[illegible]rator Theory: Advances and Applications
Vol. 23

Editor:
I. Gohberg
Tel Aviv University
Ramat Aviv, Israel

Editorial Office:
School of Mathematical Sciences
Tel Aviv University
Ramat Aviv, Israel

OT 23:
Operator Theory: Advances and Applications
Vol. 23

Birkhäuser Verlag
Basel · Boston · Stuttgart

W. Greenberg, C. van der Mee,
V. Protopopescu

Boundary Value Problems in Abstract Kinetic Theory

1987

Birkhäuser Verlag
Basel · Boston · Stuttgart

Authors Adresses:
William Greenberg
Virginia Polytechnic Institute
& State University
BLACKSBURG, Va. 24061
U.S.A.

Cornelis van der Mee
Department of Mathematics
& Computer Science
Clarkson University
POTSDAM, N.Y. 13676
U.S.A.

Vladimir Protopopescu
Engineering Physics & Mathematics Division
Oak Ridge National Laboratory
OAK RIDGE, Tennessee 37831
U.S.A.

Library of Congress Cataloging in Publication Data

Greenberg, William, 1941–
Boundary value problems in abstract kinetic theory
(Operator theory, advances and applications ; v. 23)
Bibliography: p
Includes index
1. Boundary value problems. 2. Initial value problems
3. Matter, Kinetic theory of. I. Title. II. Series.
III. van der Mee, Cornelis. IV. Protopopescu, Vladimir.
QA379.G735 1987 515.3'5 87-29981
ISBN 3-7643-1765-5

CIP-Kurztitelaufnahme der Deutschen Bibliothek

Greenberg, William:
Boundary value problems in abstract kinetic
theory/William Greenberg; Cornelis van der Mee;
Vladimir Protopopescu. – Basel ; Boston ; Stuttgart
Birkhäuser, 1987
(Operator theory ; Vol. 23)
ISBN 3-7643-1765-5
NE: van der Mee, Cornelis:; Protopopescu, V.:; GT

Printed in Germany
ISBN 3-7643-1765-5
ISBN 0-8176-1765-5

TABLE OF CONTENTS

Chapter VI
EQUIVALENCE OF DIFFERENTIAL AND INTEGRAL FORMULATIONS

Chapter VII
SEMIGROUP FACTORIZATION AND RECONSTRUCTION

Chapter VIII
ALBEDO OPERATORS, H-EQUATIONS AND REPRESENTATION OF SOLUTIONS

Chapter IX
APPLICATIONS OF THE STATIONARY THEORY

Chapter X
INDEFINITE STURM-LIOUVILLE PROBLEMS

Chapter XI
TIME DEPENDENT KINETIC EQUATIONS: METHOD OF CHARACTERISTICS

Chapter XII
TIME DEPENDENT KINETIC EQUATIONS: SEMIGROUP APPROACH

Chapter XIII
APPLICATIONS OF THE INITIAL VALUE PROBLEM

PREFACE

This monograph is intended to be a reasonably self-contained and fairly complete exposition of rigorous results in abstract kinetic theory. Throughout, abstract kinetic equations refer to (an abstract formulation of) equations which describe transport of particles, momentum, energy, or, indeed, any transportable physical quantity. These include the equations of traditional (neutron) transport theory, radiative transfer, and rarefied gas dynamics, as well as a plethora of additional applications in various areas of physics, chemistry, biology and engineering.

The mathematical problems addressed within the monograph deal with existence and uniqueness of solutions of initial-boundary value problems, as well as questions of positivity, continuity, growth, stability, explicit representation of solutions, and equivalence of various formulations of the transport equations under consideration. The reader is assumed to have a certain familiarity with elementary aspects of functional analysis, especially basic semigroup theory, and an effort is made to outline any more specialized topics as they are introduced.

Over the past several years there has been substantial progress in developing an abstract mathematical framework for treating linear transport problems. The benefits of such an abstract theory are twofold: (i) a mathematically rigorous basis has been established for a variety of problems which were traditionally treated by somewhat heuristic distribution theory methods; and (ii) the results obtained are applicable to a great variety of disparate kinetic processes. Thus, numerous different systems of integrodifferential equations which model a variety of kinetic processes are themselves modelled by an abstract operator equation on a Hilbert (or Banach) space. The general results so obtained are equally applicable to problems which range from neutron transport in nuclear reactors and polarized light transfer through planetary atmospheres, to reaction-diffusion processes in solutions, electron scattering in semiconductors and metals, and cell growth in tumors. We attempt to present herein a description of the methods and the history of this abstract generalization of kinetic equations, at the same time discussing an extensive list of concrete physical applications.

The presentation divides into four subjects, which may be read somewhat independently. We outline the division below.

Stationary problems Differential methods	CHAPTER II CHAPTER III CHAPTER IV CHAPTER V
Stationary problems Convolution methods	CHAPTER VI CHAPTER VII CHAPTER VIII
Time dependent problems	CHAPTER XI CHAPTER XII
Applications	CHAPTER IX CHAPTER X CHAPTER XIII

There are some exceptions within the indicated divisions, however. Chapter X on Sturm–Liouville equations develops aspects of the abstract theory. Section III.3 details evaporation models. Section VIII.5 on adding methods is equally relevant to differential and convolution formulations, although the method for representing solutions is certainly integral. We have made a strenuous effort to separate the abstract theory from the various applications. For example, Chapter IX, the longest chapter in the monograph, consists entirely of examples of stationary transport problems, and is cross–referenced throughout the text.

Four years ago, Birkhäuser Verlag published a book [211] covering certain aspects of linear kinetic theory. One may legitimately question whether the publication of a second one is timely and opportune. We would point out that, in the last years, many new – mostly abstract – results have been obtained which could not have been contained in [211]. Specifically, Chapters IV, V, VII, VIII, X and XI, and parts of Chapters III, VI, IX and XII are based on recent developments. The topics discussed in Chapters II and III have largely been tackled in [211], but not in the generality and abstract setting adopted here. Also, the time dependent theory in [211] is devoted to the neutron transport equation in an L_1-setting with vacuum boundary conditions. In Chapters XI – XIII we shall study the initial value problem for general abstract kinetic equations and a wide class of boundary conditions.

We are indebted to the Department of Mathematics at Virginia Polytechnic Institute and State University and to C. Wayne Patty, chairman, for providing substantial secretarial services during the preparation of this manuscript, and for their hospitality during the visit of one of us (vdM). We are also indebted to J. W. Hovenier and his group at the Department of Physics and Astronomy of the Free University of Amsterdam for financial support and a variety of discussions on radiative transfer with one of us (vdM) and to the Engineering Physics and Mathematics Division at O.R.N.L. for its supportive attitude toward this project. We would like to thank G. Busoni, G. Frosali, A. Ganchev, 蔡 东 康, J. Voigt and W. Walus for various suggestions on improvements in the manuscript. We especially thank R. Beals, C. Cercignani and P. F. Zweifel for having thoroughly and enlighteningly discussed with us many important matters included in this book. Parts of the book have been written and discussed while we, together and separately, have visited the University of Florence, and for this we are particularly indebted to V. Boffi and to the Istituto Matematico "U. Dini" and the Istituto di Matematica Applicata "G. Sansone". We are much obliged to M. Williams for solving a variety of technical problems associated with the reproduction of the manuscript. Finally, we are appreciative of the encouragement given us by I. Gohberg, and the assistance of Birkhäuser Verlag, in carrying out this project.

The research contained herein was supported in part by the U. S. Department of Energy under grants DE-AS05 83ER10711-1 and DE-AC05 84OR21400, and by the National Science Foundation under grants DMS 83-12451 and DMS 85-01337.

Chapter I

ELEMENTS OF LINEAR KINETIC THEORY

1. Introduction

In this monograph we present a rigorous exposition of boundary value problems for an extensive class of kinetic equations. These equations describe the transport of particles or radiation through a host medium (possibly the vacuum) in the region Ω with velocities or frequencies belonging to a subset $V \subset \mathbb{R}^n$, and the solutions represent particle densities as a function of position and velocity, or radiative intensities as a function of position and frequency. At the boundary $\partial\Omega$ of the spatial domain Ω, particles or radiation may be incident from outside and may be exiting the medium, and reflection and absorption processes may be specified. One may also deal with particle or radiative sources within the region Ω.

The basic assumption underlying all kinetic equations to be studied here is that the particle flux or radiation does not interact with itself in a many-body sense, but only with the given host medium or with its own thermal equilibrium distribution in a perturbative way, and with the boundary under a linear reflection law. That is to say, we will treat only linear or linearized kinetic equations. The distinction between systems which are genuinely linear, such as neutron transport and radiative transfer in scattering atmospheres, in which the neutron-neutron and photon-photon cross sections are for all practical purposes zero, and linearized systems, such as rarefied gas dynamics, will play no role in modeling the corresponding equations by abstract operator equations. We shall analyze stationary boundary value problems, where the solution does not depend on the time t, as well as time dependent problems, where both initial values and boundary conditions must be specified. The techniques for these two types of problems, however, will differ completely, and, as we shall see, for stationary problems we will essentially be confined to spatially homogeneous, force free, one dimensional models.

Before we give additional details on the nature of the problems to be treated in the ensuing chapters, let us present typical kinetic equations from an important field of application, namely neutron transport in a fissionable medium. We emphasize that neutron transport theory, although historically the area in which some of the techniques to be presented were first exploited, is only one of many areas where linear kinetic equations play an important role. Such kinetic equations are also fundamental in gas dynamics, in radiative transfer, and in electron and phonon transport, and arise in numerous other applications – from traffic modeling to medical physiology. Indeed, it is precisely the similar mathematical structure of equations from so many different areas which makes an abstract theory particularly rewarding.

In neutron transport, the quantity of interest is the neutron (angular) density ψ as a function of position $\mathbf{r}\in\Omega$ and velocity $\mathbf{v}\in V$. In the diffusion of neutrons through a fissionable medium, it is reasonable to assume that there are no neutron–neutron collisions, since the neutron density is typically less than 10^{-12} times the atomic density of the medium. Let $\psi(\mathbf{r},\mathbf{v},t)d^3r d^3v$ represent the number of neutrons in a volume element d^3r about $\mathbf{r}$ whose velocities lie in the element $\mathbf{v}+d^3v$ of velocity space. Then the change at time t in the number of neutrons with velocity in $\mathbf{v}+d^3v$ which are located in a small volume Δ about the point $\mathbf{r}$ is given by the balance equation

$$d\left[\int_\Delta \psi(\mathbf{r},\mathbf{v},t)d^3r\right]d^3v dt \;=\; -\text{ (net number flowing out of } \partial\Delta) - \text{(number of velocity } \mathbf{v} \text{ suffering collisions in } \Delta) + \text{(number of secondaries of velocity } \mathbf{v} \text{ produced in } \Delta) + \text{(number of velocity } \mathbf{v} \text{ produced in } \Delta \text{ by sources).}$$

The first term on the right is a free streaming term obtained by integration about the surface $\partial\Delta$ of Δ, or via Green's theorem, as a divergence

$$d^3v dt\int_{\partial\Delta}\mathbf{v}\psi(\mathbf{r},\mathbf{v},t)\cdot\hat{n}_o d^2r \;=\; d^3v dt\int_\Delta \nabla\cdot\mathbf{v}\psi(\mathbf{r},\mathbf{v},t)d^3r,$$

where $\hat{n}_o$ is the unit outer normal on $\partial\Delta$.

The second term can be written in terms of the macroscopic total cross section $\sigma(\mathbf{r},\mathbf{v})$, which represents a variety of neutron–nuclear reactions, including elastic scattering, inelastic scattering, radiative capture and fission,

$$d^3v dt\int_\Delta v\sigma(\mathbf{r},\mathbf{v})\psi(\mathbf{r},\mathbf{v},t)d^3r.$$

The third term may be written in terms of the differential collision cross section $\sigma(\hat{\mathbf{v}}\to\mathbf{v},\mathbf{r})$,

$$d^3v\,dt \int_\Delta \int_V \hat{v}\psi(\mathbf{r},\hat{\mathbf{v}},t)\sigma(\hat{\mathbf{v}}\to\mathbf{v},\mathbf{r})d^3\hat{v}\,d^3r,$$

and the fourth term can be described in terms of the angular source density $q(\mathbf{r},\mathbf{v},t)$:

$$d^3v\,dt \int_\Delta q(\mathbf{r},\mathbf{v},t)d^3r.$$

Collecting all this leads to the neutron transport equation

$$\frac{\partial\psi}{\partial t}(\mathbf{r},\mathbf{v},t) + \mathbf{v}\cdot\nabla\psi(\mathbf{r},\mathbf{v},t) =$$

$$= -\ v\sigma(\mathbf{r},v)\psi(\mathbf{r},\mathbf{v},t) + \int \sigma(\hat{\mathbf{v}}\to\mathbf{v},\mathbf{r})\hat{v}\psi(\mathbf{r},\hat{\mathbf{v}},t)d^3\hat{v} + q(\mathbf{r},\mathbf{v},t). \tag{1.1}$$

If the system is plane symmetric, then the cross section will depend only on one spatial variable, x say, and its angular dependence only on $\Omega\cdot\hat{\mathbf{x}} \equiv \cos\,\theta \equiv \mu$ or on $\Omega\cdot\hat{\Omega}$:

$$\frac{\partial\psi}{\partial t}(x,v,\mu,t) + v\mu\frac{\partial\psi}{\partial x}(x,v,\mu,t) =$$

$$= -\ v\sigma(x,v)\psi(x,v,\mu,t) + \int \hat{v}\sigma(v,\hat{v},\hat{\Omega}\cdot\Omega,x)\psi(x,\hat{v},\hat{\mu},t)d^3\hat{v} + q(x,v,\mu,t).$$

An important approximation is the **one speed approximation** or **constant cross section approximation.** In this approximation, $\sigma(\hat{\mathbf{v}}\to\mathbf{v},\mathbf{r}) = \sigma(\Omega\cdot\Omega,\mathbf{r},v)\delta(v-\hat{v})/v^2$ and $\sigma(\mathbf{r},\mathbf{v}) = \sigma(\mathbf{r},v)$. Defining $k(\Omega\cdot\hat{\Omega},\mathbf{r},v) = \sigma(\Omega\cdot\hat{\Omega},\mathbf{r},v)/\sigma(\mathbf{r},v)$, the equation for the integrated angular density

$$\psi(\mathbf{r},\Omega,t) = \int_{v_1}^{v_2} v^2\psi(\mathbf{r},\mathbf{v},t)dv$$

in terms of the integrated angular source density $q(\mathbf{r},\Omega,t)$ becomes

$$\frac{\partial\psi}{\partial t}(\mathbf{r},\Omega,t) + v\Omega\cdot\nabla\psi(\mathbf{r},\Omega,t) =$$

$$= -\ v\sigma(\mathbf{r},v)\psi(\mathbf{r},\Omega,t) + v\sigma(\mathbf{r},v)\int \psi(\mathbf{r},\hat{\Omega},t)k(\Omega\cdot\hat{\Omega},\mathbf{r},v)d\hat{\Omega} + q(\mathbf{r},\Omega,t).$$

Consider, finally, a time independent boundary value problem with plane symmetry and the one speed approximation. By replacing x with $\int^x \sigma(\hat{x})d\hat{x}$, but still labeling the new variable x, and assuming that the kernel $k(\Omega\cdot\hat{\Omega},\mathbf{r},v)$ is independent of position, we have, after some spherical harmonics algebra,

$$\mu\frac{\partial\psi}{\sigma x}(x,\mu) + \psi(x,\mu) =$$

$$= \tfrac{1}{2}c \sum_{\ell=0}^{\infty} (2\ell+1)a_\ell P_\ell(\mu)\int_{-1}^{1} \psi(x,\hat{\mu})P_\ell(\hat{\mu})d\hat{\mu} + q(x,\mu). \tag{1.2}$$

Eqs. (1.1) and (1.2) are examples of the sort we wish to discuss here. Along with (1.1) are provided an initial condition and boundary conditions at the spatial boundary (and possibly at the boundary of the velocity space). In particular, the initial-boundary value problem (1.1) requires the specification of the neutron distribution ψ_0 at time t=0,

$$\psi(x,\xi,0) = \psi_0(x,\xi),$$

as well as the "incoming flux" φ at the boundary,

$$\psi(x,\xi,t) = \varphi(x,\xi,t)$$

for $x\in\partial\Omega$ and $\xi\cdot n(x)<0$ with $n(x)$ the unit outer normal at $x\in\partial\Omega$. Similarly, since the stationary problem (1.2) is in plane symmetry, one will be interested in boundary conditions corresponding to half space geometry $0\leq x<\infty$ and slab geometry $0\leq x\leq\tau<\infty$. (The full space problem $-\infty<x<\infty$ is uninteresting technically as well as physically.) For example, for the slab problem related to (1.2), typical "incoming flux" boundary conditions are

$$\psi(x=0,\mu) = \varphi_+(\mu), \quad \mu>0,$$

$$\psi(x=\tau,\mu) = \varphi_-(\mu), \quad \mu<0,$$

for given $\varphi_\pm$, representing a flux φ_+ entering the slab at x=0 and a flux φ_- entering the slab at $x=\tau$. For the half space problem, the boundary conditions might be

$$\psi(x=0,\mu) = \varphi_+(\mu), \quad \mu>0,$$

$$\limsup_{x\to\infty} \int_{-1}^{1} |\psi(x,\mu)|^2 d\mu < \infty.$$

In neutron transport, boundary reflections are not relevant. However, to study kinetic equations in gas dynamics and radiative transfer, it will be necessary to include reflection at the boundaries.

This discussion was not intended to be a rigorous "derivation" of a kinetic equation (e.g., regularity conditions necessary for existence of traces and various changes of integration order have not been considered), but rather to familiarize the reader with the sort of equation which will be of interest in the sequel.

Let us now consider the abstract formulation of problems such as the neutron transport equations (1.1) and (1.2). Consider first the abstract Banach space equation

$$\frac{\partial \psi}{\partial t} + \mathbf{v} \cdot \nabla \psi = - A\psi + q$$

defined on $X = L_p(\Omega \times \mathbb{R}^3,\ d^3r\ d^3v)$, which models the time dependent equation (1.1). Here, A is the operator

$$(Ag)(\mathbf{r},\mathbf{v}) = v\sigma(\mathbf{r},\mathbf{v})g(\mathbf{r},\mathbf{v}) - \int \sigma(\hat{\mathbf{v}} \rightarrow \mathbf{v},\mathbf{r})\hat{v}g(\mathbf{r},\hat{\mathbf{v}})d^3\hat{v},$$

$\psi(t) \in X$, and appropriate initial-boundary values have to be posed. When the total time derivative $\frac{D}{Dt} \equiv \frac{\partial}{\partial t} + \mathbf{v} \cdot \nabla$ is formulated on the left hand side of the equation, the method of characteristics suggests itself as a natural approach. In this case the spatial domain Ω and the boundary conditions themselves may be time dependent. On the other hand, one may include the streaming term $\mathbf{v} \cdot \nabla \psi$ on the right hand side,

$$\frac{\partial \psi}{\partial t} = - \mathbf{v} \cdot \nabla \psi - A\psi + q,$$

whence a perturbative semigroup approach is obvious, if it can be shown that the streaming operator is the generator of a semigroup. (For the definition of a semigroup, refer to section 3.)

To formulate the half space stationary problem (1.2), let us consider the Hilbert space $H = L_2([-1,1],d\mu)$ with respect to Lebesgue measure $d\mu$, and define the bounded operators

$$(Tg)(\mu) = \mu g(\mu),$$

$$(Ag)(\mu) = g(\mu) - \frac{c}{2}\sum_{\ell=0}^{\infty}(2\ell+1)a_\ell P_\ell(\mu)\int_{-1}^{1} g(\hat{\mu})P_\ell(\hat{\mu})d\hat{\mu},$$

$$(Q_+g)(\mu) = \begin{cases} g(\mu), & \mu>0, \\ 0, & \mu<0. \end{cases}$$

Then the boundary value problem can be written

$$T\frac{\partial\psi}{\partial x} = -A\psi + q, \tag{1.3a}$$

$$Q_+\psi(x=0) = \varphi_+, \tag{1.3b}$$

$$\lim_{x\to\infty} \|\psi(x)\| < \infty, \tag{1.3c}$$

for an H-valued fuction $\psi:\mathbb{R}_+\to H$. Defining the operator $K \equiv T^{-1}A$, Eq. (1.3a) has the form

$$\frac{d\psi}{dx} = -K\psi + q, \tag{1.4}$$

which may appear to be a semigroup problem. However, at the "initial" value x=0 the solution is specified only in the forward (incoming or incident flux) direction; the backward (outgoing) flux must be derived from the equation itself. For this reason, such boundary value problems are referred to as "forward-backward" equations. Actually, the derivation of the outgoing flux at the boundary x=0 completely solves the problem, in the sense that standard semigroup theory provides the solution of the kinetic equation at interior points once the solution at the boundary is known. Thus, the major part of our effort will be devoted to producing the outgoing flux from a specified ingoing flux. This is sometimes referred to as the albedo problem.

The equations discussed above are examples of neutron transport. There are many other similar equations, describing phenomena in physics, engineering, chemistry, and biology, as well as from more diverse applications: economics, sociology, etc. The main feature of these equations is that they consist of a linear operator $\frac{\partial}{\partial t} + \mathbf{v}\cdot\nabla$ representing streaming and a (linear or nonlinear) collision operator A. That is to say, the kinetic equation can be written

$$\frac{\partial\psi}{\partial t} + \mathbf{v}\cdot\nabla\psi = -A\psi + q,$$

and the specific features of the model are contained primarily in the details of the collision operator A.

In this monograph, we will consider only linear collision operators A. Chapters II through X will be devoted to a formulation of the stationary equation in plane parallel geometry, with $\psi(x,\cdot)$ a Hilbert space valued or Banach space valued function of the position coordinate x. The abstract equation will be written in the form

$$T\frac{\partial\psi}{\partial x} = -A\psi + q \tag{1.5}$$

with the operators T and A possibly unbounded linear operators. The following three chapters will consider time dependent equations essentially of the form

$$\frac{\partial\psi}{\partial t} + \mathbf{v}\cdot\nabla\psi = -A\psi + q \tag{1.6}$$

on Banach spaces $L_p(\Lambda,d\mu)$, $1\le p<\infty$. We will include results relating to the inclusion of an external force field and to time dependent boundary conditions and spatial domains.

2. Historical development

Around 60 B.C., elaborating on the Greek atomists the Roman poet and philosopher Titus Lucretius Carus enunciated a descriptive and crude kinetic theory, a world of indestructible atoms ("primordia" or "elementa") moving without cease. Even the notion of stationary states did not escape him: "Great armies cover wide fields in maneuvers .... and yet there is a place on the peaks from which these armies appear to be motionless." [252]

Although Lucretius had a profound influence on Horace and Virgil, his scientific writings were quickly forgotten and it was not until the nineteenth century that kinetic theory was reformulated, then of course on a mathematical and experimental foundation. The high point of that development was the Boltzmann equation, proposed by Ludwig Boltzmann [48] in 1872 based on heuristic arguments and gradually recognized as a keystone of future developments. The Boltzmann equation is deceptively simple in appearance, a semilinear integrodifferential equation consisting of a linear differential operator representing pure streaming of particles and a quadratic integral operator describing the collision process. Quite remarkably, virtually every aspect of the equation has led to problems which, more than 100 years later, are still unsolved [158, 299]. These include the existence (and uniqueness) of solutions, their explicit representations and properties, numerical algorithms, and even the rigorous derivation of the equation from the principles of statistical mechanics.

Beginning with the work of van Kampen [371] and Case [68] in the 1950's, the singular eigenfunction method has become a workhorse in the computational analysis of stationary linear kinetic equations. Consider, for example, Eq. (1.5) with q=0 and half space boundary conditions

$$\frac{\partial\psi}{\partial x} = -K\psi,$$

$$Q_+\psi = \varphi_+,$$

$$\lim_{x\to\infty} \|\psi(x)\| < \infty,$$

where we have written $K = T^{-1}A$, and assume for simplicity that the spectrum $\sigma(K)$ of K is real. Then standard separation of variables suggests that elementary solutions have the form $\exp\{-\lambda x\}\phi_{\alpha,\lambda}$, and that the boundary value problem may be solved by a superposition of such solutions, where the coefficients are to be obtained by imposing the boundary condition. That is to say, suppose the distributional functions $\phi_{\alpha,\lambda}$ are continuum eigenfunctions of K,

$$K\phi_{\alpha,\lambda} = \lambda\phi_{\alpha,\lambda},$$

for $\lambda\in\sigma(K)$ and α a label representing possible degeneracies of the spectrum. If the (continuum) eigenfunctions of K are complete, then the solution of the boundary value problem at x=0 can be expanded

$$\psi(0) = \int_{-\infty}^{\infty} \sum_{\alpha=1}^{N_\lambda} A_{\psi(0)}(\alpha,\lambda)\phi_{\alpha,\lambda}\,d\rho(\lambda),$$

for $\rho(\lambda)$ a Borel measure on $\sigma(K)$, and one may expect that

$$\psi(x) = \int_{-\infty}^{\infty} \sum_{\alpha=1}^{N_\lambda} \exp\{-\lambda x\} A_{\psi(0)}(\alpha,\lambda)\phi_{\alpha,\lambda} d\rho(\lambda), \tag{2.1}$$

will be the solution of the boundary value problem. Of course, the difficulty is that the boundary value $\psi(0)$ is not known in its entirety, but only a projected part $Q_+\psi(0) = \varphi_+ \in \text{Ran } Q_+$. Moreover, boundedness of the solution as $x\to\infty$ would seem to require that the expansion in (2.1) involve only eigenfunctions with Re $\lambda \geq 0$. Indeed, what is needed, for half space problems, is a **half range completeness theorem**, guaranteeing that the set of functions $\{Q_+\phi_{\alpha,\lambda} : \text{Re } \lambda\geq 0\}$ is complete on a dense subspace D of Ran Q_+, the range of the projection Q_+. Then if $\varphi_+ \in D$ is expanded as

$$\varphi_+ = \int_0^{\infty} \sum_{\alpha=1}^{N_\lambda} A_{\varphi_+}(\alpha,\lambda)Q_+\phi_{\alpha,\lambda} d\hat{\rho}(\lambda)$$

the solution of (1.5) may be written

$$\psi(x) = \int_0^{\infty} \sum_{\alpha=1}^{N_\lambda} \exp\{-\lambda x\}A_{\varphi_+}(\alpha,\lambda)\phi_{\alpha,\lambda} d\hat{\rho}(\lambda). \tag{2.2}$$

Note that the measure $\hat{\rho}$ on $\sigma(K)\cap\mathbb{R}_+$ in (2.2) is not the same as the spectral measure ρ in (2.1). Further, the expansion in (2.2) involves the functions $\phi_{\alpha,\lambda}$ and not $Q_+\phi_{\alpha,\lambda}$; physically, this corresponds to the validity of the solution represented by (2.2) in both incoming and outgoing directions.

Although the Case–van Kampen eigenfunction expansion procedure was a keystone in the development of linear kinetic theory, the method suffers from some serious drawbacks. First of all, the analysis depends upon a study of the properties of the eigenfunctions for the specific model under consideration, and thus is not easily adaptable to the study of the abstract problem. Of even more concern, the proof of a half range completeness theorem and the derivation of the measure $\hat{\rho}$ require careful manipulation of eigenfunctions which exist in general only in a distributional sense. Although this difficulty may in principle be surmounted, in practice in the literature it always results in heuristic treatment of the completeness and orthogonality "theorems" (but see [221] for a study of isotropic neutron transport in terms of rigged Hilbert spaces). Indeed, the seminal paper of Case on solutions of the linearized Vlasov equation [67] has been a standard in plasma theory since it appeared in 1959, yet only recently was it noted [16] that the result derived for certain modes is incorrect.

Finally, it should be pointed out that the derivation of the measure $\hat{\rho}$ needed for a half range completeness theorem is nontrivial, and, in fact, is the central problem in the analysis. It requires the factorization of a function basic to the particular kinetic equation under study into a pair of factors, each analytic on a half plane. This is the so-called Wiener–Hopf factorization problem, or in the case of a matrix function, the Riemann factorization problem.

In an effort to provide a mathematically more satisfactory treatment of continuum eigenfunctions, Larsen and Habetler [241] proposed in 1972 that the analysis might be carried out in an effective manner using contour integration to obtain the necessary spectral projections. Their ideas have, over the past decade, been expanded to generate a rigorous and methodical treatment of the class of models which had been subject to continuum eigenfunction methods (see, for example, [56, 240] and reviews in [168, 407]). This technique is referred to in the literature as the resolvent integration method.

Consider for simplicity the case when K has a bounded inverse with spectrum $\sigma(K^{-1})$ on a simple Jordan curve. The principal idea in this method is to obtain a family of spectral projections associated with the operator K^{-1} by integrating the resolvent $(z-K^{-1})^{-1}$ about a contour C surrounding $\sigma(K^{-1})$ and then contracting the coutour C to $\sigma(K^{-1})$. More precisely, for a dense set D of vectors in the Banach space X, one must estimate the uniformity of convergence of the limits $R_g^{\pm}(\lambda,K^{-1}) \equiv \lim_{\delta\to 0\pm} (\lambda+\delta-K^{-1})^{-1}g$ for $\lambda\in\sigma(K^{-1})$ and $g\in D$. If the limits exist in an appropriate (weak) sense, then general properties of spectral operators of scalar type (cf. [105, 109]) will insure that the integral

$$E_g(\Gamma) = \frac{1}{2\pi i}\int_\Gamma \{R_g^+(\lambda,K^{-1}) - R_g^-(\lambda,K^{-1})\}\, d\lambda, \quad \Gamma\subset\sigma(K^{-1}),$$

gives the spectral resolution of K^{-1}. In this case the expansion (2.1) is an evident application of the spectral calculus:

$$\psi(x) = \int_{-\infty}^{\infty} \exp\{-x/\lambda\}\, dE_{\psi(0)}(\lambda).$$

If K does not have a bounded inverse, then it is possible to consider a bounded function of K. For example, one might choose to study the resolvent $(z-(K+i)^{-1})^{-1}$ in order to obtain a functional calculus for K.

The resolvent integration method does not reduce the difficult problem of dealing with a (Wiener–Hopf or Riemann matrix) factorization problem to obtain a half range expansion. Nor does it give a great deal of insight into the nature of the abstract problem. However, it does provide an elementary and mathematically consistent way to handle the distributional functions arising in the Case–van Kampen method. Indeed, it was by resolvent integration that the difficulty in the formula for the Vlasov modes was (quite easily) recognized.

During the same period that the Larsen–Habetler work appeared, Hangelbroek [181] proposed a rather different approach to these problems. Consider the case when T and A are bounded and self adjoint on an abstract Hilbert space H, with $\sigma(A) \subset (0,\infty)$. Since the operator $A^{1/2}KA^{-1/2} \equiv A^{1/2}T^{-1}A^{1/2}$ is transparently self adjoint, if the similarity transformation induced by $A^{1/2}$ is used to define an equivalent inner product on H, then K itself will be self adjoint, and the spectral theorem for self adjoint operators will provide the machinery to construct a spectral family $\{E(\Gamma) : \Gamma \subset \mathbb{R}$ Borel$\}$ for K, and thus to obtain the completeness theorem used in (2.1). This observation does not deal with the heart of the problem, namely "half range completeness." However, writing $P_\pm \equiv E(\mathbb{R}_\pm)$, it was soon realized that the existence of the (Wiener Hopf or Riemann matrix) factorization, and therefore of half range completeness, is equivalent to certain matching properties of the projections $P_\pm$ and $Q_\pm$. This then provided a new approach to dealing with these kinetic equations, an approach naturally suited to abstract generalization which replaces a difficult problem in analytic function theory with a straightforward characterization of noncommuting projections. A detailed study of stationary equations utilizing this functional analytic approach will be the topic of Chapters III – VI.

Already at the beginning of this century, a connection between certain transport type integrodifferential equations and (scalar) convolution equations had been made. In particular, we note the Schwarzschild–Milne integral equation [265], formulated for describing the transfer of light through a stellar atmosphere, and the early theory of Wiener–Hopf equations [392]. In the 1940's and 1950's the elaboration of invariant imbedding theory in radiative transfer, especially the work of Ambarzumian [6], Chandrasekhar [89], Sobolev [340] and Busbridge [61], produced a plethora of explicit solution formulas for scalar equations (Note, however, the vector valued polarized light equation with Rayleigh scattering, also solved in [89]). The result of these investigations has been the complete solution of the scalar equation of radiative transfer in terms of solutions of nonlinear integral equations, by Busbridge [61] for the half space and Sobolev [341] and Hovenier [200] for the slab, a performance not matched using the Case–van Kampen method. Only in the last two decades have these studies been extended to more general matrix and operator equations, and consequently to the treatment of more diverse applications in kinetic theory. The work of Maslennikov [259] and Feldman [116] is particularly to be noted. The main reason for the emergence of these generalizations was the development, from 1958 to 1974, of the theory of vector valued convolution equations and operator Wiener–Hopf factorizations. We mention the work of Gohberg, Krein, Feldman, Semençul and Heinig [139, 140, 141, 151] on the theory of convolution equations, and the articles of Gohberg and Leiterer [138, 148, 149, 150] on the theory of Wiener–Hopf factorizations. In the years after 1980 these methods facilitated a merger of the invariant imbedding tradition with the semigroup approach initiated by Hangelbroek. The outcome has included a proof of the equivalence of the integrodifferential and integral formulations, explicit representations of solutions, and the construction of the semigroup apparatus for non self adjoint kinetic models.

Chapters VI–VIII are devoted to the convolution equations approach. A proof is presented of the equivalence of a class of differential equations of the type (1.5) with boundary conditions and convolution equations of the form

$$\psi(x) - \int_0^\infty \mathcal{H}(x-y)B\psi(y)dy = \omega(x), \quad 0<x<\infty.$$

Explicit representations of solutions are derived in terms of solutions of certain operator valued nonlinear integral equations, both for finite and semi-infinite domains, and their connection to Wiener–Hopf factorizations. Existence and uniqueness results are then transferred from the Hilbert space setting to a Banach space setting with the help of a Fredholm argument.

Paradoxically, in some respects the time dependent linear problem is easier to study than the stationary problem. That is because the time evolution problem is unidirectional and not forward–backward, and, indeed, the construction of the initial–boundary data, which is the hard part of the stationary theory, is obviously unnecessary in the time dependent setting.

Substantial progress in understanding the properties of solutions of the initial–boundary value problem for time dependent kinetic equations dates from the development of semigroup theory in the 1950's. Among the first to apply rigorous methods to obtain existence and uniqueness results for specific linear kinetic equations were Lehner and Wing [246], whose strategy of proving the transport operator to generate a strongly continuous semigroup became a model for treating wellposedness of such problems. Motivated by the desire to develop time dependent theory in the physically more interesting L_1–setting, Birkhoff [44] and Vidav [375] introduced Banach lattice theory to the study of these equations, obtaining also the long time behavior of solutions. In fact, a guiding principle was that transport operators should have dominant eigenvalues which determine the asymptotic behavior, and for this reason compact perturbation methods were introduced. In the 1970's this strategy was further elaborated upon; we especially mention Voigt [382] and Greiner [171] for their contributions in this direction. An offspring of this approach was the development and application of positive semigroup theory after 1979, mainly by the Tübingen group (see [102, 173]), with its obvious implications for the asymptotic spectrum of the transport operator. In Chapter XII we shall give a thorough treatment of the semigroup method. We shall establish the unique solvability of a large class of time dependent linear kinetic equations describing an autonomous transport system, and shall discuss certain aspects of positive semigroups and the asymptotics of solutions of time dependent kinetic equations.

For initial value problems with time dependent phase space, boundary reflection or collision operators, semigroup theory is not applicable. Recently, extending a program espoused by Bardos [23], the classical method of characteristics has been applied in the abstract setting of vector fields of the type appearing in kinetic theory [37]. In Chapter XI we shall apply the method of characteristics to an extensive class of kinetic equations, where we shall incorporate divergence free force fields and arbitrary collision operators that allow separation into a scalar gain term and a bounded loss term.

The bulk of the monograph relates to abstract kinetic theory, which leans heavily on a variety of functional analytic methods involving semigroups, selfadjoint operators, Banach lattices, convolution equations and vector fields. However, three chapters are devoted to a large number of applications from a variety of fields. In Chapter IX we shall present some details on a number of model problems from the stationary theory of radiative transfer in planetary atmospheres, neutron transport, rarefied gas dynamics and phonon and electron transport. In Chapter X we shall focus on stationary problems arising in the study of kinetic equations of Fokker–Planck type. Finally, in Chapter XIII we shall discuss several specific applications of time dependent kinetic theory. These involve some well known equations from neutron transport theory, as well as the lesser known Spencer–Lewis equation, the runaway electron problem and a problem from cell growth dynamics.

3. Semigroups

We collect here some quite elementary results on linear operators and semigroup theory. For a more complete account of semigroup theory we refer to the monographs by Pazy [301], Davies [97], Hille and Phillips [194], Kato [213], and Krasnoselskii et al. [224]

Let X be a (real or complex) Banach space and $S:X\to X$ a linear operator on X with domain D(S). Then S is **closed** if $\{x_n\}\subset D(S)$, $x_n\to x\in X$, and $Sx_n\to y\in X$ imply $x\in D(S)$ and $Sx=y$, and is **closable** if it has a closed extension. All operators will henceforth be assumed densely defined and closed, unless otherwise specified.

The bounded operator S is **compact** if the image under S of every bounded sequence in X has a convergent subsequence. This includes, for example, the finite rank operators, i.e., operators S with image Ran S of finite dimension. The special role of compact operators is exhibited by the Fredholm alternative. We give also an analytic version of this result.

Fredholm Alternative: If S is compact, then either $(I-S)^{-1}$ is a bounded operator on X or $Sx=x$ has a nonzero solution $x\in X$.

Analytic Fredholm Theorem: If D is an open connected subset of $\mathbb{C}$, X a complex Banach space, and S(z) an analytic operator valued function such that S(z) is a compact operator for each $z\in D$, then either $(I-S(z))^{-1}$ is a bounded operator on X for no $z\in D$ or there is a subset $D_0\subset D$ with no limit points in D such that $(I-S(z))^{-1}$ is a bounded operator on X for all $z\in D\setminus D_0$, analytic in $D\setminus D_0$ with finite rank residues at $z\in D_0$, and $S(z)x=x$ has a nonzero solution $x\in X$ at each $z\in D_0$

For S an operator on a complex Banach space X, we say that $\lambda\in\mathbb{C}$ belongs to the **resolvent set** $\rho(S)$ of S if $\lambda I-S$ is injective on D(S) and has X as its range; the Closed Graph Theorem then implies that the inverse $(\lambda I-S)^{-1}$ is a bounded operator. On the open set $\rho(S)$ (when nonempty) the resolvent $(\lambda I-S)^{-1}$ depends analytically on λ. The complement of $\rho(S)$ in the complex plane is referred to as the **spectrum** $\sigma(S)$ of S. If $\lambda I-S$ is not injective on D(S), then λ is called an **eigenvalue** of S and a nonzero vector $x\in D(S)$ satisfying $Sx=\lambda x$ is called an **eigenvector** of S corresponding to the eigenvalue λ. Those nonzero vectors $x\in D(S^n)$ which are not eigenvectors of S but satisfy $(\lambda I-S)^n x=0$ for some $n\geq 2$ are called **generalized**

eigenvectors of S corresponding to the eigenvalue λ. The eigenvalues of S form the **point spectrum** $\sigma_p(S)$. A complex number $\lambda \in \sigma(S)$ which is not an eigenvalue is in the **continuous spectrum** $\sigma_c(S)$ if $\lambda I-S$ maps D(S) onto a dense submanifold of X, and is in the **residual spectrum** $\sigma_r(S)$ if $\lambda I-S$ fails to have a dense range. The union of the point spectrum and the continuous spectrum forms the **approximate point spectrum** $\sigma_{ap}(S)$; $\lambda \in \sigma_{ap}(S)$ if and only if there exists a sequence $\{x_n\}_{n=1}^{\infty}$ of vectors in X of unit norm such that $(\lambda I-S)x_n \to 0$. The **spectral radius** of S is defined by $r(S) = \sup \{|\lambda| : \lambda \in \sigma(S)\}$.

A $\mathbf{C_0}$**-semigroup** on X is a family $\{U(t)\}_{t\geq 0}$ of bounded operators such that

(i) $U(0) = I$,

(ii) $U(s)U(t) = U(s+t), \quad s,t\geq 0$,

(iii) the mapping $t \to U(t)$ is strongly continuous.

If $\|U(t)\|\leq 1$ for all $t\geq 0$, then the C_0-semigroup is said to be a **contraction semigroup**. If $A:X\to X$ is defined by

$$Ax = \lim_{t\to 0} t^{-1}(U(t)-I)x,$$

with $D(A) = \{x\in X : \lim_{t\to 0} t^{-1}(U(t)-I)x \text{ exists}\}$, then A is said to be the **generator** (or **infinitesimal generator**) of the semigroup U. It is not hard to see that A will be closed and densely defined. By the **type** $\omega_0(U)$ of the semigroup U we mean the infimum (possibly $-\infty$) of all $\omega \in \mathbb{R}$ satisfying $\|U(t)\| \leq Me^{\omega t}$ for some $M>0$ and all $t>0$. We have $\omega_0(U) = \lim_{t\to\infty} \frac{\log \|U(t)\|}{t}$ and $\{\lambda \in \mathbb{C} : \operatorname{Re} \lambda > \omega_0(U)\} \subset \rho(A)$.

The connection between the generator and the semigroup is given by the two formulas

$$(\lambda I-A)^{-1} = \int_0^{\infty} e^{-\lambda t} U(t)dt,$$

valid for $\operatorname{Re} \lambda > \omega_0(U)$, and

$$U(t)x = \lim_{n\to\infty} (I+\tfrac{t}{n}A)^n x, \quad x\in X,$$

and justifies the "symbolic" equality $U(t) = \exp\{tA\}$.

If U(t) is the restriction to the positive real axis of an analytic family of operators T(z) for z in the open sector $S_\theta = \{z : |\arg z|\leq\theta\}$ which satisfy the semigroup property $T(z+\hat{z})=T(z)T(\hat{z})$ for all $z,\hat{z}\in S_\theta$ and which converge strongly to

$T(0)=I$ as $z \to 0$ in every sector $S_{\hat{\theta}}$ with $\hat{\theta}<\theta$, then $U(t)$ is called an **analytic semigroup**. Such semigroups share a number of useful properties with the exponentiation of a positive (Hilbert space) operator, of which they are a generalization. In particular, $U(t)$ maps X into $D(A)$ and $t^n A^n U(t)$ is bounded uniformly in t for $0<t<N$ and integer $n>0$.

An operator A on a Banach space X with dual space X^* is said to be **accretive** if $\mathrm{Re}\langle x^*, Ax\rangle \geq 0$ for each $x \in D(A)$ and some $x^* \in X^*$ satisfying $\langle x^*, x\rangle = \|x\|^2 = \|x^*\|^2$. The most important theorems characterizing generators of semigroups are due to Hille, Phillips, Yosida, and Lumer.

Hille–Phillips Theorem [304]: The operator A generates a C_0-semigroup iff there exists $\omega, M>0$ such that $\lambda \notin \sigma(A)$ and $\|(\lambda I-A)^{-n}\| \leq M(\lambda-\omega)^{-n}$ for all $\lambda>\omega$, $n \in \mathbb{Z}_+$.

Hille–Yosida Theorem [194, 399]: The operator A generates a contraction semigroup iff $\lambda \notin \sigma(A)$ and $\|(\lambda I-A)^{-1}\| \leq \lambda^{-1}$ for all $\lambda>0$.

Lumer–Phillips Theorem [253, 305]: The operator A generates a contraction semigroup iff A is accretive and $\mathrm{Ran}(\lambda I-A)=X$ for some $\lambda>0$.

Hille–Phillips–Yosida Theorem [194, 400]: The operator A generates an analytic semigroup iff there exists $0<\theta\leq\frac{1}{2}\pi$ and for each $\hat{\theta}<\theta$ constants $M,\omega>0$ such that $z-\omega \notin \sigma(A)$ and $\|((z-\omega)I-A)^{-1}\| \leq Mr(z,\hat{\theta})^{-1}$ for all z with $|\arg z|<\frac{1}{2}\pi+\hat{\theta}$, where $r(z,\hat{\theta}) = \inf\{|z-y| : |\arg y| \geq \frac{1}{2}\pi+\hat{\theta}\}$.

The simplest perturbation result for C_0-semigroups is the invariance of the generator property under bounded perturbations. Indeed, if A is the generator of a C_0-semigroup $\{U_0(t)\}_{t\geq 0}$ and if the semigroup generated by $A+B$, for a bounded operator B, is denoted by $\{U(t)\}_{t\geq 0}$, then U_0 and U are related by

$$U(t) = U_0(t) + \int_0^t U_0(t-s)BU(s)ds,$$

or by the **Hille–Dyson–Phillips expansion**,

$$U(t) = \sum_{n=0}^{\infty} U^{(n)}(t)$$

with

$$U^{(n)}(t) = \int_0^t U^{(n-1)}(t-s)BU_0(s)ds,$$

$$U^{(0)}(t) = U_0(t).$$

The importance of semigroups in mathematical physics lies in the fact that they provide the solution operators for Cauchy problems of the form

$$\frac{\partial\psi}{\partial t} = A\psi + q, \tag{3.1a}$$

$$\psi(0) = \varphi. \tag{3.1b}$$

More precisely,

THEOREM 3.1. Suppose $q:\mathbb{R}_+\to X$ is strongly continuous on $[0,\infty)$ and strongly continuously differentiable on $(0,\infty)$. Then for each $\varphi\in D(A)$ there exists a function $\psi:\mathbb{R}_+\to D(A)$ which is strongly continuous on $[0.\infty)$, strongly continuously differentiable on $(0,\infty)$, and satisfies (3.1) iff A is the generator of a C_0-semigroup U. In this case the solution of (3.1) is

$$\psi(t) = U(t)\varphi + \int_0^t U(t-s)q(s)ds.$$

As was pointed out in Section 1, time dependent kinetic equations (with time independent generator and boundary conditions) can be viewed as the study of the related semigroups, provided the spatial and velocity domains and the collision and boundary processes do not depend on time. However, the situation for the stationary equations is quite different. Writing (1.3a) as

$$\frac{\partial\psi}{\partial x} = -T^{-1}A\psi + \hat{q},$$

one finds that $-K=-T^{-1}A$ does not generate a semigroup. Indeed, it will turn out that K is a **dichotomous** operator, i.e., the Banach space X has a direct sum decomposition $X = X_\ell \oplus X_r$ with the property that $-K|_{X_r}$ is the generator of a semigroup on X_r and $K|_{X_\ell}$ is the generator of a semigroup on X_ℓ. The relationship between the decomposition of X and the boundary conditions on ψ will be the principal subject of study for the stationary equations.

4. Positive cones and Banach lattices

In this section we shall introduce a series of standard results on positive cones and on linear operators that leave them invariant. These results will be used in subsequent chapters primarily to determine when the solutions of certain kinetic equations are nonnegative.

Given a real Banach space X, a **positive cone** in X is a nonempty closed subset K of X that satisfies the following properties:

(i) K is closed with respect to addition, i.e., $x+y \in K$ whenever $x \in K$ and $y \in K$.

(ii) K is closed with respect to multiplication by a nonnegative scalar constant, i.e., $\lambda x \in K$ whenever $\lambda \in [0,\infty)$ and $x \in K$.

(iii) The zero vector is the only element $x \in X$ for which x and $-x$ belong to K.

A positive cone in X defines a partial order on X by $y \geq x$ whenever $y-x \in K$, whence $\{x \in X : x \geq 0\} = K$. We call K a **solid cone** if it has a nonempty interior, i.e., if there exist $x \in K$ and $\varepsilon > 0$ such that $\{y \in X : \|x-y\|_X < \varepsilon\} \subset K$. We call K a **reproducing cone** if every element of X is the difference of two elements of K, i.e., $\{x-y : x \in K, y \in K\} = X$. Finally, we call K a **normal cone** in X if there exists a constant M such that $\|x\| \leq M\|y\|$ whenever x and $y-x$ belong to K.

It is easily seen that every solid cone is reproducing. A very simple example of a solid cone is provided by vectors with nonnegative entries in $\mathbb{R}^n$. If $(\Sigma,d\mu)$ is a measure space, then the nonnegative functions in $L_p(\Sigma,d\mu)$ form a reproducing cone, which is solid if $p=\infty$. If Σ is a compact Hausdorff space, then the nonnegative functions in the space $C(\Sigma)$ of real continuous functions on Σ with supremum norm form a solid cone in $C(\Sigma)$. More generally, if Σ is a Tychonoff space, then the nonnegative functions in the space $C(\Sigma)$ of bounded real continuous functions on Σ with supremum norm form a solid cone in $C(\Sigma)$. All of these cones are normal with $M=1$ for the constant.

Consider a real Banach space X with reproducing cone K, and let X^* be the dual of X. Then the positive linear functionals on X, i.e., those $\varphi \in X^*$ satisfying $\varphi(x) \geq 0$ for all $x \in K$, form a positive cone in X^*, which is referred to as the **dual cone** to K and denoted by K^*. It should be emphasized that K is taken to be reproducing so as to guarantee that K^* is a cone in X^*. One may prove that K^* is a normal cone in X^*, and that K^* is a reproducing cone if and only if K is a normal cone. We call a vector $x \in K$ **strictly positive**, if $\varphi(x) > 0$ for every $0 \neq \varphi \in K^*$. In some publications strictly positive vectors are called interior vectors. This is related to the fact that in Banach spaces with solid cones the strictly positive vectors are precisely the points of the interior of K. It is easily seen that the cone of nonnegative functions in $L_p(\Sigma,d\mu)$ is the dual of the cone of nonnegative functions in $L_q(\Sigma,d\mu)$, where $q=p/(p-1)$ and $1<p\leq\infty$. In these spaces a nonnegative function f is strictly positive if its set of zeros has μ-measure zero.

Given a real Banach space X, its **complexification** is defined to be the vector space $X_{\mathbb{C}}$ of ordered pairs (x,y) of vectors $x,y \in X$, with (x,y) written as x+iy, endowed with the norm

$$\|x+iy\| = \sup_{0\leq\theta\leq 2\pi} \{\cos\theta \;\|x\| + \sin\theta \;\|y\|\},$$

and with scalar multiplication defined by $\lambda(x+iy) = (ux-vy)+i(vx+uy)$ for $\lambda \in \mathbb{C}$ with $u=\text{Re}\lambda$ and $v=\text{Im}\lambda$. The real space X is contained in $X_{\mathbb{C}}$ by identifying $x \in X$ with x+i0 in $X_{\mathbb{C}}$. The notions defined above related to positive cones in Banach spaces carry over in an obvious fashion to their complexifications. It is common to utilize the complexification of a real Banach space without explicit distinction in the notation.

Positive cones in Banach spaces were first systematically studied by Krein and Rutman in the late 1930's. The main results were published in [229], primarily for solid cones. The extension to reproducing cones, and even more general ones, was made by Krasnoselskii and co-workers (see [223], Chapter 1). The relationship between normal and reproducing for dual cones is due to Ando [7], who generalized a result by Krein [225] for solid cones.

A **real Banach lattice** is a real Banach space X with reproducing and normal cone K that satisfies the following two conditions:

(i) Every pair of vectors $x,y \in X$ has a supremum $z=\max\{x,y\}$, i.e., $z\geq x$ and $z\geq y$, while $\hat{z}\geq x$ and $\hat{z}\geq y$ imply $\hat{z}\geq z$.

(ii) The norm of X is monotonic with respect to the order induced by K, i.e., if $0\leq x\leq y$, then $\|x\| \leq \|y\|$.

As a result every pair of vectors $x,y \in X$ has an infimum, namely $w=-\sup\{-x,-y\}$. Also, for every $x \in X$ one may define the absolute value $|x| \in K$ as $\max\{x,-x\}$. Then $|x|=x$ if and only if $x \in K$, while $\| |x| \| = \|x\|$. Moreover, there exist unique vectors x_+ and x_- in K such that $x_+ + x_- = |x|$ and $x_+ - x_- = x$. By an **ideal** in X we mean a (not necessarily closed) linear submanifold I of X satisfying the condition $\{x \in X : |x| \leq |y| \text{ for some } y \in I\} = I$.

The complexification of a real Banach lattice is called a **complex Banach lattice**. All of the examples of Banach spaces with cone provided above are in fact Banach lattices. A comprehensive account of the theory of Banach lattices can be found in the monographs by Schaefer [324], Zaanen [402], and Aliprantis and Burkinshaw [3].

Given a (real or complex) Banach lattice X, one may construct the dual cone K^* and the corresponding partial order on X^*. Then for every $\varphi \in K^*$ we have $|\varphi(x)| \leq \varphi(|x|)$ for any $x \in X$, whence $\|\varphi\| = \sup\{\varphi(x): x \in K, \|x\|=1\}$, $\varphi \in K^*$. As a result, if $0 \leq \varphi \leq \psi$ in X^*, then $\|\varphi\| \leq \|\psi\|$. One may then derive that X^* with the above partial order is a Banach lattice, called the **dual Banach lattice** to X.

If $\{U(t)\}_{t \geq 0}$ is a semigroup defined on a Banach lattice X, then U is said to be a **positive semigroup** if $U(t)x \geq 0$ for all $t>0$, whenever $x \geq 0$. Positive semigroups have important applications connected to their ergodic properties (cf. [120, 278]). In Section XII.4, a systematic study of the relationship between the spectral properties of a positive semigroup and its generator and the asymptotic behavior of the solution of the corresponding Cauchy problem will be presented.

A triumph of the theory of positive cones and Banach lattices has been the spectral theory of positive operators. The origin of this theory is Perron and Frobenius' work on positive eigenvalues of nonnegative matrices (cf. [123, 124, 303]) and its generalization to integral operators by Jentzsch [203]. Let X be a (real or complex) Banach space with cone K, and let A be a linear operator with domain X. Then A is called a **positive operator in the lattice sense** (or simply positive when there is no chance of confusion with the notion of positive in the Hilbert space sense) if $A[K] \subset K$. It is easily seen that every positive operator is bounded. Given $0 \neq u_0 \in K$, we call A u_0**-positive**, if for every $0 \neq x \in K$ there exist $m \in \mathbb{N}$ and $\alpha, \beta \in (0,\infty)$ such that $\alpha u_0 \leq A^m x \leq \beta u_0$. We call A **irreducible** if $\{0\}$ and X are the only closed invariant ideals of A. Otherwise we call A **reducible**. Every u_0-positive operator A is irreducible; indeed, it is immediate to see that the only closed invariant ideal I of A is the principal ideal $I = \{x \in X : |x| \leq \beta u_0 \text{ for some } \beta \in (0,\infty)\}$. This ideal then coincides with X, due to the fact that the cone of X is reproducing and normal.

The next result, as well as Theorems 4.3 and 4.4, are due to Krasnoselskii ([223], Chapter 2), extending earlier results of Krein and Rutman [229] for solid cones. There are various generalizations to operators whose spectral radii are isolated eigenvalues (cf. [324, 402]). Actually, the result is usually formulated for compact operators; the extension to power compact operators (i.e., A^n compact for some $n \in \mathbb{N}$) is straightforward.

THEOREM 4.1. Let A be a power compact positive operator on a Banach space X with a reproducing cone. Then either the spectral radius of A vanishes or the spectral radius of A is a positive eigenvalue with at least one corresponding eigenvector in K. The adjoint operator A^* has the same property.

For power compact operators one may conclude that the spectral radius belongs to the spectrum. That this is also true for general positive operators can be seen from the following result of Karlin ([212], Theorem 4).

THEOREM 4.2. Let X be a Banach space with reproducing and normal cone K. Then the spectral radius of every positive operator on X belongs to its spectrum.

The next results have generalizations to irreducible operators (cf. [324, 402]) and related results for positive semigroups and the resolvents of their generators.

THEOREM 4.3. Let X be a (real or complex) Banach space with cone K, and let A be a u_0-positive operator on X. Then we have the following statements:

(i) The spectral radius r(A) of A is a positive eigenvalue of A. This eigenvalue is algebraically simple and the corresponding eigenvector x_0 belongs to K and satisfies $\gamma u_0 \leq x_0 \leq \delta u_0$.

(ii) The only eigenvalue of A on the circle $\{\lambda \in \mathbb{C} : |\lambda| = r(A)\}$ is $\lambda = r(A)$.

(iii) The only eigenvalue of A to which corresponds at least one eigenvector in K is $\lambda = r(A)$.

THEOREM 4.4. Let X be a (real or complex) Banach space with reproducing cone K, and let A be a u_0-positive operator on X with spectral radius r(A). Consider the vector equation $(I - cA)x = y$, where $y \in K$. Then the following statements hold true:

(i) For $0 \le c < r(A)^{-1}$ there is a unique solution $x \in K$ for every $y \in K$, which is given by the absolutely convergent series

$$x = \sum_{n=0}^{\infty} c^n A^n y.$$

(ii) For $c = r(A)^{-1}$ and $y \in K$ there is no solution $x \in K$ unless $y = 0$. In that case all the solutions in K are positive multiples of the positive eigenvector corresponding to the eigenvalue $r(A)$.

(iii) For $c > r(A)^{-1}$ and $y \in K$ there do not exist any solutions $x \in K$ unless $y = 0$. In this case $x = 0$ is the only solution in K.

We conclude this chapter with a result of Nelson [282] which generalizes somewhat the previous theorems.

THEOREM 4.5. Let X be a Banach lattice and B a positive power compact operator on X, and suppose $0 < c_1 < c_2 < \ldots$ are a finite or countably infinite set of numbers for which $(I - cB)h = 0$ with $c = c_i$ has a nonzero positive solution. Then for $c \ge 0$ and $k \ge 0$ the linear equation $(I - cB)h = k$ has a solution $h \ge 0$ if and only if $(k, |\varphi|) = 0$ for every $\varphi \in X^*$ such that $(I - c_i B^*)^p \varphi = 0$ for some $c_i \le c$ and $p \in \mathbb{N}$.

Chapter II

STRICTLY DISSIPATIVE KINETIC MODELS

1. Introduction and historical development

In the next several chapters we shall develop an existence and uniqueness theory for the Hilbert space boundary value problem

$$T\psi'(x) = -A\psi(x) + q(x), \quad 0<x<\infty, \tag{1.1}$$

$$Q_+\psi(0) = \varphi_+, \tag{1.2}$$

$$\limsup_{x\to\infty} \|\psi(x)\| < \infty, \tag{1.3}$$

where T and A are self adjoint operators on a complex Hilbert space H, the null space Ker T $= \{0\}$, and Q_+ is the orthogonal projection of H onto the maximal T-invariant subspace on which T is positive. The meaning of $T\psi'$ and of a solution to (1.1)–(1.3) will be made precise below. Because of the mathematical techniques to be employed, it is convenient to study separately several different classes of collision operators A. In this chapter we shall assume that A is strictly positive; namely, its spectrum $\sigma(A)\subset(0,\infty)$. In the next chapter we shall relax this assumption to allow $\sigma(A)\subset[0,\infty)$ with A Fredholm. Finally, in Chapter IV we shall allow A to have a finite dimensional negative part. Aside from mathematical considerations, these classes of collision operators represent different types of physical models. Strictly positive collision operators describe transport in dissipative media, such as subcritical neutron transport and radiative transfer in absorbing media. Positive collision operators with Ker A non-trivial are typical of transport in systems with conservation laws, such as those in rarefied gas kinetics. Finally, non-positive collision operators occur in kinetic equations relevant to supercritical media, as in neutron transport in multiplying media.

Within this chapter, the diversity of mathematical techniques, and, indeed, the results themselves will warrant a further division into subclasses depending upon the collision operator A being a compact perturbation of the identity (Section 2), a bounded operator (Section 3), or an unbounded operator (Section 4). An additional complication, requiring somewhat sensitive treatment, is introduced if the operator T is unbounded.

At this time, before concluding with some historical comments, it seems opportune to sketch the method which will dominate this and subsequent chapters. In order to keep matters simple, we will assume A (self adjoint) strictly positive and bounded, T self adjoint, bounded and injective. Then the homogeneous boundary value problem to be solved, with $q(x)\equiv 0$, may be written as

$$\frac{\partial \psi}{\partial x} = -K\psi, \quad 0<x<\infty, \tag{1.4}$$

$$Q_+\psi(0) = \varphi_+, \tag{1.5}$$

$$\limsup_{x\to\infty} \|\psi(x)\| < \infty, \tag{1.6}$$

where $K = T^{-1}A$. We shall define, in later sections, precisely what is meant by a solution of (1.4)-(1.6). At present we will require $\psi:[0,\infty)\to H$ continuous on $[0,\infty)$, (strongly) differentiable on $(0,\infty)$, and satisfying (1.4)-(1.6). To deal with the inhomogeneous boundary value problem, where $q(x)\not\equiv 0$, one has only to prove the existence of a particular solution.

At first glance, the solution of such a problem for general vectors $\varphi_+\in \text{Ran}\, Q_+$ may appear not to exist. Indeed, in typical cases the spectrum $\sigma(T)$ of T satisfies $\sigma(T)\supset(-\varepsilon,\varepsilon)$, $\varepsilon>0$, and A is a compact perturbation of the identity, so that the spectrum $\sigma(K)$ of K contains unbounded subsets of both left and right real half-axes, and K certainly does not generate a semigroup on H. Of course, the relevant point is that the initial datum (1.5) is not the specification of the initial state $\psi(0)$, but rather only a projected part of $\psi(0)$, the "incoming flux" $Q_+\psi(0)$. The formal solution

$$\psi(x) = e^{-xK}\,\psi(0) \tag{1.7}$$

will be a solution of the boundary value problem if and only if $\psi(0)$ belongs to the subspace associated with the "positive part" of K and (1.5) is satisfied.

Let us be more precise. Suppose H may be decomposed as

$$H = H_+ \oplus H_- \tag{1.8}$$

in such a way that $\mp K$ generate bounded semigroups on $H_{\pm}$. We let $P_{\pm}$ be the complementary projections $P_{\pm}:H\to H_{\pm}$ and we define the complementary projection $Q_{-}=I-Q_{+}$. Q_{-} is related to the homogeneous left half space problem

$$\frac{\partial\psi}{\partial x} = -K\psi, \qquad -\infty<x<0 , \tag{1.9}$$

$$Q_{-}\psi(0) = \varphi_{-}, \tag{1.10}$$

$$\limsup_{x\to-\infty} \|\psi(x)\| < \infty. \tag{1.11}$$

Then $\psi(x)$ as given in (1.7) will be a solution of the right half space boundary value problem if $\psi(0)\in \mathrm{Ran}\, P_{+}$ and $Q_{+}\psi(0) = \varphi_{+}\in \mathrm{Ran}\, Q_{+}$. Likewise, it will be a solution of the left half space boundary value problem if $\psi(0)\in \mathrm{Ran}\, P_{-}$ and $Q_{-}\psi(0) = \varphi_{-}\in \mathrm{Ran}\, Q_{-}$.

The **albedo operator** E is defined by $E: \varphi_{\pm}\to\psi(0)$, where $\psi(x)$ is the solution of the corresponding homogeneous half space boundary value problem (if such a solution exists). Both left and right homogeneous half space problems are treated simultaneously precisely so that the albedo operator will be defined on all of $\mathrm{Ran}\, Q_{+} \oplus \mathrm{Ran}\, Q_{-} = H$. Since the solution of the homogeneous half space problems is given by

$$\psi(x) = e^{-xK}\, E\varphi_{\pm}, \tag{1.12}$$

the existence of solutions for all $\varphi_{\pm}\in Q_{\pm}[H]$ (uniqueness is guaranteed by the dissipativity assumption on $-A$) is equivalent to the existence of the linear operator $E:H\to H$.

The albedo operator may be defined abstractly as a linear operator $E:H\to H$ satisfying:

$$\text{(i)}\qquad Q_{\pm}EQ_{\pm} = Q_{\pm}, \tag{1.13}$$

$$\text{(ii)}\qquad P_{\mp}EQ_{\pm} = 0. \tag{1.14}$$

The first condition merely assures that the solution (1.12) satisfies the boundary condition (1.5) or (1.10). The second condition guarantees $E\varphi_{\pm}\in \mathrm{Ran}\, P_{\pm}$. In the terminology of transport theory, these conditions imply that if $f\in \mathrm{Ran}\, Q_{+}$ is an incoming flux for a right half space problem, then Ef will be the corresponding total (incoming plus reflected) flux, and if $f \in \text{Ran } Q_-$ is an incoming flux for a left half space problem, then Ef will be the corresponding total flux. Assuming, for the moment, that such an albedo operator exists, let us find an explicit representation for it. Using (i) and (ii) above gives easily

$$P_\pm E = P_\pm E(Q_+ + Q_-) = P_\pm EQ_\pm = EQ_\pm \tag{1.15}$$

and

$$Q_\pm P_\pm EQ_\pm = Q_\pm . \tag{1.16}$$

Then, adding the ± equations in (1.16) and utilizing (1.15) yields

$$Q_+P_+EQ_+ + Q_-P_-EQ_- = (Q_+P_+ + Q_-P_-)E = I. \tag{1.17}$$

Indeed, the existence of E, and therefore the existence of solutions of the half space problems for **all** boundary data in Ran $Q_\pm$, will follow at once from the bijectivity of

$$V = Q_+P_+ + Q_-P_- . \tag{1.18}$$

Most of the work to be carried out in the present and the next two chapters will be devoted to proving the invertibility of V.

We have glossed over the decomposition (1.8). In terms of the (equivalent) inner product

$$(h,k)_A = (Ah,k) \tag{1.19}$$

on H one sees easily that K is self adjoint with respect to this inner product. Then the Spectral Theorem provides the decomposition and the projections $P_\pm$, and only the analysis of V remains.

When A is strictly positive and a compact perturbation of the identity, the scenario above may be carried out rigorously without additional complications. A non-trivial null space Ker $A \neq \{0\}$ complicates the geometry somewhat; however, a trick introduced by Beals [32] and van der Mee [360] enables one to carry out the necessary bookkeeping in a convenient fashion. Already for more general bounded (but still positive) A, a more serious complication arises, as the proof of the invertibility of V on H breaks down. We shall rectify this difficulty by seeking a solution of the boundary value problem in a larger Hilbert space $H_T \supset H$; in this sense we will be admitting weaker solutions. Both A unbounded and T unbounded introduce increasing complications, and when the dissipativity assumption on $-A$ is relaxed, it will be convenient to view the inner product (1.19) as an indefinite metric on a Krein space.

As discussed in the previous chapter, van Kampen [371] and Case [68] developed the singular eigenfunction expansion method to solve half space boundary value problems. Despite the widespread use of this method, even at present, many mathematicians remained unconvinced of the validity of such computations because of the non-rigorous treatment of the continuous spectrum involved in the eigenfunction expansion. This problem was alleviated somewhat by the resolvent integration approach introduced in 1971 by Larsen [239, 241]. The method became a popular tool in deriving explicit representations of solutions of various kinetic equations and has been applied far beyond the range of strictly dissipative models. In 1973 Hangelbroek [180, 181] introduced an operator theoretic approach to linear transport equations which, with further developments by a number of authors, has provided the framework upon which the abstract theory of the next few chapters will be constructed.

Hangelbroek's work was extended to isotropic neutron transport in conservative media by Lekkerkerker [248] and to the scalar BGK model equation by Kaper [207]. Other applications followed in anisotropic one-speed neutron transport. Lekkerkerker [249] studied the case of a degenerate scattering law but did not include a proof of the invertibility of V, though it can be constructed using the method of [184]. A proof covering both degenerate and non-degenerate scattering is due to van der Mee [359] and is presented also in Chapter 6 of [25]. We note also a proof for special non-degenerate cases [182]. Finally, a somewhat weaker existence and uniqueness result follows from the abstract theory of boundary value problems presented by Beals [32].

In recent years the class of half space problems that can be treated rigorously has been vastly extended by the emergence of so-called abstract kinetic models, where the boundary value problem (1.1)-(1.3) is studied for a large class of operators T and A. Concrete problems in neutron transport, radiative transfer, rarefied gas dynamics et al. then arise as natural applications. Overall, two variants of such a theory can be distinguished with different possibilities of generalization. The first one of these variants was developed by Beals [32, 33] and extended by Greenberg et al. [161, 166], and has proved particularly successful in situations where A is a differential operator (see [34]; also Chapter X). The second one of these variants was developed by van der Mee [359, 360] and applies to models where A is a compact perturbation of the identity.

Beals [32] considered an arbitrary (bounded or unbounded) injective self adjoint operator T and an arbitrary bounded strictly positive A (as well as non-injective A with special Jordan structure). The novelty in Beals' approach was to consider two additional Hilbert space extensions, namely the completion H_T of D(T) with respect to the inner product

$$(h,k)_T = (|T|h,k) \tag{1.20}$$

and the completion H_S of D(A) with respect to the inner product

$$(h,k)_S = (|A^{-1}T|h,k)_A = (T(P_+-P_-)h,k). \tag{1.21}$$

It is then immediate that the orthogonal projections $Q_\pm$ of H onto the maximal T-positive/negative T-invariant subspaces admit continuous extensions from D(T) to H_T, while the orthogonal (with respect to the inner product (1.19)) projections $P_\pm$ associated with the spectral decomposition of $T^{-1}A$ allow continuous extensions from $D(A^{-1}T)$ to H_S. The second new element of Beals' approach was the proof of the equivalence of the inner products (1.20) and (1.21) on D(T), thereby enabling the natural identification $H_T \simeq H_S$. The invertibility of the operator $V = Q_+P_+ + Q_-P_-$ on H_T then arises as a corollary. As a result, Eqs. (1.1)–(1.3) were shown to be uniquely solvable in the space H_T, which is an extension of D(T).

Van der Mee [359, 360] considered an arbitrary bounded injective self adjoint operator T and a positive operator A which is a compact perturbation of the identity satisfying the weak regularity condition

$$\exists\alpha>0: \text{Ran}(I-A) \subset \text{Ran}\,|T|^{\alpha}. \tag{1.22}$$

Under these hypotheses he proved the invertibility of V and therefore the unique solvability of Eqs. (1.1)–(1.3) on the given Hilbert space H. Although this result implies Beals' result in H_T (see Section 3), it provides more detailed information about the solution, useful in further developments (see Chapters VI to VIII), but the price one must pay is its applicability to a more restricted class of problems. The invertibility proof for V relies on the Fredholm alternative, where the proof of Ker V = {0} is borrowed from [184]. The compactness of I–V, needed to implement the

Fredholm alternative, follows from the identity $I - V = (Q_- - Q_+)(P_+ - Q_+)$ in combination with an estimate applied to a resolvent integral representation for the projections P_+ and Q_+. The estimate requires Eq. (1.22) as a tool. In Section 2 we shall also provide an extension of this approach to unbounded T, due to Greenberg et al. [165]

On generalizing the Beals method to include unbounded operators A, the proof of the equivalence of the inner products (1.20) and (1.21), and thus the proof of the natural identification $H_T \simeq H_S$, appeared to be the major obstacle. A one sentence argument by Beals [32] intended to establish this result for the electron scattering equation (see Section X.6) was generally considered overconcise. Finally, Beals [34] proved $H_T \simeq H_S$ and the unique solvability for a large class of problems (1.1)–(1.3), where T is multiplication by a (sufficiently regular) function and A is a positive self adjoint Sturm–Liouville differential operator. In this way the unique solvability of the electron scattering problem was settled. (We remark that a previous and detailed proof of $H_T \simeq H_S$, formulated for a specific Sturm–Liouville example, namely the Fokker–Planck equation, appeared in [35].) Recently Curgus [96] observed that the identification $H_T \simeq H_S$ is equivalent to infinity being a regular critical point of $T^{-1}A$ in a suitable indefinite inner product on H_T (see Section IV.1 for the terminology), and used this observation to obtain an alternative proof.

A somewhat different route was followed by Greenberg et al. [166] Arguing that $E = V^{-1}$ is the relevant operator rather than V (since E appears in the solution and V does not), they proved that $E = V^{-1}$ admits a continuous extension from $V[D(A^{1/2})]$ to a bounded operator from H_T into the completion of D(A) with respect to the sum of the inner products (1.20) and (1.21), and so they circumvented the issue of establishing whether $H_T \simeq H_S$ or not. A technical difficulty still is a surjectivity assumption on V. The approach of [166] was subsequently extended to unbounded T (see [160]), but now the theory only goes through with the additional assumption of the essential self adjointness of $A^{-1}T$ on the space $D(A^{1/2})$ endowed with the complete inner product (1.19).

Rigorous results for multiplying media (A with negative part) have been obtained only more recently. Ball and Greenberg [21] utilized Krein space theory to study the neutron transport equation for isotropic scattering and the one speed approximation. A theory for the abstract equation with T bounded has been given by Greenberg and van der Mee [163]. In concluding this historical overview, we note that the last reference provided also an analysis of the connection between existence and uniqueness and the asymptotic behavior of solutions for both conservative and multiplying media.

The following notation is more or less standard up through Chapter X. H is a complex Hilbert space, T is a self adjoint injective operator on H, and A is a self adjoint Fredholm operator on H. $Q_\pm$ are the orthogonal projections of H onto the maximal T-invariant subspaces on which T is positive/negative. $Q_\pm$ will also be used for the natural extensions of the orthogonal projections to $H_T \supset D(T)$.

2. Strong solutions

In this section A will be a strictly positive operator which is a compact perturbation of the identity satisfying the regularity condition

$$\exists \alpha>0: \operatorname{Ran}(I-A) \subset \operatorname{Ran} |T|^{\alpha} \cap D(|T|^{1+\alpha}). \tag{2.1}$$

We shall consider the boundary value problem

$$(T\psi)'(x) = -A\psi(x) + q(x), \quad 0<x<\infty, \tag{2.2}$$

$$Q_+\psi(0) = \varphi_+, \tag{2.3}$$

$$\|\psi(x)\|_H = O(1) \ (x\to\infty). \tag{2.4}$$

As we shall illustrate in Chapter IX, boundary value problems of the above type, with T and A subject to these hypotheses, occur frequently in one-speed and symmetric multigroup neutron transport, radiative transfer and rarefied gas dynamics. For bounded T the results of this section are due to van der Mee [359, 360], except for Lemma 2.7 which was derived by Hangelbroek and Lekkerkerker [184]. The generalization to unbounded T is due to Greenberg et al. [165]

We shall define a solution of Eqs. (2.2)-(2.4) for any $\varphi_+ \in Q_+[D(T)]$ to be a continuous function $\psi:[0,\infty)\to H$ with values in D(T) such that $T\psi$ is strongly differentiable on $(0,\infty)$ and Eqs. (2.2)-(2.4) are satisfied. Since ψ takes values in the original space H, and not an enlargement thereof, ψ is called a **strong** solution of the boundary value problem (cf. Sections 3 and 4). The inhomogeneous term $q:[0,\infty)\to H$ will be bounded and uniformly Hölder continuous. Toward the end of this section we will define a different type of (strong) solution which will allow for a wider class of boundary data (for unbounded T), namely $\varphi_+ \in Q_+[H]$.

Let us consider the inner product (1.19). Due to the strict positivity and boundedness of A, this inner product is equivalent to the original inner product. Furthermore, $A^{-1}T$ is an (unbounded) self adjoint operator on H relative to this inner product. Indeed, if $h \in D(T) \to (A^{-1}Th,k)_A = (Th,k)$ is bounded, then this implies that $k \in D(T^*) = D(T) = D(A^{-1}T)$. We write H_A for the vector space H endowed with the inner product (1.19) and define $P_\pm$ as the H_A-orthogonal projections of H onto the maximal $A^{-1}T$-positive/negative $A^{-1}T$-invariant subspaces. Since, by assumption, T and $A^{-1}T$ have zero null spaces, $P_\pm$ (and $Q_\pm$) are pairs of complementary projections. Actually, one can say more. $P_\pm$ (and $Q_\pm$) are invariant on D(T) and are bounded complementary projections on the complete inner product space D(T) with graph norm defined by

$$(h,k)_{GT} = (h,k) + (Th,Tk). \tag{2.5}$$

The self adjointness of $A^{-1}T$ with respect to the inner product (1.19) allows the machinery of the Spectral Theorem to be introduced. It is then evident that the restrictions of $e^{\mp xT^{-1}A}P_\pm$ to Ran $P_\pm$ are bounded analytic semigroups on Ran $P_\pm$, whose infinitesimal generators are the (unbounded) inverses of the restrictions of $\mp A^{-1}T$ to Ran $P_\pm$, and therefore differentiable on $(0,\infty)$ with respect to the operator norm topology. From the injectivity of A, we have

$$\lim_{x\to\infty} \| e^{\mp xT^{-1}A} P_\pm h \| = 0, \qquad h \in H. \tag{2.6}$$

Exploiting the invariance of the semigroups on D(T), we may also demonstrate that the semigroups restricted to D(T) are bounded analytic semigroups on Ran $P_\pm \cap D(T)$ relative to the topology generated by the graph norm (2.5).

We shall first state a result on Bochner integration, due to Hille and contained in [104], which is straightforward to prove and will be crucial in the present section as well as in Chapters VI and VII.

LEMMA 2.1. Let R be a closed linear operator and let $\varphi:(a,b)\to D(R)\subset H$ be a vector function satisfying the integrability conditions

$$\int_a^b \|\varphi(x)\| dx < \infty,$$

$$\int_a^b \|R\varphi(x)\|dx < \infty.$$

Then the integral $\int_a^b \varphi(x)dx \in D(R)$ and $R\int_a^b \varphi(x)dx = \int_a^b R\varphi(x)dx.$

We next apply this lemma to derive a representation of the solution of the homogeneous boundary value problem.

LEMMA 2.2. $\psi(x)$ is a solution of the homogeneous boundary value problem (2.2)–(2.4), where $q(x)\equiv 0$, if and only if

$$\psi(x) = e^{-xT^{-1}A}\, h, \qquad 0<x<\infty, \tag{2.7}$$

for some $h\in \text{Ran}\, P_+\cap D(T)$ with $Q_+h=\varphi_+$. All such solutions are strongly differentiable on $(0,\infty)$ and vanish at infinity with respect to the graph topology of $D(T)$ as well as the original topology of H.

Proof: Suppose $\psi:[0,\infty)\to D(T)$ is a solution of the homogeneous boundary value problem. Using the facts that $\psi:[0,\infty)\to H$ is continuous, A is bounded and $(T\psi)' = -A\psi$, we see that the function $(T\psi)'$ is bounded and continuous on $(0,\infty)$. Since, for $0<\varepsilon<x<\infty$,

$$T\psi(x)-T\psi(\varepsilon) = \int_\varepsilon^x (T\psi)'(y)dy = -A\int_\varepsilon^x \psi(y)dy,$$

it follows that $T\psi$ is continuous on $(0,\infty)$ and satisfies $\|T\psi(x)\|_H=O(x)$ $(x\to\infty)$. For all $\text{Re}\lambda<0$ and $m>0$, we have

$$\int_\varepsilon^m e^{x/\lambda}\psi(x)dx \in D(T), \qquad T\int_\varepsilon^m e^{x/\lambda}\psi(x)dx = \int_\varepsilon^m e^{x/\lambda}T\psi(x)dx,$$

as a consequence of the previous lemma, whence

$$0 = \lambda\int_\varepsilon^m e^{x/\lambda}\{(T\psi)'(x) + A\psi(x)\}dx =$$

$$= [\lambda e^{x/\lambda}T\psi(x)]_{x=\varepsilon}^m - (T-\lambda A)\int_\varepsilon^m e^{x/\lambda}\psi(x)dx. \tag{2.8}$$

Notice that $T\psi(x)$ has a strong limit as $x\downarrow 0$; using that ψ is continuous on $[0,\infty)$ and T is a closed operator, the limit is, in fact, $T\psi(0)$, and (2.8) is valid for $\varepsilon=0$. Noting also that $\lambda \epsilon^{x/\lambda} T\psi(x)$ vanishes as $x\to\infty$, we have

$$\int_0^\infty e^{x/\lambda}\psi(x)dx = \lambda(\lambda - A^{-1}T)^{-1}A^{-1}T\psi(0) = G(\lambda)$$

for all non-real λ in the left half plane. Since $G(\lambda)$ has an analytic continuation to the left half plane, the same is true for $(\lambda - A^{-1}T)^{-1}\psi(0) = \lambda^{-2}G(\lambda) + \lambda^{-1}\psi(0)$. We may therefore conclude that $\psi(0)\in \text{Ran } P_+ \cap D(T)$, and Eq. (2.2) with $\psi(0)$ given a priori is an initial value problem on Ran P_+ whose solution must have the form (2.7). The relation $Q_+\psi(0) = \varphi_+$ is immediate, as is the proof of the converse argument. ∎

It is clear that the operator $V = Q_+P_+ + Q_-P_-$ leaves invariant $D(T)$. We shall establish the compactness of $I-V$ on H and on $D(T)$. First we shall state three technical lemmas from operator theory. The first of these is a consequence of the norm closedness of the algebra of compact operators, and the second is a moment inequality, which follows easily from the Spectral Theorem and Hölder's inequality. The third was proved by Krein and Sobolevskii [231]; we will sketch a proof presented by Krasnoselskii et al. [224]

LEMMA 2.3. The integral of a (norm) continuous compact operator-valued function with integrable norm is a compact operator.

LEMMA 2.4. Let A be a positive definite self adjoint operator. Then $\|A^\tau x\| \le \|Ax\|^\tau \|x\|^{1-\tau}$ for $x\in D(A)$ and any $\tau\in(0,1)$.

LEMMA 2.5. Let A be a positive definite self adjoint operator and B a closed operator satisfying $D(A)\subset D(B)$ and $\|Bx\| \le k\|Ax\|^\tau \|x\|^{1-\tau}$ for $x\in D(A)$ and some $\tau\in(0,1)$. Then $D(A^\delta)\subset D(B)$ and $\|Bx\| \le k_0 \|A^\delta x\|$ for $x\in D(A^\delta)$ and all $\delta > \tau$.

Proof: We may see easily that the vector function

$$t^{-\delta}B(tI + A)^{-1}x = BA^{-1}t^{-\delta}A(tI + A)^{-1}x$$

is continuous for $t>0$ and all $x\in D(A)$. Let us estimate:

$$\int_0^\infty t^{-\delta}\|B(tI+A)^{-1}x\|dt \le k\int_0^\infty t^{-\delta}\|A(tI+A)^{-1}x\|^{\tau}\|(tI+A)^{-1}x\|^{1-\tau}dt \le$$

$$\le k\int_0^\infty t^{-\delta}(a+t)^{\tau-1}dt\|x\| = k_1\|x\|$$

for some $a>0$, where we have used that $1>\delta>\tau$. Thus $\int_0^\infty t^{-\delta}(tI+A)^{-1}xdt$ $\epsilon D(B)$ and

$$B\int_0^\infty t^{-\delta}(tI+A)^{-1}xdt = \int_0^\infty t^{-\delta}B(tI+A)^{-1}dt.$$

Using the representation

$$A^{-\alpha} = \pi^{-1}\sin\alpha\pi\int_0^\infty t^{-\alpha}(tI+A)^{-1}dt,$$

which is obtained from the Spectral Theorem with some manipulation of integrands, it follows that $D(A^{\delta})\subset D(B)$ and $\|BA^{-\delta}x\|\le k_1\|x\|$. ∎

LEMMA 2.6. The operator P_+-Q_+ is compact on H and the restriction of P_+-Q_+ to D(T) is compact on D(T) (endowed with the inner product (2.5)). Moreover, $(P_+-Q_+)[H]\subset D(T)$.

Proof: We will prove first that P_+-Q_+ is compact on H and $(P_+-Q_+)[H]\subset D(T)$. Let $\Delta_1=\Delta(\varepsilon,M)$ denote the oriented curve composed of the straight lines from $-i\varepsilon$ to $-i$, from $-i$ to $M-i$, from $M+i$ to i, and from $+i$ to $+i+\varepsilon$. Let $\Delta_2=\Delta(M)$ denote the oriented curve composed of the straight lines from $M-i$ to $+\infty-i$ and from $+\infty+i$ to $M+i$. Write $\Delta=\Delta_1\cup\Delta_2$ with the orientation inherited from Δ_1 and Δ_2. We recall that the projections P_+ and Q_+ are bounded on H and on D(T) endowed with the graph inner product (2.2). We have the integral representations

$$P_+ = \lim_{\varepsilon\to 0}\frac{1}{2\pi i}\int_\Delta(\lambda-S)^{-1}d\lambda,$$

$$Q_+ = \lim_{\varepsilon\to 0}\frac{1}{2\pi i}\int_\Delta(\lambda-T)^{-1}d\lambda,$$

where the limits are in the strong topology and $S=A^{-1}T$. Let

$$P_+^{(1)} = \lim_{\varepsilon\to 0}\frac{1}{2\pi i}\int_{\Delta_1}(\lambda-S)^{-1}d\lambda,\quad P_+^{(2)} = \frac{1}{2\pi i}\int_{\Delta_2}(\lambda-S)^{-1}d\lambda,$$

and

$$Q_+^{(1)} = \lim_{\varepsilon\to 0} \frac{1}{2\pi i}\int_{\Delta_1} (\lambda-T)^{-1}d\lambda, \quad Q_+^{(2)} = \frac{1}{2\pi i}\int_{\Delta_2} (\lambda-T)^{-1}d\lambda.$$

Then one has

$$P_+-Q_+ = (P_+^{(1)}-Q_+^{(1)}) + (P_+^{(2)}-Q_+^{(2)}). \tag{2.9}$$

We will show that $P_+^{(1)}-Q_+^{(1)}$ and $P_+^{(2)}-Q_+^{(2)}$ are compact on H, and $(P_+^{(1)}-Q_+^{(1)})[H]\subset D(T)$ as well as $(P_+^{(2)}-Q_+^{(2)})[H]\subset D(T)$.

Consider first

$$P_+^{(1)}-Q_+^{(1)} = \lim_{\varepsilon\to 0} \frac{1}{2\pi i}\int_{\Delta_1} \{(\lambda-S)^{-1}-(\lambda-T)^{-1}\}d\lambda. \tag{2.10}$$

We shall see that this limit can be taken in the norm topology. We exploit the regularity condition (2.1) and obtain from the Closed Graph Theorem the existence of a bounded operator D such that $B=|T|^{\alpha}D$. Then, for non real λ,

$$(\lambda-S)^{-1}-(\lambda-T)^{-1} = (\lambda-T)^{-1}(S-T)(\lambda-S)^{-1} = (\lambda-T)^{-1}BS(\lambda-S)^{-1} =$$

$$= (\lambda-T)^{-1}|T|^{\alpha}D\, S(\lambda-S)^{-1},$$

which shows that $(\lambda-S)^{-1}-(\lambda-T)^{-1}$ is a compact operator on H. Next, since S is self adjoint on H with respect to the inner product (1.19), we may use the Spectral Theorem to derive the norm estimate

$$\|S(i\mu-S)^{-1}\|_{\mathcal{L}(H_A)} \le \sup_{t\in\mathbb{R}} \left|\frac{t}{i\mu-t}\right| \le 1.$$

But the inner products on H and H_A are equivalent, and thus also are the $\mathcal{L}(H)$ and $\mathcal{L}(H_A)$ norms, so there is a constant c_0 such that $\|S(i\mu-S)^{-1}\|_{\mathcal{L}(H)}\le c_0$.

Likewise, from the Spectral Theorem,

$$\| |T|^{\alpha}(i\mu-T)^{-1}\|_{\mathcal{L}(H)} \le \sup_{t\in\mathbb{R}} |\frac{t^{\alpha}}{i\mu-t}| \le c_{\alpha}|\mu|^{\alpha-1}.$$

Thus

$$\|(\int_{\Delta(\varepsilon,M)}-\int_{\Delta(\gamma,M)})[(\lambda-S)^{-1}-(\lambda-T)^{-1}]d\lambda\|_{\mathcal{L}(H)} \le 2\|D\|_{\mathcal{L}(H)}c_0c_{\alpha}\int_{\varepsilon}^{\gamma}\mu^{\alpha-1}d\mu,$$

which shows that the limit (2.10) exists in the operator norm topology, and consequently proves the compactness of $P_+^{(1)}-Q_+^{(1)}$.

Since the vector functions $[(\lambda-S)^{-1}-(\lambda-T)^{-1}]x$ and $T[(\lambda-S)^{-1}-(\lambda-T)^{-1}]x = T(\lambda-T)^{-1}BS(\lambda-S)^{-1}x$ for $x\in H$ are bounded and continuous on Δ_1, we have

$$\frac{1}{2\pi i}\int_{\Delta_1}[(\lambda-S)^{-1}-(\lambda-T)^{-1}]xd\lambda \in D(T)$$

and

$$T(\frac{1}{2\pi i}\int_{\Delta_1}[(\lambda-S)^{-1}-(\lambda-T)^{-1}]xd\lambda) = \frac{1}{2\pi i}\int_{\Delta_1}T[(\lambda-S)^{-1}-(\lambda-T)^{-1}]xd\lambda.$$

Now, note that

$$\|(\int_{\Delta(\varepsilon,M)}-\int_{\Delta(\gamma,M)})T[(\lambda-S)^{-1}-(\lambda-T)^{-1}]d\lambda\|_{\mathcal{L}(H)} \le 2c_0\|B\|_{\mathcal{L}(H)}|\varepsilon-\gamma|$$

implies the existence of the limit

$$\lim_{\varepsilon\to 0}\frac{1}{2\pi i}\int_{\Delta(\varepsilon,M)}T[(\lambda-S)^{-1}-(\lambda-T)^{-1}]xd\lambda$$

in the operator norm topology. Therefore, by the closedness of T,

$$(P_+^{(1)}-Q_+^{(1)})x = \lim_{\varepsilon\to 0}\frac{1}{2\pi i}\int_{\Delta_1}[(\lambda-S)^{-1}-(\lambda-T)^{-1}]xd\lambda \in D(T),$$

which proves the inclusion $(P_+^{(1)}-Q_+^{(1)})[H] \subset D(T)$.

Next let us consider

$$P_+^{(2)}-Q_+^{(2)} = \frac{1}{2\pi i}\int_{\Delta_2}[(\lambda-S)^{-1}-(\lambda-T)^{-1}]d\lambda.$$

Since, for non-real λ, $(\lambda-S)^{-1}-(\lambda-T)^{-1} = (\lambda-T)^{-1}BS(\lambda-S)^{-1}$ is compact, it is sufficient to show the integrability of this operator. We can rewrite the operator in the following form with $C=BA^{-1}$:

$$(\lambda-S)^{-1}-(\lambda-T)^{-1} = (\lambda-T)^{-1}BS(\lambda-T)^{-1}[(\lambda-T)(\lambda-S)^{-1}] =$$

$$= (\lambda-T)^{-1}CT(\lambda-T)^{-1}[(\lambda-T)(\lambda-S)^{-1}].$$

Evidently $\operatorname{Ran}(\lambda-S)^{-1}=D(\lambda-T)$ and by the Closed Graph Theorem $(\lambda-T)(\lambda-S)^{-1}$ is a bounded operator on H. In fact, we will show that the norm of this operator is uniformly bounded for $\lambda\in\Delta_2$. We easily derive the identity $(\lambda-T)(\lambda-S)^{-1} = I+BS(\lambda-S)^{-1} = I+CT(\lambda-S)^{-1}$ By virtue of the estimate $\|(\lambda-S)^{-1}\|_{\mathcal{L}(H)}\le c_0$ for $\lambda\in\Delta_2$, it is sufficient to show that CT is bounded on D(T)=D(S). But by the regularity condition (2.1), Ran C = Ran B $\subset D(|T|^{1+\alpha}) \subset D(T)$ and then by the Closed Graph Theorem, the operator TC is bounded on H, thus $CT \subset (TC)^*$ is bounded on D(T). Finally, for any $\lambda\in\Delta_2$ we have $\|(\lambda-T)(\lambda-S)^{-1}\|_{\mathcal{L}(H)} \le 1+\|(TC)^*\|_{\mathcal{L}(H)}\, c_0$, which provides a λ-uniform bound, as claimed.

Therefore it is sufficient to show the integrability of $F(\lambda) = (\lambda-T)^{-1}CT(\lambda-T)^{-1}$. Let Q_0 be a spectral projection belonging to the spectral decomposition of the self adjoint operator T such that the resolvent set of the restriction of T to the range of $Q_1=I-Q_0$ contains a real neighborhood of zero. We can decompose $F(\lambda)$ as follows:

$$F(\lambda) = (\lambda-T)^{-1}|T|^{-\omega}Q_1|T|^{\omega}C|T|^{1+\nu}|T|^{-1-\nu}T(\lambda-T)^{-1}Q_1 +$$

$$+ (\lambda-T)^{-1}|T|^{-\omega}Q_1|T|^{\omega}CT(\lambda-T)^{-1}Q_0 +$$

$$+ (\lambda-T)^{-1}Q_0C|T|^{1+\nu}|T|^{-1-\nu}T(\lambda-T)^{-1}Q_1 +$$

$$+ (\lambda-T)^{-1}Q_0CT(\lambda-T)^{-1}Q_0,$$

where $\nu=\frac{1}{2}\alpha$ and $2\omega>\max\{1+\alpha,2-\alpha\}$, and we may choose $\omega<1+\frac{1}{2}\alpha$. Note that $\nu+\omega>1$. For $\lambda\in[M\pm i, \infty\pm i)$ we have the following estimates:

$$\|(\lambda-T)^{-1}|T|^{-\omega}Q_1\|_{\mathcal{L}(H)} \le \text{const. } (\operatorname{Re}\lambda)^{-\omega},$$

$$\| |T|^{-1-\nu} T(\lambda-T)^{-1} Q_1 \|_{\mathcal{L}(H)} \leq \text{const. } (\mathrm{Re}\lambda)^{-\nu},$$

$$\|(\lambda-T)^{-1} Q_0\|_{\mathcal{L}(H)} \leq \text{const. } (\mathrm{Re}\lambda)^{-1},$$

$$\|T(\lambda-T)^{-1} Q_0\|_{\mathcal{L}(H)} \leq \text{const. } (\mathrm{Re}\lambda)^{-1}.$$

Moreover, since Ran C = Ran B $\subset D(|T|^{1+\omega}) \subset D(|T|^{1+\alpha}) \subset D(|T|^{1+\nu}) \subset D(|T|^{\omega})$, both $|T|^{\omega}C$ and $(C|T|^{1+\nu})^* = |T|^{1+\nu}C$ are bounded, thus also $C|T|^{1+\nu}$ (on $D(|T|^{1+\nu})$). So we must consider $|T|^{\omega}C|T|^{1+\nu}$.

Fix $\sigma \in (0,1)$. As $C|T|^{1+\alpha}$ is bounded on $D(|T|^{1+\alpha})$, there exists a constant k such that $\|Ch\| \leq k\| |T|^{-1-\alpha}h\|$ for all $h \in D(|T|^{-1-\alpha}) = \mathrm{Ran}(|T|^{1+\alpha})$. Then, by Lemma 2.4, we have $\| |C|^{\sigma}h\| \leq k^{\sigma}\| |T|^{-1-\alpha}h\|^{\sigma}\|h\|^{1-\sigma}$ for all $h \in D(|T|^{-1-\alpha})$. Hence, by Lemma 2.5, $\| |C|^{\sigma}h\| \leq k_0\| |T|^{-\delta(1+\alpha)}h\|$ for all $h \in D(|T|^{-\delta(1+\alpha)})$ and $\delta>\sigma$. Thus $|T|^{\delta(1+\alpha)}|C|^{\sigma}$ and $|C|^{\sigma}|T|^{\delta(1+\alpha)}$ are bounded. For $\delta=\frac{\omega}{1+\alpha}$ and $\delta=\frac{1+\nu}{1+\alpha}$, respectively, and $\sigma=½$ we recover as bounded operators $|T|^{\omega}|C|^{½}$ and $|C|^{½}|T|^{1+\nu}$. Then, using the polar decomposition $B=U|B|$, we can represent $|T|^{\omega}C|T|^{1+\nu}$ as a composition of bounded operators; one has

$$|T|^{\omega}C|T|^{1+\nu} = |T|^{\omega}|C|^{½}U|C|^{½}|T|^{1+\nu}.$$

Now we can estimate the norm of $F(\lambda)$:

$$\|F(\lambda)\|_{\mathcal{L}(H)} \leq \text{const.}((\mathrm{Re}\lambda)^{-\nu-\omega}+(\mathrm{Re}\lambda)^{-1-\omega}+(\mathrm{Re}\lambda)^{-1-\nu}+(\mathrm{Re}\lambda)^{-2}) \leq$$

$$\leq \text{const. } (\mathrm{Re}\lambda)^{-s},$$

where $s=\min\{\nu+\omega,1+\omega,1+\nu,2\}$. This estimate, along with the uniform boundedness of $(\lambda-T)(\lambda-S)^{-1}$ for $\lambda \in \Delta_2$, shows the integrability of $(\lambda-S)^{-1}-(\lambda-T)^{-1}$ on Δ_2 and completes the proof of the compactness of $P_+^{(2)}-Q_+^{(2)}$.

Let $x \in H$. Note that $[(\lambda-S)^{-1}-(\lambda-T)^{-1}]x \in D(T)$ for any $\lambda \in \Delta_2$. In order to prove that $(P_+^{(2)}-Q_+^{(2)})x \in D(T)$ it is sufficient to show that $T[(\lambda-S)^{-1}-(\lambda-T)^{-1}]x$ is Bochner integrable on Δ_2 (see Lemma 2.1). Since

$$\|T[(\lambda-S)^{-1}-(\lambda-T)^{-1}]x\| = \|T(\lambda-T)^{-1}BS(\lambda-S)^{-1}x\| =$$

$$= \|T(\lambda-T)^{-1}CT(\lambda-T)^{-1}[(\lambda-T)(\lambda-S)^{-1}x]\| \leq$$

$$\leq (1+\|(TC)^*\|_{\mathcal{L}(H)})\| T(\lambda-T)^{-1}CT(\lambda-T)^{-1}\|_{\mathcal{L}(H)}\|x\|,$$

it is sufficient to prove the integrability of $\|T(\lambda-T)^{-1}CT(\lambda-T)^{-1}\|_{\mathcal{L}(H)}$ on Δ_2. But this can be done in the same way as in the case of $\|F(\lambda)\|_{\mathcal{L}(H)}$, the only change being that one must use the factorization $|T|^{-1-\omega}|T|^{1+\omega}$ instead of $|T|^{-\omega}|T|^{\omega}$ in the decomposition of $TF(\lambda)$. Then (2.9) implies that P_+-Q_+ is compact on H and $(P_+-Q_+)[H]\subset D(T)$.

It remains to prove that the restriction of P_+-Q_+ to D(T) is compact with respect to the graph norm (2.5). Let us write $\hat{P}_+=AP_+A^{-1}$. Then, clearly, $\hat{P}_+-Q_+=P_+-Q_++P_+C-BP_+-BP_+C$ is compact in H. Moreover, for $h\in D(T)$, $(\hat{P}_+-Q_+)Th=T(P_+-Q_+)h$. Using the compactness of $\hat{P}_+-Q_+$ and this intertwining property, one can show the compactness of the restriction of P_+-Q_+ to D(T) with respect to the graph norm. Indeed, put $L=P_+-Q_+$ and $\hat{L}=\hat{P}_+-Q_+$, and observe that $L[D(T)]\subset D(T)$ and $TL=\hat{L}T$ on D(T). Let $\{h_n\}_{n=1}^{\infty}$ be a sequence in D(T) that is bounded with respect to (2.5). Then $\{h_n\}_{n=1}^{\infty}$ and $\{Th_n\}_{n=1}^{\infty}$ are bounded in the H-norm. Since L and $\hat{L}$ are compact, the sequences $\{Lh_n\}_{n=1}^{\infty}$ and $\{\hat{L}Th_n\}_{n=1}^{\infty}$ have convergent subsequences $\{Lh_{n_k}\}_{k=1}^{\infty}$ and $\{\hat{L}Th_{n_k}\}_{k=1}^{\infty}$ in H. Using the intertwining property in combination with (2.5), we obtain the convergence with respect to (2.5) of the subsequence $\{Lh_{n_k}\}_{k=1}^{\infty}$, whence L is compact. This completes the proof of the lemma. ∎

For bounded T the proof of the next lemma was given in [184].

LEMMA 2.7. The operator V has zero null space.

Proof: Suppose $Vh=0$ for some $h\in H$. Then $Q_+P_+h=-Q_-P_-h=0$ yields (cf. the previous lemma)

$$P_+h = (Q_--P_-)P_+h = (P_+-Q_+)P_+h \in D(T),$$

$$P_-h = -(P_+-Q_+)P_-h \in D(T),$$

whence $h=P_+h + P_-h \in D(T)$. Thus $\operatorname{Ker} V \subset D(T)$. We now obtain

$$(TP_+h, P_+h) \le 0,$$

$$(TP_+h, P_+h) = (A^{-1}TP_+h, P_+h)_A \ge 0,$$

in the first case because $P_+h \in \operatorname{Ran} Q_-$ and in the second case because $P_+h \in \operatorname{Ran} P_+$. Thus we conclude that $P_+h=0$ and in a similar way derive $P_-h=0$. ■

By virtue of the simple but important identity

$$I-V = Q_-P_+ + Q_+P_- = Q_-P_+ - Q_+P_+ + Q_+ = (Q_- - Q_+)(P_+ - Q_+) \tag{2.11}$$

and Lemma 2.6, we may conclude that $I-V$ is compact (in both topologies under consideration). Now with the above lemma, the Fredholm alternative gives us the principal result of this section for $q(x)\equiv 0$. The extension to $q(x)\not\equiv 0$ is then direct. Let us define the operator $\mathcal{H}_S(x)$ by

$$\mathcal{H}_S(x) = \begin{cases} +T^{-1}Ae^{-xT^{-1}A}P_+, & x>0 \\ -T^{-1}Ae^{-xT^{-1}A}P_-, & x<0. \end{cases}$$

THEOREM 2.8. The operator V is invertible. The boundary value problem (2.2)–(2.4) is uniquely solvable for each $\varphi_+ \in Q_+[D(T)]$ and each bounded uniformly Hölder continuous function $q:[0,\infty)\to H$ such that $\| |T|^\gamma q(x)\| = O(1)$ $(x\to\infty)$ for some $\gamma>0$, and the solution is

$$\psi(x) = e^{-xT^{-1}A}E(\varphi_+ - Q_+\chi(0)) + \chi(x), \quad 0\le x<\infty, \tag{2.12}$$

where $E=V^{-1}$ and

$$\chi(x) = \int_0^\infty \mathcal{H}_S(x-y)A^{-1}q(y)dy. \tag{2.13}$$

Proof: The proof for $q(x)\equiv 0$ is immediate from the previous two lemmas. If there is an inhomogeneous term, we will follow the line of reasoning prevailing in Section VI.3. For this reason we shall only sketch the proof and refer the reader to that section for additional details.

In order to interpret $\chi(x)$ as a Bochner integral, we rewrite it as

$$\chi(x) = \{I-e^{-xT^{-1}A}\}A^{-1}q(x) + \int_0^\infty \mathcal{H}_S(x-y)A^{-1}\{q(y)-q(x)\}dy,$$

where the second term at the right hand side is a well-defined absolutely convergent Bochner integral. This may be seen from the uniform Hölder continuity of q on $[0,\infty)$. The boundedness of q implies the boundedness of χ. On premultiplying (2.13) by $S=A^{-1}T$, we obtain the absolutely convergent Bochner integral

$$S\chi(x) = \int_0^\infty S\mathcal{H}_S(x-y)A^{-1}q(y)dy,$$

which is readily proved strongly differentiable with derivative $-\chi(x)+A^{-1}q(x)$. It should be noted that Lemma 2.1 implies $\chi(x)\in D(S)$ if $q(x)\in D(S)$. ∎

We have established an existence and uniqueness theory for half space problems, where A is a strictly positive compact perturbation of the identity satisfying the regularity assumption (2.1). It is possible to seek solutions of the boundary value problem for all $\varphi_+\in Q_+[H]$ rather than just $\varphi_+\in Q_+[D(T)]$. However, in this case it seems necessary to reformulate the problem slightly. The differential equation (2.2) is replaced by

$$T(\psi'(x)) = -A\psi(x) + q(x), \qquad 0<x<\infty, \tag{2.14}$$

and a solution is defined to be a continuous function $\psi:[0,\infty)\to H$ which is continuously differentiable on $(0,\infty)$ such that $\psi'(x)\in D(T)$ for $0<x<\infty$, and which satisfies (2.14)-(2.3)-(2.4). One should then assume that $q(x)\in \mathrm{Ran}\, S$ for all $x\in[0,\infty)$, while $S^{-1}q(x)$ is bounded and uniformly Hölder continuous on $[0,\infty)$ with $\| |S|^{\gamma-1}q(x)\|=O(1)$ $(x\to\infty)$. Then the analog of Theorem 2.8 may be proved in the same manner, but of course unique solvability will be obtained for each $\varphi_+\in Q_+[H]$. We shall see in Section VI.3 that the differentiability of $T\psi$ leads to an equivalent vector-valued convolution equation, but the differentiability of ψ does not.

3. Bounded collision operators

In this section A will be a bounded strictly positive operator. We shall drop all compactness assumptions on A. It will turn out that under these conditions the boundary value problem (1.1)-(1.3) associated with T and A is not necessarily well posed on the Hilbert space H or the domain D(T) of T. We shall therefore construct a Hilbert space extension H_T on which the problem is well posed. In this way we shall recover a result of Beals [32]. However, we shall follow a different route to these results and avoid the use of a second Hilbert space extension (as was done in his derivation).

First, let us introduce some pairs of complementary projections. We recall the projections $Q_\pm$ associated with T, which are, of course, invariant on D(T):

$$TQ_\pm h = Q_\pm Th, \qquad h \in D(T). \tag{3.1}$$

As in the previous section, we see that $A^{-1}T$ is self adjoint on H relative to the inner product (1.19), which is equivalent to the given inner product on H, since A is bounded and strictly positive. The operator TA^{-1} is self adjoint with respect to the second inner product

$$(h,k)_{A^{-1}} = (A^{-1}h,k), \tag{3.2}$$

which is again equivalent. Now let $P_\pm$ be the orthogonal projections of the Hilbert space H onto the maximal $A^{-1}T$-positive/negative $A^{-1}T$-invariant subspaces (where the positivity and orthogonality notions concern (1.19)), while $\hat{P}_\pm$ are the orthogonal projections of H onto the maximal TA^{-1}-positive/negative TA^{-1}-invariant subspaces (where the positivity and orthogonality properties relate to (3.2)). Finally, let us note that $\mp T^{-1}A$ and $\mp AT^{-1}$ generate bounded analytic semigroups on Ran $P_\pm$ and Ran $\hat{P}_\pm$, respectively. Using the invariance of $P_\pm$ and $\hat{P}_\pm$ on $D(A^{-1}T)=D(T)$ and $D(AT^{-1})=\text{Ran } T$, respectively, we obtain the intertwining relation:

$$T(e^{\mp xT^{-1}A}P_\pm h) = e^{\mp xAT^{-1}}\hat{P}_\pm(Th), \qquad h \in D(T). \tag{3.3}$$

We now define the Hilbert space H_T as the completion of D(T) with respect to the inner product (1.20). If T is bounded, then H can be imbedded in H_T in a natural way, but for unbounded T such natural imbedding does not exist. We shall develop the existence and uniqueness theory for the boundary value problem (1.1)–(1.3) in the space H_T. In order to implement this extension we have to continue certain operators on H from their restrictions on D(T) to bounded operators on H_T. This will be achieved using a proposition originally due to Krein [226], which appears as Theorem 1.2 of [41].

PROPOSITION 3.1: Let R and $\hat{R}$ be two bounded operators on H such that R leaves invariant D(T) and satisfies the intertwining property

$$TRh = \hat{R}Th, \quad h \in D(T). \tag{3.4}$$

Then the restriction of R to D(T) has a continuous extension to H_T, while

$$\|R\|_{H_T} \leq \max\{\|R\|_H, \|\hat{R}\|_H\}. \tag{3.5}$$

Proof: First we turn the domain of T into a Hilbert space by introducing the graph inner product (2.5). Using the intertwining property (3.4) we have the estimate

$$(Rh,Rh)_{GT} = (Rh,Rh) + (\hat{R}Th,\hat{R}Th) \leq \|R\|^2\|h\|^2 + \|\hat{R}\|^2\|Th\|^2 \leq$$

$$\leq [\max\{\|R\|,\|\hat{R}\|\}]^2 \ \|h\|^2_{GT},$$

implying that the restriction of R to D(T) is bounded with respect to (2.5). We also observe that $\hat{R}^*$, with the adjoint relative to H, leaves invariant D(T) and satisfies the intertwining property

$$T\hat{R}^*h = R^*Th, \quad h \in D(T). \tag{3.6}$$

Repeating the above estimate we find that the restriction of $\hat{R}^*$ to D(T) is bounded with respect to (2.5) with norm estimate

$$\|\hat{R}^*\|_{GT} \leq \max\{\|\hat{R}^*\|,\|R^*\|\} = \max\{\|R\|,\|\hat{R}\|\}.$$

Next, put $L=(Q_+-Q_-)\hat{R}^*(Q_+-Q_-)$. Then L leaves invariant D(T) and

$$\|L\|_{GT} \leq \max\{\|R\|,\|\hat{R}\|\}.$$

Furthermore, for all $h,k \in D(T)$ we have

$$(Lh,k)_T = (|T|Lh,k) = (R^*|T|h,k) = (h,Rk)_T, \tag{3.7}$$

which means that L and R are adjoints with respect to H_T. (We shall see shortly that L and R are bounded on H_T.) Fix $h \in D(T)$ with $\|h\|_T=1$, and let $S_n=\|(LR)^n h\|_T^2$. Using the symmetry of LR with respect to (1.20), for real λ we obtain the inequality

$$0 \le \|(LR)^{n-1}h + \lambda(LR)^{n+1}h\|_T^2 = S_{n-1} + 2\lambda S_n + \lambda^2 S_{n+1},$$

which implies that $S_n^2 \le S_{n-1}S_{n+1}$ for $n \in \mathbb{N}$. Since $S_0=1$, we find

$$S_1 \le (S_2/S_1) \le (S_3/S_2) \le (S_4/S_3) \le \dots,$$

whence

$$S_n \ge S_1 S_{n-1} \ge S_1^2 S_{n-2} \ge \dots \ge S_1^n, \quad n=1,2,3,\dots,$$

and so $S_1 \le S_n^{1/n}$.

On the other hand, using the estimate

$$2\|h\|_T^2 = 2(|T|h,h) \le 2\|Th\|.\|h\| \le \|h\|^2+\|Th\|^2 = \|h\|_{GT}^2 \tag{3.8}$$

for $h \in D(T)$, we obtain

$$S_n = \|(LR)^n h\|_T^2 \le \tfrac{1}{2}\|(LR)^n h\|_{GT}^2 \le \tfrac{1}{2}\|(LR)^n\|_{GT}^2\|h\|_{GT}^2 \le$$

$$\le \tfrac{1}{2}\|R\|_{GT}^{2n}\|L\|_{GT}^{2n}\|h\|_{GT}^2 \le \tfrac{1}{2}[\max\{\|R\|,\|\hat{R}\|\}]^{4n}\|h\|_{GT}^2,$$

implying

$$S_1 \le S_n^{1/n} \le 2^{-1/n}[\max\{\|R\|,\|\hat{R}\|\}]^4\|h\|_{GT}^{2/n}.$$

Letting $n \to \infty$ we find

$$\|LRh\|_T \le [\max\{\|R\|,\|\hat{R}\|\}]^2,$$

where h is an arbitrary vector in D(T) with $\|h\|_T=1$. Since D(T) is a dense linear subspace of H_T, we must conclude that LR extends to a bounded operator on H_T with norm $\le [\max\{\|R\|,\|\hat{R}\|\}]^2$. Thus for $h \in D(T)$ we must have

$$\|Rh\|_T^2 = (Rh,Rh)_T = (LRh,h)_T \le \|LR\|_T \|h\|_T^2,$$

which implies the proposition. ∎

The above proposition has far reaching ramifications for the boundary value problems under consideration. Using (3.1) we may see that $Q_\pm$ extend to orthogonal projections on H_T. The analogous equation for $P_\pm$ yields that $P_\pm$ extend to bounded projections on H_T. Moreover, the families $\{\exp(\mp xT^{-1}A)P_\pm : 0 \le x < \infty\}$ have continuous extensions to H_T which form a bounded set in H_T-operator norm. To see that they, in fact, form analytic semigroups on $P_\pm[H_T]$ requires more effort. First, the proposition implies the analyticity in x of both families on the open right half plane, as well as the semigroup property. It therefore remains to prove the identities

$$\lim_{x\to\infty} \|e^{\mp xT^{-1}A}P_\pm h\|_T = 0, \quad h \in H_T, \tag{3.9}$$

$$\lim_{x\to 0,\ |\arg x|\le\delta} \|e^{\mp xT^{-1}A}P_\pm h - P_\pm h\|_T = 0, \quad h \in H_T, \tag{3.10}$$

where $0 \le \delta < \frac{1}{2}\pi$. Both of these identities hold true for $h \in D(T)$ in the norm induced by the inner product (2.5). Thus, as a consequence of (3.8), they are also valid for $h \in D(T)$ in H_T-norm (i.e., as they are stated). However, since the families $\{\exp(\mp xT^{-1}A)P_\pm : 0 \le x < \infty\}$ are bounded in the H_T-operator norm and D(T) is dense in H_T, the identities (3.9) and (3.10) are true for all $h \in H_T$, which settles the analyticity of the semigroups. We will write $K \supset T^{-1}A$ for the operator which gives the infinitesimal generators (within ± sign) of the extensions of $\exp\{\mp xT^{-1}A\}P_\pm$. It will be seen in the next section that K may also be obtained as an appropriate closure of $T^{-1}A$.

In order to establish the unique solvability of the boundary value problem in an appropriate functional formulation, we first prove

THEOREM 3.2. The operator $V=Q_+P_+ + Q_-P_-$ extends to a bounded invertible operator on H_T.

Proof: Since the projections $Q_\pm$ and $P_\pm$ extend to bounded operators on H_T, the same property must be true for V. One easily computes that

$$2V-I = (2Q_+P_+-P_+) + (2Q_-P_--P_-) = (2Q_+-I)P_+ + (2Q_--I)P_- =$$

$$= (Q_+-Q_-)(P_+-P_-). \tag{3.11}$$

Then for all $h \in D(T)$ we must have

$$((2V-I)h,h)_T = (|T|(Q_+-Q_-)(P_+-P_-)h,h) = (T(P_+-P_-)h,h) =$$

$$= (|A^{-1}T|h,h)_A \geq 0,$$

where $|A^{-1}T|$ denotes the absolute value of $A^{-1}T$ with respect to (1.19). Hence,

$$2(Vh,h)_T \geq \|h\|_T^2, \quad h \in D(T).$$

Thus V is a bounded strictly positive self adjoint operator on H_T. ■

It remains to link the invertibility of V on H_T to the unique solvability of the boundary value problem

$$\psi'(x) = -K\psi(x) + \hat{q}(x), \quad 0<x<\infty, \tag{3.12}$$

$$Q_+\psi(0) = \varphi_+, \tag{3.13}$$

$$\|\psi(x)\|_T = O(1) \ (x\to\infty), \tag{3.14}$$

where $\hat{q}(x)=A^{-1}q(x)$. We define a **(weak)** solution of the boundary value problem (3.12)-(3.13) for any $\varphi_+ \in Q_+[H_T]$ to be a continuous function $\psi:[0,\infty)\to H_T$ such that ψ is continuously differentiable on $(0,\infty)$ with values in D(K) and Eqs. (3.12)-(3.14) are satisfied. Putting $E=V^{-1}$ on H_T, one would expect unique solutions of the homogeneous boundary value problem of the form

$$\psi(x) = e^{-xK}E\varphi_+, \quad 0 \leq x < \infty. \tag{3.15}$$

THEOREM 3.3. For every $\varphi_+ \in Q_+[H_T]$ the homogeneous boundary value problem (3.12)–(3.14), where $q(x) \equiv 0$, has a unique solution, which is given by (3.15). If $\hat{q}:[0,\infty) \to H_T$ takes values in $D(K)$ and $K\hat{q}$ is uniformly Hölder continuous, then $\psi(x)$ is given by

$$\psi(x) = e^{-xK}E(\varphi_+ - Q_+\chi(0)) + \chi(x)$$

with

$$\chi(x) = \int_0^\infty \mathcal{H}_S(x-y)A^{-1}q(y)dy.$$

Proof: Let $\psi:[0,\infty) \to H_T$ be a solution of the boundary value problem (3.12)–(3.14) with $q(x) \equiv 0$. Then for $0 < \varepsilon < m < \infty$ and $\mathrm{Re}\lambda < 0$ we have

$$0 = \lambda \int_\varepsilon^m e^{x/\lambda}\{\psi'(x) + K\psi(x)\}dx =$$

$$= [\lambda e^{x/\lambda}\psi(x)]_{x=\varepsilon}^m - (I - \lambda K) \int_\varepsilon^m e^{x/\lambda}\psi(x)dx.$$

Here $\int_\varepsilon^m \psi(x)dx \in D(K)$ and $K\int_\varepsilon^m \psi(x)dx = \int_\varepsilon^m K\psi(x)dx$, as a consequence of the continuous differentiability of ψ on $(0,\infty)$. We easily prove, using that $\|\psi(x)\|_T = O(1)$ $(x \to \infty)$, that for all non-real λ in the left half plane

$$\int_0^\infty e^{x/\lambda}\psi(x)dx = -\lambda(I - \lambda K)^{-1}\psi(0). \tag{3.16}$$

Because of the definition of K and the analyticity of (3.16) in the left half plane, we obtain $\psi(0) \in P_+[H_T]$. The boundary condition $Q_+\psi(0) = \varphi_+$ and the invertibility of V on H_T imply $\psi(0) = E\varphi_+$, where $E = V^{-1}$. The solution (3.15) follows by solving the evolution equation $\psi' = -K\psi$ with initial value $\psi(0) = E\varphi_+$ on $P_+[H_T]$. The extension to the inhomogeneous case follows the proof of Theorem 2.8. ∎

We notice that the solution of Eqs. (3.12)–(3.14) is a function with values in H_T, even if $\varphi_+ \in Q_+[D(T)]$. The solution has all its values in D(T) only if $\varphi_+ \in Q_+V[D(T)] = VP_+[D(T)]$. It is not known whether the half space problem is well posed in a functional formulation only involving D(T).

4. Unbounded collision operators

We will consider now A an (unbounded) self adjoint strictly positive Fredholm operator. Because of complications introduced by the unboundedness of T, we will assume either that T is bounded, or more generally that $D(A^{1/2}) \subset D(T)$. As we shall see, the boundary value problem (1.1)–(1.3) will be transferred to yet another Hilbert space setting H_S, with the boundary values $\psi(0) \in H_T \cap H_S$. That is to say, the albedo operator E will map $E: H_T \to H_T \cap H_S$. Moreover, although E will turn out to be continuous from H_T into either H_T or H_S, its inverse $V = E^{-1}$ will in general be unbounded. Special emphasis will be given to conditions under which H_T and H_S can be identified.

Let us assume that $D(A) \cap D(T)$ is dense in H, and let H_A be the completion of D(A) with respect to the inner product (1.19). We will always view H_A as a linear submanifold of H to be identified with $D(A^{1/2})$. Assuming $A^{-1}T$ to be essentially self adjoint with respect to the H_A inner product, we denote its unique self adjoint extension by S. Then $H_A \cap D(T) = D(A^{-1}T) \subset D(S) \subset H_A$. The essential self adjointness assumption is obviously satisfied if $H_A \subset D(T)$, since in this case the Closed Graph Theorem implies that $T: H_A \to H$ is bounded and $S = A^{-1}T$ is bounded self adjoint on H_A. We now define by H_{TS} the completion of $H_A \cap D(T)$ with respect to the inner product

$$(h,k)_{TS} = (|T|h,k) + (|S|h,k)_A. \tag{4.1}$$

Write $Q_\pm$ for the H–orthogonal projections onto the positive/negative spectral subspaces of T and $P_\pm$ for the H_A–orthogonal projections onto the positive/negative spectral subspaces of S. Then $S = |S|W$ and $|T| = TU$, where $W = Q_+ - Q_-$ and $U = P_+ - P_-$. We define H_T as the completion of H_{TS} with respect to the inner product (1.20), and H_S as the completion of H_{TS} with respect to the inner product

$$(h,k)_S = (|S|h,k). \tag{4.2}$$

The continuous and dense embeddings

$$H_A \longrightarrow \begin{cases} \longrightarrow H_{TS} \longrightarrow \begin{cases} \longrightarrow H_S \\ \longrightarrow H_T \end{cases} \\ \longrightarrow H \end{cases}$$

for $H_A \subset D(T)$ with the additional continuous and dense embedding $H \longrightarrow H_T$ for T bounded, and the continuous and dense embeddings

$$H_A \cap D(T) \longrightarrow \begin{cases} \longrightarrow H_{TS} \longrightarrow \begin{cases} \longrightarrow H_S \\ \longrightarrow H_T \end{cases} \\ \longrightarrow H_A \longrightarrow H \end{cases}$$

for unbounded T are evident. It is also evident that $H_A \cap D(T)$ is densely imbedded in H_S and in H_T. Clearly S and $P_\pm$ extend from D(S) to self adjoint operators on H_S, and T and $Q_\pm$ extend from D(T) to self adjoint operators on H_T.

Let us now give a precise statement of the boundary value problem. For the sake of convenience we will take $q(x) \equiv 0$. Given $\varphi_\pm \in Q_\pm[H_T]$, a (**weak**) solution of Eqs. (1.1)–(1.3)/(1.9)–(1.11) is a continuous function $\psi:[0,\infty) \rightarrow H_S$ having its values $\psi(x) \in D(S^{-1}) \cap H_S$ for $0<x<\infty$, which is continuously H_S-differentiable on $(0,\infty)$, has $\psi(0) \in H_{TS}$, and satisfies (1.2)–(1.3)/(1.10)–(1.11) along with

$$\psi'(x) = - K\psi(x) + \hat{q}(x). \tag{4.3}$$

We have written K for S^{-1} on H_S and $\hat{q} = T^{-1}q$.

As in the case of bounded A, we are choosing boundary data φ_+ from $Q_+[H_T]$ and demanding that the total boundary flux $\psi(0) \in H_T$. However, because of the (possibly) singular behavior of the generator K on H_T, we must seek solutions in H_S. In effect, this problem did not arise in the previous section because H_T and H_S can be identified when A is bounded. The albedo operator now may be viewed as a map from H_T into either H_T or H_S, since the definition of a solution $\psi(x)$ requires that Ran $E \subset H_{TS}$.

Thus, we define an albedo operator E to be a linear operator $E:H_T \rightarrow H_{TS}$ such that

$$\text{(i)} \quad Q_\pm E Q_\pm = Q_\pm \tag{4.4}$$

(ii) $P_{\mp}EQ_{\pm} = 0$ (4.5)

on H_T. Note that an immediate consequence of (4.5), if an albedo operator exists, is the invariance of Ran E under $P_{\pm}$.

The equivalence of the existence of solutions to the homogeneous boundary value problems (1.1)–(1.3)/(1.9)–(1.11) for all $\varphi_{\pm} \in Q_{\pm}[H_T]$ and the existence of an albedo operator is transparent. Indeed, assuming $\psi(x;\varphi_+)$ is the solution corresponding to φ_+, and defining E: $\varphi_+ \to \psi(0;\varphi_+)$, only the validity of (4.5) might be questioned. However, for Re $\lambda<0$ and Im $\lambda\neq0$, the (Bochner) integral $\int_0^\infty e^{x/\lambda}\psi(x;\varphi_+)dx$ is H_S-absolutely convergent, and

$$(\lambda-S)\int_0^\infty e^{x/\lambda}\ \psi(x;\varphi_+)dx = \lambda S\psi(0;\varphi_+)$$

after an integration by parts. Thus

$$\int_0^\infty e^{x/\lambda}\ \psi(x;\varphi_+)dx = \lambda S(\lambda-S)^{-1}\psi(0;\varphi_+)$$

has an H_S-analytic continuation to the open left half plane and therefore $\psi(0;\varphi_+)\in P_+[H_S]$, or, equivalently, (4.5) holds true.

Motivated by (1.15), let us define $V_0 = Q_+P_+ + Q_-P_- : H_{TS}\to H_T$ with $D(V_0) = \{f\in H_{TS} : P_{\pm}f\in H_{TS}\}$. Thus we also have $V_0 f = \frac{1}{2}(WU+I)f$ for all $f\in D(V_0)\subset H_{TS}$. Moreover, V_0 is closed as an operator from H_{TS} into H_T. Indeed, if $\{h_n\}_{n=1}^\infty$ is a sequence in $D(V_0)$ such that $h_n\to h$ and $V_0h_n\to g$, then $P_+h_n\to(Q_+-Q_-)g+Q_-h$ and $P_-h_n\to(Q_--Q_+)g+Q_+h$ in H_T. Since also $P_{\pm}h_n\to P_{\pm}h$ in H_S, the two limits of $\{P_{\pm}h_n\}_{n=1}^\infty$ coincide; they are limits in the H_{TS}-norm and hence $h\in D(V_0)$ and $V_0h=g$, which proves the statement.

Let us consider first the case when $H_A\subset D(T)$. We have

LEMMA 4.1. If $H_A\subset D(T)$. Then Ker $V_0=\{0\}$, and, as an operator on H_T, V_0 is closed, symmetric and positive.

Proof: The lemma will follow from the identity

$$2(V_0f,f)_T = \|f\|_T^2 + \|f\|_S^2, \quad f\in D(V_0). \tag{4.6}$$

Suppose $f\in D(V_0)$ and $g\in H_A$. We have

$$2(V_0 f,g)_T - (f,g)_T = (Uf, Wg)_T = (TUf, |T|g)_{T^{-1}},$$

where the unitary equivalence $|T|:H_T \to H_{T^{-1}}$ has been utilized. Here $H_{T^{-1}}$ is the completion of Ran $T \subset H$ with respect to the inner product

$$(h,k)_{T^{-1}} = (|T|^{-1}h,k).$$

If $H_A \subset D(T)$, then the unitary equivalence $S:H_S \to H_{S^{-1}}$, where $H_{S^{-1}}$ is the completion of Ran $S \subset H_A$ with respect to the inner product

$$(h,k)_{S^{-1}} = (|S|^{-1}h,k),$$

and the fact that $SUf \in H_A$ give

$$(f,g)_S = (Sf,Sg)_{S^{-1}} = (|S|^{1/2}Sf, |S|^{1/2}Sg)_A = (SUf,g)_A = (TUf,|T|g)_{T^{-1}},$$

which completes the proof. ∎

LEMMA 4.2. If $H_A \subset D(T)$, then Ran $V_0 = H_T$ if and only if there exists an albedo operator, and then $E_0 = V_0^{-1}$ is the unique albedo operator.

Proof: Assume Ran $V_0 = H_T$. By construction, $E_0:H_T \to H_{TS}$. If $g \in H_T$ and we write $E_0 Q_\pm g = h_\pm$, then $Q_\pm g = Q_+P_+h_\pm + Q_-P_-h_\pm$, whence $Q_\mp P_\mp h_\pm = 0$. But we also have $V_0 P_\mp h_\pm = Q_\mp P_\mp h_\pm = 0$, so $P_\mp h_\pm \in \text{Ker } V_0$ and, by injectivity, $P_\mp h_\pm = 0$. We have shown that $P_\mp E_0 Q_\pm = 0$ on H_T. From $V_0 E_0 = I$ and the result just obtained, we have $Q_\pm = Q_\pm P_\pm E_0 Q_\pm = Q_\pm E_0 Q_\pm$ on H_T. Thus E_0 is an albedo operator.

Suppose E_0 is an albedo operator. Since we have $P_\pm E Q_\pm = E Q_\pm$ on H_T, Ran $E \subset \{f \in H_{TS} : P_\pm f \in H_T\}$. Following Section 1, we derive easily the intertwining relation $P_\pm E = E Q_\pm$ on H_T, and thus $Q_\pm P_\pm E Q_\pm = Q_\pm$. Adding together the ± equations gives $V_0 E = I$ on H_T, from which the lemma follows. ∎

We now derive the main results of this section.

THEOREM 4.3. If $H_A \subset D(T)$, then the following statements are equivalent:

(i) The boundary value problems (1.1)–(1.3)/(1.9)–(1.11) with $q(x)\equiv 0$ are solvable for all $\varphi_\pm \in Q_\pm[H_T]$.

(ii) There exists an albedo operator $E: H_T \to H_{TS}$.

(iii) The operator V_0 has dense range in H_T.

(iv) The operator V_0 is a self adjoint operator on H_T.

Proof: The equivalence of (i) and (ii) is clear from previous considerations. Note that E exists as a bounded operator on H_T. That is immediate from the estimate

$$2(V_0 h,h)_T \geq \|h\|_S^2, \quad h \in D(V_0). \tag{4.7}$$

The boundedness of E and the closedness of V_0 imply the closedness of Ran $V_0 \subset H_T$, and hence the equivalence of (iii) and (ii). Let V be the Friedrichs extension of the positive symmetric operator V_0. As we must have

$$2(Vh,h)_T \geq \|h\|_T^2, \quad h \in D(V), \tag{4.8}$$

the extension V coincides with V_0 if (iii) is valid, whence V_0 is self adjoint. Conversely, the self adjointness of V_0 along with (4.8) gives (iii). ■

COROLLARY 4.4. If E exists, then it is injective, and $E: H_T \to H_S$ and $E: H_T \to H_T$. Moreover, the boundary value problems (1.1)–(1.3)/(1.9)–(1.11) are uniquely solvable for each $\varphi_\pm \in Q_\pm[H_T]$.

THEOREM 4.5. If $H_A \subset D(T)$, then the following statements are equivalent:

(i) $V_0: H_T \to H_T$ is bounded.

(ii) $V_0: H_S \to H_T$ is bounded.

(iii) The norms $\|\cdot\|_T$ and $\|\cdot\|_S$ are equivalent: $H_T \simeq H_S$.

If any of these, then V_0 is invertible and $E = V_0^{-1}$ is the albedo operator.

Proof: If (i) is satisfied, then (4.7) implies

$$2\|V_0\|_{H_T} \|h\|_T^2 \geq \|h\|_S^2.$$

Now following an estimate of Beals [32], we have

$$\|h\|_T^2 = (Th,Wh) = (Uh,Wh)_S \leq \|h\|_S \|h\|_T \left[2\|V_0\|_{H_T}\right]^{1/2}$$

for $h \in H_A \subset H$, which implies $H_T \simeq H_S$. In a similar fashion, if (ii) is satisfied, then

$$2\|V_0\| \ \|h\|_S \|h\|_T \geq \max \{2\|h\|_T^2, 2\|h\|_S^2\},$$

which gives $H_T \simeq H_S$ directly. Conversely, $H_T \simeq H_S$ implies $D(V_0)=H_T$, by definition, and then (i)–(ii) by the Closed Graph Theorem.

If (i)–(ii)–(iii) are fulfilled, then the self adjointness of V_0 along with the identity (4.6) complete the proof of the theorem. ■

The next corollary indicates why one did not need the space H_S in the previous section. Although for bounded A we need not have $H_A \subset D(T)$, all previous arguments can still be carried out if A is bounded.

COROLLARY 4.6. If A is bounded, then $H_T \simeq H_S$.

Proof: If A is bounded, we may identify H_A and H, and the operators $P_\pm$ and V_0 will then be bounded on H while leaving invariant D(T). ■

In applications of the present theory to physical models, one seems to have $H_T \simeq H_S$. This is evident in models for which A is bounded, which covers most of radiative transfer, neutron transport and rarefied gas dynamics. In Chapter X we will encounter an extensive class of Sturm–Liouville diffusion equations, where T is multiplication by an indefinite weight and A is a Sturm–Liouville differential operator. The general proof for this case that $H_T \simeq H_S$ has been given by Beals [34]. An alternative proof has recently been given by Curgus [96]. On the other hand, Kaper et al. [209] have constructed an example, even with bounded T, where $H_T \not\simeq H_S$. For this the assumptions of Theorem 4.3 are still satisfied; we shall present the example below.

Let us define the matrices

$$T_n = \begin{bmatrix} 0 & 1 \\ 1 & 0 \end{bmatrix}, \quad A_n = \begin{bmatrix} n & 0 \\ 0 & n^3 \end{bmatrix}.$$

Next we consider the Hilbert space $H = \bigoplus_{n=1}^{\infty} \mathbb{C}^2$, on which we define the operators

$$T = \bigoplus_{n=1}^{\infty} T_n, \qquad A = \bigoplus_{n=1}^{\infty} A_n.$$

Then T is bounded and self adjoint with $\sigma(T)=\{-1,1\}$, and A is unbounded and strictly positive self adjoint; we have

$$S = \bigoplus_{n=1}^{\infty} S_n, \qquad Q_\pm = \bigoplus_{n=1}^{\infty} Q_{\pm,n}, \qquad P_\pm = \bigoplus_{n=1}^{\infty} P_{\pm,n},$$

where

$$S_n = \begin{bmatrix} 0 & n^{-1} \\ n^{-3} & 0 \end{bmatrix}$$

and

$$Q_{\pm,n} = \frac{1}{2}\begin{bmatrix} 1 & \pm 1 \\ \pm 1 & 1 \end{bmatrix}, \qquad P_{\pm,n} = \frac{1}{2}\begin{bmatrix} 1 & \pm n \\ \pm n^{-1} & 1 \end{bmatrix}.$$

We now easily derive that $|T|$ is the identity operator on H and hence $H_T=H$. However, $|S|$, which can be factorized as $S(P_+-P_-)$, has the form

$$|S| = \bigoplus_{n=1}^{\infty} |S_n|,$$

where

$$|S_n| = \frac{1}{n^2}\begin{bmatrix} 1 & 0 \\ 0 & 1 \end{bmatrix}.$$

Hence H_S properly extends $H_T=H$, and we have $H_S \neq H_T$. On computing $V=Q_+P_++Q_-P_-$ one sees easily that V is unbounded on H.

Chapter III.

CONSERVATIVE KINETIC MODELS

1. Preliminary decompositions and reductions

In the previous chapter we studied boundary value problems in half space geometry of the type

$$T\psi'(x) = -A\psi(x), \qquad 0<x<\infty, \tag{1.1a}$$

$$Q_+\psi(0) = \varphi_+, \tag{1.2a}$$

$$\|\psi(x)\| = O(1) \ (x\to\infty), \tag{1.3a}$$

and

$$T\psi'(x) = -A\psi(x), \qquad -\infty<x<0, \tag{1.1b}$$

$$Q_-\psi(0) = \varphi_-, \tag{1.2b}$$

$$\|\psi(x)\| = O(1) \ (x\to-\infty) \tag{1.3b}$$

with A strictly positive. Depending on the specific assumptions on T and A, different variants of the boundary value problems were defined. In each case, the above problems were reformulated as invertibility problems for the operator

$$V = Q_+P_+ + Q_-P_-, \tag{1.4}$$

on a suitable Hilbert space, where the projections $P_\pm$ associated with the operator

$A^{-1}T$ are analogs of the projections $Q_{\pm}$ associated with the operator T. The result is that the problems (under appropriate hypotheses) are solvable, indeed are uniquely solvable, and, even more, are uniquely solvable if the boundary condition at infinity (1.3a)/(1.3b) is replaced by

$$\lim_{x\to\pm\infty} \|\psi(x)\| = 0, \tag{1.5}$$

or

$$\exists n \in \mathbb{N}: \ \|\psi(x)\| = O(|x|^n) \ (x\to\pm\infty). \tag{1.6}$$

In fact, solvability for every pair of vectors $\varphi_{\pm} \in \operatorname{Ran} Q_{\pm}$ is precisely equivalent to the invertibility of V.

In the previous chapter we have also dealt with the inhomogeneous boundary value problem, where a term q(x) is present on the right hand side of the differential equation. As the inhomogeneous half space problem amounts to construction of a particular solution once the homogeneous version is solved, the introduction of such a term q(x) does not yield a challenging problem. In fact, we will shortly introduce a transformation which will modify the half space problem in such a way that the resultant kinetic equation will be represented on an infinite dimensional subspace with a strictly positive collision operator, and we may proceed as in Chapter II. The remaining finite dimensional problem will be essentially trivial. For this reason, in this chapter we shall confine ourselves to the homogeneous version of the boundary value problem.

Requiring Ker A={0} excludes from consideration many physically important problems, such as most linearized gas kinetics equations (where the existence of conservation laws results in the collision operator A having a nontrivial kernel), the neutron transport equations precisely at criticality, and radiative transfer problems for non-absorbing media. In this chapter we will generalize the results of the previous chapter to the case where A is (still) positive, but with nontrivial kernel, and will study conditions at ±∞ corresponding to (1.5) and to (1.6) with n=0 and n=1. The existence of a nontrivial kernel for the collision operator will cause a number of difficulties – even the proper definition of the projections $P_{\pm}$ is not a priori clear – and the equivalence of solvability with invertibility of V will be lost. Indeed, depending on the boundary condition at ±∞, both non-uniqueness and non-existence of solutions are possible.

For R a linear operator defined on the Hilbert space H and λ in the spectrum of R, $\lambda \in \sigma(R)$, the root linear manifold $Z_\lambda(R)$ is defined by

$$Z_\lambda(R) = \{h \in H : h \in D(R^n),\ (R-\lambda I)^n h = 0 \text{ for some } n \in \mathbb{N}\}.$$

Of course, if R is normal, then the Spectral Theorem guarantees that $Z_\lambda(R)$ is precisely the eigenspace corresponding to the eigenvalue λ. More generally, $Z_\lambda(R)$ consists of the eigenvectors and generalized eigenvectors associated with the "Jordan blocks" corresponding to $\lambda \in \sigma(R)$.

Let us write K_0 for the H-closure of $T^{-1}A$. The definition and study of the projections $P_\pm$ will be accomplished by proving the decompositions

$$Z_0(K_0) \oplus Z_0(K_0^*)^\perp = H, \tag{1.7a}$$

$$Z_0(K_0^*) \oplus Z_0(K_0)^\perp = H. \tag{1.7b}$$

On $Z_0(K_0^*)^\perp$ one may define the inner product $(\cdot,\cdot)_A$. We then denote by H_A the Hilbert space obtained by taking the orthogonal direct sum of $Z_0(K_0)$ (endowed with some inner product) and the $(\cdot,\cdot)_A$-completion of $Z_0(K_0^*)^\perp$ (constructed as a submanifold of H, which is possible because A has closed range). Since, as will be proved separately, $Z_0(K_0)$ is finite dimensional, all inner products thus constructed are equivalent. Then the $(\cdot,\cdot)_A$-closure of the restriction of $T^{-1}A$ to $Z_0(K_0^*)^\perp$ is the inverse of an injective self adjoint operator with respect to the $(\cdot,\cdot)_A$ inner product. This inverse can be used to construct the projections $P_\pm$ on $Z_0(K_0^*)^\perp$. By replacing A with a **strictly** positive self adjoint operator A_β coinciding with A on $Z_0(K_0^*)^\perp$, the inverse of the restriction of $T^{-1}A$ to $Z_0(K_0^*)^\perp$ extends to an operator $A_\beta^{-1}T$ on H, as do the projections $P_\pm$. In this way the boundary value problems can be reduced to four subproblems:

(i) To solve Eqs. (1.1)-(1.3) in a suitable functional formulation with A replaced by A_β. Since A_β is strictly positive, this problem can be dealt with using the methods of Chapter II.

(ii) To solve an evolution equation of the form $\psi_0'(x) = -T^{-1}A\psi_0(x)$ on $Z_0(K_0)$. As this space will be finite dimensional, the equation is trivially solvable.

(iii) To match the boundary conditions (1.2) and (1.3), (1.5) or (1.6). Essentially this involves an analysis of $Z_0(K_0)$ only.

(iv) To assure that the solution is independent of the alteration of A.

If solutions of the original problem (1.1)–(1.3) in fact exist, we expect to write them in the form (say for T,A bounded):

$$\psi(x) = e^{-xT^{-1}A}PE\varphi_+ + e^{-xT^{-1}A}\psi_0(0),$$

where P is the projection of H onto $Z_0(K_0^*)^\perp$ along $Z_0(K_0)$, the albedo operator E arises as the inverse of the operator V_β defined as in Eq. (1.4), $\psi_0(0) \in \operatorname{Ker} A$, and E must be constructed in such a way that $\operatorname{Ran} EQ_+ \cap Z_0(K_0) \subset \operatorname{Ker} A$. For Eqs. (1.1)–(1.2)–(1.6), the solution will have the form

$$\psi(x) = e^{-xT^{-1}A}PE\varphi_+ + (I-xT^{-1}A)\psi_0(0)$$

with $\psi_0(0) \in Z_0(T^{-1}A)$.

The decomposition (1.7) and the projections $PP_\pm$ were first applied by Lekkerkerker [248] to one speed neutron transport with isotropic scattering. The abstract generalization was developed by van der Mee [360] and further refined by Greenberg et al. [160, 166] Under more restrictive assumptions on the structure of $Z_0(T^{-1}A)$, such a procedure was applied already to transport problems by Beals [32], who more recently formulated his own version of the above program (cf. [34]). We shall now outline this procedure in more detail.

Let us assume T is injective self adjoint and A is non-negative self adjoint and Fredholm. In addition we will assume that $Z_0(T^{-1}A) \subset D(T)$ and $D(A) \cap D(T) \subset H$ densely. We recall that for a self adjoint operator A to be Fredholm means that $\dim \operatorname{Ker} A < \infty$ and Ran A is closed. We will see shortly that also $\dim Z_0(K_0) < \infty$.

LEMMA 1.1. We have $\operatorname{Ker} K_0 = \operatorname{Ker} A$ and $Z_0(K_0) = Z_0(T^{-1}A)$.

Proof: Let $h \in \operatorname{Ker} K_0$. Then there exists a sequence $\{h_n\}_{n=1}^\infty$ in $D(T^{-1}A) \cap (\operatorname{Ker} A)^\perp$ such that $h_n \to h$ and $T^{-1}Ah_n \to 0$. Put $k_n = T^{-1}Ah_n$. Then, for A^{-1} the bounded operator from Ran A into $(\operatorname{Ker} A)^\perp$ inverting A, we find $h_n = A^{-1}Tk_n$. Since $A^{-1}T \subset (TA^{-1})^*$ and TA^{-1} is densely defined, the operator $A^{-1}T$ is closable and therefore $h=0$. Thus $\operatorname{Ker} K_0 \subset \operatorname{Ker} A$, thereby proving they coincide. Next, suppose that $K_0 f = g \in \operatorname{Ker} A$. Then there exists a sequence $\{f_n\}_{n=1}^\infty$ in

$D(T^{-1}A)\cap(\mathrm{Ker}\ A)^{\perp}$ such that $f_n \to f$ and $g_n = T^{-1}Af_n \to g$. Therefore, $g_n \in D(A^{-1}T)$ (as A^{-1} has its range in $(\mathrm{Ker}\ A)^{\perp}$), $g_n \to g$ and $A^{-1}Tg_n \to f$, whence g is in the domain of the closure of the operator $A^{-1}T$, and

$$\left[\overline{A^{-1}T}\right] g = f.$$

Now $g \in D(T)$, by assumption, while, for all $\ell \in \mathrm{Ker}\ A$,

$$(Tg,\ell) = \lim_{n\to\infty} (T^{-1}Af_n, T\ell) = \lim_{n\to\infty} (f_n, A\ell) = 0,$$

whence $Tg \in (\mathrm{Ker}\ A)^{\perp} = D(A^{-1})$. Thus $A^{-1}Tg = \left[\overline{A^{-1}T}\right] g = f$, implying that $f \in Z_0(T^{-1}A)$. Finally, if $K_0 e = d \in \mathrm{Ker}(T^{-1}A)^n$ for some $n \geq 2$, we may repeat the previous argument to obtain $e \in \mathrm{Ker}(T^{-1}A)^{n+1}$. Thus $Z_0(K_0) = Z_0(T^{-1}A)$. ■

LEMMA 1.2. If $h \in Z_0(T^{-1}A)$, then there exists $k \in Z_0(T^{-1}A)$ such that $T^{-1}Ah = k$ and $T^{-1}Ak = 0$.

Proof: Put $k = T^{-1}Ah$ and $\ell = T^{-1}Ak$ for $\ell \in \mathrm{Ker}\ A$. Using $k,h,\ell \in D(A)$ and $\ell,k \in D(T)$, we have

$$(Ak,k) = (T\ell,k) = (\ell,Tk) = (\ell,Ah) = (A\ell,h) = 0.$$

Since A is positive self adjoint, we find $Ak=0$. Using that $Ak=T\ell$ and $\mathrm{Ker}\ T=\{0\}$, we obtain $\ell=0$, which establishes the lemma. ■

This lemma was published in [360] for A a compact perturbation of the identity and T bounded and in [166] for A unbounded. It means that the Jordan chains of $T^{-1}A$ at the zero eigenvalue have at most length two, and thus implies $\dim Z_0(K_0) \leq 2 \dim \mathrm{Ker}\ A$.

PROPOSITION 1.3. One has

$$T[Z_0(K_0)] = Z_0(K_0^{*}),$$

$$A[Z_0(K_0^{*})^{\perp}\cap D(A)] = Z_0(K_0)^{\perp} = \overline{T[Z_0(K_0^{*})^{\perp}\cap D(T)]},$$

and the following decompositions hold true:

$$Z_0(K_0) \oplus Z_0(K_0^*)^{\perp} = H, \tag{1.8a}$$

$$Z_0(K_0^*) \oplus Z_0(K_0)^{\perp} = H. \tag{1.8b}$$

Proof: Let us first prove the identity

$$T[Z_0(K_0)] = Z_0(K_0^*). \tag{1.9}$$

If $h \in \mathrm{Ker}\ A$, then $Th \in D(AT^{-1}) \subset D(K_0^*)$ and $K_0^* Th = Ah = 0$, whence $Th \in \mathrm{Ker}\ K_0^*$. Similarly, if $f \in D(T^{-1}A)$ and $T^{-1}Af \in \mathrm{Ker}\ A$, then $f \in Z_0(T^{-1}A) \subset D(T)$. Thus, $Tf \in D(AT^{-1})$ and $K_0^* Tf = Af = TK_0 f$, where $f \in Z_0(K_0)$. Since

$$Z_0(K_0) = Z_0(T^{-1}A) = \mathrm{Ker}\ (T^{-1}A)^2,$$

we obtain

$$T[Z_0(K_0)] \subset Z_0(K_0^*).$$

To prove the converse, let us first show that $D(K_0^*) \subset \mathrm{Ran}\ T$. Take $g \in D(K_0^*)$. Then $h \to (T^{-1}Ah, g)$ extends from $D(T^{-1}A)$ to a bounded linear functional on H, and therefore $z \to (T^{-1}z, g)$ extends from $\mathrm{Ran}\ A \cap \mathrm{Ran}\ T$ to a bounded linear functional on Ran A. Because

$$\dim \frac{\mathrm{Ran}\ T}{\mathrm{Ran}\ A \cap \mathrm{Ran}\ T} = \dim \frac{\mathrm{Ran}\ A + \mathrm{Ran}\ T}{\mathrm{Ran}\ A} \leq \dim \frac{H}{\mathrm{Ran}\ A} < \infty,$$

we can define $z \to (T^{-1}z, g)$ as a linear functional on Ran T, which is bounded with respect to the H-norm. Since T^{-1} is self adjoint on H, we have $g \in D(T^{-1}) = \mathrm{Ran}\ T$, which proves the statement.

Now take $g \in \mathrm{Ker}\ K_0^*$. Then $g \in D(T^{-1})$, and for all $\ell \in D(T^{-1}A)$ we have $(T^{-1}g, A\ell) = (g, T^{-1}A\ell) = 0$. Thus, we also have $T^{-1}g \in (\mathrm{Ran}\ A)^{\perp} = \mathrm{Ker}\ A$, and therefore $T^{-1}g \in \mathrm{Ker}(T^{-1}A)$.

Suppose, by induction hypothesis, that for some $n \geq 2$

$$T^{-1}[\mathrm{Ker}\,(K_0^*)^{n-1}] \subset \mathrm{Ker}\,(T^{-1}A)^{n-1}.$$

Take $g \in \mathrm{Ker}\,(K_0^*)^n$. Then $g \in D(K_0^*) \subset \mathrm{Ran}\,T$ and $K_0^* g = T\ell$ for some $\ell \in D(T^{-1}A)^{n-1}$. Now observe that $T\ell \in D(AT^{-1})$ and $K_0^* T\ell = (AT^{-1})T\ell = A\ell$. Since $A\ell \in D(K_0^*) \subset \mathrm{Ran}\,T$, we have

$$T^{-1}A\ell \in T^{-1}[\mathrm{Ker}\,(K_0^*)^{n-2}] \subset \mathrm{Ker}\,(T^{-1}A)^{n-2},$$

whence $\ell \in \mathrm{Ker}\,(T^{-1}A)^{n-1}$. Finally, because $g \in D(K_0^*)$ and $h \longrightarrow (T^{-1}Ah,g)$ extends from $D(T^{-1}A)$ to a bounded linear functional on H, we also have that $h \longrightarrow (Ah,T^{-1}g)$ extends to a bounded linear functional on H. By virtue of the self adjointness of A, the latter in turn implies that $T^{-1}g \in D(A)$, and therefore $g \in D(AT^{-1})$. But then we have $AT^{-1}g = K_0^* g \in \mathrm{Ker}\,(T^{-1}A)^{n-1}$, whence $T^{-1}g \in \mathrm{Ker}\,(T^{-1}A)^n$, which we intended to prove. Consequently, we have established (1.9).

Next take $h \in Z_0(K_0) \cap Z_0(K_0^*)^\perp$. Then $Z_0(K_0) = Z_0(T^{-1}A)$ and (1.9) imply $(Th,k) = 0$ for all $k \in Z_0(T^{-1}A)$, and thus $Th \in Z_0(T^{-1}A)^\perp \subset (\mathrm{Ker}\,A)^\perp = \mathrm{Ran}\,A$. This in turn implies the existence of some $\ell \in Z_0(T^{-1}A)$ such that $h = T^{-1}A\ell$, so $(A\ell,\ell) = (Th,\ell) = 0$, $Th = A\ell = 0$, and $h = 0$. Hence, $Z_0(K_0) \cap Z_0(K_0^*)^\perp = \{0\}$. In a similar way, if $f \in Z_0(K_0^*) \cap Z_0(K_0)^\perp$, then $f = Th$ for some $h \in Z_0(K_0) \cap Z_0(K_0^*)^\perp = \{0\}$ and therefore $f = 0$. Thus also $Z_0(K_0^*) \cap Z_0(K_0)^\perp = \{0\}$.

We now use some simple dimension arguments. Obviously, for $d = \dim Z_0(K_0)$, $d^* = \dim Z_0(K_0^*)$, $c = \mathrm{codim}\, Z_0(K_0)^\perp$ and $c^* = \mathrm{codim}\, Z_0(K_0^*)^\perp$ one has

$$c = d, \qquad c^* = d^*, \qquad d = d^* \tag{1.10}$$

(see (1.9) for the last one). We then have $d = c^*$, $d^* = c$, and the decompositions (1.8) are clear.

Finally, we observe that T maps $Z_0(K_0)$ onto $Z_0(K_0^*)$ and $Z_0(K_0^*)^\perp \cap D(T)$ into $Z_0(K_0)^\perp$. Since T has dense range, we have, using (1.8),

$$\overline{T\{Z_0(K_0^*)^\perp \cap D(T)\}} = Z_0(K_0)^\perp. \tag{1.11}$$

Analogously, we observe that A maps $Z_0(K_0^*)^\perp \cap D(A)$ into $Z_0(K_0)^\perp$ and maps $Z_0(K_0) \subset D(A)$ into $Z_0(K_0^*)$, while $\mathrm{Ker}\,A \subset Z_0(K_0)$. Since Ker A has finite dimension and A has closed range, we must have

$$A\{Z_0(K_0^*)^\perp \cap D(A)\} = Z_0(K_0)^\perp,$$

which completes the proof. ∎

The decompositions (1.8) will now enable us to reduce the boundary value problems (1.1)-(1.2)-(1.3)/(1.5)/(1.6) with given A to one with strictly positive A. In fact, this reduction follows immediately from the following proposition.

PROPOSITION 1.4. Let β be an invertible operator on $Z_0(T^{-1}A)$ satisfying

$$(T\beta h,h) \geq 0, \quad h \in Z_0(T^{-1}A). \tag{1.12}$$

Let P denote the projection of H onto $Z_0(K_0^*)^\perp$ along $Z_0(T^{-1}A)$ and define A_β by

$$D(A_\beta) = D(A), \qquad A_\beta h = T\beta^{-1}(I-P)h + APh. \tag{1.13}$$

Then A_β is strictly positive self adjoint and satisfies

$$A_\beta^{-1}T = \beta \oplus \left[T^{-1}A\,|_{Z_0(K_0^*)^\perp}\right]^{-1}. \tag{1.14}$$

Moreover, if A is a compact perturbation of the identity satisfying the condition

$$\exists\alpha>0:\ \mathrm{Ran}(I-A) \subset \mathrm{Ran}\,|T|^\alpha \cap D(|T|^{3+\alpha}), \tag{1.15}$$

then A_β is a compact perturbation of the identity satisfying the condition

$$\exists\alpha>0:\ \mathrm{Ran}(I-A_\beta) \subset \mathrm{Ran}\,|T|^\alpha \cap D(|T|^{1+\alpha}). \tag{1.16}$$

Proof: From (1.13) we easily derive (1.14). Moreover, for $g \in D(A)$ we have

$$(A_\beta g,g) = (APg,Pg) + (T\beta^{-1}(I-P)g,(I-P)g) \geq 0,$$

where we used (1.12) for $h=\beta^{-1}(I-P)g$. Since $\sigma(A) \subset \{0\} \cup [\varepsilon,\infty)$ for some $\varepsilon>0$ and $Z_0(T^{-1}A)$ has finite dimension, we must have strict positivity of A_β from the obvious triviality of its kernel.

Next, let A be a compact perturbation of the identity satisfying (1.15) for certain $0<\alpha<1$. Since $A_\beta - A = (A_\beta - A)(I-P)$ has finite rank, A_β must be a compact perturbation of the identity too. Furthermore,

$$I-A_\beta = (I-A) - (A_\beta - A)(I-P) = (I-A) - T(\beta^{-1} - T^{-1}A)(I-P),$$

and therefore

$$\mathrm{Ran}(I-A_\beta) \subset \mathrm{Ran}\ |T|^\alpha. \tag{1.17}$$

Also, using that $\mathrm{Ran}(I-A) \subset D(|T|^{2+\alpha})$, we find, for $h \in Z_0(T^{-1}A)$ and $g=T^{-1}Ah$,

$$h = (I-A)h + Tg = (I-A)h + T(I-A)g \in D(|T|^{2+\alpha}),$$

which implies

$$\mathrm{Ran}(I-A_\beta) \subset D(|T|^{1+\alpha}). \tag{1.18}$$

From (1.17) and (1.18) we obtain (1.16). ■

If A is a bounded operator on H, so is A_β, while $A_\beta^{-1}T$ is a self adjoint injective operator on H_A. Here we endow H_A with the (complete) inner product

$$(h,k)_{A_\beta} = (A_\beta^{1/2}h, A_\beta^{1/2}k). \tag{1.19}$$

We note first that, since the β-dependence of A_β is isolated on the finite dimensional subspace $Z_0(K_0)$, the completion of D(A) with respect to $(\cdot,\cdot)_{A_\beta}$ is independent of β and may be denoted by H_A for each β. The boundedness and strict positivity of A_β imply the coincidence of H_A and H as sets and the equivalence of (1.19) to the original inner product of H. We may then define $P_\pm$ as the H_A- orthogonal projections of H onto maximal $A_\beta^{-1}T$-positive/negative $A_\beta^{-1}T$-invariant subspaces. If P is the projection of H onto $Z_0(K_0^*)^\perp$ along $Z_0(K_0)$, the decomposition (1.14) makes it clear that PP_+, PP_- and $I-P$ form a family of complementary projections, independent of β and commuting with $T^{-1}A$.

If A is unbounded, then A_β is unbounded as well and the above analysis is more involved. By Lemma 1.1, the finite dimensional subspace $Z_0(T^{-1}A) \subset D(A)$, and therefore the projection P leaves invariant H_A, its restriction to H_A is bounded, and it has null space $Z_0(T^{-1}A)$ and range $Z_0(K_0^{\,*})^{\perp} \cap H_A$. If T is bounded, $A_\beta^{-1}T$ is self adjoint on H_A and $P_\pm$ may be defined as the H_A-orthogonal projections of H_A onto maximal $A_\beta^{-1}T$- positive/negative $A_\beta^{-1}T$-invariant subspaces. If T is unbounded, let K be the H_A-closure of $T^{-1}A$. Since H_A is continuously imbedded in H, we have $T^{-1}A \subset K \subset K_0$ and $D(K) = D(K_0) \cap H_A$. Moreover, Lemma 1.1 implies

$$Z_0(T^{-1}A) = Z_0(K) = Z_0(K_0) = \mathrm{Ker}\{(T^{-1}A)^2\}, \tag{1.20}$$

while Proposition 1.3 gives rise to the $(\cdot,\cdot)_{A_\beta}$-orthogonal decomposition

$$Z_0(T^{-1}A) \oplus \{Z_0(K_0^{\,*})^{\perp} \cap H_A\} = H_A. \tag{1.21}$$

We then obtain for the H_A-closure of $A_\beta^{-1}T$ the operator

$$K_\beta^{-1} = \overline{A_\beta^{-1}T}^{(H_A)} = \beta \oplus (K|_{Z_0(K_0^*)^{\perp} \cap H_A})^{-1}.$$

It is easily observed that this operator is closed and symmetric on H_A (with respect to $(\cdot,\cdot)_{A_\beta}$). Let us choose a self adjoint extension S_β (provided it exists) and define $P_\pm$ as the H_A-orthogonal projections of H_A onto the maximal S_β-positive/negative S_β-invariant subspaces. Again, PP_+, PP_- and I−P are a family of β-independent complementary bounded projections on H_A.

For various cases the next proposition was established in [160, 166, 360].

PROPOSITION 1.5. The subspaces

$$M_\pm = \{PP_\mp[H_A] \oplus Q_\pm[H]\} \cap Z_0(T^{-1}A) \tag{1.22}$$

satisfy the condition

$$\pm(Tf,f) > 0, \quad 0 \neq f \in M_\pm, \tag{1.23}$$

while

$$M_+ \oplus M_- = Z_0(T^{-1}A). \tag{1.24}$$

Proof: For $f \in M_+$ there exist $g \in PP_-[H_A]$ and $h \in Q_+[H]$ such that $f=g+h$. Then

$$0 \le (Th,h) = (Tf,f) + (Tg,g) - (Tf,g) - (Tg,f).$$

Since $Tf \in Z_0(K_0^*)$ and $g \in Z_0(K_0^*)^\perp$, we have $(Tf,g)=0$, while $f \in Z_0(T^{-1}A)$ and $Tg \in Z_0(T^{-1}A)^\perp$ imply $(Tg,f)=0$. Also,

$$(Tg,g) = (A_\beta^{-1}Tg,g)_{A_\beta} \le 0.$$

Thus, $(Tf,f) \ge 0$. However, $(Tf,f)=0$ would imply $(Th,h)=(Tg,g)=0$, and therefore $h=g=0$, which in turn would imply $f=0$. Hence, $(Tf,f)>0$ for all $0 \ne f \in M_+$ and Eq. (1.23) is clear. It is also clear that

$$M_+ \cap M_- = \{0\}. \tag{1.25}$$

Note that $e \in PP_-[H_A]$ and $k \in PP_+[H_A]$ imply $(Te,k)=(A_\beta^{-1}Te,k)_{A_\beta}=0$ and so $(TPP_-[H_A])^\perp \supset PP_+[H_A]$. From the identity (cf. (1.9))

$$T[M_\pm] = \{TPP_\mp[H_A] \oplus TQ_\pm[H]\} \cap Z_0(K_0^*),$$

we find

$$(T[M_\pm])^\perp = \overline{\{(TPP_\mp[H_A])^\perp \cap (TQ_\pm[H])^\perp\} + Z_0(K_0^*)^\perp} =$$

$$= \overline{[\{PP_\pm[H_A] \oplus Z_0(K)\} \cap Q_\mp[H]] + Z_0(K_0^*)^\perp} = M_\mp \oplus Z_0(K_0^*)^\perp,$$

where we have used the finite dimensionality of $Z_0(K_0)$ and the finite co-dimensionality of $Z_0(K_0^*)^\perp$. Thus,

$$\overline{T(M_+ + M_-)} = \{(T[M_+])^\perp \cap (T[M_-])^\perp\}^\perp = \{Z_0(K_0^*)^\perp\}^\perp = Z_0(K_0^*).$$

Using the finite dimensionality of M_+ and M_- and (1.8) we find

$$M_+ + M_- = Z_0(T^{-1}A),$$

which together with (1.25) implies (1.24). ■

PROPOSITION 1.6. The subspaces

$$M_\pm \cap \text{Ker } A = \{PP_\mp[H_A] \oplus Q_\pm[H]\} \cap \text{Ker } A \tag{1.26}$$

satisfy the direct sum decomposition

$$\text{Ker } A = \{M_+ \cap \text{Ker } A\} \oplus \{M_- \cap \text{Ker } A\} \oplus T^{-1}A[Z_0(K)], \tag{1.27}$$

while

$$\{M_\pm \cap \text{Ker } A\} \oplus T^{-1}A[Z_0(K)]$$

is a maximal subspace of $Z_0(K)$ on which $\pm(Th,h)$ is nonnegative.

Proof: Let us apply the $(T\cdot,\cdot)$-orthogonal decomposition (1.24) as well as the inclusion $T^{-1}A[Z_0(K)] \subset \text{Ker } A$ and compute that

$$(T[M_+ \cap \text{Ker } A])^\perp \cap \text{Ker } A = [M_- \cap \text{Ker } A] \oplus T^{-1}A[Z_0(K)]. \tag{1.28}$$

Since in this equality the left hand side is the $(T\cdot,\cdot)$-orthogonal complement in Ker A of a subspace on which $(Th,h)>0$ for nonzero h and the right hand side is a subspace of Ker A on which (Th,h) is nonpositive, we clearly have (1.28). ■

If we combine Proposition 1.6 with the construction in the statement of Proposition 1.4, we can derive an identity which has important ramifications for the existence of bounded solutions of the half space problem. Indeed, let us choose a maximal subspace N_+ of $Z_0(K)$ on which $(Th,h)\geq 0$, and a maximal subspace N_- of $Z_0(K)$ on which $(Th,h)<0$ for nonzero h. Then obviously

$$N_+ \oplus N_- = Z_0(K).$$

Further, the above proposition shows that one may take $N_+ \subset \mathrm{Ker}\, A$. On defining $\beta h = \pm h$ for $h \in N_\pm$ and extending β linearly to all of $Z_0(K)$, we can construct A_β in such a way that

$$\mathrm{Ran}\, P_+ = PP_+[H_A] \oplus N_+ \subset PP_+[H_A] \oplus \mathrm{Ker}\, A. \tag{1.29}$$

The positivity of $T\beta$ on $Z_0(K)$ then is most easily seen from the identity

$$(T\beta h,h) = (Th_+,h_+) - (Th_-,h_-) + (Th_+,h_-) - (Th_-,h_+) \geq 0$$

for all $h = h_+ - h_-$ with $h_\pm \in N_\pm$. It should be observed that (Th_+,h_-) need not vanish, since Ker A may contain a nontrivial subspace of $T^{-1}A[Z_0(K)]$.

Let us next construct the completion of $D(A) \cap D(T)$ with respect to the inner product

$$(h,k)_{S_\beta} = (|T|h,k) + (|S_\beta|h,k)_{A_\beta}.$$

Again, because the β-dependence of A_β is isolated on the finite dimensional subspace $Z_0(T^{-1}A)$, this Hilbert space does not depend on β and may be denoted by H_{TS}. From H_{TS} we then obtain H_T as the completion with respect to $(h,k)_T = (|T|h,k)$ and H_S as the completion with respect to $(h,k)_{S_\beta} = (|S_\beta|h,k)_{A_\beta}$. The latter does not depend on β. The projections $P_\pm$ and P can all be extended to bounded projections on H_S, resulting in the triple of complementary projections PP_+, PP_-, $I-P$ on H_S. As it turns out, the space H_T is defined as in the previous chapter, with the projections $Q_\pm$ extending as orthogonal projections on H_T.

COROLLARY 1.7. The subspaces

$$M_\pm^{(S)} = \{PP_\mp[H_S] \oplus Q_\pm[H_T]\} \cap Z_0(T^{-1}A) \tag{1.30}$$

have the property

$$\pm(Tf,f) > 0, \quad 0 \neq f \in M_\pm^{(S)},$$

and therefore coincide with $M_\pm$. The analogous statement applies to the intersection of these spaces with Ker A.

2. Boundary value problems

In this section we will analyze the boundary value problems (1.1)–(1.2) along with a condition at infinity, namely, one of

$$\lim_{x \to \pm\infty} \|\psi(x)\| = 0, \tag{2.1}$$

$$\|\psi(x)\| = O(1) \quad (x \to \pm\infty), \tag{2.2}$$

$$\|\psi(x)\| = O(|x|) \quad (x \to \pm\infty). \tag{2.3}$$

(The upper/lower signs are to be taken with (1.1a)–(1.2a)/(1.1b)–(1.2b).) Actually, by virtue of Lemma 1.2, (2.3) is equivalent to (1.6). In Chapter II, these boundary value problems were studied assuming (2.2), and in all cases the unique solution satisfied (2.1). Indeed, the result would have been the same even with the (apparently) more general boundary condition (1.6), as is evident from standard semigroup theory. We shall now generalize these results to non-negative A with nontrivial kernel. However, it will be necessary to distinguish carefully among the boundary conditions at infinity.

We shall first suppose that A is a compact perturbation of the identity, T bounded and $\mathrm{Ran}\,(I-A) \subset \mathrm{Ran}\,|T|^{\alpha}$ for some $0<\alpha<1$. The definition of a solution of the boundary value problem is given in the second paragraph of Section II.2. Writing $\psi_1 = P\psi$ and $\psi_0 = (I-P)\psi$, Eq. (1.1a) may be decomposed as follows:

$$(T\psi_1)'(x) = -A\psi_1(x), \qquad 0<x<\infty, \tag{2.4}$$

and

$$\psi_0{}'(x) = -T^{-1}A\psi_0(x), \qquad 0<x<\infty. \tag{2.5}$$

The second equation is an evolution equation on the finite dimensional zero root linear manifold $Z_0(T^{-1}A)$, and therefore admits the elementary solution

$$\psi_0(x) = e^{-xT^{-1}A}\psi_0(0) = (I-xT^{-1}A)\psi_0(0), \tag{2.6}$$

where we have used Lemma 1.2. Clearly, on imposing condition (2.1), (2.2), or (2.3), we must require that $\psi_0(0)=0$, $\psi_0(0)\in\text{Ker } A$ or $\psi_0(0)\in Z_0(T^{-1}A)$, respectively.

We consider next Eq. (2.4). Let us adjoin to this equation the dummy equation

$$(T\phi_0)'(x) = -A_\beta\phi_0(x), \quad 0<x<\infty, \tag{2.7}$$

on $Z_0(T^{-1}A)$, whose solution is not of concern, as it will be projected out shortly. However, defining $\phi=\phi_0+\psi_1$, we can now combine Eqs. (2.4) and (2.7) to obtain

$$(T\phi)'(x) = -A_\beta\phi(x), \quad 0<x<\infty. \tag{2.8}$$

Referring to Section II.2, we may write the solution as

$$\phi(x) = \exp\{-xT^{-1}A_\beta\}Eg_+, \quad 0\le x<\infty, \tag{2.9}$$

where $g_+\in Q_+[H]$ and E is the inverse of $V=Q_+P_++Q_-P_-$. Then the solution of (1.1a) is given by

$$\psi(x) = \exp\{-xT^{-1}A\}PEg_+ + \psi_0(x), \quad 0\le x<\infty, \tag{2.10}$$

where $\psi_0(x)$ is given by (2.6), $\psi_0(0)$ determines which boundary condition at infinity is met, and the boundary condition (1.2a) is still to be satisfied. More precisely, depending on whether (2.1), (2.2) or (2.3) is imposed, we must find $g_+\in Q_+[H]$ and $\psi_0(0)\in\{0\}$, Ker A or $Z_0(T^{-1}A)$, respectively, satisfying

$$Q_+(PEg_+ + \psi_0(0)) = \varphi_+. \tag{2.11}$$

In case T is unbounded and A is a compact perturbation of the identity satisfying the somewhat more restrictive condition

$$\exists\alpha>0\colon \text{Ran}(I-A) \subset \text{Ran}\,|T|^\alpha\cap D(|T|^{3+\alpha}), \tag{2.12}$$

we seek solutions for boundary data $\varphi_+\in Q_+[D(T)]$. Again we find that all such solutions have the form (2.10), where we must find $g_+\in Q_+[D(T)]$ and $\psi_0(0)\in\{0\}$,

Ker A or $Z_0(T^{-1}A)$, satisfying (2.11). For solutions of the type described at the end of Section II.2, one only demands $g_+ \in Q_+[H]$.

Next we consider the case when A is bounded and seek solutions in the Hilbert space H_T (cf. Section II.3). T may be either bounded or unbounded, with $Z_0(T^{-1}A) \subset D(T)$. By Proposition II 3.1, P and I−P extend to H_T. Then (the unbounded operator) $T^{-1}A_\beta$ extends to H_T and may be treated as in Section II.3; a closed extension of $T^{-1}A$ to H_T is obtained by $T^{-1}A=(T^{-1}A_\beta)P + (T^{-1}A)(I-P)$. If E is the bounded inverse of V on H_T, which exists according to the results of Section II.3 applied to T and A_β, we obtain again $\psi(x)$ in the form (2.10), where we must find $g_+ \in Q_+[H_T]$ and $\psi_0(0)$ precisely as in the previous cases.

Finally, let us consider the most general case, when A is unbounded and we seek solutions in H_S for initial data $\varphi_+ \in Q_+[H_T]$. As in Section II.4, we assume that $Z_0(T^{-1}A) \subset D(T)$ and T satisfies $H_A \subset D(T)$ if T is unbounded. As a result, the techniques of that section can be applied to the transformed problem for $\varphi_+ \in Q_+[H_T]$.

In summary, we are considering three functional formulations of the boundary value problem:

(i) A a compact perturbation of the identity satisfying (2.12) with solutions in H for $\varphi_+ \in Q_+[D(T)]$, or for $\varphi_+ \in Q_+[H]$.

(ii) A bounded with solutions in H_T for $\varphi_+ \in Q_+[H_T]$.

(iii) A unbounded, V_0 surjective, and $H_A \subset D(T)$ if T is unbounded, with solutions in H_S for $\varphi_+ \in Q_+[H_T]$.

In all cases it is assumed that $Z_0(T^{-1}A) \subset D(T)$ and $D(T) \cap D(A)$ are dense in H if T is unbounded. As there is complete symmetry between left and right half space problems, we will write results for the right half space problem only.

We give first a precise meaning to measures of non-existence and non-uniqueness.

Definition: The measure of non-completeness γ_+^0 for Eqs. (1.1)-(1.2)-(2.1), the measure of non-completeness γ_+ for Eqs. (1.1)-(1.2)-(2.2) and the measure of non-completeness γ_+^1 for Eqs. (1.1)-(1.2)-(2.3) are the codimensions in Ran Q_+ of the subspaces of boundary values $\varphi_+ \in \text{Ran } Q_+$ for which these problems are solvable. The measure of non-uniqueness δ_+^0 for Eqs. (1.1)-(1.2)-(2.1), the measure of non-uniqueness δ_+ for Eqs. (1.1)-(1.2)-(2.2) and the measure of non-uniqueness δ_+^1 for Eqs. (1.1)-(1.2)-(2.3) are the dimensions of the solution spaces of the corresponding homogeneous problems.

For historical reasons, the codimensionality of the subspace of boundary values at x=0 for which the boundary value problem is well posed is called the "measure of non-completeness", rather than "measure of non-existence", although the latter nomenclature would correspond more naturally to the related term "measure of non-uniqueness".

THEOREM 2.1. In each of the above functional formulations, the boundary value problem

$$T\psi'(x) = -A\psi(x), \quad 0<x<\infty, \tag{2.13}$$

$$Q_+\psi(0) = \varphi_+, \tag{2.14}$$

$$\lim_{x\to\infty} \|\psi(x)\| = 0, \tag{2.15}$$

has at most one solution for every φ_+, and the measure of non-completeness γ_+^0 for solutions of this problem coincides with the maximal number of linearly independent vectors $g_1,...,g_k \in \text{Ker } A$ satisfying

$$\text{(a)} \ (Tg_i,g_j) = 0, \quad 1\le i,j\le k, \quad i\ne j, \tag{2.16}$$

$$\text{(b)} \ (Tg_i,g_i) \ge 0, \quad 1\le i\le k. \tag{2.17}$$

THEOREM 2.2. In each of the above functional formulations, the boundary value problem

$$T\psi'(x) = -A\psi(x), \quad 0<x<\infty, \tag{2.18}$$

$$Q_+\psi(0) = \varphi_+, \tag{2.19}$$

$$\|\psi(x)\| = O(1) \ (x\to\infty), \tag{2.20}$$

has at least one solution for every φ_+, and the measure of non-uniqueness δ_+ for solutions of this problem coincides with the maximal number of linearly independent vectors $h_1,\ldots,h_\ell \in \mathrm{Ker}\, A$ satisfying

$$\text{(a)}\ (Th_i,h_j) = 0, \qquad 1\le i,j\le \ell, \quad i\ne j, \tag{2.21}$$

$$\text{(b)}\ (Th_i,h_i) < 0, \qquad 1\le i\le \ell. \tag{2.22}$$

THEOREM 2.3. In each of the above functional formulations, the boundary value problem

$$T\psi'(x) = -A\psi(x), \qquad 0<x<\infty, \tag{2.23}$$

$$Q_+\psi(0) = 0, \tag{2.24}$$

$$\|\psi(x)\| = O(x) \ (x\to\infty), \tag{2.25}$$

has at least one solution, and the measure of non-uniqueness δ_+^1 for solutions of this problem coincides with the maximal number of linearly independent vectors $e_1,\ldots,e_m \in Z_0(T^{-1}A)$ satisfying

$$\text{(a)}\ (Te_i,e_j) = 0, \qquad 1\le i,j\le m, \quad i\ne j, \tag{2.26}$$

$$\text{(b)}\ (Te_i,e_i) < 0, \qquad 1\le i\le m. \tag{2.27}$$

COROLLARY 2.4. The solution of (2.13)–(2.15), if it exists, is given by

$$\psi(x) = \exp\{-xK\}PEg_+, \tag{2.28}$$

where g_+ is the unique solution of

$$Q_+PEg_+ = \varphi_+. \tag{2.29}$$

The solutions of (2.18)–(2.20) have the form

$$\psi(x) = \exp\{-xK\}PEh_+ + h_0, \tag{2.30}$$

where $h_0 \in \{\mathrm{Ran}\ PP_+ \oplus \mathrm{Ran}\ Q_-\} \cap \mathrm{Ker}\ A$ and h_+ is the unique solution of

$$Q_+PEh_+ + Q_+h_0 = \varphi_+. \tag{2.31}$$

The solutions of (2.23)-(2.25) have the form

$$\psi(x) = \exp\{-xK\}PEf_+ + (I-xT^{-1}A)f_0, \tag{2.32}$$

where $f_0 \in \{\mathrm{Ran}\ PP_+ \oplus \mathrm{Ran}\ Q_-\} \cap Z_0(T^{-1}A)$ and f_+ is the unique solution of

$$Q_+PEf_+ + Q_+f_0 = \varphi_+. \tag{2.33}$$

Here the operator K is to be interpreted as an appropriate extension of $T^{-1}A$, and likewise Ran Q_+, according to the context, as $Q_+[H]$, $Q_+[D(T)]$ or $Q_+[H_T]$.

Proof of Theorems 2.1, 2.2 and 2.3: Consider first the basic vector equation (2.11), where we require the solution to satisfy the conditions (1.2a) and (2.3). We must then find $\psi_0(0) \in Z_0(K)$ and $g_+ \in$ Ran Q_+ such that (2.11) is fulfilled. Using Proposition 1.4 we may construct β in such a way that

$$\mathrm{Ran}\ P_+ \subset \mathrm{Ran}\ PP_+ \oplus Z_0(K).$$

Then the vector function

$$\psi(x) = e^{-xK}PE\varphi_+ + (I-xK)(I-P)E\varphi_+, \quad 0 \leq x < \infty,$$

is a solution, and the existence of a solution to the boundary value problem (1.1a)-(1.1b)-(2.3) is established.

Next, we recall that the results of the third chapter remain unchanged if the boundedness condition is replaced by the boundary condition (2.25). Using the reduction to strictly positive A_β, we find the formula (2.32) for ψ where $f_+ = PEg_+$ for a vector $g_+ \in \mathrm{Ran}\ Q_+$ and $f_0 \in Z_0(T^{-1}A)$ such that $(f_+ + f_0) \in \mathrm{Ran}\ Q_-$. Therefore

$$f_0 \in [\mathrm{Ran}\ PP_+ \oplus \mathrm{Ran}\ Q_-] \cap Z_0(T^{-1}A) = M_-. \tag{2.34}$$

Conversely, if f_0 satisfies (2.34), we can find a unique $f_+ \in \text{Ran } PP_+$ so that $f_+ + f_0 \in \text{Ran } Q_-$, whence (2.32) will provide a solution of (2.23)–(2.25). We have thus proved Theorem 2.3. At the same time we have shown the existence of a solution of the boundary value problem (1.1a)–(1.2a)–(2.2). Indeed, it is possible to construct within Ker A a maximal subspace of $Z_0(K)$ on which (Th,h) is nonnegative (see Proposition 1.6). Such a subspace will then be a complement of M_- in $Z_0(K)$, and hence it is possible to subtract from every solution of (1.1a)–(1.2a)–(2.3) a solution of (2.23)–(2.25) as to obtain a solution of (1.1a)–(1.2a)–(2.2). Using (1.29) one may see that

$$\psi(x) = e^{-xK}PE\varphi_+ + (I-P)E\varphi_+$$

is a solution of the boundary value problem (1.1a)–(1.2a)–(2.2), where $E=(Q_+P_++Q_-P_-)^{-1}$.

To study the uniqueness of this problem, let us suppose that ψ is a solution of (2.18)–(2.20) corresponding to $\varphi_+=0$. Then ψ has the form (2.10) with $PEg_+ + \psi_0(0) \in \text{Ran } Q_-$ for some $g_+ \in \text{Ran } Q_+$ and $\psi_0(0) \in \text{Ker } A$. Therefore,

$$\psi_0(0) \in [\text{Ran } PP_+ \oplus \text{Ran } Q_-] \cap \text{Ker } A = M_- \cap \text{Ker } A,$$

where we are using the notation of Proposition 1.5. Conversely, if $\psi(0) \in M_- \cap \text{Ker } A$, we can find vectors $g_+ \in \text{Ran } Q_+$ and $h_- \in \text{Ran } Q_-$ such that $\psi_0(0)=PEg_+-h_-$, thereby leading to a solution of (2.18)–(2.20) with $\varphi_+=0$. Hence, the measure of non-uniqueness of solutions of the boundary value problem (2.18)–(2.20) coincides with the dimension of the subspace $M_- \cap \text{Ker } A$. However, $M_- \cap \text{Ker } A$ is a subspace of Ker A maximal with regard to vectors h satisfying $(Th,h)<0$ for $h \neq 0$ (see Proposition 1.6). We shall see in Section IV.1 that the dimension of such a maximal subspace does not depend on the specific choice of the subspace. Therefore, the dimension of $M_- \cap \text{Ker } A$ equals the dimension of the subspace spanned by h_i satisfying (2.21)–(2.22), which completes the proof of Theorem 2.2.

The uniqueness question for solutions of the boundary value problem (2.13)–(2.15) follows from the uniqueness of solutions of the problem

$$T\varphi'(x) = -A_\beta\varphi(x), \quad 0<x<\infty, \tag{2.35}$$

$$Q_+\varphi(0) = \varphi_+, \tag{2.36}$$

$$\lim_{x\to\infty} \|\varphi(x)\| = 0, \tag{2.37}$$

which was proved in Chapter II. In fact, every solution of (2.13)–(2.15) has its initial value $\psi(0)$ in Ran PP_+ and therefore must be a solution of (2.35)–(2.37).

To analyze existence for Eqs. (2.13)–(2.15), we consider the possibility of finding $g_+ \in \text{Ran } Q_+$ such that

$$\varphi_+ = Q_+PEg_+ = PEg_+ - Q_-PEg_+,$$

or, equivalently,

$$\varphi_+ \in (\text{Ran } PP_+ \oplus \text{Ran } Q_-)\cap\text{Ran } Q_+. \tag{2.38}$$

Let us compute the orthogonal complement:

$$[T\{(\text{Ran } PP_- \oplus \text{Ran } Q_+)\cap\text{Ran } Q_-\}]^\perp = [(T[\text{Ran } PP_-])^\perp\cap\text{Ran } Q_-] \oplus \text{Ran } Q_+ =$$

$$= [(\text{Ran } PP_+ \oplus Z_0(K))\cap\text{Ran } Q_-] \oplus \text{Ran } Q_+.$$

Using the inclusion

$$\{0\} \subset \text{Ran } PP_+\cap\text{Ran } Q_- \subset \text{Ran } P_+\cap\text{Ran } Q_- \subset \text{Ker } V = \{0\}$$

and the density of the subspaces

$$\text{Ran } PP_+ \oplus Z_0(T^{-1}A) + \text{Ran } Q_- \supset \text{Ran } P_+ + \text{Ran } Q_- \supset \text{Ran } V,$$

we find that the orthogonal complement of the subspace $(\text{Ran } PP_+\oplus\text{Ran } Q_-)\cap\text{Ran } Q_+$ in Ran Q_+ has the same dimension as $M_+ = (\text{Ran } PP_-\oplus\text{Ran } Q_+)\cap Z_0(T^{-1}A)$. In Proposition 1.6, we saw that M_+ is a subspace of $Z_0(T^{-1}A)$ maximal with regard to vectors h satisfying $(Th,h)\geq 0$. Arguing as before, we obtain that the dimension of M_+ equals the maximal number of linearly independent vectors $x_1,\dots,x_k$ such that $(Tx_i,x_j)=0$ for $i\neq j$ and $(Tx_i,x_i)>0$. The dimension of this subspace is the co-dimension in Ran Q_+ of the closure of the set of vectors φ_+ for which Eqs. (2.13)–(2.15) are solvable. This completes the proof of Theorem 2.1. ∎

COROLLARY 2.5. The conclusions of Theorem 2.3 and Corollary 2.4 remain unchanged if (2.25) is replaced by

$$\exists n \geq 1: \ \|\psi(x)\| = O(|x|^n) \ (x \to \infty). \tag{2.39}$$

For A a compact perturbation of the identity, T bounded and (2.12) satisfied, Theorem 2.2 and 2.3 were established by van der Mee [360]. All three results were obtained in more general settings by Greenberg et al. [160, 162, 166]. Earlier treatments of these problems (generally for specific models, but see [32] for abstract models of this type) either considered A strictly positive or a situation where $T^{-1}A[Z_0(T^{-1}A)] = \text{Ker } A$, whence unique solvability of the boundary value problem (2.18)–(2.20) was assured. It appears that this may have been responsible for introducing an unfortunate misconception into the literature, wherein it is claimed that invertibility of $V = Q_+P_+ + Q_-P_-$ is equivalent to half-range completeness, thus completely disregarding the boundary value problem leading to the operator V.

3. Evaporation models

A few years ago Arthur and Cercignani [15] conjectured that the non-existence of a steady flow in a gas beyond the speed of sound would show up in the linearized BGK model for a rarefied gas as a non-completeness result. These authors applied resolvent integration techniques to the linearized BGK equation

$$(v+d)\frac{\partial\psi}{\partial x}(x,v) + \psi(x,v) =$$

$$= \pi^{-\frac{1}{2}}\int_{-\infty}^{\infty} \{1 + 2v\hat{v} + 2(v^2 - \tfrac{1}{2})(\hat{v}^2 - \tfrac{1}{2})\}\psi(x,\hat{v})e^{-\hat{v}^2}\,d\hat{v}, \tag{3.1}$$

with $0 < x < \infty$, $-\infty < v < \infty$, subject to the boundary condition

$$\lim_{x\to\infty} \pi^{-\frac{1}{2}}\int_{-\infty}^{\infty} |\psi(x,v)|^2 e^{-v^2}\,dv = 0 \tag{3.2}$$

at infinity, and an incoming flux boundary condition at $x=0$ reflecting conservation of mass and energy. (As a result of the BGK linearization, ψ represents the perturbation from an equilibrium distribution.) Here $d \geq 0$ is the drift velocity with

$d=(3/2)^{1/2}$ corresponding to the speed of sound of the gas. They showed that the existence of solutions, valid for $d^2<3/2$, did in fact break down for $d^2>3/2$. Later, Siewert and Thomas [334] reproduced these results by reducing the analysis to a suitable Riemann–Hilbert problem, and obtained analytical solutions. They subsequently applied their technique to a more complicated model incorporating transverse velocity variations and a three dimensional scattering kernel [335].

In this section we will follow the analysis of Greenberg and van der Mee [162] and derive non–existence results for the corresponding abstract kinetic equations from the general theory of the previous sections. To a large extent this is no more than a restatement of results obtained in Section 2. However, it seems worthwhile to present the abstract theory in a fashion more oriented toward existence results for restricted sets of initial data, as this abstract analysis provides a powerful method of efficiently determining regimes of existence and non–existence for specific models. For example, the Arthur–Cercignani results for the model described by Eq. (3.1) can be obtained at a fraction of the effort required by the earlier analyses. An additional application of these abstract results will be presented in Section IX.5.

For the sake of convenience we will consider the case where A is a bounded non–negative Fredholm operator, T a (bounded or unbounded) injective self adjoint operator, and the abstract equation is studied on the Hilbert space extension H_T of D(T). We then seek a solution in H_T of the boundary value problem (2.13)–(2.15).

PROPOSITION 3.1. Let k be the maximal number of linearly independent vectors $h_1,\ldots,h_k$ in Ker A satisfying the conditions

$$\text{(a)}\ (Th_i,h_j) = 0, \quad i,j\leq k, \quad i\neq j, \tag{3.3}$$

$$\text{(b)}\ (Th_i,h_i) < 0, \quad i\leq k, \tag{3.4}$$

and put $m=\dim\{\mathrm{Ker}\ A\cap \mathrm{Ran}\ Q_-\}$. Then the vectors $\varphi_+\in Q_+[\mathrm{Ker}\ A]$ for which the boundary value problem (2.13)–(2.15) is solvable form a linear subspace of dimension k–m.

Proof: According to Theorem 2.1, there is at most one solution of Eqs. (2.13)–(2.15), which will be of the form

$$\psi(x) = e^{-xK}g, \quad 0\leq x<\infty, \tag{3.5}$$

precisely if there is a $g \in \text{Ran } PP_+$ with $Q_+g = \varphi_+ \in Q_+[\text{Ker } A]$. Thus $\varphi_+ = Q_+\alpha$ for some $\alpha \in \text{Ker } A$ with $g - \alpha \in [\text{Ran } PP_+ \oplus \text{Ker } A] \cap \text{Ran } Q_-$. This implies that

$$\alpha \in [\text{Ran } PP_+ \oplus \text{Ran } Q_-] \cap \text{Ker } A = M_- \cap \text{Ker } A$$

and g is the unique (for fixed α) vector in Ran PP_+ satisfying $g - \alpha \in \text{Ran } Q_-$. Thus, the number of linearly independent $\alpha \in \text{Ker } A$ such that (2.13)–(2.15) with $\varphi_+ = Q_+\alpha$ is solvable coincides with $\dim\{M_- \cap \text{Ker } A\} = k$ (cf. the proof of Theorem 2.2).

Finally, in order to obtain the number of linearly independent $\varphi_+ \in Q_+[\text{Ker } A]$ for which (2.13)–(2.15) is solvable, we must correct the estimate above by the degree of non-uniqueness in constructing $\alpha \in \text{Ker } A$ from $\varphi_+ \in Q_+[\text{Ker } A]$. This correction is evidently $m = \dim\{\text{Ker } A \cap \text{Ran } Q_-\}$, which completes the proof of the proposition. ■

In the example of Eq. (3.1), if $H = L_2(\mathbb{R}, \pi^{-1/2} e^{-v^2} dv)$ with

$$Tf(v) = (v+d)f(v),$$

$$Af(v) = f(v) - \pi^{-1/2} \int_{-\infty}^{\infty} \{1 + 2v\hat{v} + 2(v^2 - \tfrac{1}{2})(\hat{v}^2 - \tfrac{1}{2})\} e^{-\hat{v}^2} f(\hat{v}) d\hat{v},$$

then $h_1 = dv - v^2$, $h_2 = 1$, $h_3 = v^2 - \tfrac{1}{2}$ is a basis of the type indicated in Proposition 3.1, and since $(Th_1, h_1) = \tfrac{1}{2}d(d^2 - 3/2)$, $(Th_2, h_2) = d > 0$, $(Th_3, h_3) = 5d/4 > 0$, it is evident that the measure of non-completeness is 2 for $d^2 < 3/2$ and 3 for $d^2 \geq 3/2$. If we consider $\varphi_+ \in Q_+[\text{Ker } A]$ given by

$$\varphi_+(v) = \rho_0 + c_1 v + T_0(v^2 - \tfrac{1}{2}), \quad v \geq -d,$$

then for $d^2 < 3/2$ there will be unique values of the "density perturbation" ρ_0 and the "temperature perturbation" T_0 for each $c_1 \in \mathbb{R}$ such that $\varphi_+ \in Q_+[M_- \cap \text{Ker } A]$. This result corresponds to conservation of mass and energy (but not momentum). If $d^2 \geq 3/2$, then $\varphi_+ \in Q_+[M_- \cap \text{Ker } A]$ if and only if $\rho_0 = T_0 = c_1 = 0$.

Note that, in general, if the operator T has the form $T = T_0 + dI$ for $d > 0$ and if $m = \dim(\text{Ker } A \cap \text{Ran } Q_-)$ is independent of d, then the number $k - m$ of Proposition 3.1 as a function of d is monotonically non-increasing and reaches zero for d sufficiently large. In fact, we must have $(Th, h) = (T_0 h, h) + d(h, h)$ for every $h \in \text{Ker } A$. This may be viewed as an abstract generalization of the supersonic breakdown of linear systems predicted by Arthur and Cercignani.

4. Reflective boundary conditions

The development of abstract half space theory has so far been limited to "autonomous" incoming fluxes, where there is no reflection at the boundary $x=0$. This is a severe restriction in applications related to gas dynamics and Brownian motion, where reflection at the boundary is generally present. Reflective boundary conditions are common as well in radiative transfer.

A typical reflective boundary condition for gas dynamics or Brownian motion in half space geometry requires

$$\psi(0,v) = \alpha\psi(0,-v) + \beta\int_{-\infty}^{0} \sigma(\hat{v}\to v)\psi(0,\hat{v})d\hat{v} + \varphi_+(v), \qquad 0<v<\infty, \tag{4.1}$$

and an appropriate boundary condition at $x=\infty$. Here $\alpha\epsilon[0,1]$ and $\beta\epsilon[0,1-\alpha]$ are the accommodation coefficients for specular and diffuse reflection, respectively, and the surface scattering kernel $\Sigma(\hat{v}\to v)$, defined by

$$\Sigma(\hat{v}\to v) = (v/|\hat{v}|)\, e^{-\frac{1}{2}(v^2-\hat{v}^2)}\sigma(\hat{v}\to v), \qquad \hat{v}<0<v,$$

satisfies the following positivity, reciprocity and normalization conditions:

$$\text{(i)} \quad \Sigma(\hat{v}\to v) \geq 0, \qquad \hat{v}<0<v, \tag{4.2}$$

$$\text{(ii)} \quad ve^{-\frac{1}{2}v^2}\Sigma(-v\to-\hat{v}) = |\hat{v}|\, e^{-\frac{1}{2}\hat{v}^2}\Sigma(\hat{v}\to v), \qquad \hat{v}<0<v, \tag{4.3}$$

$$\text{(iii)} \quad \int_0^{\infty} \Sigma(\hat{v}\to v)dv = 1, \qquad \hat{v}<0. \tag{4.4}$$

For a detailed discussion of surface scattering kernels we refer to the monograph of Cercignani [84].

In the following, we shall formulate an abstract generalization of such problems. We assume T is a (bounded or unbounded) injective self adjoint operator on a complex Hilbert space H, A a positive self adjoint Fredholm operator, with the following properties:

(i) $Z_0(T^{-1}A) \subset D(T)$,
(ii) $T^{-1}A$ essentially is self adjoint on H_A,
(iii) $D(T) \cap D(A)$ is dense in H, H_S and H_T, and the H_S- and H_T-inner products are equivalent on $D(T) \cap D(A)$.

This third condition enables one to implement the analysis entirely on H_T, and is valid, for example, for A a Sturm-Liouville differential operator and T multiplication by an indefinite weight function (cf. Section X.1).

To formulate the abstract reflective boundary condition at x=0, we assume the existence of an inversion symmetry (or signature operator) $J:H \to H$ and a surface reflection operator $R:Q_+[H] \to Q_+[H]$ such that

(iv) $J = J^* = J^{-1}$, $JT = -TJ$, $JA = AJ$,
(v) $TRh = \hat{R}Th$ for some bounded operator $\hat{R}:H \to H$,
(vi) $(|T|Rh,Rh) \le (|T|h,h)$ for all $h \in Q_+[D(T)]$.

Condition (iv) implies that $J[D(T)] \subset D(T)$ and $J[D(A)] \subset D(A)$, and (v) establishes that R is bounded on $Q_+[D(T)]$. The last condition represents the fact that the "current" reflected at the boundary should not exceed the incident current.

The identities $JQ_\pm = Q_\mp J$ may be exploited to extend R to H with RJ=JR. Further, R extends to a contraction on H_T and J extends to a self adjoint unitary operator on H_T. For the model (4.1), the operators J and R may be taken as $(Jh)(v) = h(-v)$, $-\infty < v < \infty$, and

$$(Rh)(v) = \alpha h(v) + \beta \int_{-\infty}^{0} \sigma(\hat{v} \to v) h(-\hat{v}) d\hat{v}, \quad 0 \le v < \infty.$$

In treating these stationary kinetic equations, we have separated the operator describing boundary reflecting processes into a factor J accounting for pure inversion of the direction and a factor R accounting for the remaining reflection processes. The result is a decomposition into two operators, which are usually self adjoint in H_T. When treating time dependent problems in Chapters XI to XIII, we do not make such a distinction. The main reason is that in the time dependent setting one generally does not assume plane parallel geometry, which results in the loss of self adjointness properties of the reflection operator.

The abstract boundary value problem may be written

$$T\psi'(x) = -A\psi(x), \quad 0<x<\infty, \tag{4.5}$$

$$Q_+\psi(0) = RJQ_-\psi(0) + \varphi_+, \tag{4.6}$$

$$\|\psi(x)\|_T = O(1) \ (x\to\infty), \tag{4.7}$$

where conditions (i)-(vi) will be assumed throughout this section.

PROPOSITION 4.1. The boundary value problem (4.5)-(4.7) with $\varphi_+ \in Q_+[H_T]$ has a solution in H_T of the form

$$\psi(x) = \exp\{-xK\}PE\mathbb{S}_R^{-1}\varphi_+, \quad 0\le x<\infty, \tag{4.8}$$

where P is the continuous extension to H_T of the projection along $Z_0(T^{-1}A)$ onto $Z_0(K^*)^\perp$, E is the albedo operator, and $\mathbb{S}_R$ is an invertible operator on H_T defined by

$$\mathbb{S}_R = I + RJ(I-E). \tag{4.9}$$

Proof: Defining

$$\mathbb{S}_R = (Q_+-RJQ_-)EQ_+ + (Q_--RJQ_+)EQ_- \tag{4.10}$$

and using the identity II (1.13) yields easily the expression (4.9). The albedo operator satisfies the inclusion

$$\text{Ran } EQ_+ \subset \text{Ran } PP_+ \oplus \text{Ker } A. \tag{4.11}$$

Because of the equivalence of the inner products in H_T and H_S, E is a strictly positive self adjoint operator on H_T with spectrum $\sigma(E)\subset(0,2)$ (cf. Eq. II (4.7)). Thus $\|I-E\|_{H_T}<1$ and

$$\|I-\mathbb{S}_R\|_{H_T} \le \|R\|_{H_T}\|J\|_{H_T}\|I-E\|_{H_T} < 1,$$

which yield the invertibility of $\mathbb{S}_R$. Since, by (4.14), $(I-P)E\mathbb{S}_R^{-1}\varphi_+ \in \text{Ker } A$, it is immediate that the function in Eq. (4.8) satisfies (4.5) with boundary condition (4.7). We may then compute

$$(Q_+ - RJQ_-)E\mathfrak{S}_R^{-1}\varphi_+ = (Q_+\mathfrak{S}_R)\mathfrak{S}_R^{-1}\varphi_+ = \varphi_+,$$

and the proposition follows. ■

THEOREM 4.2. Define M_-^R by

$$M_-^R = [\text{Ran } PP_+ \oplus \text{Ker}(Q_+ - RJQ_-)] \cap Z_0(T^{-1}A).$$

Then the measure of non-uniqueness for solutions of the boundary value problem (4.5)-(4.7) coincides with $\dim(M_-^R \cap \text{Ker } A)$.

Proof: The solutions of the boundary value problem for $\varphi_+=0$ are of the form

$$\psi(x) = e^{-xK}h_+ + h_0, \quad 0 \le x < \infty,$$

where $h_+ \in \text{Ran } PP_+$, $h_0 \in \text{Ker } A$ and

$$(Q_+ - RJQ_-)(h_+ + h_0) = 0.$$

Now $g \in \text{Ran } PP_+ \cap \text{Ker}(Q_+ - RJQ_-)$ implies $g = Eg_+$ with $g_+ \in \text{Ran } Q_+$ and $\mathfrak{S}_R g_+ = 0$. The invertibility of $\mathfrak{S}_R$ then gives $g=0$. Therefore, we have $h_0 \in [\text{Ran } PP_+ \oplus \text{Ker}(Q_+ - RJQ_-)] \cap \text{Ker } A$ and $h_+ + h_0 \in \text{Ker}(Q_+ - RJQ_-)$, which proves the theorem. ■

COROLLARY 4.3. Under the hypothesis $\text{Ker}(Q_+ - RJQ_-) \cap \text{Ker } A = \{0\}$, the boundary value problem (4.5)-(4.7) has measure of non-uniqueness k, where k is the maximal number of linearly independent vectors $h_1, \ldots h_k \in \text{Ker } A$ satisfying

(a) $(Th_i, h_j) = 0, \quad i \ne j,$

(b) $(Th_i, h_i) < 0, \quad i \le k.$

In particular, the hypothesis is satisfied if $\|R\|_{H_T} < 1$.

Proof: Let us prove first that $(Th,h)\le 0$ if $h\in M_-^R$. We write $h=f+g$ with $f\in \mathrm{Ran}\ PP_+$ and $g\in \mathrm{Ker}(Q_+-RJQ_-)$. Since $h\in Z_0(T^{-1}A)$ and $f\in Z_0(K^*)^\perp$, we have $(Th,f)=0$. Hence,

$$(Th,h) + (Tf,f) = (Tg,g) = \|Q_+g\|_T^2-\|Q_-g\|_T^2 = \|RJQ_-g\|_T^2-\|Q_-g\|_T^2 \le$$

$$\le -(1-\|R\|_{H_T}^2)\|Q_-g\|_T^2.$$

Since $(Tf,f)=(A_\beta^{-1}Tf,f)_{A_\beta}\ge 0$, we must have $(Th,h)\le 0$.

Note that if $\|R\|_{H_T}<1$, then $(Th,h)<0$ with $h\ne 0$ as above. For $(Th,h)=0$ implies $f=0$ and either $\|R\|_{H_T}=1$ or $Q_-g=0$. But the latter would give $Q_+g=RJQ_-g=0$ and so $g=0$, whence $h=0$.

Now we compute (orthogonal complements to be taken in H):

$$(TM_-^R)^\perp = [(T\mathrm{Ran}\ PP_+)^\perp\cap \mathrm{Ran}\ T^{-1}(Q_+-Q_-J\hat{R}^*)]+Z_0(K^*)^\perp =$$

$$= [(\mathrm{Ran}\ PP_-\oplus Z_0(T^{-1}A))\cap \mathrm{Ran}(Q_++Q_-J\hat{R}^*)T^{-1}]+Z_0(K^*)^\perp =$$

$$= \{[\mathrm{Ran}\ PP_-\oplus \mathrm{Ran}(Q_++Q_-J\hat{R}^*)]\cap Z_0(T^{-1}A)\}\oplus Z_0(K^*)^\perp.$$

We have used the fact that $\mathrm{Ran}(Q_++Q_-J\hat{R}^*)$ is closed in H, since $Q_++Q_-J\hat{R}^*$ is a bounded projection on H. Writing $h=f+g\in Z_0(T^{-1}A)$ with $f\in \mathrm{Ran}\ PP_-$ and $g=(Q_++Q_-J\hat{R}^*)\ell$, we obtain

$$(Th,h) + (Tf,f) = (Tg,g) = \|Q_+g\|_T^2 - \|Q_-g\|_T^2 = \|Q_+\ell\|_T^2 - \|Q_-J\hat{R}^*\ell\|_T^2 \ge$$

$$\ge (1-\|\hat{R}^*\|_{H_T})\|Q_+\ell\|_T^2,$$

which is non-negative since

$$0 \le (\hat{R}^*h,\hat{R}^*h)_T = (R^*Th,(Q_+-Q_-)\hat{R}^*h) = ((Q_+-Q_-)h,R(Q_+-Q_-)\hat{R}^*h)_T \le$$

$$\le \|h\|_T\|\hat{R}^*h\|_T$$

and therefore $\|\hat{R}^*h\|_T \le \|h\|_T$. Thus $(Th,h)\ge 0$ for $h\in(TM_-^R)^\perp\cap Z_0(K)$, with strict positivity if $\|R\|_{H_T}<1$. Since M_-^R has the property that $(Th,h)<0$ for $0\neq h\in M_-^R$, it must be maximal in this respect, and likewise within Ker A for the corresponding subspace $M_-^R\cap\text{Ker } A$. ∎

COROLLARY 4.4. For $R=I$ (i.e., purely specular reflection), the measure of non-uniqueness for solutions of the boundary value problem (4.5)-(4.7) is the dimension of the subspace $\{h\in\text{Ker } A : Jh=h\}$.

Proof: Note that any vector $h\in H_T$ satisfying $Q_+h=JQ_-h$ has the property

$$(Th,h) = \|Q_+h\|_T^2 - \|Q_-h\|_T^2 = \|Q_+h\|_T^2 - \|JQ_-h\|_T^2 = 0.$$

If $h\in\text{Ran } PP_+\oplus\text{Ker } A$, then $h=f+g$ for $f\in\text{Ran } PP_+$ and $g\in\text{Ker } A$. Now $h=Jh$, and so $f=Jf$, $g=Jg$. But $Jf\in\text{Ran } PP_-$ and $\text{Ran } PP_+\cap\text{Ran } PP_-=\{0\}$ give $f=0$. Thus $h=Jh\in\text{Ker } A$. Conversely, if $h=Jh\in\text{Ker } A$, then $Q_+h=Q_+Jh=JQ_-h$. ∎

We have presented an existence and uniqueness theory for half space boundary value problems with reflection at one surface. For strictly positive self adjoint A such a theory was provided by van der Mee [367], and extended to non-negative A by van der Mee and Protopopescu [369], using the techniques of this section. Earlier, Beals and Protopopescu [35, 36] established the existence of solutions for the Fokker-Planck equation (4.5) with a general reflection law, but an extra condition on the incoming flux φ_+ was imposed to ensure existence. This condition is now known to be unnecessary.

Guiraud [177, 178] has derived existence and uniqueness results for the stationary Boltzmann equation with sufficiently regular scattering term and for rather arbitrary reflective boundary conditions. These results were improved by Maslova [260, 261]. We shall return to the one dimensional linearized Boltzmann equation in Section IX.5.

Chapter IV

NON-DISSIPATIVE AND NON-SYMMETRIC KINETIC MODELS

1. Indefinite inner product spaces

In previous chapters we developed the existence and uniqueness theory for boundary value problems of the form

$$T\psi'(x) = -A\psi(x), \quad 0<x<\infty, \tag{1.1}$$

$$Q_+\psi(0) = \varphi_+, \tag{1.2}$$

with an appropriate condition at $x=\infty$, either

$$\|\psi(x)\| = O(1) \ (x\to\infty) \tag{1.3}$$

or

$$\lim_{x\to\infty} \|\psi(x)\|=0, \tag{1.4}$$

as well as the corresponding problem for the interval $-\infty<x\leq 0$. The operators T and A were assumed self adjoint on a complex Hilbert space H, with T injective and A non-negative Fredholm. Further, Q_+ denoted the orthogonal projection of H onto the maximal T-positive, T-invariant subspace, and the specific functional formulation of the problem depended on the boundedness properties of T and A. The idea was to strive for semigroup type solutions of the form

$$\psi(x) = \exp\{-xT^{-1}A\}E\varphi_+, \quad 0\leq x<\infty, \tag{1.5}$$

where the range of the operator E belonged to a suitable $T^{-1}A$-invariant subspace. (In certain functional formulations we had to specify this more carefully; see Sections II.3 and II.4.) In order to find such a subspace, we employed the Spectral Theorem, since, at least for strictly positive A, (an extension of) the operator $T^{-1}A$ was similar to a self adjoint operator. For A with a nontrivial kernel, we were able to carry out a similar analysis after splitting off a finite dimensional root linear manifold containing Ker A, and to obtain results on related problems involving reflective boundaries.

In Sections 2 and 3 we will develop an existence and uniqueness theory for boundary value problems of the type indicated above with T bounded and A having a finite dimensional negative part. To accomplish this, we will exploit a Spectral Theorem for self adjoint operators on indefinite inner product spaces. Although the stationary theory of kinetic models with multiplying media obviously is not important for physically relevant models in half space geometry, the results obtained herein will play an important role in the analysis of the abstract problem in slab geometries, and models with multiplying media are of considerable importance in such finite geometries. In Section 4 we will consider the case in which A has a nonnegative real part.

In analogy with the case of non-negative A, our approach aims at obtaining a reduction of the problem to four subproblems:

(i) Eqs. (1.1)-(1.2) with A replaced by a finite dimensional perturbation A_β which is strictly positive.

(ii) An elementary evolution equation on a finite dimensional subspace.

(iii) The matching of boundary conditions.

(iv) An analysis of the effects of the perturbation.

As we shall see, part (ii) of the program will have to be modified due to spectrality problems related to indefinite inner product spaces.

The most elementary example of the boundary value problem under study arises from neutron transport in a multiplying medium. Under conditions of isotropic scattering and fission and the one speed approximation, we have the stationary neutron transport equation

$$\mu \frac{\partial \psi}{\partial x}(x,\mu) + \psi(x,\mu) = \frac{c}{2}\int_{-1}^{1} \psi(x,\hat{\mu})d\hat{\mu}, \qquad 0<x<\infty, \tag{1.6}$$

for $\psi(x,\cdot)\in L_2([-1,1],d\mu)$, and we may define

$$Th(\mu) = \mu h(\mu),$$

$$Ah(\mu) = h(\mu) - \frac{c}{2}\int_{-1}^{1} h(\hat{\mu})d\hat{\mu}.$$

Under the multiplying medium condition $c>1$, it was observed by Ball and Greenberg [21] that $T^{-1}A$ is self adjoint with respect to an indefinite inner product. This observation was used to determine the spectral decomposition of $T^{-1}A$, although the boundary value problem (1.1)-(1.2) remained unstudied. An analysis of the abstract boundary value problem was made by Greenberg and van der Mee [163] along the lines to be described in this chapter.

In the remainder of this section we will summarize some standard material on indefinite inner product spaces. For this theory we refer in particular to the monographs of Bognár [46] and Gohberg et al. [146, 147] The latter contains results primarily in a finite dimensional setting. In the most general formulation developed to date (i.e., for definitizable operators) the spectral theory for operators self adjoint on indefinite inner product spaces was largely advanced through the pioneering work of Krein and Langer (e.g., [228, 237]). The state of affairs up to 1978 has been reviewed by Azizov and Iokhvidov [19]. For a complete proof of the Spectral Theorem one may refer to a recent article of Langer [238]. A new proof of the Spectral Theorem was recently given by Bognár [47].

Let H be a complex vector space and $[\cdot,\cdot]$ a symmetric sesquilinear form on H, with respect to which H is complete. The space H with its indefinite metric is called a (complete) indefinite inner product space. In general, such spaces do not have sufficient structure to be of interest from the point of view of operator theory. Of considerably more interest are certain subclasses of these, Krein spaces and Π_κ spaces.

An indefinite inner product space H is a **Krein space** if H has an $[\cdot,\cdot]$-orthogonal decomposition $H = H^+ \oplus H^-$ such that the metric $[\cdot,\cdot]$ is complete and (strictly) positive definite on H^+, and complete and (strictly) negative definite on H^-. If one of these subspaces (generally taken to be H^-) is finite dimensional of dimension κ, then H is called a Π_κ **space** (or **Pontrjagin space**). Throughout we will assume that the dimension of the maximal negative subspace of a Π_κ-space is κ. If a linear operator J is defined by $Jh = \pm h$ for $h \in H^\pm$, then

$$(h,k) = [Jh,k] \tag{1.7}$$

defines a complete (positive definite) inner product on H; i.e., H with $(\cdot,\cdot)$ is a Hilbert space, and J is unitary and self adjoint with respect to $(\cdot,\cdot)$. The above decomposition is called a fundamental decomposition of the Krein space H. Such a decomposition is never unique, though the dimension κ is.

A subspace M of an indefinite inner product space is strictly positive/strictly negative if $0 \neq x \in M$ implies $\pm[x,x]>0$, and positive/negative if $\pm[x,x]\geq 0$. It is neutral if $[x,x]=0$, and non-degenerate if $0 \neq x \in M$ implies the existence of some $y \in M$ with $[x,y]\neq 0$. We call M maximal positive if $M \subset N$ and N positive imply M=N, and similarly for the maximality of neutral and negative subspaces.

We note that any indefinite inner product space has a fundamental decomposition as an orthogonal direct sum of a strictly positive, a strictly negative and a neutral subspace.

The following results will be used frequently (see, for example, [46]).

PROPOSITION 1.1. In an arbitrary inner product space, any positive linear manifold is contained in a maximal positive subspace. Any two maximal positive subspaces have the same dimension. Both statements are true with "positive" replaced by "strictly positive", "negative" or "strictly negative".

COROLLARY 1.2. If M is a non-degenerate (closed) subspace, then $M \cap M^{\perp}=\{0\}$ and $H=M \oplus M^{\perp}$. If M is a maximal positive subspace, then $M^{\perp}$ is a strictly negative subspace. This is also true with positive replaced by "strictly positive", "negative" or "strictly negative" and "strictly negative" replaced by "negative", "strictly positive" or "positive", respectively. If H is a Krein space and M is a (closed) non-degenerate subspace, then $M^{\perp}$ is non-degenerate.

In the previous chapter we have used these results on several occasions. If A is non-negative, the zero root linear manifold $Z_0(K)$ is a finite dimensional indefinite inner product space with respect to the inner product

$$[h,k] = (Th,k). \tag{1.8}$$

Proposition III 1.6 can be reformulated as follows: M_+ and M_- are maximal strictly positive and maximal strictly negative subspaces of $Z_0(K)$, respectively, while the inner product (1.8) is non-degenerate. In the existence and uniqueness theory of Sections III.2 to III.4 an important role is played by bases of subspaces of Ker A, which consist of vectors h with $[h,h]<0$ or vectors h with $[h,h]\geq 0$. In fact, all of these results can be formulated in terms of dimensions of maximal strictly negative and maximal positive subspaces of Ker A. The more general results for indefinite self adjoint A obtained in this chapter are valid also for non-strictly positive self adjoint A, thereby reformulating the existence and uniqueness results of Chapter III in terms of these dimensions.

Let us now turn to the spectral theory of self adjoint operators on a Krein space. First we remark that all topological notions related to a Krein space H are connected with the strong (or weak) topology of the Hilbert space (with inner product (1.7)) associated with the given Krein space. For instance, $\{h_n\}_{n=1}^{\infty}$ converges to h in the strong sense means that $\lim_{n\to\infty} (h_n-h,h_n-h) = 0$, where $(\cdot,\cdot) = [J\cdot,\cdot]$ is defined by (1.7). By a **self adjoint operator** on a Krein space H we mean a (closed) linear operator R on H, with domain D(R), satisfying the conditions

(i) $[Rh,k] = [h,Rk]$ for all $h,k \in D(R) \subset H$.

(ii) $D(R) = \{k \in H : h \mapsto [Rh,k]$ extends from D(R) to a continuous functional on H$\}$.

We remark that this is the natural extension to Krein spaces of the usual definition for self adjoint operators on Hilbert spaces. In general, one may not expect a Spectral Theorem for self adjoint Krein space operators. As a matter of fact, if R_0 is a bounded linear operator on the Hilbert space H_0 and if we make $H=H_0 \oplus H_0$ into a Krein space by putting

$$[(h_1,h_2), (k_1,k_2)] = (h_1,k_2) + (h_2,k_1),$$

then the operator $R=R_0 \oplus R_0^*$, where R_0^* is the adjoint of R_0, will be self adjoint with respect to $[\cdot,\cdot]$, but in general its invariant subspace structure will not allow a Spectral Theorem. Thus some restriction is necessary. It turns out that a self adjoint operator on a Π_κ-space always allows a Spectral Theorem. A more general version, for so-called definitizable operators, is due to Langer [237] (see [238] for an accessible proof). It generalizes both the result for Π_κ-spaces (derived first in [228]) and the result for positive operators (due to Krein and Smuljan [230]). A new and most simple proof for positive operators was recently given by Bognár [47].

Let us formulate the Spectral Theorem, restricting ourselves to bounded self adjoint operators. By a (bounded) **positive operator** on a Krein space H we mean a (bounded) operator R satisfying $[Rh,h] \geq 0$ for all $h \in H$. Such an operator necessarily is self adjoint. A (bounded) **definitizable operator** on a Krein space H will be a (bounded) self adjoint operator R such that $\rho(R)$ is positive for some polynomial ρ. Such a polynomial is called a definitizing polynomial for S. It is easily shown (cf. [46], Theorem IX 7.3) that every (bounded) self adjoint operator on a Π_κ-space is definitizable.

Every (bounded) definitizable operator S admits a decomposition $S=S_1 \oplus S_2$, where S_1 and S_2 are defined on the $[\cdot,\cdot]$-orthogonal subspaces H_1 and H_2, respectively, such that

(i) the spectrum of S_1 is real.

(ii) H_2 has finite dimension and the real spectrum of S_2 is empty.

In this case the spectrum of S_2 is symmetric with respect to the real line, also as far as its multiplicities and Jordan structure are concerned. In particular, if H is a Π_κ-space, then the dimension of H_2 is at most 2κ, while H_1 is a Π_ℓ-space with $\ell=\kappa-\frac{1}{2}\dim H_2$. For all these results we refer to Chapter IX of [46]. In consideration of the above, we may assume that there is only real spectrum in formulating the Spectral Theorem for definitizable operators.

THEOREM 1.3. Let R be a (bounded) definitizable operator on the Krein space H with real spectrum only. Then there exists a unique projection valued function F on the real line with the following properties:

(i) F is monotonically non-decreasing in the sense that

$$[F(t)h,h] \leq [F(s)h,h], \quad t<s, \quad h\in H,$$

except if the interval $[t,s]$ contains one of the finitely many (real) **critical points** $\alpha_1,...,\alpha_r$; at these points and for all $t<\alpha_i<s$ there exists a vector h_i such that $[F(t)h_i,h_i] > [F(s)h_i,h_i]$.

(ii) F is strongly right and left continuous, except possibly at the critical points; moreover, $F(t)\to 0$ and $F(s)\to I$ as $t\to-\infty$ and $s\to+\infty$ in the strong sense.

(iii) for all t,s which are not critical points we have $F(t)F(s) = F(s)F(t) = F(\min\{t,s\})$.

(iv) for every closed interval $[t,s]$ not containing a critical point we have

$$\int_{t_+}^{s_+} z\,dF(z)h = R(F(s)-F(t))h, \quad h\in H,$$

where the integral is defined as the strong limit of Riemann sums.

(v) for all $t<s$ which are not critical points the projection $F(t)-F(s)$ commutes with R and the restriction of R to its range has its spectrum inside $[t,s]$.

The critical points necessarily appear as zeros of a definitizable polynomial of R, though not all such zeros need to be critical points. The eigenvalues of R are precisely those points t where F(t) is not strongly continuous, the real resolvent set of R consists of those t where F(t) is locally constant, and the continuous spectrum of R consists of all t where F(t) is strongly continuous but not locally constant. There is no residual spectrum of R. The function F is called the resolution of the identity associated with R.

Except for the critical points, definitizable operators share many properties with self adjoint operators on Hilbert spaces. For the critical points we have the following proposition (cf. [228, 237, 238]).

PROPOSITION 1.4. For a critical point α of a (bounded) definitizable operator R on a Krein space H the following statements are equivalent:

(i) $F(\cdot)$ has strong left and right limits at α.

(ii) $F(\cdot)$ is bounded in a deleted neighborhood of α.

(iii) for sufficiently small $\varepsilon>0$ and $M_\varepsilon=\mathrm{Ran}\,(F(\alpha+\varepsilon)-F(\alpha-\varepsilon))$ we have

$$\{h\in M_\varepsilon : [h,k]=0 \text{ for all } k\in M_\varepsilon\} = \{0\}.$$

If H is a Π_κ-space, then we have the fourth equivalent condition:

(iv) α is an eigenvalue and

$$\{h\in Z_\alpha(R) : [h,k]=0 \text{ for all } k\in Z_\alpha(R)\} = \{0\}.$$

Critical points satisfying the above equivalent conditions are called **regular**. If these conditions are not satisfied, they are called **irregular**. It is clear that we may extend conditions (iii) and (v) of Theorem 1.3 to all t,s which are not irregular critical points and condition (iv) to closed intervals $[t,s]$ which do not contain irregular critical points.

With respect to Π_κ-spaces we mention two important results for later use.

First, $Z_\alpha(R) = \text{Ker}\,(R-\alpha I)^{2\kappa+1}$ for $\alpha \in \mathbb{R}$; i.e., Jordan chains for real eigenvalues have length at most $2\kappa+1$ (see [46], Theorem IX 4.9). Further, every (bounded) self adjoint operator on a Π_κ-space has a closed invariant subspace which is maximal positive (see [310]).

2. Reduction to a strictly dissipative kinetic model

In this section we reduce the boundary value problems (1.1)-(1.2)-(1.3)/(1.4) to boundary value problems of the same form, where A is replaced by a strictly positive self adjoint perturbation of A of finite rank. Here, and in the next section also, we assume T and A are self adjoint operators on the Hilbert space H, T is bounded injective and A is (possibly unbounded) Fredholm with spectrum intersecting $(-\infty,0]$ at finitely many points representing eigenvalues of finite multiplicity. By imposing the boundedness condition on T we remove many of the technical difficulties encountered in Section III.4, while enabling the application of the theory of bounded (rather than unbounded) definitizable operators.

Let $\lambda_1,\ldots,\lambda_m$ be the negative eigenvalues of A. Then we have the decomposition

$$D(A) = \bigoplus_{i=1}^{m} Z_{\lambda_i}(\bar{A}) \oplus Z_0(A) \oplus \{Z_{(0,\infty)}(A) \cap D(A)\}, \tag{2.1}$$

where $Z_\lambda(A)$ is the λ root linear manifold of A and $Z_{(0,\infty)}(A)$ is the orthogonal complement of the m+1 other constituent subspaces in (2.1). We note that $Z_{(0,\infty)}(A)$ is a maximal A-strictly positive, A-invariant subspace of H. Let us define H_A as the completion of D(A) with respect to the indefinite metric

$$\{h,\ell\} = (Ah,\ell). \tag{2.2}$$

Then H_A is an indefinite inner product space with fundamental decomposition

$$H_A = H_A^- \oplus H_A^0 \oplus H_A^+, \tag{2.3}$$

where

$$H_A^- = \bigoplus_{i=1}^{m} Z_{\lambda_i}(A), \qquad H_A^0 = Z_0(A), \qquad H_A^+ = Z_{(0,\infty)}(A) \cap H_A. \tag{2.4}$$

It should be noted that H_A may be identified with H if A is bounded. Otherwise, H_A is a proper dense submanifold of H. If Ker A = {0}, then H_A is a Π_κ space, where κ is the sum of the multiplicities of the negative eigenvalues of A. The injectivity of A and the Fredholm assumption guarantee that $S=A^{-1}T$ is bounded self adjoint with respect to (2.2). Then S has at most κ non-real eigenvalues (multiplicities taken into account) occurring in complex conjugate pairs with pairwise coinciding Jordan structures, while the length of a Jordan chain for a real eigenvalue of S does not exceed $2\kappa+1$. There is a resolution of the identity for the real part of the spectrum of S, possibly with finitely many critical points at certain eigenvalues (see Theorem 1.3).

If A has non-zero kernel, then H_A with metric (2.2) is not even a Krein space, and the spectral analysis of S is somewhat more circuitous. In this case we will use the fact that S is invariant on $Z_0(AT^{-1})^\perp$, which is a Π_κ space under (2.2), though κ need not coincide either with the number of negative eigenvalues or with the number of nonpositive eigenvalues of A.

Definition: The operator A will be called T-regular if for each $0 \neq \alpha \in \mathbb{C}$, $Z_\alpha(S)$ is non-degenerate with respect to the indefinite inner product (2.2), and if $Z_0(S)$ is finite dimensional and non-degenerate with respect to the indefinite inner product (1.8).

The condition on $Z_0(S)$ is sometimes called T-regularity at zero [163]. The T-regularity assumption precisely guarantees that the critical points in the spectral resolution of S are regular, and thus that the resolution of the identity of S is bounded on the real line (cf. Proposition 1.4). If A has a nontrivial kernel, T-regularity at zero is essential for extending S from $Z_0(AT^{-1})^\perp$ to H_A in such a way as to preserve the applicability of the Spectral Theorem. Actually, the reduction

$$H = Z_0(K) \oplus Z_0(K^*)^\perp \tag{2.5}$$

for $K = T^{-1}A$ and the deformation of A to a non-singular operator A_β in the fashion of Proposition III 1.4 are still available, by virtue of the T-regularity. What is required, however, is a decomposition of H into K-invariant subspaces in such a way that A can be deformed to a strictly positive operator. Such a decomposition is provided by the following proposition.

PROPOSITION 2.1. Let A be T-regular and write $K=T^{-1}A$ and $K^*=AT^{-1}$. Then there exist a K-invariant finite dimensional subspace $Z(K)\subset D(K)$ and a K-invariant finite codimensional subspace $Z(K^*)^\perp$ with the following properties:

(i) $T[Z(K)] = Z(K^*)$, $\overline{T[Z(K^*)^\perp]} = Z(K)^\perp$.

(ii) $A[Z(K^*)^\perp \cap D(A)] = Z(K)^\perp$.

(iii) $Z(K)\oplus Z(K^*)^\perp = Z(K^*)\oplus Z(K)^\perp = H$.

(iv) $Z(K)\oplus[Z(K^*)^\perp \cap H_A] = H_A$.

(v) $Z(K^*)^\perp$ is strictly positive with respect to (2.2).

(vi) $Z(K)$ is non-degenerate with respect to (1.8).

Moreover, if P denotes the projection of H onto $Z(K^*)^\perp$ along $Z(K)$ and β is an invertible positive operator on $Z(K)$ (with respect to (1.8)), then the operator

$$A_\beta = T\beta^{-1}(I-P) + AP \tag{2.6}$$

with $D(A_\beta) = D(A)$ is strictly positive self adjoint and Fredholm, and

$$A_\beta^{-1}T = \beta \oplus \left[T^{-1}A\big|_{Z(K^*)^\perp}\right]^{-1}. \tag{2.7}$$

Proof: Observe that $Z_0(K^*)^\perp \cap H_A$ is a Π_κ-space for certain κ. Let $\bar{\kappa}$ be the sum of the algebraic multiplicities of K in the open upper (or lower) half plane. Because the real part of $S=A^{-1}T$ on $Z_0(K^*)^\perp$ does not have irregular critical points, the subspaces $Z_\lambda(K)$, with λ running through all non-zero critical points, are $\Pi_{\kappa(\lambda)}$-spaces, where

$$\sum_\lambda \kappa(\lambda) = \kappa - \bar{\kappa}. \tag{2.8}$$

For every critical point λ we choose a $T^{-1}A$-invariant subspace M_λ of $Z_\lambda(K)$, which is maximal strictly negative (see the last paragraph of Section 1; strict negativity is by virtue of the finite dimensionality). Now we extend M_λ to the $T^{-1}A$-invariant subspace N_λ of $Z_\lambda(K)$ spanned by the maximal Jordan chains of $T^{-1}A$ corresponding to the eigenvalue λ which have at least one vector in M_λ. Then

$$\dim N_\lambda \le (2\kappa(\lambda)+1)\dim M_\lambda = (2\kappa(\lambda)+1)\kappa(\lambda) < \infty. \tag{2.9}$$

Next we put

$$Z(K) = \{\bigoplus_{\tau} Z_{\tau}(K)\} \oplus \{\bigoplus_{\lambda} N_{\lambda}\} \oplus Z_0(K), \tag{2.10}$$

where τ runs through the nonreal eigenvalues and λ through the nonzero critical points. In this way we find Z(K) to have finite dimension. Obviously, Z(K) is $T^{-1}A$-invariant and contained in D(A).

If we take the orthogonal complements of both sides of Eq. (2.10), we get

$$Z(K)\cap(T[Z(K)])^{\perp} =$$

$$= \bigcap_{\mathrm{Im}\tau>0} \{W_{\tau}\cap(T[W_{\tau}])^{\perp}\} \cap \{Z_0(K)\cap(T[Z_0(K)])^{\perp}\} \bigcap_{\lambda} \{N_{\lambda}\cap(T[N_{\lambda}])^{\perp}\},$$

where $W_{\tau}=Z_{\tau}(K)\oplus Z_{\bar{\tau}}(K)$. For all $h,k\in W_{\tau}$ we have

$$(h,k)_A = (Th,T^{-1}Ak) = [h,T^{-1}Ak] = 0.$$

The non-degeneracy of W_{τ} with respect to (2.2) gives $W_{\tau}\cap(T[W_{\tau}])^{\perp} = \{0\}$. From the T-regularity of A at zero we obtain $Z_0(K)\cap(T[Z_0(K)])^{\perp} = \{0\}$. Since N_{λ} contains the maximal strictly negative subspace M_{λ} of $Z_{\lambda}(K)$, we cannot have $N_{\lambda}\cap(T[N_{\lambda}])^{\perp}\neq\{0\}$, as otherwise for h in this intersection we would have

$$(h,k)_A = (Th,T^{-1}Ak) = [h,T^{-1}Ak] = 0$$

for all $k\in N_{\lambda}$. But this would imply that $M_{\lambda}\oplus\{N_{\lambda}\cap(T[N_{\lambda}])^{\perp}\}$ is a negative subspace of $Z_{\lambda}(K)$ (with respect to (2.2)), which properly contains M_{λ}, thereby contradicting the maximality of M_{λ}. Hence, $N_{\lambda}\cap(T[N_{\lambda}])^{\perp} = \{0\}$. We therefore conclude that

$$Z(K)\cap(T[Z(K)])^{\perp} = \{0\}, \tag{2.11}$$

which is equivalent to the non-degeneracy of Z(K) with respect to (1.8).

From $K = T^{-1}A$ and $K^* = AT^{-1}$ we obtain the properties (ii) and (iv) of the theorem, while (i) and (iii) follow easily from (2.11). It is clear also that Z(K) and $Z(K^*)^{\perp}$ are orthogonal with respect to (2.2). Since $Z(K)\cap Z_0(K^*)^{\perp}$ contains

$$\bigoplus_{\mathrm{Im}\ \tau>0} Z_\tau(K) \oplus \{\bigoplus_\lambda M_\lambda\}$$

as a negative subspace of dimension $\bar{\kappa}+\sum_\lambda \kappa(\lambda)=\kappa$ and $Z_0(K^*)^\perp \cap H_A$ is a Π_κ-space (both with respect to the indefinite metric (2.2)), it follows then that the subspace $(T[Z(K)])^\perp$ must be strictly positive with respect to (2.2), which yields (v).

Finally, let P denote the projection of H onto $Z(K^*)^\perp$ along Z(K), and define A_β by (2.6). Then Eq. (2.7) is easily verified, while

$$(A_\beta h,h) = [\beta^{-1}(I-P)h,(I-P)h] + (Ph,Ph)_A, \quad h\in D(A).$$

Since condition (v) holds true, the strict positive self adjointness of A_β then follows easily. ∎

COROLLARY 2.2. If I−A is compact with Ran $(I-A)\subset$Ran $|T|^\alpha$ for some $0<\alpha<1$, then $I-A_\beta$ is compact with Ran $(I-A_\beta)\subset$Ran $|T|^\alpha$.

Proof: Since A_β is a finite rank perturbation of A, it is necessary only to show that $Z_\lambda(K) \subset$ Ran $|T|^\alpha$ for every eigenvalue λ of K. If $Ax=\lambda Tx$, then $x=(I-A)x+\lambda Tx \in$ Ran $|T|^\alpha$, which shows Ker $(K-\lambda) \subset$ Ran $|T|^\alpha$. Let us assume that Ker $(K-\lambda)^n \subset$ Ran $|T|^\alpha$ and $(K-\lambda)^{n+1}y=0$. Then for $z\in$Ker $(K-\lambda)^n$ one has $(A-\lambda T)y=Tz$, and thus $y=(I-A)y + \lambda Ty + Tz \in$ Ran $|T|^\alpha$. Hence, Ker $(K-\lambda)^{n+1} \subset$ Ran $|T|^\alpha$. ∎

We have proved that Z(K) is non-degenerate with respect to the indefinite inner product (1.8). The next proposition indicates examples of strictly positive and negative subspaces of Z(K). The proof is completely analogous to the proof of Proposition III 1.6.

PROPOSITION 2.3. The subspaces

$$M_\pm = [\mathrm{Ran}\ PP_\mp \oplus \mathrm{Ran}\ Q_\pm] \cap Z(K)$$

are maximally strictly positive/strictly negative in Z(K) with respect to (1.8). We have $(M_\pm)^\perp = T[M_\mp] \oplus Z(K)^\perp$, and $Z(K) = M_+ \oplus M_-$ is a fundamental decomposition of Z(K) with respect to the indefinite inner product (1.8).

We recall that $Q_\pm$ are the orthogonal projections of H onto the maximal T-positive/negative, T-invariant subspaces with respect to the Hilbert space inner product on H. Let us construct analogous projections for $S_\beta=A_\beta^{-1}T$. We note first that S_β is a bounded self adjoint operator on H_A with respect to the Hilbert space inner product obtained from the completion of

$$(x,y)_{A_\beta} = (A_\beta x,y), \quad x,y \in D(A). \tag{2.12}$$

Since the operators A_β coincide on a subspace of H of finite codimension $(T[Z(K)])^\perp$, all inner products (2.12) are equivalent on H_A. Let $P_\pm$ be the orthogonal projections of H_A onto the maximal S_β-positive/negative, S_β-invariant subspaces, all with respect to (2.12). These projections are complementary, because S_β has zero null space. The projection P of H onto $(T[Z(K)])^\perp$ along Z(K) also maps H_A onto $(T[Z(K)])^\perp \cap H_A$ along Z(K). As the restriction of S_β to $(T[Z(K)])^\perp$ does not depend on β, neither do $PP_\pm$, and PP_+, PP_-, I−P are a set of disjoint complementary projections on H_A.

In the next section we shall study the boundary value problems (1.1)−(2.1)/(2.2).

3. Existence and uniqueness theory

For the conservative case developed in Chapter III, the boundary value problem was studied by replacing the collision operator A with a strictly positive operator A_β. The existence and uniqueness of solutions, depending upon the boundary condition at infinity, was entirely determined by the structure of the generalized eigenspace $Z_0(K)$ of $K=T^{-1}A$.

For the case of multiplying media, we will follow an analogous procedure. A finite rank perturbation A_β of A which is strictly positive and which respects the decomposition of H into K-reducing subspaces Z(K) and $Z(K^*)^\perp$ is provided by Proposition 2.1. In this case, however, the existence of negative eigenvalues of A may result in both imaginary and complex eigenvalues for $K=T^{-1}A$, and the structure of all of the corresponding generalized eigenspaces, along with $Z_0(K)$, play an important role in analyzing existence and uniqueness properties. With this in mind, let us define

$$K_\pm^0 = \bigoplus_{\mathrm{Re}\,\lambda>0} Z_{\pm\lambda}(K),$$

$$K_\pm = \bigoplus_{\mathrm{Re}\,\lambda>0} Z_{\pm\lambda}(K) \bigoplus_{\mathrm{Re}\,\lambda=0} \mathrm{Ker}(K-\lambda),$$

$$L_\pm^0 = \bigoplus_{\mathrm{Re}\,\lambda\geq 0} Z_{\pm\lambda}(K),$$

$$L_\pm = \bigoplus_{\mathrm{Re}\,\lambda>0} Z_{\pm\lambda}(K) \bigoplus_{\mathrm{Re}\,\lambda=0} (K-\lambda)Z_\lambda(K),$$

where λ runs through the eigenvalues of the restriction $K|_{Z(K)}$ of K to the subspace Z(K).

We will now see to what extent one may develop an existence and uniqueness theory for the boundary value problems (1.1)-(1.2)-(1.3)/(1.4) under the assumption that A is T-regular with finite dimensional negative part. We consider three functional formulations:

(i) The operator A is a compact perturbation of the identity satisfying

$$\mathrm{Ran}\,(I-A) \subset \mathrm{Ran}\,|T|^\alpha \tag{3.1}$$

for some $0<\alpha<1$. Solutions are to be found in H (cf. Section II.2).

(ii) The operator A is bounded. Solutions are to be found in H_T (cf. Section II.3).

(iii) The operator A is unbounded, while the H_T- and H_S-topologies coincide on H_A. Solutions are to be found in H_T (cf. Section II.4).

We recall that P is the projection of H onto $Z(K^*)^\perp$ along Z(K). P extends to a bounded projection on H_T with kernel Z(K) in formulation (ii), and to a bounded projection on H_S with kernel Z(K) in formulation (iii), and, as is our custom, we use the same notation for P and its extensions. Similarly, $P_\pm$ extend to complementary projections on H_T and on H_S for the appropriate formulations. By virtue of Proposition 1.1, the boundary value problem decomposes into the finite dimensional equation

$$\psi_0'(x) = -T^{-1}A\psi_0(x) \tag{3.2}$$

on Ran (I-P) and an equation on Ran P which (following the strategy of Chapter III) may be replaced by

$$\varphi'(x) = -T^{-1}A_\beta\varphi(x). \tag{3.3}$$

Then, on adding boundary conditions of the type (1.2) and (1.3)/(1.4) and solving (3.2), (3.3), we obtain $\psi(x) = \exp\{-xK\}\psi(0)$ as the solution of (1.1)-(1.2)-(1.3)/(1.4), where

$$\psi(0) \in ([\mathrm{Ran}\ PP_+]\oplus K_+)\cap H_T \tag{3.4}$$

if one imposes condition (1.3), and

$$\psi(0) \in ([\mathrm{Ran}\ PP_+]\oplus K_+^0)\cap H_T \tag{3.5}$$

if one imposes condition (1.4), and where the intersection with H_T should be omitted if I-A is compact satisfying (3.1). Thus, the boundary value problem has been reduced entirely to conditions (1.2) and (1.3)/(1.4) on $\psi(0)$.

We shall now give a characterization of the measures of non-uniqueness and non-completeness for the boundary value problem. We recall that the measures of non-uniqueness δ_+ and δ_+^0, defined in Section III.2, are the dimensions of the solution spaces of the homogeneous problem with appropriate boundary condition at infinity, and the measures of non-completeness γ_+ and γ_+^0 are the codimensions of the subspaces of boundary values at x=0 for which the boundary value problem is well posed.

THEOREM 3.1. The measures of non-uniqueness δ_+ and δ_+^0 for the boundary value problems (1.1)-(1.2)-(1.3) and (1.1)-(1.2)-(1.4)) are given by

$$\delta_+ = \dim\ \{[\mathrm{Ran}\ PP_+\oplus \mathrm{Ran}\ Q_-] \cap K_+ = \dim\ (M_-\cap K_+) \tag{3.6a}$$

and

$$\delta_+^0 = \dim\ \{[\mathrm{Ran}\ PP_+\oplus \mathrm{Ran}\ Q_-] \cap K_+^0\} = \dim\ (M_-\cap K_+^0). \tag{3.6b}$$

Proof: The measures of non-uniqueness are determined by considering the corresponding homogeneous problems. In this case, the condition on $\psi(0)$ for the problem (1.1)-(1.2)-(1.3) becomes $\psi(0) \in [\mathrm{Ran}\ PP_+\oplus K_+] \cap \mathrm{Ran}\ Q_-$, and for the problem (1.1)–(1.2)–(1.4) becomes $\psi(0) \in [\text{Ran } PP_+ \oplus K_+^0] \cap \text{Ran } Q_-$. On considering the modified problem, where A is replaced by A_β and $V=Q_+P_+ + Q_-P_-$ is defined in terms of the spectral projections $P_\pm$ of $T^{-1}A_\beta$, it follows that Ran PP_+ and Ran Q_- have zero intersection. The theorem now follows easily from the triviality of this intersection. ■

THEOREM 3.2. The measures of non-completeness γ_+ and γ_+^0 for the boundary value problems (1.1)–(1.2)–(1.3) and (1.1)–(1.2)–(1.4)) are given by

$$\gamma_+ = \dim \{[\text{Ran } PP_- \oplus \text{Ran } Q_+] \cap L_-\} = \dim (M_+ \cap L_-) \tag{3.7a}$$

and

$$\gamma_+^0 = \dim \{[\text{Ran } PP_- \oplus \text{Ran } Q_+] \cap L_-^0\} = \dim (M_+ \cap L_-^0). \tag{3.7b}$$

Proof: Since $\varphi_+ = \psi(0) - Q_-\psi(0) \in \text{Ran } Q_+$, it is evident that we must require

$$\varphi_+ \in \{[\text{Ran } PP_+ \oplus K_+] + \text{Ran } Q_-\} \cap \text{Ran } Q_+$$

for problem (1.1)–(1.2)–(1.3), and

$$\varphi_+ \in \{[\text{Ran } PP_+ \oplus K_+^0] + \text{Ran } Q_-\} \cap \text{Ran } Q_+$$

for problem (1.1)–(1.2)–(1.4). The indicated ranges must be interpreted, according to the functional formulation, as submanifolds of H, H_T or H_S.

We will consider γ_+ only, as the proof for γ_+^0 proceeds mutatis mutandis with K_+ replaced by K_+^0. We proceed first for the case I–A compact satisfying (3.1), i.e., for solutions in H. It is clear that

$$\gamma_+ = \dim \frac{\text{Ran} Q_+}{\{[\text{Ran} PP_+ \oplus K_+] + \text{Ran} Q_-\} \cap \text{Ran} Q_+} = \dim \frac{H}{[\text{Ran} PP_+ \oplus K_+] + \text{Ran} Q_-}.$$

We may use the inclusions

$$H \supset [\text{Ran } PP_+ \oplus Z(K)] + \text{Ran } Q_- \supset \text{Ran } P_+ + \text{Ran } Q_- \supset V[H] = H$$

and obtain

$$\gamma_+ = \dim \frac{[\text{Ran}PP_+ \oplus Z(K)] + \text{Ran}Q_-}{[\text{Ran}PP_+ \oplus K_+] + \text{Ran}Q_-} = \dim \frac{[\text{Ran}PP_- \oplus L_-] \cap \text{Ran}Q_+}{\text{Ran}PP_- \cap \text{Ran}Q_+},$$

and thus

$$\gamma_+ = \dim \{[\text{Ran } PP_- \oplus L_-] \cap \text{Ran } Q_+\} = \dim \{M_+ \cap L_-\}.$$

For the other functional formulations, the intermediate steps in Eq. (3.7) must be taken somewhat differently, but again we obtain the above expressions for γ_+ for the most general case of A unbounded, provided H_T and H_S coincide naturally. ∎

In [163] it was stated erroneously that $M_\pm \cap N$ is a maximal strictly positive/negative subspace of N with respect to (1.8), where N is one of the subspaces K_+, L_-, K_+^0 and L_-^0. As a matter of fact, these spaces are strictly positive/negative, but not necessarily maximal in this respect. In general, the right hand sides of (3.6) and (3.7) may be bounded in terms of the sign characteristics of the restriction of $T^{-1}A$ to Z(K), i.e., in data which one may readily derive from the matrix representation of this (finite dimensional) operator with respect to a special basis. (For sign characteristics we refer to [144, 145, 146, 315].) If A is positive self adjoint, then the maximality statement is satisfied and the measures of non-uniqueness and non-completeness are given by these explicit expressions.

4. Nonsymmetric collision operators

In the last several chapters the selfadjointness of the collision operator A has played a key role in the development of existence and uniqueness results for stationary abstract kinetic equations. Recently, Ganchev et al. [126, 128] have indicated how the theory of these chapters might be extended to a class of nonsymmetric collision operators. The techniques employed in the arguments utilize results from the perturbation theory of C_0 bisemigroups, and thus lie somewhat between the differential equations theory of these chapters and the convolution equations theory developed in Chapters VI and VII. Nevertheless, we feel it opportune to sketch the ideas at this time, drawing on one basic result contained in Chapter VI.

In what follows, the selfadjointness of A will be replaced by an accretiveness assumption:

$$\text{Re } A \equiv \tfrac{1}{2}(A+A^*) \geq 0.$$

We will consider the case T injective and selfadjoint, and $B = I-A$ compact and satisfying the regularity assumption $\text{Ran } B \subset D(|T|^\alpha) \cap \text{Ran } |T|^\gamma$ for some $\alpha>1$ and $\gamma>0$. In addition, for simplicity we will make the dissipative medium assumption $\text{Ker } A = \text{Ker } (\text{Re } A) = \{0\}$. Nonsymmetric collision operators typically occur in multiphase transport, in which the symmetry of the collision process is broken, for example, in multigroup neutron transport, in radiative transfer of polarized light, and in the dynamics of rarefied gas mixtures (see Chapter IX for the various applications).

Although the study of left and right half space boundary value problems in this and the preceding two chapters has led to the construction of semigroups acting, repectively, to the left and to the right, it is convenient at this time to develop more carefully the notion of a bisemigroup. By a **C_0 bisemigroup** $E(t)$ on a Banach space X we will mean a function E from $\mathbb{R}\setminus\{0\}$ into $L(X)$, the bounded operators on X, with the properties

(i) $E(t)E(s) = \pm E(t+s)$ if $\text{sgn}(t)=\text{sgn}(s)=\pm 1$ and $E(t)E(s) = 0$ if $\text{sgn}(t)=-\text{sgn}(s)$.

(ii) $E(\cdot)$ is strongly continuous.

(iii) $\Pi_+ + \Pi_- = I$, where $\Pi_\pm \equiv \text{s-}\lim_{\pm t \downarrow 0} (\pm E(t))$.

The bounded projections $\Pi_\pm$ are called separating projectors, and $\Pi_+\Pi_-=0=\Pi_-\Pi_+$. From (iii) we may see that $\pm E(t)\Pi_\pm$, $\pm t\geq 0$, are C_0 semigroups on $\text{Ran } \Pi_\pm$. An operator S is the generator of $E(t)$ if $\Pi_\pm$ leaves $D(S)$ invariant and $S\Pi_\pm h=\Pi_\pm Sh$, for all $h\in D(S)$, and if $E(t)=\pm\exp(-tS)\Pi_\pm$, $\pm t>0$. If the Laplace transform of $E(t)$ exists, then it is the resolvent of S on the imaginary axis, i.e., $(S-\lambda)^{-1}h = \int_{-\infty}^{\infty} e^{\lambda t}E(t)h\,dt$ for $\text{Re}\lambda=0$ and $h\in X$. The bisemigroup will be called bounded analytic, strongly decaying analytic, or exponentially decaying analytic if the semigroups $\pm E(t)\Pi_\pm$, $\pm t>0$, have the respective properties. Here by a strongly decaying (bi)semigroup we mean a (bi)semigroup that converges to zero at infinity in the strong sense. For a systematic study of bisemigroups we refer to [27].

Following Ganchev and Greenberg [126], we have the following lemma, which follows easily from the properties of bounded analytic semigroups.

LEMMA 4.1. Suppose that S generates a bounded analytic semigroup $\exp(-tS)$ and that zero is in the spectrum of S. Then the semigroup is strongly decaying if and only if zero is in the continuous spectrum of S.

Now let us consider the case at hand. Evidently, the self adjoint operator T^{-1} on the Hilbert space H is generator of a strongly decaying, analytic bisemigroup E(t) with separating projectors given by the spectral projections $Q_{\pm}$. We wish to conclude that $K = T^{-1}A$ generates a bisemigroup as well. The essential observation is that K has no eigenvalues on the imaginary axis.

THEOREM 4.2. K generates an analytic bisemigroup $E^{\times}(t)$ with separating projectors $P_{\pm}$. For any $t\in\mathbb{R}\setminus\{0\}$ we have that $E(t)-E^{\times}(t)$ and $Q_{\pm}-P_{\pm}$ are compact. The bisemigroup $E^{\times}(t)$ is strongly decaying. If $\sigma(T^{-1})$ has a gap at zero (i.e., T is bounded), then $E^{\times}(t)$ is exponentially decaying.

Proof: We sketch the proof; for additional details, see Section VII.2 Consider the operator valued function $\mathbb{k}(t)=T^{-1}E(t)B$ defined for $t\in\mathbb{R}\setminus\{0\}$. The regularity condition on B assures that $\mathbb{k}(\cdot)\in L_1(L(X))_{-\infty}^{\infty}$, the space of norm integrable operator valued functions on the real line. Define the bounded operator $\mathcal{L}$ on the spaces $L_p(H)_{-\infty}^{\infty}$ of L_p Bochner integrable vector valued functions on the real line, or on $C(H)_{-\infty}^{\infty}$, the space of norm continuous vector valued functions on the real line, by

$$(\mathcal{L}\psi)(t) = \int_{-\infty}^{\infty} \mathbb{k}(t-s)\psi(s)ds.$$

Then the Laplace transform of the operator $(I-\mathcal{L})$ is given by

$$W(\lambda) = I - \int_{-\infty}^{\infty} e^{\lambda t}\mathbb{k}(t)dt = (\lambda-S)^{-1}(\lambda-S^{\times}) = I + \lambda^{-1}(S^{-1}-\lambda^{-1})^{-1}B$$

for $\mathrm{Re}\lambda=0$. From the assumptions on B it follows that $W(\lambda)$ has a bounded inverse on the extended imaginary axis. We may then apply the Bochner-Phillips Theorem (Theorem VI 2.2) to conclude that the operator $I-\mathcal{L}$ on $L_p(H)_{-\infty}^{\infty}$ or $C(H)_{-\infty}^{\infty}$ is invertible with inverse $I+\mathcal{L}^{\times}$, where

$$(\mathcal{L}^{X}\psi)(t) = \int_{-\infty}^{\infty} \Bbbk^{X}(t-s)\psi(s)ds.$$

We claim that $E^{X}(t)h=(I+\mathcal{L}^{X})E(t)h$, for $h\in X$ and $t\in\mathbb{R}\setminus\{0\}$, is the bisemigroup generated by K. First let us check that $E^{X}(t)$ defined above is a bisemigroup. Fix some $s>0$ and $h\in H$ and define $\psi(t)=E^{X}(t+s)h$ if $t>0$ and $\psi(t)=0$ if $t<0$. For $t>0$ we have

$$(I-\mathcal{L})\psi(t) = E^{X}(t+s)h - \int_{s}^{\infty}\Bbbk(t+s-r)E^{X}(r)hdr =$$

$$= (I-\mathcal{L})E^{X}(t+s)h + \int_{-\infty}^{s}\Bbbk(t+s-r)E^{X}(r)hdr,$$

since for $r<s$ we have $t+s-r>0$ and $\Bbbk(t+s-r)=E(t)\Bbbk(s-r)$, while for $r>s$ we have $0=E(t)\Bbbk(s-r)$. We can rewrite the above as

$$(I-\mathcal{L})\psi(t) = E(t+s)h + E(t)\mathcal{L}E^{X}(S)h = E(t)(E(s) + \mathcal{L}E^{X}(s))h = E(t)E^{X}(s)h.$$

Therefore $\psi(t)=E^{X}(t)E^{X}(s)h$ for $t>0$. For $t<0$ we have

$$(I-\mathcal{L})\psi(t) = -\int_{s}^{\infty}\Bbbk(t+s-r)E^{X}(r)hdr = E(t)\int_{-\infty}^{\infty}\Bbbk(s-r)E^{X}(r)hdr =$$

$$= E(t)(E(s)+\mathcal{L}E^{X}(s))h = E(t)E^{X}(s)h.$$

Combining these cases and recalling the definition of $\psi(t)$, we get $E^{X}(t)E^{X}(s)=E^{X}(t+s)$ for $t,s>0$ and $E^{X}(t)E^{X}(s)=0$ for $t<0$, $s>0$.

It is easy to see that $E^{X}(t)$ is strongly continuous and bounded. To check that $P_{+}+P_{-}=I$, note that the jump of $E^{X}(t)$ at $t=0$ is equal to the jump of $E(t)$, i.e.,

$$(P_{+}+P_{-})h = E^{X}(+0)h - E^{X}(-0)h = E(+0)h - E(-0)h = (Q_{+}+Q_{-})h = h.$$

Therefore $E^{X}(t)$ is a bounded C_0 bisemigroup.

Because $T^{-1}-K$ is T^{-1}-relatively compact, the spectrum of K outside of $\mathbb{R}$ consists of isolated eigenvalues that can accumulate only on $\sigma(T^{-1})$. Therefore, we may take the Laplace transform of $E^{X}(t)$, and we get

$$(K-\lambda)^{-1}h = \int_{-\infty}^{\infty} e^{\lambda t}E^{X}(t)hdt, \quad \mathrm{Re}\lambda=0, \quad \lambda\neq 0, \quad h\in H.$$

Hence, K is the generator of $E^{X}(t)$ and $P_{\pm}$ are positive/negative spectral projectors for

K.

The compactness of

$$E^{x}(t) - E(t) = \int_{-\infty}^{\infty} k(t-s)E^{x}(s)ds,$$

and consequently of $P_{\pm}-Q_{\pm}$, follows from the fact that the (Bochner) integral of an integrable compact operator valued function is compact (Lemma II 2.3). Because $D(K^{-1})=D(A^{-1}T)=D(T)$ is dense, we get that zero is either in the resolvent set or in the continuous spectrum of K, and hence that E^{x} is strongly decaying. If the spectrum of T^{-1} has a gap at zero, then it is immediate that E(t), and thus $k(t)$, is exponentially decaying. Then it can be shown that $k^{x}(t)$ will be exponentially decaying, implying the exponential decay of $E^{x}(t)$. This completes the proof of the theorem. ∎

Now we may consider Eqs. (1.1)–(1.3). By a solution we will understand a continuous function $\psi:[0,\infty)\to D(T)$, such that $T\psi(x)$ is strongly differentiable for $x\in(0,\infty)$ and the equation and boundary conditions are satisfied. We may show, as in Chapter II, that every solution has the form $\psi_h(x)=\exp(-xK)P_+h$, $x\geq 0$, for some $h\in D(T)$ such that $Q_+P_+h=\varphi_+$, and, moreover, that (1.4) is satisfied.

The unique solvability of the boundary value problem is equivalent to the fact that Q_+ maps $P_+D(T)$ one-to-one onto $Q_+D(T)$. Introducing the operator $V=Q_+P_++Q_-P_-$, we have unique solvability if V maps D(T) one-to-one onto D(T). Denoting by $E=V^{-1}$ the albedo operator, we will have $P_+h=E\varphi_+$.

LEMMA 4.3. The projectors $P_{\pm}$ and the operator V leave D(T) invariant. The operator (I−V) is compact on H and also on D(T) equipped with the T graph norm.

Proof: We can write $I-V=(Q_--Q_+)(P_+-Q_+)$. According to Theorem 4.2 we have $P_+-Q_+ = \int_{-\infty}^{\infty}k(-y)E^{x}(y)dy$ and it is a compact operator, hence (I−V) is compact in H. The regularity assumption on B with $\alpha>1$ assures that $Tk(\cdot)\in L_1(L(H))_{-\infty}^{\infty}$. Therefore, $(P_+-Q_+)h\in D(T)$ for all $h\in H$. This implies that $P_{\pm}D(T)\subset D(T)$ and $VD(T)\subset D(T)$.

Let us define $\hat{P}_{\pm}^{*}=AP_{\pm}A^{-1}$. Obviously, we have that $\hat{P}_{+}^{*}-Q_+ = (P_+-Q_+)+P_+BA^{-1}-BP_+-BP_+BA^{-1}$ is a compact operator on H. For $h\in D(T)$ we have $\hat{P}_{+}^{*}Th = AP_+A^{-1}Th = A(A^{-1}T)P_+h = TP_+h$, and so $T(P_+-Q_+)h = (\hat{P}_{+}^{*}-Q_+)Th$. This, along with the compactness of $(\hat{P}_+^*-Q_+)$, implies that P_+-Q_+, and hence $(I-V)$, is compact on $D(T)$ in the T-graph norm. ■

LEMMA 4.4. If $h\in D(T)$, then $\psi_h(\cdot)\in L_2(H)_0^\infty$, $\psi_h(x)\in D(T)$ for all $x\geq 0$, and $\|T\psi_h(x)\|\to 0$ as $x\to\infty$.

Proof: Because we have an analytic bisemigroup, the derivative $\dot{\psi}_h(x) = -T^{-1}A\psi_h(x)$ exists and obviously belongs to $D(T)$. Since $P_+h\in D(T)$, we have that $\psi_h(x) = P_+h + \int_0^x\dot{\psi}_h(y)dy$ is in $D(T)$.

Using the shorthand notation $\omega_h(\cdot)$ for the bisemigroup generated by T^{-1} applied to a vector h, one has that $h\in D(T)$ implies $\omega_h(\cdot)\in L_2(H)_{-\infty}^\infty$ (the space of Bochner square integrable H valued functions). For, using the Spectral Theorem,

$$\int_{-\infty}^\infty \|\omega_h(x)\|^2dx = \int_{-\infty}^0 \|e^{-xT^{-1}}Q_-h\|^2dx + \int_0^\infty\|e^{-xT^{-1}}Q_+h\|^2dx =$$

$$= \int_{-\infty}^0 dx\int_{-\infty}^0 \exp\{-2\tfrac{x}{\mu}\}dF_{h,h}(\mu) + \int_0^\infty dx\int_0^\infty \exp\{-2\tfrac{x}{\mu}\}dF_{h,h}(\mu) =$$

$$= \int_{-\infty}^\infty \tfrac{1}{2}|\mu|\,dF_{h,h}(\mu) = \tfrac{1}{2}(|T|h,h) < \infty.$$

But $\omega_h(\cdot)\in L_2(H)_{-\infty}^\infty$ implies $\psi_h(\cdot)\in L_2(H)_{-\infty}^\infty$. Then, noting the equality $T\psi_h(x) = TP_+h-A\int_0^x\psi_h(y)dy$ and taking $x\to\infty$, we get

$$\lim_{x\to\infty} T\psi_h(x) = TP_+h - A\int_0^\infty\psi_h(y)dy = TP_+h - AA^{-1}TP_+h = 0,$$

which completes the proof. ■

LEMMA 4.5. Ker $V = \{0\}$.

Proof: First note that Ker $V\subset D(T)$. Indeed, let $Vh=0$. Then $h=(I-V)h\in D(T)$, where the inclusion is assured by the first lemma. Because we have Ker V = (Ran $P_+\cap$Ran Q_-)$\oplus$(Ran $P_-\cap$Ran Q_+), the proof will be completed if we can show that Ran $Q_\mp\cap$Ran $P_\pm\cap D(T)=0$.

Suppose that $h\in$ Ran $P_+\cap$Ran $Q_-\cap D(T)$. Adapting an argument of [396], one has

$$-(2(\mathrm{Re}A)\psi_h(x),\psi_h(x)) = -(A\psi_h(x),\psi_h(x)) - (\psi_h(x),A\psi_h(x)) =$$

$$= (\frac{d}{dx}T\psi_h(x),\psi_h(x)) + (\psi_h(x),\frac{d}{dx}T\psi_h(x)) = \frac{d}{dx}(T\psi_h(x),\psi_h(x)).$$

But $-(Th,h)\geq 0$, because $h\in \text{Ran } Q_-$. Using the accretiveness of A and Lemma 4.4 we have the estimate

$$0 \geq \lim_{\tau\to\infty} \int_0^\tau -(2(\text{Re}A)\psi_h(x),\psi_h(x))dx = \lim_{\tau\to\infty} \int_0^\tau \frac{d}{dx}(T\psi_h(x),\psi_h(x))dx =$$

$$= \lim_{\tau\to\infty} (T\psi_h(\tau),\psi_h(\tau)) - (T\psi_h(+0),\psi_h(+0)) = -(Th,h) \geq 0.$$

Therefore $(Th,h)=0$. But T is injective and negative definite on Ran Q_-, whence we get the equality $h=0$. This shows that $\text{Ran } P_+\cap \text{Ran } Q_-\cap D(T)=\{0\}$. Analogously $\text{Ran } P_-\cap \text{Ran } Q_+\cap D(T)=\{0\}$ holds, and hence we have proved that V is injective. ■

THEOREM 4.6. For every $\varphi_+\in Q_+D(T)$ the boundary value problem (1.1)–(1.3) has a unique solution, which is given by $\psi(x)=\exp(-xK)E\varphi_+$. The solution is decaying at infinity and is also square integrable in x. If T is a bounded operator, then the solution is exponentially decaying.

Proof: Because V maps D(T) into itself and I–V is compact in D(T) in the T graph norm, the injectivity of V implies that V maps D(T) onto D(T). Therefore the operator $E=V^{-1}$ is a bounded operator in D(T) with the T graph norm, and obviously for $\varphi_+\in Q_+D(T)$ we have $E\varphi_+\in P_+D(T)$ and $Q_+E\varphi_+=\varphi_+$. ■

As in Chapter II, if T is an unbounded operator, the boundary value problem can be interpreted in another way, in which case the boundary value φ_+ may satisfy $\varphi_+\in Q_+H$ rather than just $\varphi_+\in Q_+D(T)$. With an appropriate redefinition of solution, we again have unique solvability for every $\varphi_+\in Q_+H$, and the decay of all solutions at infinity; only the square integrability of the solution may be lost.

If A has a nontrivial kernel, a somewhat more detailed analysis along the lines of this section, but utilizing a K–invariant decomposition of H as in Chapter II, leads to the characterization of the boundary value problem in terms of measures of non–uniqueness and measures of non–completeness. So far, this generalization seems to require some restrictions on Ker A, which are, however, satisfied in various radiative transfer models. We refer to [128] for these later developments.

Chapter V.

KINETIC EQUATIONS ON FINITE DOMAINS

1. Slab geometry

In the previous three chapters we have analyzed in detail the existence and uniqueness theory for the abstract differential equation

$$T\psi'(x) = -A\psi(x), \qquad 0<x<\tau, \tag{1.1}$$

with $\tau=\infty$, with an incoming flux boundary condition at $x=0$, possibly with reflection at the surface, and with one of several possible boundedness conditions at $x=\infty$. Physically speaking, this equation models transport processes in the half space $\{(x,y,z)\in\mathbb{R}^3 : x\geq 0\}$ which are uniform in the transverse (y−z) directions. In this chapter we will present an analysis of the abstract equation (1.1) on the finite interval $[0,\tau]$, $\tau<\infty$, which represents **slab geometry**. That is to say, we now wish to study transport processes on the slab $\{(x,y,z)\in\mathbb{R}^3 : 0\leq x\leq\tau\}$ which are again uniform in the transverse directions. The appropriate boundary conditions are

$$Q_+\psi(0) = \varphi_+, \tag{1.2a}$$

$$Q_-\psi(\tau) = \varphi_- \tag{1.2b}$$

corresponding to incoming fluxes φ_+, φ_- at $x = 0,\tau$, respectively, or

$$Q_+\psi(0) = R_\ell JQ_-\psi(0) + \varphi_+, \tag{1.3a}$$

$$Q_-\psi(\tau) = R_r JQ_+\psi(\tau) + \varphi_-, \tag{1.3b}$$

which allows for reflection at the surfaces $x=0$ and $x=\tau$. (The surface reflection operators R_ℓ and R_r and the inversion symmetry J are defined in Section 5 below.)

Throughout this chapter, as in much of the previous chapters, we will assume T and A are self adjoint operators on a complex Hilbert space H, with T injective and A Fredholm, and $Q_\pm$ are the maximal T-positive/negative T-invariant projections on H.

Also, as before, we shall distinguish between several functional formulations, corresponding to collision operators A with different properties:

(i) A is a bounded operator, and solutions are to be found in H_T, which is the completion of D(T) with respect to an appropriate inner product.

(ii) A is a compact perturbation of the identity satisfying a regularity condition (cf. (2.7)), and solutions are to be found in H.

(iii) A may be unbounded, but a finite slab analog of $V=Q_+P_++Q_-P_-$ is essentially self adjoint on H_T. Solutions are to be found in a Hilbert space $H_S \supset D(T)\cap D(A)$.

In the next three sections we will study the existence and uniqueness theory of the boundary value problem (1.1)-(1.2) for each of the three functional formulations, and the existence of transmission and reflection operators. In the last section we will extend these results to include slab problems with reflective surfaces.

2. Boundary value problems for nonmultiplying slab media

Let us assume that T is an injective self adjoint operator, and A is a positive self adjoint Fredholm operator, both defined on the complex Hilbert space H. Positivity of the collision operator A corresponds physically to the assumption that transport takes place within a nonmultiplying medium. As for the half space problem discussed in Section III.2, we assume that the zero root linear manifold $Z_0(T^{-1}A)$ satisfies

$$Z_0(T^{-1}A) = \bigcup_{n=0}^{\infty} \mathrm{Ker}(T^{-1}A)^n \subset D(T).$$

Then we may derive the decompositions

$$Z_0(T^{-1}A)\oplus Z_0(AT^{-1})^{\perp} = H, \tag{2.1}$$

$$Z_0(AT^{-1})\oplus Z_0(T^{-1}A)^{\perp} = H, \tag{2.2}$$

and the reduction of the operator

$$T^{-1}A = T^{-1}A\,|_{Z_0(T^{-1}A)}\oplus S^{-1},$$

where S is the unique operator on the finite co-dimensional subspace $Z_0(AT^{-1})^{\perp}$ such that ASh=Th for $h\in Z_0(AT^{-1})^{\perp}\cap D(T)$. We discuss each of the functional formulations in turn.

(a) The collision operator A bounded

We observe that the inner product

$$(h,k)_A = (Ah,k), \tag{2.3}$$

is equivalent to the original inner product on H and that S is self adjoint and injective with respect to (2.3). Denoting by P the projection of H onto $Z_0(AT^{-1})^{\perp}$ along $Z_0(T^{-1}A)$, we define PP_+ (resp. PP_-) as the projection of H onto the maximal S-positive (resp. -negative) S-invariant subspace of $Z_0(AT^{-1})^{\perp}$. Here positivity and negativity relate to (2.3). Then the restriction of $-T^{-1}A$ (resp. $+T^{-1}A$) to Ran PP_+ (resp. PP_-) generates a bounded analytic semigroup. We note that $Z_0(T^{-1}A)$ has finite dimension, $Z_0(T^{-1}A)=\mathrm{Ker}(T^{-1}A)^2$ (see Lemma III 1.3), and thus we have $\exp\{-xT^{-1}A\}(I-P)h=(I-xT^{-1}A)(I-P)h$ for $h\in Z_0(T^{-1}A)$.

A similar construction can be carried out with respect to AT^{-1}. Then, using Proposition II 3.1, the projections and semigroups can be extended to H_T; we write $K\supset T^{-1}A$ for the operator which gives the infinitesimal generators (within ± sign) of the extensions of $\exp\{\mp xT^{-1}A\}P_{\pm}$. A solution of the boundary value problem (1.1)-(1.2) then is defined to be a continuous function $\psi:[0,\tau]\to H_T$ with $\psi(x)\in D(K)\subset H_T$, which is continuously differentiable on $(0,\tau)$ and satisfies (1.1)-(1.2).

Premultiplying (1.1) by PP_+, PP_- and I-P, we find easily that ψ is a solution if

$$\psi(x) = [e^{-xT^{-1}A}PP_+ + e^{(\tau-x)T^{-1}A}PP_- + (I-xT^{-1}A)(I-P)]h, \tag{2.4}$$

where

$$h = PP_+\psi(0) + PP_-\psi(\tau) + (I-P)\psi(0).$$

By fitting the boundary conditions (1.2) and writing $\varphi_+ + \varphi_- = \varphi$, we get $V_\tau h = \varphi$, where V_τ is defined on H_T by

$$V_\tau = Q_+[PP_+ + e^{\tau T^{-1}A}PP_- + (I-P)] +$$
$$+ Q_-[PP_- + e^{-\tau T^{-1}A}PP_+ + e^{-\tau T^{-1}A}(I-P)]. \tag{2.5}$$

THEOREM 2.1. In this formulation, for every $\varphi_+ \epsilon Q_+[H_T]$ and $\varphi_- \epsilon Q_-[H_T]$ there is a unique solution of the boundary value problem (1.1)–(1.2), which is given by

$$\psi(x) = [e^{-xT^{-1}A}PP_+ + e^{(\tau-x)T^{-1}A}PP_- + (I-xT^{-1}A)(I-P)]V_\tau^{-1}\varphi. \tag{2.6}$$

Here $\varphi = \varphi_+ + \varphi_-$ and V_τ is the invertible operator defined by (2.5).

Proof: In view of (2.4)–(2.5) it suffices to show that V_τ is invertible on H_T. First we restrict ourselves to invertible A. As the operators $I+\exp\{-\tau|T^{-1}A|\}$ and $I+\exp\{-\tau|AT^{-1}|\}$ are invertible on H — observe that (2.3) is an equivalent inner product on H in which $A^{-1}T$ is self adjoint, while A acts as a similarity between the two operators — a straightforward application of Proposition II 3.1 implies that $I+\exp\{-\tau|T^{-1}A|\}$ extends continuously to an invertible operator on H_T. Put

$$W_\tau = V_\tau[I+e^{-\tau|T^{-1}A|}]^{-1}.$$

Using the identity $V_\tau = V+(I-V)\exp\{-\tau|T^{-1}A|\}$, where, as in the previous three chapters, we have written $V = Q_+P_+ + Q_-P_-$, we obtain

$$W_\tau = (2V-I)[I+e^{-\tau|T^{-1}A|}]^{-1} + (I-V).$$

We then compute

$$(W_\tau h,h)_T = ((2V-I)[I+e^{-\tau|T^{-1}A|}]^{-1}h,h)_T - \tfrac{1}{2}((2V-I)h,h)_T + \tfrac{1}{2}\|h\|_T^2.$$

Defining $\Phi(z)=(1+e^{-\tau/|z|})^{-1/2}$ and employing the identity $2V-I=(Q_+-Q_-)(P_+-P_-)$, we find, for $h\in D(T)$,

$$(W_\tau h,h)_T = \| |A^{-1}T|^{1/2}\Phi(A^{-1}T)h\|_A^2 - \tfrac{1}{2}\| |A^{-1}T|^{1/2}h\|_A^2 + \tfrac{1}{2}\|h\|_T^2 \geq \tfrac{1}{2}\|h\|_T^2,$$

since $\Phi(z)\geq\frac{1}{2}$. Hence, W_τ is an invertible strictly positive self adjoint operator on H_T.

Next, let us drop the invertibility assumption on A. Following the general procedure developed in Section III.1, we extend the restriction of A to $Z_0(AT^{-1})^\perp$ to a strictly positive self adjoint operator A_β on H satisfying $A_\beta^{-1}T=\beta\oplus S$, where β is invertible on $Z_0(T^{-1}A)$. We then define $P_\pm$ as the (extension to H_T of the) orthogonal projections of H onto the maximal $A_\beta^{-1}T$-positive/negative $A_\beta^{-1}T$-invariant subspaces. Here orthogonality is related to the equivalent inner product (2.3) with A_β instead of A. Define

$$V_{\tau,\beta} = Q_+[P_+ + \exp\{\tau T^{-1}A_\beta\}P_-] + Q_-[P_- + \exp\{-\tau T^{-1}A_\beta\}P_+] =$$

$$= V + (I-V)\exp\{-\tau|T^{-1}A_\beta|\}.$$

Then $V_{\tau,\beta}$ is invertible on H_T and $V_{\tau,\beta}^{-1}V_\tau$ is a finite rank perturbation of the identity. Thus it suffices to show that V_τ has zero null space. Indeed, let $V_\tau h=0$ and write $[h,k]=(Th,k)$. Then (2.6) implies $Q_+[PP_+ + \exp\{\tau T^{-1}A\}PP_- + (I-P)]h=0$ and $Q_-[PP_- + \exp\{-\tau T^{-1}A\}PP_+ + (I-\tau T^{-1}A)(I-P)]h=0$, whence

$$0 \geq [PP_+h,PP_+h] + [\exp\{\tau T^{-1}A\}PP_-h,\exp\{\tau T^{-1}A\}PP_-h] + [h_0,h_0]$$

and

$$0 \leq [PP_-h,PP_-h] + [\exp\{-\tau T^{-1}A\}PP_+h,\exp\{-\tau T^{-1}A\}PP_+h] + [h_\tau,h_\tau],$$

where $h_0=(I-P)h$ and $h_\tau=(I-\tau T^{-1}A)(I-P)h$. On subtracting these we obtain

$$0 \leq [(I-\exp\{-2\tau T^{-1}A\})PP_+h,PP_+h] - [(I-\exp\{2\tau T^{-1}A\})PP_-h,PP_-h] +$$

$+ \ 2\tau(Ah_0,h_0).$

Thus, $(I-\exp\{-2\tau T^{-1}A\})PP_+h=0$, $(I-\exp\{2\tau T^{-1}A\})PP_-h=0$, and $A(I-P)h=0$, which implies that $h \in \operatorname{Ker} A$. But for such vectors we have $h=V_\tau h=0$, whence V_τ has zero kernel. ■

The above result was obtained for the case of injective and certain non-injective A by Beals [32]. Except for the last part of the proof, which is taken from [360], the arguments presented here are new.

(b) The collision operator A a compact perturbation of the identity

We shall next investigate the boundary value problem (1.1)–(1.2) for the case where T is a (bounded or unbounded) injective self adjoint operator and A is a positive self adjoint operator which is a compact perturbation of the identity satisfying

$$\exists 0<\gamma<1: \quad \operatorname{Ran}(I-A) \subset \operatorname{Ran}|T|^{\gamma} \cap D(|T|^{2+\gamma}). \tag{2.7}$$

As in Section II.2 it is possible to distinguish between two types of solutions, and the results contained herein are applicable to both, with appropriate restrictions on boundary conditions and the class of solutions. We shall define a solution of the boundary value problem for $\varphi_+ \in Q_+[D(T)]$ and $\varphi_- \in Q_-[D(T)]$ to be a continuous function $\psi:[0,\tau] \to H$ with values in D(T) such that $T\psi$ is a strongly differentiable on $(0,\tau)$ and Eqs. (1.1)–(1.2) are fulfilled.

As in the previous subsection, we consider the bounded analytic semigroups which are connected with the operator $T^{-1}A$ restricted to $PP_+[H]$, $PP_-[H]$ and $(I-P)[H]$, and the bounded analytic semigroups which are connected with the operator AT^{-1} restricted to $\hat{P}\hat{P}_+[H]$, $\hat{P}\hat{P}_-[H]$ and $(I-\hat{P})[H]$. Then, in the same manner as before, we obtain (2.4)–(2.5), where V_τ is a bounded operator on H leaving invariant D(T). For every $\varphi=\varphi_+ + \varphi_- \in D(T)$ the boundary value problem (1.1)–(1.2) has a unique solution if and only if the restriction of V_τ to D(T) is invertible on D(T).

THEOREM 2.2. In this formulation, for every $\varphi_+ \in Q_+[D(T)]$ and $\varphi_- \in Q_-[D(T)]$ there is a unique solution of the boundary value problem (1.1)–(1.2), which is given by (2.6), where V_τ has a restriction to D(T) which is invertible on D(T).

Proof: Since $D(T)\subset H_T$, we obtain $(\mathrm{Ker}\ V_\tau)\cap D(T)=\{0\}$ from Theorem 2.1. Using that PP_+-Q_+ and PP_--Q_- are compact operators on H (cf. Lemma II 2.6; the result can easily be extended to non-invertible A), we find

$$I-V_\tau = \tau Q_-T^{-1}A(I-P) + Q_+(Q_--PP_-) + Q_-(Q_+-PP_+) +$$
$$+ Q_+(Q_--PP_-)e^{\tau T^{-1}A}PP_- + Q_-(Q_+-PP_+)e^{-\tau T^{-1}A}PP_+,$$

which is a compact operator. We may prove in the same way that the operator $\hat{V}_\tau$ satisfying the intertwining relation $TV_\tau=\hat{V}_\tau T$ on D(T) is compact. From this identity and the compactness of V_τ and $\hat{V}_\tau$ we can easily derive the compactness of $V_\tau|_{D(T)}$ on the Hilbert space D(T) with (complete) inner product

$$(h,k)_{GT} = (h,k) + (Th,Tk), \tag{2.8}$$

using the reasoning of the last paragraph of the proof of Lemma II 2.6. Hence, $V_\tau|_{D(T)}$ is an invertible operator on D(T).

It remains to prove that $\mathrm{Ker}\ V_\tau\subset D(T)$, but this easily follows, using that Q_+ and Q_- leave invariant D(T), $Z_0(T^{-1}A)\subset D(T)$ and $(PP_\pm-Q_\pm)[H]\subset D(T)$. We conclude that V_τ has zero null space and $I-V_\tau$ is compact on H, which implies its invertibility on H. ■

In Section II.2 we have introduced an alternate notion of solution in which boundary data may be chosen in H rather than in D(T). Here we define the analogous notion for slabs. Given $\varphi_\pm\in H$, we seek a continuous function ψ on $[0,\tau]$ which is strongly differentiable on $(0,\tau)$, with derivative having its values in D(T), and which satisfies the boundary value problem. We refer to Section II.2 for additional details.

COROLLARY 2.3. In the formulation of the previous paragraph, for every $\varphi_+\in Q_+[H]$ and $\varphi_-\in Q_-[H]$ there is a unique solution of the boundary value problem (1.1)-(1.2), which is given by (2.6), where V_τ is invertible on H.

(c) The collision operator A unbounded

We will now consider a bounded self adjoint operator T with zero null space, and an (unbounded) self adjoint operator A which is positive and Fredholm. Under these assumptions we shall analyze the boundary value problem (1.1)–(1.2) in the Hilbert space setting H_S of Section II.4, with boundary values $\psi(0)\epsilon H_T$ and $\psi(\tau)\epsilon H_T$. We shall focus on the technicalities involved in dropping the boundedness of A, and assume A is injective. The injectivity condition may, in fact, be removed by a construction parallelling that discussed in Section III.1.

Let us define the Hilbert space $H_A=D(A^{1/2})\subset H$ with inner product $(h,k)_A = (A^{1/2}h,A^{1/2}k)$. On H_A the bounded operator $S=A^{-1}T$ is self adjoint, and so is $K=S^{-1}\supset T^{-1}A$. Next, let us define by H_{TS} the completion of H_A with respect to the inner product

$$(h,k)_{TS} = (|T|h,k) + (|S|h,k)_A.$$

By H_S we shall denote the completion of H_{TS} with respect to the inner product

$$(h,k)_S = (|S|h,k)_A = (T(P_+-P_-)h,k)_A. \tag{2.9}$$

Here $P_\pm$ denote the maximal positive/negative projections associated with S on H_A, and the same symbols are used for their extensions to H_S. The completion of H_{TS} with respect to the inner product

$$(h,k)_T = (|T|h,k) \tag{2.10}$$

will be denoted by H_T. As before, $Q_\pm$ extend to orthogonal projections on $H_T\supset H$ and will be specified by the same symbols.

Let us give a precise statement of the boundary value problem (1.1)–(1.2). Given $\varphi_+\epsilon Q_+[H_T]$ and $\varphi_-\epsilon Q_-[H_T]$, a solution of Eqs. (1.1)–(1.2) is a continuous function $\psi:[0,\tau]\to H_S$ which is H_S-differentiable on $(0,\tau)$ with $\psi(0)\epsilon H_T$ and satisfies (1.1)–(1.2). As in the case of bounded A, we are choosing boundary data $\varphi_+\epsilon Q_+[H_T]$ and $\varphi_-\epsilon Q_-[H_T]$ and demanding that the "total boundary fluxes" $\psi(0)\epsilon H_{TS}$ and $\psi(\tau)\epsilon H_{TS}$. However, because of the (possibly) singular behavior of the operator K, we must seek solutions in H_S. We note that this problem did not arise previously because $H_T\cong H_S$ when A is bounded (see Corollary II 4.6).

On premultiplying Eq. (1.1) by $P_\pm$, solving the resulting equations and adding their solutions, we arrive at the expressions

$$\psi(x) = e^{-xT^{-1}A}P_+\psi(0) + e^{(\tau-x)T^{-1}A}P_-\psi(\tau), \quad 0<x<\tau, \tag{2.11}$$

where the semigroups are the continuous extensions to H_S of corresponding semigroups on H_A. In order to obtain a vector equation relating $P_+\psi(0)$ and $P_-\psi(\tau)$ to the boundary data φ_+ and φ_-, we define an operator $V_{\tau,0}$ with domain $D(V_{\tau,0}) = \{h\in H_S : [P_\pm+\exp\{\pm\tau T^{-1}A\}P_\mp]h\in H_{TS}\}$ by

$$V_{\tau,0}h = \{Q_+[P_++e^{\tau T^{-1}A}P_-] + Q_-[P_-+e^{-\tau T^{-1}A}P_+]\}h.$$

The boundary value problem (1.1)-(1.2) has at least one solution for given $\varphi=\varphi_++\varphi_-\in H_T$ if and only if there exists $h\in D(V_{\tau,0})$ such that $V_{\tau,0}h=\varphi$. The corresponding solutions are then given by (2.11), where $P_+\psi(0)=P_+h$ and $P_-\psi(\tau)=P_-h$. Existence and uniqueness for the boundary value problem thus amounts to the invertibility of $V_{\tau,0}$ as an operator from its domain onto H_T.

In order to address this issue, we introduce the operator $Z_{\tau,0}$ with domain $D(Z_{\tau,0}) = \{h\in H_S : [I+\exp\{-\tau|T^{-1}A|\}]^{-1}h\in D(V_{\tau,0})\}$ by

$$Z_{\tau,0} = V_{\tau,0}[I+\exp\{-\tau|T^{-1}A|\}]^{-1}.$$

Since $I+\exp\{-\tau|T^{-1}A|\}$ is bounded and invertible on H_S, the invertibility of $V_{\tau,0}$ is equivalent to the invertibility of $Z_{\tau,0}$, where both operators are viewed as acting from their respective domains into H_T. As in Section II.4 we define V_0 with $D(V_0) = \{h\in H_{TS} : P_\pm h\in H_T\}$ by $V_0 = (Q_+P_++Q_-P_-)$.

Let us first derive the analog of Eq. (II 4.6).

LEMMA 2.4. For all $h\in D(Z_{\tau,0})$ and $g\in H_{TS}$ we have

$$2(Z_{\tau,0}h,g)_T = (h,g)_T + (\Psi_\tau(A^{-1}T)h,g)_S,$$

where $\Psi_\tau(z) = (1-\exp\{-\tau/z\})(1+\exp\{-\tau/z\})^{-1}$. Moreover, $D(Z_{\tau,0})\subset H_{TS}$.

Proof: We compute, on $D(V_{\tau,0})$,

$$2V_{\tau,0} - [I+e^{-\tau|T^{-1}A|}] =$$

$$= (2Q_+-I)[P_+ + e^{\tau T^{-1}A}P_-] + (2Q_--I)[P_- + e^{-\tau T^{-1}A}P_+] =$$

$$= (Q_+-Q_-)\{[P_+ + e^{\tau T^{-1}A}P_-] - [P_- + e^{-\tau T^{-1}A}P_+]\}.$$

Notice that both sides map an arbitrary vector of $D(V_{\tau,0})$ into H_T. Hence, on $D(Z_{\tau,0})$,

$$2Z_{\tau,0}-I = (Q_+-Q_-)(P_+-P_-)\Psi_\tau(A^{-1}T).$$

For $h\in D(Z_{\tau,0})$ we have $(P_+-P_-)\Psi_\tau(A^{-1}T)h\in H_{TS}$. Therefore, for $h\in D(Z_{\tau,0})$ the vectors $Z_{\tau,0}h$ and $(Q_+-Q_-)(P_+-P_-)\Psi_\tau(A^{-1}T)h$ belong to H_T, and therefore $h\in H_{TS}$. We also remark that $H_A \subset H_T$. Thus, for all $g\in H_A$,

$$2(Z_{\tau,0}h,g)_T - (h,g)_T = ((Q_+-Q_-)(P_+-P_-)\Psi_\tau(A^{-1}T)h,g)_T.$$

Following precisely the method to prove Eq. (II 4.6), we derive

$$2(Z_{\tau,0}h,g)_T - (h,g)_T = (T(P_+-P_-)\Psi_\tau(A^{-1}T)h, |T|g)_{T^{-1}} =$$

$$= (T(P_+-P_-)\Psi_\tau(A^{-1}T)h,g) = (S(P_+-P_-)\Psi_\tau(A^{-1}T)h,g)_A = (\Psi(A^{-1}T)h,g)_S,$$

which settles the desired equality for $g\in H_A$. The more general case follows by continuous extension to H_{TS}. ∎

Lemma 2.4 implies that Ker $Z_{\tau,0}=\{0\}$ and therefore that every solution of Eqs. (1.1)–(1.2) is unique. Indeed, since $\sigma(A^{-1}T)\subset[-M,M]$ for some finite M, we have $0<\Psi_\tau(M)\le\Psi_\tau(z)\le 1$ for $z\in\sigma(A^{-1}T)$. Thus $(h,k)_{S,\tau} = (\Psi_\tau(A^{-1}T)h,k)_S$ is an equivalent inner product on H_S, while $(h,k)_{TS,\tau} = (h,k)_T + (\Psi_\tau(A^{-1}T)h,k)_S$ is an equivalent inner product on H_{TS}. Hence, $Z_{\tau,0}h=0$ implies $(h,g)_{S,\tau}=0$ for all $g\in H_{TS}$, which in turn implies $h=0$.

LEMMA 2.5. $Z_{\tau,0}$ is a closed symmetric operator on H_T.

Proof: Let $\{h_n\}_{n=1}^\infty$ be a sequence in $D(Z_{\tau,0}) \subset H_{TS}$ satisfying $h_n \to h$ and $Z_{\tau,0}h_n \to f$ in H_T-norm. Then, for every $g\in H_{TS}$ we have

$$2(Z_{\tau,0}h_n,g)_T = (h_n,g)_T + (\Psi_\tau(A^{-1}T)h_n,g)_S,$$

whence, by taking $n\to\infty$, we see that $h\in H_{TS}$ and

$$2(f,g)_T = (h,g)_T + (\Psi_\tau(A^{-1}T)h,g)_S.$$

Indeed, the above hypotheses, the inclusion $D(Z_{\tau,0}) \subset H_{TS}$ and the completeness of the $(\cdot,\cdot)_{TS,\tau}$ inner product on H_{TS} imply that $\{h_n\}_{n=1}^{\infty}$ is a weak Cauchy sequence in H_{TS}. Hence, $h_n \to \hat{h}$ in the weak topology of H_{TS} for some $\hat{h}\in H_{TS}$. Since this topology is weaker than the H_T-norm topology, we have $\hat{h}=h$ and consequently $h\in H_{TS}$. On the other hand, using the identity

$$\lim_{n\to\infty} \|(P_+-P_-)\Psi_\tau(A^{-1}T)h_n - (Q_+-Q_-)(2f-h)\|_T = 0$$

and $(P_+-P_-)\Psi_{\tau,0}(A^{-1}T)h\in H_S$, we must conclude that the latter vector belongs to H_{TS}. Hence, $(P_\pm+\exp\{\pm\tau T^{-1}A\}P_\mp)[I+\exp\{-\tau|T^{-1}A|\}]^{-1}h \in H_{TS}$, and therefore $h\in D(Z_{\tau,0})$ and $Z_{\tau,0}h=f$. Substituting $g\in D(Z_{\tau,0})$ in Lemma 2.4 we see that $Z_{\tau,0}$ is symmetric in H_T. ■

As a consequence of Lemma 2.4, we have $2(Z_{\tau,0}h,h)_T \geq \|h\|_T^2$ for $h\in D(Z_{\tau,0})$. Thus $Z_{\tau,0}$ has closed range in H_T and $Z_{\tau,0}^{-1}$ acts as a bounded operator from Ran $Z_{\tau,0}$ ($\subset H_T$) into H_T. Since we also have, for some $\delta>0$,

$$2(Z_{\tau,0}h,h)_T \geq \delta(\|h\|_T^2+\|h\|_S^2) \geq 2\delta\|h\|_T\|h\|_S, \quad h\in D(Z_{\tau,0}),$$

we also find that $Z_{\tau,0}^{-1}$ acts as a bounded operator from Ran $Z_{\tau,0}\subset H_T$ into H_S. We obtain easily an analog of Theorem II 4.3.

THEOREM 2.6. The following statements are equivalent:

(i) For all $\varphi_+\in Q_+[H_T]$ and $\varphi_-\in Q_-[H_T]$ the boundary value problem (1.1)-(1.2) has a (unique) solution.

(ii) Ran $Z_{\tau,0} = H_T$.

(iii) $Z_{\tau,0}[H_A]$ is a dense subspace of H.

(iv) $Z_{\tau,0}$ is a self adjoint operator on H_T.

In a similar way we obtain an analog of Theorem II 4.5.

THEOREM 2.7. The following statements are equivalent:

(i) $Z_{\tau,0}$ is a bounded operator on H_T.
(ii) $Z_{\tau,0}$ is a bounded invertible operator on H_T.
(iii) $Z_{\tau,0}$ is a bounded operator from H_S into H_T.
(iv) $Z_{\tau,0}$ is a bounded invertible operator from H_S into H_T.
(v) The Hilbert spaces H_T and H_S allow a natural identification.

Proof: It is sufficient to observe that $(\cdot,\cdot)_{S,\tau}$ represents a complete inner product on H_S. Therefore, we may repeat the proof of Theorem II 4.5 completely, using this inner product instead of $(\cdot,\cdot)_S$. ∎

The latter result implies that either the operators $Z_{\tau,0}$ and V_0 are bounded invertible on H_T for all τ or unbounded for all τ. Thus, by Theorem II 4.6, the operators $Z_{\tau,0}$ are bounded on H_T if A is bounded. As we shall see in Chapter X, one may also identify H_T and H_S naturally in the case of a Sturm–Liouville diffusion equation, and consequently the operators V_0 and $Z_{\tau,0}$ are bounded for this case too.

We may modify these results for the case when Ker A is nontrivial in the manner of Chapter III, leading to a modified operator $Z_{\tau,0}$. This is accomplished with the help of a strictly positive self adjoint auxiliary operator A_β, which coincides with A on $Z_0(AT^{-1})^\perp$. In this way we may prove uniqueness of solutions, but existence again depends on the completeness property Ran $Z_{\tau,0}=H_T$.

3. Boundary value problems for multiplying slab media

In this section we study the abstract boundary value problem (1.1)–(1.2) relevant to transport in a multiplying medium confined within the slab $[0,\tau]$ and with bounded collision operator A, following closely the analysis of Greenberg and Walus [167, 386]. The multiplying character of the medium manifests itself mathematically in allowing A to have a finite dimensional negative part. That is to say, we shall let A be a bounded self adjoint Fredholm operator on H satisfying conditions (i) and (ii) on

$Z_0(T^{-1}A)$ specified below and such that the spectrum of A in $(-\infty,0]$ consists of a finite number of eigenvalues of finite multiplicity. We shall let T be a self adjoint injective operator on H, and for simplicity we assume T is bounded.

Let us introduce an (indefinite) inner product on H defined by (see Section IV.1)

$$[f,g]_A = (Af,g). \tag{3.1}$$

We will denote by H_A the space H equipped with the $[\cdot,\cdot]_A$-inner product. If Ker $A=\{0\}$, then H_A is a Π_κ-space where κ is the sum of the multiplicities of the negative eigenvalues of A. One can easily show that the operator $T^{-1}A$ is self adjoint with respect to the indefinite inner product (3.1). Then $T^{-1}A$ has at most 2κ nonreal eigenvalues (multiplicities taken into account) occurring in complex conjugate pairs with pairwise coinciding Jordan structures, while the length of a Jordan chain for a real eigenvalue of $T^{-1}A$ does not exceed $2\kappa+1$. Moreover, there is a resolution of the identity for the real spectrum of $T^{-1}A$, possibly with finitely many critical points at certain eigenvalues (see Theorem IV 1.3). If A has a nontrivial kernel, then H_A is no longer a Π_κ-space, nor a Krein space; however, the decomposition of H to be developed will allow the spectral analysis of the operator $T^{-1}A$ to be carried out.

In order to exclude the existence of so-called irregular critical points, we shall assume the following (cf. Proposition IV 1.4).

(i) $Z_0(T^{-1}A)$ is finite dimensional and nondegenerate with respect to the indefinite inner product $[f,g]_T=(Tf,g)$; i.e., if $f\in Z_0(T^{-1}A)$ and $(Tf,g)=0$ for all $g\in Z_0(T^{-1}A)$, then $f=0$.

(ii) for any real nonzero λ, $Z_\lambda(T^{-1}A)$ is nondegenerate with respect to the indefinite inner product $[\cdot,\cdot]_A$.

Under these assumptions we obtain a decomposition of the Hilbert space H into $T^{-1}A$ invariant subspaces in such a way that A can be deformed to a strictly positive operator by a perturbation of finite rank. Such a decomposition is provided by Proposition IV 2.1, which we reformulate here for the sake of completeness.

PROPOSITION 3.1. There exists a $T^{-1}A$-invariant finite dimensional subspace $Z(T^{-1}A)$ of H with the following properties:

(i) $Z(T^{-1}A) \oplus (T[Z(T^{-1}A)])^{\perp} = H$,

(ii) The subspace $(T[Z(T^{-1}A)])^{\perp}$ is $T^{-1}A$-invariant and strictly positive with respect to $[\cdot,\cdot]_A$,

(iii) The constituent subspaces in the decomposition (i) are $[\cdot,\cdot]_A$-orthogonal.

Moreover, if P denotes the projection of H onto $(T[Z(T^{-1}A)])^{\perp}$ along $Z(T^{-1}A)$ and β is an invertible $[\cdot,\cdot]_T$-positive operator on $Z(T^{-1}A)$, then the bounded operator $A_\beta = AP + T\beta^{-1}(I-P)$ is strictly positive with respect to the H-inner product and Fredholm. Here, $[h,k]_T = (Th,k)$.

Since A_β is strictly positive on H and A_β^{-1} is bounded, the inner product $(\cdot,\cdot)_{A_\beta}$ defined by $(f,g)_{A_\beta} = (A_\beta f,g)$ is equivalent to the original inner product on H. We will denote by H_{A_β} the space H endowed with the $(\cdot,\cdot)_{A_\beta}$-inner product. It is clear that its topology does not depend on β, and that the operator S_β defined by $S_\beta = A_\beta^{-1}T$ is bounded, injective and self adjoint on H_{A_β}. We define $P_\pm$ as the orthogonal complementary projections of H_{A_β} onto the maximal S_β-positive/negative S_β-invariant subspaces. Note that $P_\pm$ depend on β. However, since the decomposition (i) reduces the operator S_β, and $S_\beta|_{(T[Z(T^{-1}A)])^{\perp}}$ does not depend on β, the operators PP_+, PP_- and $I-P$ form a set of β-independent complementary projections on H.

As in Section II.4 we first introduce H_{TS_β} as the completion of $H_A \cap D(T)$ with respect to the inner product

$$(h,k)_{TS_\beta} = (|T|h,k) + (|S_\beta|h,k)_{A_\beta}.$$

Next, let H_{S_β} be the completion of H_{TS_β} with respect to the inner product

$$(f,g)_{S_\beta} = (|S_\beta|f,g)_{A_\beta}.$$

As $Z(T^{-1}A)$ is finite dimensional, the topologies of H_{TS_β} and H_{S_β} do not depend on β. In fact, one can show that H_{S_β} is topologically isomorphic to H_T, the completion of H with respect to the inner product $(f,g)_T = (|T|f,g)$ (cf. Corollary II 4.6). Therefore we will suppress the subscripts β in H_{TS_β} and H_{S_β} and write H_{TS} and H_S, respectively.

Since the extension of $T^{-1}A_\beta$ to the space H_S is self adjoint, one can define the projections $P_\pm$ and the contraction semigroups $\exp\{\mp xT^{-1}A_\beta\}P_\pm$ with the help of the Spectral Theorem. Moreover, one can extend $P_\pm$ and P to bounded projections acting in H_S; we will denote by K the resulting extension of $T^{-1}A$ to H_S. Solving Eq. (1.1) on the subspaces Ran PP_+, Ran PP_- and Ran(I−P)=Z($T^{-1}A$) as in the previous section, we obtain, for the solution of the boundary value problem (1.1)–(1.2),

$$\psi(x) = [e^{-xK}PP_+ + e^{(\tau-x)K}PP_- + e^{-xT^{-1}A}(I-P)]h, \tag{3.2}$$

for $V_\tau h = \varphi$, where $\varphi=\varphi_+ + \varphi_-$ and V_τ is defined in (2.5). Therefore the unique solvability of the boundary value problem is equivalent to the bounded invertibility of the operator V_τ on H_S. The proof of invertibility utilizes the following estimate on the operator $V_{\tau,\beta}$ defined in Section 2 (a).

LEMMA 3.2. For any $\tau>0$ the operator $V_{\tau,\beta}$ is invertible on H_S and

$$\| V_{\tau,\beta}^{-1} \|_{\mathcal{L}(H_S)} \le \| V^{-1} \|_{\mathcal{L}(H_S)} (1 - \| V^{-1} - I \|_{\mathcal{L}(H_S)})^{-1}, \tag{3.3}$$

where $V=Q_+P_+ + Q_-P_-$ has a bounded inverse on H_S satisfying $\| V^{-1}-I \|_{\mathcal{L}(H_S)}<1$.

Proof: The following proof is adapted from an argument of Beals [32]. We first recall that the Hilbert spaces H_S and H_T can be identified in a natural way by means of the equivalence of their norms on a suitable common dense subspace (see Corollary II 4.6). As a result $V=Q_+P_+ + Q_-P_-$ establishes a topological isomorphism of H_S onto H_T. Let $W=I-V= Q_+P_- + Q_-P_+$. Then, for any $h \in H_{A_\beta}$, we have the useful identity

$$\|Vh\|_T^2 - \|Wh\|_T^2 = ((V-W)(V+W)h,h)_T = ((2V-I)h,h)_T = (T(P_+-P_-)h,h) =$$

$$= (S_\beta(P_+-P_-)h,h)_{A_\beta} = \|h\|^2_{S_\beta}. \tag{3.4a}$$

In a similar way we prove that

$$\|V^*k\|^2_{S_\beta} - \|W^*k\|^2_{S_\beta} = \|k\|^2_T. \tag{3.4b}$$

Here we have exploited the identity

$$((2V-I)h,k)_T = (h,k)_{S_\beta}$$

to derive the equalities

$$V^* = P_+Q_+ + P_-Q_-,$$

$$W^* = I^* - V^* = -(P_+Q_- + P_-Q_+),$$

where the adjoints are operators from H_T into H_S.

In order to proceed to the proof of the inequality (3.3), note that

$$\|V^{-1}-I\|_{\mathcal{L}(H_S)} = \|V^{-1}W\|_{\mathcal{L}(H_S)} = \|W^*(V^*)^{-1}\|_{\mathcal{L}(H_S)}.$$

Moreover, from (3.4) it follows that

$$\|W^*(V^*)^{-1}h\|^2_{S_\beta} = \|h\|^2_{S_\beta} - \|(V^*)^{-1}h\|^2_T.$$

Since V^* is a topological isomorphism, $\|(V^*)^{-1}h\|_T \geq C\|h\|_{S_\beta}$ for some constant $C>0$. Finally it is clear that $C\leq 1$ and

$$\|V^{-1}-I\|_{\mathcal{L}(H_S)} = \|W^*(V^*)^{-1}\|_{\mathcal{L}(H_S)} \leq (1-C^2)^{1/2} < 1.$$

Using simple algebra we show

$$V_{\tau,\beta} = V\,[I+(V^{-1}-I)\exp\{-\tau\,|T^{-1}A_\beta|\}].$$

Then since

$$\|(V^{-1}-I)\exp\{-\tau\,|T^{-1}A_\beta|\}\|_{\mathcal{L}(H_S)} \leq \|V^{-1}-I\|_{\mathcal{L}(H_S)} < 1,$$

the inequality (3.3) follows by direct computation. ∎

THEOREM 3.3. There exists $\tau_c>0$ such that for all $0<\tau<\tau_c$ and every $\varphi_+\in Q_+[H_S]$ and $\varphi_-\in Q_-[H_S]$ the boundary value problem (1.1)–(1.2) has a unique solution. The solution is given by

$$\psi(x) = [e^{-xK}PP_+ + e^{(\tau-x)K}PP_- + e^{-xT^{-1}A}(I-P)]V_\tau^{-1}\varphi, \tag{3.5}$$

where $\varphi=\varphi_++\varphi_-$ and V_τ is the invertible operator on H_T defined by (2.5).

Proof: We will show that, for sufficiently small τ,

$$\| V_\tau - V_{\tau,\beta} \|_{\mathcal{L}(H_S)} < (\| V_{\tau,\beta}^{-1} \|_{\mathcal{L}(H_S)})^{-1}, \tag{3.6}$$

which guarantees the bounded invertibility of V_τ. Simple algebra shows that

$$V_\tau - V_{\tau,\beta} = Q_+[I-e^{\tau\beta^{-1}}](I-P)P_- + Q_-[I-e^{-\tau\beta^{-1}}](I-P)P_+ +$$

$$+ Q_-[e^{-\tau(T^{-1}A \mid Z(T^{-1}A))_-}I](I-P).$$

Therefore,

$$\| V_\tau - V_{\tau,\beta} \|_{\mathcal{L}(H_S)} \leq c_1(e^{\tau \|\beta^{-1}\|}-1) + c_2(e^{\tau \|T^{-1}A \mid Z(T^{-1}A)\|}-1) \equiv f(\tau)$$

for constants $c_1, c_2 \in \mathbb{R}$. Clearly $f(\tau) \to 0$ as $\tau \to 0$. Let

$$\tau_c = \sup_{\tau>0} \{\tau : f(\tau) < (1-\| V^{-1}-I\|_{\mathcal{L}(H_S)})\| V^{-1}\|^{-1}_{\mathcal{L}(H_S)}\}.$$

Now using Lemma 3.2 one shows that the inequality (3.6) holds true for any $\tau \in (0,\tau_c)$. ■

In the case in which the collision operator A is a compact perturbation of the identity satisfying the regularity condition Ran B $\subset$ Ran$|T|^\gamma \cap D(|T|^n)$ for some $\gamma>0$ and all $n \in \mathbb{N}$, one can improve the statement of Theorem 3.3. In particular, the function $z \to V_z - I$ is compact operator valued (on H) and analytic in the open right half plane. Then, using the analytic version of the Fredholm alternative, one can infer the existence of a discrete subset Δ of the open right half plane such that for any real $\tau \in \Delta$ the boundary value problem (1.1)–(1.2) with $\varphi_\pm = 0$ has at least one nontrivial solution, which is given by (3.2), where h is a nontrivial solution of $V_\tau h = 0$. Further, if $V_z - I$ is a trace class operator, which is the case, for example, if I−A is a finite rank perturbation of the identity, then Δ can be characterized as the set of solutions of det $[V_z] = 0$ with Re $z > 0$.

4. Reflection and transmission operators

Reflection and transmission operators for abstract kinetic equations were studied in detail by van der Mee [362]. They were introduced there as the functional analytic counterparts of the reflection and transmission functions which play an important role in radiative transfer (see, for instance, [89, 342]). In [164] they were studied further in connection with an abstract differential equation with reflecting boundary conditions. In both of these articles these operators were analyzed in the original Hilbert space H. In [183] Hangelbroek introduced similar operators to investigate the monotonicity of reflection and transmission as a function of slab diameter.

In the present and the next section we will restrict ourselves to bounded nonnegative collision operators A. Let us define unique **reflection operators** $R_{+\tau}$ and $R_{-\tau}$ and **transmission operators** $T_{+\tau}$ and $T_{-\tau}$, corresponding to the boundary value problem (1.1)–(1.2), satisfying the conditions

$$\psi(0) = (R_{+\tau}+T_{-\tau})\varphi, \qquad \psi(\tau) = (R_{-\tau}+T_{+\tau})\varphi, \tag{4.1}$$

$$R_{\pm\tau}Q_{\mp} = 0, \qquad T_{\pm\tau}Q_{\mp} = 0. \tag{4.2}$$

We study these operators first on H_T. Using (2.6) we find easily the expressions

$$R_{+\tau} = [PP_{+} + e^{\tau T^{-1}A}PP_{-} + (I-P)]V_{\tau}^{-1}Q_{+}, \tag{4.3}$$

$$R_{-\tau} = [PP_{-} + e^{-\tau T^{-1}A}PP_{+} + (I-\tau T^{-1}A)(I-P)]V_{\tau}^{-1}Q_{-}, \tag{4.4}$$

$$T_{+\tau} = [PP_{-} + e^{-\tau T^{-1}A}PP_{+} + (I-\tau T^{-1}A)(I-P)]V_{\tau}^{-1}Q_{+}, \tag{4.5}$$

$$T_{-\tau} = [PP_{+} + e^{\tau T^{-1}A}PP_{-} + (I-P)]V_{\tau}^{-1}Q_{-}. \tag{4.6}$$

Using the boundary conditions (1.2) (or Eq. (2.5) directly) we obtain $Q_{\pm}R_{\pm\tau}=Q_{\pm}$ and $Q_{\mp}T_{\pm\tau}=0$, from which it is evident that $T_{\pm\tau}$ leaves invariant $Q_{\pm}[H_T]$ and $R_{\pm\tau}$ is a projection on H_T with kernel $Q_{\mp}[H_T]$.

These identities may also be found in a different way. It is seen immediately that $R_{+\tau}\varphi$ and $T_{+\tau}\varphi$ are the values at $x=0$ and $x=\tau$ of the unique solution of the boundary value problem (1.1)-(1.2) with $\varphi_+=Q_+\varphi$ and $\varphi_-=0$, and $R_{-\tau}\varphi$ and $T_{-\tau}\varphi$ are the values at $x=\tau$ and $x=0$ of the solution of this boundary value problem for $\varphi_+=0$ and $\varphi_-=Q_-\varphi$. Then (4.2) and the properties mentioned in the sentence following (4.6) are immediate. In the present definition the transmission operators account for the fluxes transmitted through the slab, while the reflection operators account for incident plus reflected fluxes.

THEOREM 4.1. The operators $R_{+\tau}+R_{-\tau}$, $T_{+\tau}$ and $T_{-\tau}$ are self adjoint on H_T, and $\|[R_{+\tau}+R_{-\tau}]-I\|_{H_T}<1$.

Proof: Let us denote by ψ the solution of Eqs. (1.1)-(1.2) with boundary data φ, and by $\hat{\psi}$ the solution of these equations with boundary data $\hat{\varphi}$. Putting $N=Q_+-Q_-$, we compute

$$\frac{d}{dx}(N\psi(x),\ \hat{\psi}(\tau-x))_T = -(NK\psi(x),\ \hat{\psi}(\tau-x))_T + (\psi(x),\ NK\hat{\psi}(\tau-x))_T = 0,$$

implying

$$(N[R_{+\tau}+T_{-\tau}]\varphi,[R_{-\tau}+T_{+\tau}]\hat{\varphi})_T = (N[R_{-\tau}+T_{+\tau}]\varphi,[R_{+\tau}+T_{-\tau}]\hat{\varphi})_T. \tag{4.7}$$

For the transmission operator $T_{+\tau}$ we have

$$(T_{+\tau}\varphi,\hat{\varphi})_T = (NT_{+\tau}Q_+\varphi,Q_+\hat{\varphi})_T = (NT_{+\tau}Q_+\varphi,Q_+R_{+\tau}Q_+\hat{\varphi})_T =$$

$$= (NR_{+\tau}Q_+\varphi,T_{+\tau}Q_+\hat{\varphi})_T = (NQ_+\varphi,T_{+\tau}\hat{\varphi})_T = (\varphi,T_{+\tau}\hat{\varphi})_T,$$

where we have employed Ran $(I-R_{+\tau})=$Ran Q_- and $Q_-T_{+\tau}=0$. The selfadjointness of $T_{-\tau}$ is proved similarly.

Next let us consider the reflection operators $R_{+\tau}$ and $R_{-\tau}$. A careful computation yields

$$((I-R_{+\tau})\varphi,\hat{\varphi})_T = -(N(I-R_{+\tau})\varphi,[R_{-\tau}+T_{+\tau}]Q_-\hat{\varphi})_T = -(N\varphi,[R_{-\tau}+T_{+\tau}]Q_-\hat{\varphi})_T +$$

$$+ (N[R_{+\tau}+T_{-\tau}]\varphi,[R_{-\tau}+T_{+\tau}]Q_-\hat{\varphi})_T - (NT_{-\tau}\varphi,[R_{-\tau}+T_{+\tau}]Q_-\hat{\varphi})_T =$$

$$= -(N\varphi,R_{-\tau}\hat{\varphi})_T - (NT_{-\tau}\varphi,[I-(I-R_{-\tau})]Q_-\hat{\varphi})_T +$$

$$+ (N[R_{-\tau}+T_{+\tau}]\varphi,[R_{+\tau}+T_{-\tau}]Q_{-}\hat{\varphi})_T =$$

$$= -(N\varphi,R_{-\tau}\hat{\varphi})_T - (N(I-R_{-\tau})\varphi,[R_{+\tau}+T_{-\tau}]Q_{-}\hat{\varphi})_T +$$

$$+ (N\varphi,[R_{+\tau}+T_{-\tau}]Q_{-}\hat{\varphi})_T + (NT_{+\tau}\varphi,[R_{+\tau}+T_{-\tau}]Q_{-}\hat{\varphi})_T + (T_{-\tau}\varphi,\hat{\varphi})_T.$$

On the right side of the last equality the second term vanishes as does the fourth term, while the third and fifth terms cancel each other as a consequence of the selfadjointness of $T_{-\tau}$ on H_T. Hence,

$$((I-R_{+\tau})\varphi,\hat{\varphi})_T = -(N\varphi,R_{-\tau}\hat{\varphi})_T = -(N\varphi,\hat{\varphi})_T + (\varphi,(I-R_{-\tau})\hat{\varphi})_T. \tag{4.8}$$

In a similar way we obtain

$$((I-R_{-\tau})\varphi,\hat{\varphi})_T = (N\varphi,\hat{\varphi})_T + (\varphi,(I-R_{+\tau})\hat{\varphi})_T. \tag{4.9}$$

On adding these we find that $R_{+\tau}+R_{-\tau}$ is self adjoint on H_T.

Let us next consider the contractiveness of $R_{+\tau}+R_{-\tau}-I$. The differential equation (1.1) gives immediately

$$-\int_0^\tau (\psi'(x),N\psi(x))_T dx = -[(\psi(x),N\psi(x))_T]_{x=0}^\tau + \int_0^\tau (\psi(x),N\psi'(x))_T dx =$$

$$= -[(\psi(x),N\psi(x))_T]_{x=0}^\tau - \int_0^\tau (\psi(x),NK\psi(x))_T dx.$$

Therefore,

$$(N\psi(0),\psi(0))_T - (N\psi(\tau),\psi(\tau))_T = 2\int_0^\tau (NK\psi(x),\psi(x))_T dx \geq \delta\,\|(I-P)\varphi\|_T^2 \geq 0,$$

where we used that P is a bounded projection on H_T. We then obtain

$$(NR_{+\tau}\varphi,R_{+\tau}\varphi)_T - (NR_{-\tau}\varphi,R_{-\tau}\varphi)_T \geq \|T_{+\tau}\varphi\|_T^2 + \|T_{-\tau}\varphi\|_T^2 + \delta\,\|(I-P)\varphi\|_T^2,$$

where we have made use of the identities (cf. (4.7))

$$(NR_{\pm\tau}\varphi,T_{\mp\tau}\varphi)_T = (NT_{\pm\tau}\varphi,R_{\mp\tau}\varphi)_T.$$

From (4.8) and (4.9) we know that the H_T-adjoint of $R_{\pm\tau}$ coincides with $R_{\mp\tau}\pm N$. This in turn leads to the equality

$$(NR_{+\tau}\varphi,R_{+\tau}\varphi)_T - (NR_{-\tau}\varphi,R_{-\tau}\varphi)_T =$$

$$= 2([R_{+\tau}+R_{-\tau}]\varphi,\varphi)_T - \|[R_{+\tau}-N]\varphi\|_T^2 - \|[R_{-\tau}+N]\varphi\|_T^2.$$

Finally, we obtain, in the partial order of self adjoint operators on H_T,

$$2\{R_{+\tau}+R_{-\tau}\} \geq T_{+\tau}^2 + T_{-\tau}^2 + \delta N(I-P) \geq 0.$$

Since $\operatorname{Ker} P=Z_0(T^{-1}A)$ has finite dimension, strict positivity of $R_{+\tau}+R_{-\tau}$ is guaranteed if $R_{+\tau}+R_{-\tau}$ has zero null space.

In order to establish this, it suffices to prove that $R_{+\tau}k=R_{-\tau}k$ implies $k=0$. For this would imply the strict positivity of $R_{+\tau}+R_{-\tau}$ and, in view of the identity $(Q_+-Q_-)[R_{+\tau}+R_{-\tau}](Q_+-Q_-) = 2I-[R_{+\tau}+R_{-\tau}]$, the strict positivity of $2I-[R_{+\tau}+R_{-\tau}]$ on H_T. Suppose that k is as indicated. Then $h=(Q_+-Q_-)k$ satisfies $[R_{+\tau}+R_{-\tau}]h=0$, and therefore

$$0 = 2([R_{+\tau}+R_{-\tau}]h,h)_T \geq ([T_{+\tau}^2+T_{-\tau}^2]h,h)_T = \|T_{+\tau}h\|_T^2 + \|T_{-\tau}h\|_T^2 \geq 0,$$

whence $T_{+\tau}h=T_{-\tau}h=0$. Then $T_{+\tau}k=T_{+\tau}h=0$ and $T_{-\tau}k=-T_{-\tau}h=0$. and so

$$k = [R_{+\tau}+T_{-\tau}]k = [PP_+ + e^{\tau T^{-1}A}PP_- + (I-P)]V_\tau^{-1}k,$$

$$k = [R_{-\tau}+T_{+\tau}]k = [e^{-\tau T^{-1}A}PP_+ + PP_- + (I-\tau T^{-1}A)(I-P)]V_\tau^{-1}k.$$

On premultiplication of these equations by the projections PP_+, PP_- and $I-P$, we get

$$PP_+k = PP_+V_\tau^{-1}k = \exp\{-\tau T^{-1}A\}PP_+V_\tau^{-1}k,$$

$$PP_-k = PP_-V_\tau^{-1}k = \exp\{\tau T^{-1}A\}PP_-V_\tau^{-1}k,$$

$$(I-P)k = (I-P)V_\tau^{-1}k = (I-\tau T^{-1}A)(I-P)V_\tau^{-1}k.$$

From these one obtains easily $PP_+V_\tau^{-1}k=0$, $PP_-V_\tau^{-1}k=0$, and $T^{-1}A(I-P)V_\tau^{-1}k=0$. Therefore $V_\tau^{-1}k\in\text{Ker }A$. However, for $\ell\in\text{Ker }A$ we have $V_\tau\ell=\ell$, whence $V_\tau^{-1}k=V_\tau(V_\tau^{-1}k)\in\text{Ker }A$. But $V_\tau[\text{Ker }A]\subset\text{Ker }A$, V_τ is injective and Ker A has a finite dimension, from which we deduce that $k\in\text{Ker }A$.

Finally, we use $T_{\pm\tau}k=0$ to obtain $PP_\pm V_\tau^{-1}Q_\pm k=0$, $PP_\pm V_\tau^{-1}Q_\mp k=0$, $(I-P)V_\tau^{-1}Q_-k=0$, and $(I-\tau T^{-1}A)(I-P)V_\tau^{-1}Q_+k=0$, which implies $V_\tau^{-1}Q_\pm k\in\text{Ker }A$ and therefore $Q_\pm k\in\text{Ker }A$. Since $V_\tau^{-1}y=y$ for all $y\in\text{Ker }A$, we must have

$$Q_+k = (I-P)V_\tau^{-1}Q_+k = (I-\tau T^{-1}A)(I-P)V_\tau^{-1}Q_+k = 0,$$

$$Q_-k = (I-P)V_\tau^{-1}Q_-k = 0,$$

whence $k=Q_+k+Q_-k=0$. This completes the proof of the theorem. ∎

Next we consider I–A compact and the reflection and transmission operators acting on the original Hilbert space H.

LEMMA 4.2. If I–A is compact and satisfies (2.7), then the operators $R_{\pm\tau}-Q_\pm$ and $T_{\pm\tau}-\exp\{\mp\tau T^{-1}\}Q_\pm$ are compact on H. Their restrictions to D(T) are compact as operators on D(T) endowed with the inner product (2.8).

Proof: Since $V_\tau^{-1}-I$ is compact and the same is true for $PP_\pm-Q_\pm$ and I–P, it suffices to show that $\exp\{\mp\tau T^{-1}A\}PP_\pm-\exp\{\mp\tau T^{-1}\}Q_\pm$ is compact. In order to prove the latter, we can repeat the proof of Lemma II 2.6 with three modifications:

(i) The result is first proved for invertible A and then extended to noninvertible A using the procedure of Section III.1.

(ii) All integrals in the proof of Lemma II 3.2 contain the additional factor $\varphi(\lambda)=e^{-\tau/\lambda}$.

(iii) The integration curve near $\lambda=0$ must be a subset of two straight lines through the origin that do not coincide with the real or the imaginary axis.

In this way we can exploit the fact that on the one hand

$$e^{-\tau T^{-1}A}PP_+ = \lim_{\epsilon\downarrow 0}\frac{1}{2\pi i}\int_{\Gamma_\epsilon} e^{-\tau/\lambda}(\lambda-S)^{-1}d\lambda$$

and

$$e^{-\tau T^{-1}}Q_+ = \lim_{\epsilon\downarrow 0} \frac{1}{2\pi i}\int_{\Gamma_\epsilon} e^{-\tau/\lambda}(\lambda-T)^{-1}d\lambda,$$

while on the other hand $\varphi(\lambda)=e^{-\tau/\lambda}$ is bounded and continuous on $|\arg\lambda|\le\frac{1}{2}\pi-\delta$ and analytic on $|\arg\lambda|<\frac{1}{2}\pi-\delta$ for any $0<\delta<\frac{1}{2}\pi$. Here Γ_ϵ is the oriented broken line consisting of the lines from $i+\infty$ to $i+\tan\delta$, from $i+\tan\delta$ to $\epsilon(i+\tan\delta)$, from $-\epsilon(i+\tan\delta)$ to $-(i+\tan\delta)$, and from $-(i+\tan\delta)$ to $-(i+\infty)$, while the limits are taken in the strong operator topology. As a result we find the desired compactness. ∎

As a consequence of the above lemma, we obtain that $R_{+\tau}+R_{-\tau}$ is a compact perturbation of the identity. We may, however, exploit the selfadjointness of this operator on H_T to prove that $\{|T|(R_{+\tau}+R_{-\tau})\}^*=|T|(R_{+\tau}+R_{-\tau})$ on D(T). (Here the adjoint relates to H). In combination with the identity $N\{R_{+\tau}+R_{-\tau}\}N = 2I-\{R_{+\tau}+R_{-\tau}\}$, we may then prove that

$$T(R_{+\tau}+R_{-\tau}) = \{2I-(R^*_{+\tau}+R^*_{-\tau})\}T, \tag{4.10}$$

where $R^*_{+\tau}+R^*_{-\tau}$ is a compact perturbation of the identity on H. The restriction of $R_{+\tau}+R_{-\tau}$ to D(T) will then be a compact perturbation of the identity on D(T), when endowed with the inner product (2.8). We have the following result.

THEOREM 4.3. If I−A is compact and satisfies (2.7), then the operator $R_{+\tau}+R_{-\tau}$ is invertible on D(T). Moreover, $\mathrm{Ran}(R_{+\tau}+R_{-\tau}-I)\subset D(T)$.

Proof: Since $R_{+\tau}+R_{-\tau}-I$ is compact on H and its restriction to D(T) is compact on D(T), it suffices to establish $\mathrm{Ker}(R_{+\tau}+R_{-\tau})\cap D(T)=\{0\}$ and $\mathrm{Ker}(R_{+\tau}+R_{-\tau})\subset D(T)$. The former follows directly from the invertibility of $R_{+\tau}+R_{-\tau}$ on H_T (cf. Theorem 4.1). In order to prove the latter, we have to show that

$$(e^{\mp\tau T^{-1}A}PP_\pm-e^{\mp\tau T^{-1}}Q_\pm)[H] \subset D(T),$$

which can be proved along the lines of Lemma II 2.6 with the above modifications. We may then exploit this property, the inclusions $(PP_\pm-Q_\pm)[H]\subset D(T)$ and $Z_0(T^{-1}A)\subset D(T)$

to obtain the inclusion. The last statement of the theorem is now immediate. ■

5. Slabs with reflective boundary conditions

In this section we analyze the boundary value problem (1.1)–(1.3), where R_ℓ and R_r are surface reflection operators for the left and right surface, respectively, and J is an inversion symmetry. The operator T will be assumed (bounded or unbounded) injective and selfadjoint, and A positive self adjoint and Fredholm. We shall take A bounded throughout.

The above problem naturally comes up in several fields of physics. In radiative transfer the boundary value problem

$$\mu\frac{\partial\psi}{\partial x}(x,\mu) + \psi(x,\mu) = \tfrac{1}{2}c\int_{-1}^{1}\psi(x,\hat{\mu})d\hat{\mu}, \quad 0<x<\tau, \quad -1\leq\mu\leq 1, \tag{5.1}$$

$$\psi(0,\mu) = \varphi_+(\mu), \quad 0\leq\mu\leq 1, \tag{5.2}$$

$$\psi(\tau,\mu) = \int_0^1 \sigma(\nu\to\mu)\psi(\tau,\nu)d\nu, \quad -1\leq\mu\leq 0, \tag{5.3}$$

gives a simple model of radiative transfer in a planetary atmosphere of finite optical thickness with sunlight incident on the top ($x=0$) and reflection by the planetary surface ($x=\tau$). The operator $(Jh)(\mu)=h(-\mu)$ then describes the inversion of the direction of a beam of light. We thus have an example of the boundary value problem (1.1)–(1.3), where $R_\ell=0$ and $\varphi_-=0$. In rarefied gas dynamics, if one describes the stationary transport of gas with full account of reflection and absorption by the walls, the BGK procedure to lowest order leads to the problem

$$v\frac{\partial\psi}{\partial x}(x,v) + \psi(x,v) = \pi^{-1/2}\int_{-\infty}^{\infty}\psi(x,\hat{v})e^{-\hat{v}^2}\,d\hat{v}, \quad 0<x<\tau, \quad v\in\mathbb{R}, \tag{5.4}$$

$$\psi(0,v) = \alpha_\ell\psi(0,-v) + \varphi_+(v), \quad 0\leq v<\infty, \tag{5.5}$$

$$\psi(\tau,v) = \alpha_r\psi(\tau,-v) + \varphi_-(v), \quad -\infty<v\leq 0, \tag{5.6}$$

where the operator $(Jh)(v)=h(-v)$ describes inversion of direction, and both boundaries $x=0$ and $x=\tau$ are partially specularly reflective.

The planetary atmosphere problem (5.1)–(5.3) was studied in an abstract setting by Greenberg and van der Mee [164]. The boundary value problem with both boundaries reflective has not been analyzed before within the present framework. We shall solve both problems in the functional formulations (i) and (ii) discussed in Section 2. Throughout we assume the existence of bounded operators R_ℓ and $\hat{R}_\ell$ on $Q_+[H]$ and R_r and $\hat{R}_r$ on $Q_-[H]$ such that R_ℓ (resp. R_r) leaves invariant $Q_+[D(T)]$ (resp. $Q_-[D(T)]$) with

$$TR_\ell = \hat{R}_\ell T, \quad TR_r = \hat{R}_r T, \tag{5.7}$$

and such that, for every $h \in Q_+[D(T)]$ and $k \in Q_-[D(T)]$, we have

$$0 \leq (|T|R_\ell h,h) \leq (|T|h,h), \tag{5.8a}$$

$$0 \leq (|T|R_r k,k) \leq (|T|k,k). \tag{5.8b}$$

The former condition implies that R_ℓ (resp. R_r) extends to a positive self adjoint contraction on $Q_+[H_T]$ (resp. $Q_-[H_T]$). The physical meaning of the latter condition is that the boundaries (e.g., the planetary surface or the walls of the gas vessel) do not increase the incident current (for radiative transfer, the incident energy). A similar condition was imposed in Section III.4 for the corresponding half space problem. We assume also the existence of an inversion symmetry J, which is an arbitrary unitary and self adjoint operator on H, leaving invariant D(T) and satisfying $JT = -TJ$ and $JA = AJ$.

Let us first analyze the "unilateral" boundary value problem, where $R_\ell = 0$.

THEOREM 5.1. The boundary value problem

$$T\psi'(x) = -A\psi(x), \quad 0<x<\tau,$$

$$Q_+\psi(0) = \varphi_+,$$

$$Q_-\psi(\tau) = R_r J Q_+\psi(\tau) + \varphi_-,$$

is uniquely solvable in the functional formulations (i) and (ii).

Proof: Clearly,

$$\psi(\tau) = T_{+\tau}Q_+\psi(0) + R_{-\tau}Q_-\psi(\tau) = T_{+\tau}\varphi_+ + R_{-\tau}Q_-\psi(\tau).$$

Substituting the second boundary condition, one obtains

$$\psi(\tau) = T_{+\tau}\varphi_+ + R_{-\tau}(R_r JQ_+\psi(\tau)+\varphi_-),$$

which implies

$$(Q_+-Q_+R_{-\tau}R_r J)Q_+\psi(\tau) = T_{+\tau}\varphi_+ + Q_+R_{-\tau}\varphi_-. \tag{5.9}$$

We extend R_r to the complete Hilbert space H (or H_T) by defining

$$Rh = JR_r JQ_+h + R_r Q_-h,$$

which yields an operator commuting with J and with H_T-norm at most one. In the same way we construct $\hat{R}$ from $\hat{R}_\ell$ and $\hat{R}_r$. We now define

$$S^{(r)}_{R,\tau} = I - Q_+R_{-\tau}RJ - Q_-R_{+\tau}RJ,$$

and write (5.9) in the form

$$S^{(r)}_{R,\tau}\, Q_+\psi(\tau) = T_{+\tau}\varphi_+ + Q_+R_{-\tau}\varphi_-,$$

where we used the commutator relation $Q_\pm S^{(r)}_{R,\tau} = S^{(r)}_{R,\tau}Q_\pm$. We may, however, write

$$S^{(r)}_{R,\tau} = I + (I-[R_{+\tau}+R_{-\tau}])RJ. \tag{5.10}$$

In view of Theorem 4.1 and the estimates $\|J\|=1$ and $\|R\|\le 1$ in H_T-norm, this operator is invertible on H_T. Since $I-[R_{+\tau}+R_{-\tau}]$ is compact on H and leaves D(T) invariant, and since Eq. (4.10) holds true for $I-(R^*_{+\tau}+R^*_{-\tau})$ a compact operator on H, it is clear that $K=(I-[R_{+\tau}+R_{-\tau}])RJ$ is a compact operator on H leaving invariant D(T) and satisfying $TK=\hat{K}T$ for the compact operator $\hat{K}=(I-[R^*_{+\tau}+R^*_{-\tau}])\hat{R}J$ on H. This means that the operator in (5.10) is a compact perturbation of the identity on H, while its restriction to D(T) is a compact perturbation of the identity on D(T) when endowed with the inner product (2.8). Since this operator is invertible on H_T, we have $[\mathrm{Ker}\ S_{R,\tau}^{(r)}] \cap D(T) = \{0\}$. Using that $\mathrm{Ran}(R_{+\tau}+R_{-\tau}-I) \subset D(T)$ (see Theorem 4.3), we find $\mathrm{Ker}\ S_{R,\tau}^{(r)} \subset \mathrm{Ran}\ (I-S_{R,\tau}^{(r)}) \subset D(T)$. Hence, $S_{R,\tau}^{(r)}$ is invertible on H and its restriction to D(T) is invertible on D(T). In the functional formulations (i) and (ii) we now obtain

$$Q_+\psi(\tau) = [S_{R,\tau}^{(r)}]^{-1}(T_{+\tau}\varphi_+ + Q_+R_{-\tau}\varphi_-).$$

This in turn implies

$$Q_-\psi(\tau) = \varphi_- + R_rJ[S_{R,\tau}^{(r)}]^{-1}(T_{+\tau}\varphi_+ + Q_+R_{-\tau}\varphi_-),$$

whence,

$$\psi(0) = R_{+\tau}Q_+\psi(0) + T_{-\tau}Q_-\psi(\tau) =$$

$$= R_{+\tau}\varphi_+ + T_{-\tau}\varphi_- + T_{-\tau}R_rJ[S_{R,\tau}^{(r)}]^{-1}(T_{+\tau}\varphi_+ + Q_+R_{-\tau}\varphi_-),$$

from which we can easily construct the unique solution. ■

Let us consider the "bilateral" boundary value problem, where R_ℓ and R_r are both nonzero. We shall give a partial solution to the problem.

THEOREM 5.2. Suppose that $\|R_\ell\|_{H_T}<1$ and $\|R_r\|_{H_T}<1$. Then the boundary value problem

$$T\psi'(x) = -A\psi(x), \quad 0<x<\tau,$$

$$Q_+\psi(0) = R_\ell JQ_-\psi(0) + \varphi_+,$$

$$Q_-\psi(\tau) = R_rJQ_+\psi(\tau) + \varphi_-,$$

is uniquely solvable in the functional formulations (i) and (ii).

Proof: Straightforward calculation yields

$$\psi(0) = R_{+\tau}Q_+\psi(0) + T_{-\tau}Q_-\psi(\tau) =$$

$$= (R_{+\tau}\varphi_+ + T_{-\tau}\varphi_-) + R_{+\tau}R_\ell JQ_-\psi(0) + T_{-\tau}R_r JQ_+\psi(\tau)$$

and

$$\psi(\tau) = T_{+\tau}Q_+\psi(0) + R_{-\tau}Q_-\psi(\tau) =$$

$$= (T_{+\tau}\varphi_+ + R_{-\tau}\varphi_-) + T_{+\tau}R_\ell JQ_-\psi(0) + R_{-\tau}R_r JQ_+\psi(\tau).$$

Let us introduce the operator

$$S_{R,\tau} = I - Q_+R_{-\tau}R_rJ - Q_-R_{+\tau}R_\ell J - T_{-\tau}R_rJ - T_{+\tau}R_\ell J.$$

We may then derive the equation

$$S_{R,\tau}(Q_+\psi(\tau)+Q_-\psi(0)) = T_{+\tau}\varphi_+ + T_{-\tau}\varphi_- + Q_+R_{-\tau}\varphi_- + Q_-R_{+\tau}\varphi_+.$$

Let us rewrite $S_{R,\tau}$ by introducing $R = R_\ell \oplus R_r$. We obtain

$$S_{R,\tau} = I + \{I-[R_{+\tau}+R_{-\tau}+T_{+\tau}+T_{-\tau}]\}RJ,$$

where $\|R\|\le 1$ and $\|J\|=1$ in H_T-norm. If one only assumes that $\|R\|\le 1$ in the H_T-norm, it would suffice to show that $Z_\tau = R_{+\tau}+R_\tau+T_{+\tau}+T_{-\tau}$ is a self adjoint operator on H_T with spectrum within (0,2) in order to establish unique solvability for the first functional formulation. However, since we can confine the spectrum only within (0,2], we must assume in general that R is a strict contraction.

In view of the identity

$$Z_\tau=[I+e^{-\tau|T^{-1}A|}]PV_\tau^{-1} + (2I-\tau T^{-1}A)(I-P)V_\tau^{-1}$$

(see(4.3)-(4.6)), we may write Z_τ as the inverse of the operator

$$W_\tau=V_\tau\{[I+e^{-\tau|T^{-1}A|}]^{-1}P + (\tfrac{1}{2}I + \tfrac{1}{4}\tau T^{-1}A)(I-P)\},$$

which generalizes the operator defined in Section 2. First we observe that

$$\{Q_+ + Q_-(I-\tau T^{-1}A)\}\,\{\tfrac{1}{2}I + \tfrac{1}{4}\tau(Q_+-Q_-)T^{-1}A\}(I-P) =$$

$$= \{\tfrac{1}{2}I + \tfrac{1}{4}\tau(Q_+-Q_-)T^{-1}A\}(I-P).$$

We then find

$$W_\tau = (Q_+-Q_-)(PP_+-PP_-)\{[I+e^{-\tau|T^{-1}A|}]^{-1} - \tfrac{1}{2}I\} + \tfrac{1}{2}I + \tfrac{1}{4}\tau(Q_+-Q_-)T^{-1}A(I-P),$$

whence

$$(W_\tau h,h)_T = (|S|\{[I+e^{-\tau|S|^{-1}}]^{-1} - \tfrac{1}{2}I\}h_1,h_1)_A + \tfrac{1}{2}\|h\|_T^2 + \tfrac{1}{4}\tau(Ah_0,h_0)$$

for $h_0=(I-P)h_0$ and $h_1=Ph$, and thus $W_\tau \geq \tfrac{1}{2}I$ on H_T. We may thus conclude that Z_τ is a self adjoint operator on H_T with spectrum in $(0,2]$. However, let us notice that

$$\psi(z) = z\{[1+e^{-\tau/z}]^{-1} - \tfrac{1}{2}\}, \qquad 0<z<\infty,$$

is a nonnegative continuous function on $(0,\infty)$ satisfying $\psi(+\infty)=\tfrac{1}{4}\tau$, while $z^{-1}\psi(z)\to\tfrac{1}{2}$ as $z\downarrow 0$. We then obtain

$$(W_\tau h,h)_T \geq \tfrac{1}{2}\|h\|_T^2 + \tfrac{1}{4}\tau(Ah_0,h_0),$$

which implies $(W_\tau h,h)_T=\tfrac{1}{2}\|h\|_T^2$ if and only if $h\in\text{Ker }A$. As a result, $\text{Ker }(2I-Z_\tau)=\text{Ker }A$. We may also conclude that

$$\|I-S_{R,\tau}\|_{H_T} \leq \|I-Z_\tau\|_{H_T}\ \|R\|_{H_T}\ \|J\|_{H_T} < 1,$$

which establishes unique solvability in the first functional formulation. The modification of the existence and uniqueness results for the second functional formulation can be implemented as in Section 2. ■

For purely specular reflection, where R=I, we easily find Ker $S_{R,\tau}$ = {h ∊ Ker A : Jh=h}, and unique solvability may, in general, be violated. However, the proof of the above theorem is valid for $\|R\|\leq 1$ if we assume that A is strictly positive self adjoint.

Chapter VI

EQUIVALENCE OF DIFFERENTIAL AND INTEGRAL FORMULATION

1. Kinetic equations in integral form

Integral forms of transport equations first appeared in radiative transfer theory at the beginning of this century. If one considers radiative transfer with isotropic scattering in a layer of finite optical thickness τ, the boundary value problem may be written (cf. Section IX.1)

$$\mu\frac{\partial\psi}{\partial x}(x,\mu) + \psi(x,\mu) = \tfrac{1}{2}c\int_{-1}^{1}\psi(x,\hat{\mu})d\hat{\mu}, \qquad 0\le x\le\tau, \tag{1.1}$$

$$\psi(0,\mu) = \varphi_{+}(\mu), \quad 0\le\mu\le 1, \tag{1.2a}$$

$$\psi(\tau,\mu) = \varphi_{-}(\mu), \quad -1\le\mu\le 0. \tag{1.2b}$$

The right hand side of Eq. (1.1) is usually called the source term. If one were to consider this term as known (which, of course, it is not), Eqs. (1.1)–(1.2) would be easily solved in the form

$$\psi(x,\mu) = e^{-x/\mu}\varphi_{+}(\mu) + \tfrac{1}{2}c\int_{0}^{x}\mu^{-1}e^{-(x-y)/\mu}\chi(y)dy \tag{1.3a}$$

for $0<\mu\le 1$, and

$$\psi(x,\mu) = e^{(\tau-x)/\mu}\varphi_{-}(\mu) - \tfrac{1}{2}c\int_{x}^{\tau}\mu^{-1}e^{-(x-y)/\mu}\chi(y)dy \tag{1.3b}$$

for $-1\le\mu<0$, where

$$\chi(x) = \int_{-1}^{1}\psi(x,\hat{\mu})d\hat{\mu}. \tag{1.4}$$

Equation (1.3) may be considered as an integral equation for the unknown function

$\psi(x,\mu)$. By integrating (1.3) with respect to the angular variable μ and adding the resulting equations, a further reduction is accomplished to the scalar integral equation of convolution type

$$\chi(x) - \tfrac{1}{2}c\int_0^\tau E_1(x-y)\chi(y)dy =$$

$$= \int_0^1 e^{-x/\mu}\varphi_+(\mu)d\mu + \int_{-1}^0 e^{(\tau-x)/\mu}\varphi_-(\mu)d\mu, \tag{1.5}$$

where

$$E_1(x) = \int_0^1 \mu^{-1}e^{-|x|/\mu}d\mu$$

is the exponential integral function. As $0<c\leq 1$, Eq. (1.5) can, in principle, be solved by iteration (see [340], for instance). Since the exponential integral function has a logarithmic singularity at $x=0$ (see [89, 255]), for numerical purposes one should prefer iterating (1.3). In radiative transfer theory this is known as the method of expansion with respect to orders of multiple scattering, since the n-th iterate accounts for the contribution to the intensity $\psi(x,\mu)$ of light scattered n times (cf. [89, 357]).

If one takes in (1.4) the limit as $\tau\to\infty$, one obtains

$$\chi(x) - \tfrac{1}{2}c\int_0^\infty E_1(x-y)\chi(y)dy = \int_0^1 e^{-x/\mu}\varphi_+(\mu)d\mu, \qquad 0\leq x<\infty, \tag{1.6}$$

which is known as the Schwarzschild-Milne integral equation (cf. [89, 342]). It was first studied by Milne [265] as early as 1921. Both convolution equations (1.5) and (1.6) were extensively analyzed in the monograph of Hopf [196] and were the major stimulus which triggered the early theory of Wiener-Hopf equations (see [392]). In the early 1940's Ambarzumian found a way to express the general solution of Eqs. (1.3) in two functions only (cf. [6]). These functions were later denoted X and Y by Chandrasekhar [89] and satisfy a coupled set of nonlinear integral equations. In the case of Eq. (1.6), one deals only with one so-called H-function. This method introduced by Ambarzumian [6] was called "invariant imbedding". It was further developed by Chandrasekhar [89], Sobolev [340], Busbridge [61] and many others. The mathematical background has been treated in detail in the monograph of Wing [397].

Theorists soon realized that radiative transfer with anisotropic scattering also allows reduction to a convolution equation, but now the kernel will be a matrix function. Similar procedures were developed for inhomogeneous media [62, 339] and polarized light

([199], for instance). Integral formulations were less salient in neutron transport and rarefied gas dynamics, presumably because of the popularity of the classical Case–van Kampen method in the 1960's. In fact, integral formulations appearing in the books of Case, de Hoffmann and Placzek [69] and Davison [98] were soon replaced by "Caseology." Radiative transfer theory was relatively unaffected by this transformation in neutron transport theory.

A rigorous, fairly general formulation of kinetic equations in integral form was given by Maslennikov [259]. Indirectly, through the work of Feldman [116], as well as through the study of H–functions (cf. [58, 216]) and criticality problems (cf. [281], for example), integral formulations reappeared in neutron transport theory. A rigorous proof of the equivalence of the abstract boundary value problems of the previous chapters to a vector–valued convolution equation is due to van der Mee [359] (also [360]). It should be emphasized that most (if not all) transport theorists have worked either with the differential form, as in the previous chapters, or with the integral form, as in the present chapter, mostly taking their equivalence for granted. Among the few to give some argument for their equivalence are Kelley and Mullikin (see [216, 274], for instance), but their arguments are unidirectional: the integral form is derived from the differential form, but not conversely. We feel the necessity of having a rigorous equivalence proof, since we intend to use the equivalence as a tool to derive properties of the abstract boundary value problems from those of the integral equations. Our objective is an integrated development of both approaches in an abstract setting.

2. Preliminaries on convolution operators

The present section is devoted to some well known and some less known properties of convolution operators. Since the seminal work of Wiener and Hopf [392], a considerable part of convolution equations theory was developed during the late 1960's and early 1970's; see Gohberg and Krein [141], Feldman [117], Gohberg and Semençul [151], Gohberg and Heinig [140], and Gohberg and Leiterer [148, 149, 150]. The major tools in these approaches are classical Fredholm theory and factorization. Not all the material is easily accessible, and for our purposes it is preferable to use (modified) Laplace transforms instead of the more usual Fourier transforms. For these reasons, we include a basic summary of the properties of convolution operators in this and the next two sections.

We first consider the full line convolution equation

$$\psi(x) - \int_{-\infty}^{\infty} k(x-y)\psi(y)dy = \omega(x), \quad -\infty<x<\infty, \tag{2.1}$$

where ψ, ω are functions from the real line to some (real or complex) Banach space X and k is a function from the real axis to the Banach algebra L(X) of bounded linear operators on X. On defining vectors ψ and ω in certain spaces of functions from $\mathbb{R}$ into X as well as the convolution operator

$$(\mathcal{L}\psi)(x) = \int_{-\infty}^{\infty} k(x-y)\psi(y)dy, \tag{2.2}$$

one can restate Eq. (2.1) as the following vector equation:

$$(I-\mathcal{L})\psi = \omega. \tag{2.3}$$

In order to formulate our assumptions on the convolution kernel k and the function spaces, we use the resources of strong measurability (always with respect to Lebesgue measure, unless explicitly stated otherwise) and Bochner integration. A short account can be found in Section 31 of [401]; more modern material appears in the monographs of Mikusinski [264] and Diestel and Uhl [104]. Given a real (possibly infinite) interval (a,b) and a (real or complex) Banach space X, we denote by $L_p(X)_a^b$ the (real or complex) Banach space of all strongly measurable functions $\psi:(a,b)\rightarrow X$ which are bounded with respect to the norm

$$\|\psi\|_p = \begin{cases} [\int_a^b \|\psi(x)\|_X^p dx]^{1/p}, & 1\leq p<\infty, \\ \text{ess sup } \{\|\psi(x)\|_X : a\leq x<b\}, & p=\infty. \end{cases}$$

We let $C(X)_a^b$ denote the subspace of $L_\infty(X)_a^b$ consisting of all bounded continuous functions $\psi:(a,b)\rightarrow X$ continuous at the endpoints a, b, as well, if they are finite. The functions $\psi\in L_1(X)_a^b$ are called Bochner integrable on (a,b) and the functions $\psi\in L_\infty(X)_a^b$ essentially bounded on (a,b).

It is easily proved (cf. [141]) that for $k\in L_1(L(X))_{a-b}^{b-a}$ the operator

$$(\mathcal{L}\psi)(x) = \int_a^b k(x-y)\psi(y)dy, \quad a<x<b,$$

is bounded on the spaces $Z=L_p(X)_a^b$ $(1\leq p\leq\infty)$ and $Z=C(X)_a^b$, in all cases with norm

$$\|\mathcal{L}\|_Z \leq \int_{a-b}^{b-a} \|k(x)\|_{L(X)}dx. \tag{2.4}$$

For $L_1(X)_a^b$ and $L_\infty(X)_a^b$ this is a straightforward estimate, from which one derives the result for $L_p(X)_a^b$ $(1<p<\infty)$ using the Riesz interpolation theorem ([224], Theorem 2.4). In order to obtain the estimate for $C(X)_a^b$ we may use the following proposition.

PROPOSITION 2.1. Let $p^{-1} + q^{-1} = 1$. If $\psi \in L_p(X)_a^b$ and $k \in L_q(L(X))_{a-b}^{b-a}$, then $\mathcal{L}\psi$ is bounded and continuous on (a,b).

Proof: Extending ψ and k to $(-\infty,\infty)$ by putting $\psi(x) = 0$ and $k(y) = 0$ for $x\in(a,b)$ and $|y|\geq|b-a|$, it is evidently sufficient to prove this proposition for $a = -\infty$ and $b = \infty$. Let us first take $1\leq p<\infty$, $\psi\in L_p(X)_{-\infty}^{\infty}$ and $k\in L_q(L(X))_{-\infty}^{\infty}$. Then there exists a sequence $\{\psi_n\}_{n=1}^{\infty}$ of strongly measurable step functions such that $\lim_{x\to\infty} \|\psi-\psi_n\|_p=0$. Therefore, using

$$\|(\mathcal{L}\psi)(x)-(\mathcal{L}\psi_n)(x)\| \leq \int_{-\infty}^{\infty} \|k(x-y)\{\psi(y)-\psi_n(y)\}\|dy \leq \|k\|_q\|\psi-\psi_n\|_p,$$

one finds that ψ is bounded continuous if all functions $\mathcal{L}\psi_n$ are bounded continuous. So let us suppose that there exists a subset E of the real line of positive measure and a fixed vector $\xi\in X$ such that

$$\psi(x) = \begin{cases} \xi, & x\in E \\ 0, & x\notin E \end{cases}$$

Then for $x_1,x_2\in\mathbb{R}$ one obtains

$$(\mathcal{L}\psi)(x_1) - (\mathcal{L}\psi)(x_2) = \int_{(E-x_1)\Delta(E-x_2)} k(-z)\xi\, dz,$$

where $A\Delta B = (A\setminus B)\cup(B\setminus A)$. Since the Lebesgue measure of the symmetric difference $(E-x_1)\Delta(E-x_2)$ vanishes for $|x_1-x_2|\to0$, the Bochner integrability of k implies that $|(\mathcal{L}\psi)(x_1)-(\mathcal{L}\psi)(x_2)|\to0$ as $|x_1-x_2|\to0$. Thus $\mathcal{L}\psi$ is continuous on $(-\infty,\infty)$ and obviously bounded. We may conclude that $\mathcal{L}\psi$ is bounded continuous if $\psi \in L_p(X)_{-\infty}^{\infty}$ and $1 \le p < \infty$. If $p=\infty$, then $q=1$ and, instead, one approximates k by step functions. ■

This result is well-known, at least for $X = \mathbb{C}^1$. It implies that $\mathcal{L}$ is bounded on $C(X)_a^b$ with norm estimate (2.4), if we have $k \in L_1(L(X))_{a-b}^{b-a}$. It will play an important role in the next two sections.

We continue our investigation of convolution operators by stating the following deep result essentially due to Bochner and Phillips [45], which plays a fundamental role in convolution equations theory. We formulate an abstract generalization of their result due to Allan [4] and Gohberg and Leiterer [149].

THEOREM 2.2. Let $k \in L_1(L(X))_{-\infty}^{\infty}$. Suppose that the operator

$$W(\lambda) = I - \int_{-\infty}^{\infty} e^{x/\lambda} k(x) dx \tag{2.5}$$

is invertible for all extended imaginary λ. Then there exists $\ell \in L_1(L(X))_{-\infty}^{\infty}$ such that

$$W(\lambda)^{-1} = I + \int_{-\infty}^{\infty} e^{x/\lambda} \ell(x) dx, \quad \mathrm{Re}\ \lambda = 0. \tag{2.6}$$

Moreover, if $k(x)$ is compact operator valued almost everywhere, then the same property holds for $\ell(x)$.

COROLLARY 2.3. If $W(\lambda)$ is invertible for all extended imaginary λ, then Eq. (2.1) is uniquely solvable in all spaces $L_p(X)_{-\infty}^{\infty}$ $(1 \le p \le \infty)$ and $C(X)_{-\infty}^{\infty}$, and the solution is given by

$$\psi(x) = \omega(x) + \int_{-\infty}^{\infty} \ell(x-y) \omega(y) dy.$$

Proof: Taking integral transforms of both sides of Eq. (2.1) and using that the Laplace transform of a convolution product of two functions is the algebraic product of the Laplace transforms of their factors, we easily obtain

$$W(\lambda) \int_{-\infty}^{\infty} e^{x/\lambda} \psi(x) dx = \int_{-\infty}^{\infty} e^{x/\lambda} \omega(x) dx, \quad \mathrm{Re}\ \lambda = 0.$$

We now use the invertibility of $W(\lambda)$ for extended imaginary λ and the identity (2.6) to obtain the corollary. ∎

Additional results for convolution equations on the half line $(0,\infty)$ and on finite intervals $(0,\tau)$ will appear in the next chapter.

3. Equivalence theorems

In this section we shall establish the equivalence of boundary value problems of the previous chapters to convolution equations. First we give a detailed proof for the finite slab problem, followed by less detailed proofs for the half space problem and a related "full space" problem. We begin by proving some estimates. The results are based to a large extent on parts of [359, 360] and the Appendix of [366]. Throughout, T will be a (possibly unbounded) self adjoint operator with zero null space and B will be a bounded operator satisfying the regularity condition

$$\exists\alpha>0: \quad \operatorname{Ran} B \subset \operatorname{Ran}|T|^{\alpha} \cap D(|T|^{1+\alpha}). \tag{3.1}$$

We shall write $A = I-B$, $\sigma(.)$ for the resolution of the identity of T, and $Q_\pm$ for the orthogonal projections onto the maximal T-positive/negative T-invariant subspaces. For $Q_\pm$ we may, of course, write $Q_+ = \sigma([0,\infty))$ and $Q_- = \sigma((-\infty,0])$. In terms of T one defines the propagator function $\mathcal{H}(x)$ by

$$\mathcal{H}(x) = \begin{cases} +T^{-1}e^{-xT^{-1}}Q_+ = \int_0^\infty t^{-1}e^{-x/t}\sigma(dt), & 0<x<\infty, \\ -T^{-1}e^{-xT^{-1}}Q_- = -\int_{-\infty}^0 t^{-1}e^{-x/t}\sigma(dt), & -\infty<x<0. \end{cases}$$

It is easily seen that the restriction of $\pm\mathcal{H}(|x|)$ to Ran $Q_\pm$ is the derivative with respect to x of the semigroup whose generator is the (unbounded) inverse of the restriction of $\mp T$ to Ran $Q_\pm$.

Let us first derive two simple lemmas.

LEMMA 3.1. We have the following estimates:

$$\int_{-\infty}^{\infty} \|\mathcal{H}(x)B\| dx < \infty,$$

$$\int_{-\infty}^{\infty} \|T\mathcal{H}(x)B\| dx < \infty.$$

Proof: If ϕ is a bounded measurable function on the real line, then $\phi(T) = \int_{-\infty}^{\infty} \phi(t)\sigma(dt)$ is defined as a strong limit of Stieltjes sums, and has the norm

$$\|\phi(T)\| = \text{ess sup } \{|\phi(t)| : t \in \sigma(T)\}.$$

This equality allows us to obtain the following norm estimates:

$$\| |T|^{\alpha}\mathcal{H}(x)\| = 0(|x|^{\alpha-1}) \quad (x \to 0),$$

$$\|T\mathcal{H}(x)\| = 0(1) \quad (x \to 0),$$

$$\| |T|^{-1-\alpha}\mathcal{H}(x)\| = 0(|x|^{-2-\alpha}) \quad (x \to \pm\infty),$$

$$\| |T|^{-\alpha}\mathcal{H}(x)\| = 0(|x|^{-1-\alpha}) \quad (x \to \pm\infty).$$

Using (3.1) the lemma is immediate. ∎

LEMMA 3.2. Let $f:(0,\infty) \to H$ be bounded and continuous. Then

$$\lim_{y \to \infty} e^{(y-x)T^{-1}} Q_{-} f(y) = 0$$

in the weak sense.

Proof: Choosing arbitrary $h \in H$ and using dominated convergence, one has

$$\lim_{y \to \infty} (e^{(y-x)T^{-1}} Q_{-} f(y), h) = \lim_{y \to \infty} \int_{0}^{\infty} e^{(x-y)/t} (\sigma(-dt) f(y), h) = 0,$$

which proves the lemma. ∎

We shall now derive the finite slab and half space equivalence results in one theorem. It will be seen that the proof actually depends only on the estimates

$\int_{-\tau}^{\tau} \|\mathcal{H}(x)B\|dx < \infty$ if τ is finite, or $\int_{-\infty}^{\infty} \|\mathcal{H}(x)B\|dx < \infty$, $\int_{-\infty}^{\infty} \|T\mathcal{H}(x)B\|dx < \infty$, if τ is infinite, rather than the stronger regularity condition (3.1).

THEOREM 3.3. Let $0<\tau\leq\infty$. Assume $\omega:[0,\tau)\to H$ is continuous, and left-continuous at $x=\tau$ if $\tau<\infty$, $\omega(x)\in D(T)$ for $0<x<\tau$, and $T\omega$ strongly differentiable on $(0,\tau)$. Then a function $\psi:(0,\tau)\to D(T)$ is a solution of the boundary value problem

$$(T\psi)'(x) = -A\psi(x) + (T\omega)'(x) + \omega(x), \quad 0<x<\tau, \tag{3.2}$$

$$\lim_{x\downarrow 0} \|Q_+\psi(x)-Q_+\omega(0)\|_H = 0, \tag{3.3a}$$

and

$$\lim_{x\uparrow \tau} \|Q_-\psi(x)-Q_-\omega(\tau)\|_H = 0 \tag{3.3b}$$

for $\tau<\infty$,

$$\|\psi(x)\|_H = O(1) \quad (x\to\infty) \tag{3.3c}$$

for $\tau=\infty$, if and only if $\psi\in L_\infty(H)_0^\tau$ and ψ satisfies the convolution equation

$$\psi(x) - \int_0^\tau \mathcal{H}(x-y)B\psi(y)dy = \omega(x), \quad 0<x<\tau. \tag{3.4}$$

Any such solution is continuous on $[0,\tau]$ if $\tau<\infty$ or on $[0,\infty)$ if $\tau=\infty$.

Proof: Assume first that $\tau<\infty$. Let $\psi:(0,\tau)\to D(T)$ be a solution to Eqs. (3.2)-(3.3), and put $\chi=\psi-\omega$. If we choose $0<x<\infty$ and pick $x_1\in(0,x)$ and $x_2\in(x,\tau)$, then

$$\int_0^{x_1} \mathcal{H}(x-y)B\psi(y)dy = \int_0^{x_1} \mathcal{H}(x-y)\{(T\chi)'(y)+\chi(y)\}dy =$$
$$= [e^{-(x-y)T^{-1}}Q_+\chi(y)]_0^{x_1}$$

and

$$\int_{x_2}^{\tau} \mathcal{H}(x-y)B\psi(y)dy = \int_{x_2}^{\tau} \mathcal{H}(x-y)\{(T\chi)'(y)+\chi(y)\}dy =$$

$$= [-e^{-(x-y)T^{-1}}Q_-\chi(y)]_{x_2}^{\tau},$$

where advantage has been taken of the differentiability of the propagator function. In fact, in the strong sense one has

$$\mathcal{H}(x-y)\{(T\chi)'(y) + \chi(y)\} = \frac{\partial}{\partial y}\mathcal{H}(x-y)T\chi(y).$$

Using the first part of Lemma 3.1 and $\psi \in L_\infty(H)_0^\tau$, we see that the left hand sides have strong limits as $x_1 \uparrow x$ and $x_2 \downarrow x$. Exploiting (3.3) and the equality $\chi = \psi - \omega$ gives (3.4).

Conversely, let $\psi \in L_\infty(H)_0^\tau$ be a solution of (3.4). Because the functions $\mathcal{H}(.)B \in L_1(L(H))_{-\tau}^\tau$ (see Lemma 3.1) and $\psi \in L_\infty(H)_0^\tau$, it is immediate from Proposition 2.1 that

$$g(x) = \int_0^\tau \mathcal{H}(x-y)B\psi(y)dy$$

depends continuously on the variable $x \in [0,\tau]$. Similarly, because the functions $T\mathcal{H}(.)B \in L_1(L(H))_{-\tau}^\tau$ (see Lemma 3.1) and $\psi \in L_\infty(H)_0^\tau$, we also find that

$$\tilde{g}(x) = \int_0^\tau T\mathcal{H}(x-y)B\psi(y)dy$$

depends continuously on the variable $x \in [0,\tau]$. Moreover, from Lemma II 2.1 we see that $g(x) \in D(T)$ and $Tg(x) = \tilde{g}(x)$ for all $0 \le x \le \tau$. Repeatedly using the lemma, we obtain, for all $\varepsilon > 0$,

$$T\{g(x+\varepsilon) - g(x)\}/\varepsilon = h_1 + h_2 + h_3 + h_4,$$

where

$$h_1 = \varepsilon^{-1}[e^{-\varepsilon T^{-1}}Q_+ - Q_+]T\int_0^x \mathcal{H}(x-y)B\psi(y)dy,$$

$$h_2 = \varepsilon^{-1}[e^{\varepsilon T^{-1}}Q_- - Q_-]T\int_x^\tau \mathcal{H}(x-y)B\psi(y+\varepsilon)dy,$$

$$h_3 = \varepsilon^{-1}\int_x^{x+\varepsilon} T\mathcal{H}(x+\varepsilon-y)B\psi(y)dy,$$

$$h_4 = -\varepsilon^{-1}\int_x^{x+\varepsilon} T\mathcal{H}(x-y)B\psi(y)dy.$$

Let us take the limit as $\varepsilon\downarrow 0$. Simple semigroup theory yields

$$h_1 \to -\int_0^x \mathcal{H}(x-y)B\psi(y)dy.$$

As $\psi = g+\omega$ and both g and ω are continuous on $[0,\tau)$, the function ψ is continuous on $[0,\tau]$. Using this together with dominated convergence (cf. [401], Section 31) and the same semigroup property, one obtains

$$h_2 \to -\int_x^\tau \mathcal{H}(x-y)B\psi(y)dy.$$

The continuity of the expressions under the integral signs implies that $h_3 \to Q_+B\psi(x)$ and $h_4 \to Q_-B\psi(x)$. Thus, Tg is strongly differentiable on $(0,\tau)$ from the right, with

$$(Tg)'(x) = -g(x) + B\psi(x).$$

Strong differentiability from the left, with the same one-sided derivative, can be proved in an analogous way. Hence, Tg is differentiable on $(0,\tau)$. From (3.4) it follows that $g = \psi-\omega$. We may conclude that ψ satisfies (3.2). The boundary conditions (3.3) follow by substitution.

Consider, finally, $\tau=\infty$. The proof that any $\psi\in L_\infty(H)_0^\infty$ satisfying Eq. (3.4) is continuous on $[0,\infty)$ and is a solution of Eqs. (3.2)–(3.3a,c) can be given in precisely the same manner as for the finite interval. Conversely, if $\psi:(0,\infty)\to D(T)$ is a solution of Eqs. (3.2)–(3.3a,c) and $x\in(0,\infty)$, we pick $x_1\in(0,x)$ and $x_2,x_3\in(x,\infty)$, put $\chi = \psi-\omega$ and compute

$$\int_0^{x_1} \mathcal{H}(x-y)B\psi(y)dy = \int_0^{x_1} \mathcal{H}(x-y)\{(T\chi)'(y)+\chi(y)\}dy =$$

$$= [e^{-(x-y)T^{-1}}Q_+\chi(y)]_0^{x_1}$$

and

$$\int_{x_2}^{x_3} \mathcal{H}(x-y)B\psi(y)dy = \int_{x_2}^{x_3} \mathcal{H}(x-y)\{(T\chi)'(y)+\chi(y)\}dy =$$
$$= [-e^{-(x-y)T^{-1}}Q_-\chi(y)]_{x_2}^{x_3}.$$

Using Lemma 3.1 and the boundedness of ψ we see that the left hand sides have limits as $x_1 \uparrow x$, $x_2 \downarrow x$ and $x_3 \to \infty$. Since $Q_+\chi(y) \to 0$ as $y \downarrow 0$ and Lemma 3.2 holds true, we obtain the convolution equation (3.4). ∎

Let us consider some special choices for ω.

(i) For $h \in D(T)$ and $\tau < \infty$, we take

$$\omega(x) = e^{-xT^{-1}}Q_+h + e^{(\tau-x)T^{-1}}Q_-h. \tag{3.5}$$

The boundary value problem equivalent to (3.4) then becomes

$$(T\psi)'(x) = -A\psi(x), \quad 0<x<\tau,$$

$$\lim_{x \downarrow 0} \|Q_+\psi(x)-Q_+h\|_H = 0,$$

$$\lim_{x \uparrow \tau} \|Q_-\psi(x)-Q_-h\|_H = 0.$$

(ii) For $f:[0,\tau] \to H$ uniformly Hölder continuous and $\tau < \infty$, we take

$$\omega(x) = \int_0^\tau \mathcal{H}(x-y)\{f(y)-f(x)\}dy + [I-e^{-xT^{-1}}Q_+-e^{(\tau-x)T^{-1}}Q_-]f(x). \tag{3.6}$$

From the Hölder continuity it follows that the above integral is a well defined Bochner integral, as

$$\|\mathcal{H}(x-y)\{f(y)-f(x)\}\| \le M|x-y|^\gamma \|\mathcal{H}(x-y)\| = O(|x-y|^{\gamma-1}) \ (|x-y| \to 0)$$

Formally one could write $\omega(x) = \int_0^\tau \mathcal{H}(x-y)f(y)dy$, but the status of the latter integral is not always clear. The boundary value problem equivalent to (3.4) now has the form

$$(T\psi)'(x) = -A\psi(x) + f(x), \qquad 0<x<\tau,$$

$$\lim_{x\downarrow 0} \|Q_+\psi(x)\|_H = 0,$$

$$\lim_{x\uparrow \tau} \|Q_-\psi(x)\|_H = 0.$$

(iii) For $Q_+h \in D(T)$ and $\tau=\infty$, the choice

$$\omega(x) = e^{-xT^{-1}}Q_+h \tag{3.7}$$

leads to the usual half space problem

$$(T\psi)'(x) = -A\psi(x), \qquad 0<x<\infty,$$

$$\lim_{x\downarrow 0} \|Q_+\psi(x)-Q_+h\|_H = 0,$$

$$\|\psi(x)\|_H = O(1) \; (x\to\infty).$$

(iv) For $f:[0,\infty)\to H$ a bounded uniformly Hölder continuous function such that $\|f(t)\|_H = O(t^{-\beta})$ $(t\to\infty)$ for some $\beta>0$ and $\tau=\infty$, the choice

$$\omega(x) = \int_0^\infty \mathcal{H}(x-y)\{f(y)-f(x)\}dy + [I-e^{-xT^{-1}}Q_+]f(x) \tag{3.8}$$

leads to the equivalent half space problem

$$(T\psi)'(x) = -A\psi(x) + f(x), \qquad 0<x<\infty,$$

$$\lim_{x\downarrow 0} \|Q_+\psi(x)\|_H = 0,$$

$$\|\psi(x)\|_H = O(1) \; (x\to\infty).$$

As before, the formal integral $\int_0^\infty \mathcal{H}(x-y)f(y)dy$ may not be well-defined in the Bochner sense.

(v) A more general class of problems is obtained by adding (3.7) and (3.8), or (3.6) and the slab analog of (3.8).

Finally, let us note that if $\omega:[0,\tau)\to H$ is continuous for any $0<\tau\leq\infty$, left continuous at τ if $\tau<\infty$, and strongly differentiable on $(0,\tau)$, and if $\omega'(x)\in D(T)$ for all $0<x<\tau$, then every solution of the boundary value problem (3.2)-(3.3) is a solution of (3.4), but the converse proof fails. One may apply this "unilateral implication" to $\omega(x)$ of the form (3.6) or (3.7), where h and Q_+h may not be contained in D(T).

The next theorem is a "full line" result, which is instrumental to the developments of the next chapter. Note that the theorem does not give an equivalence result, the implication being only in one direction.

THEOREM 3.4. Let $\omega:\mathbb{R}\to H$ be bounded and continuous, except possibly for a jump discontinuity at x=0. Suppose that $\omega(x)\in D(T)$ and $T\omega$ is strongly differentiable at all $0\neq x\in\mathbb{R}$. Then every solution $\psi\in L_\infty(H)_{-\infty}^{\infty}$ of the convolution equation

$$\psi(x) - \int_{-\infty}^{\infty} \mathcal{H}(x-y)B\psi(y)dy = \omega(x), \qquad x\in\mathbb{R}, \tag{3.9}$$

is bounded and continuous on $\mathbb{R}$, except possibly for a jump discontinuity at x=0 of size

$$\psi(0^+) - \psi(0^-) = \omega(0^+) - \omega(0^-), \tag{3.10}$$

and satisfies the vector-valued differential equation

$$(T\psi)'(x) = -A\psi(x) + (T\omega)'(x) + \omega(x), \qquad 0\neq x\in\mathbb{R}. \tag{3.11}$$

Proof: Theorem 3.4 is easily derived from Theorem 3.3 and its analogue for $(-\infty,0)$, Proposition 2.1, the essential boundedness of ψ, and Lemma 3.1, by rewriting (3.9) as

$$\psi(x) - \int_0^\infty \mathcal{H}(x-y)B\psi(y)dy = \omega(x) + \int_{-\infty}^0 \mathcal{H}(x-y)B\psi(y)dy, \qquad 0<x<\infty,$$

$$\psi(x) - \int_{-\infty}^{0} \mathcal{H}(x-y)B\psi(y)dy = \omega(x) + \int_{0}^{\infty}\mathcal{H}(x-y)B\psi(y)dy, \qquad -\infty<x<0.$$

The integrals

$$\phi(x) = \pm\int_{0}^{\pm\infty}\mathcal{H}(x-y)B\psi(y)dy, \qquad \text{sign}(x) = \mp 1,$$

in the newly obtained right hand sides satisfy the equation

$$(T\phi)'(x) + \phi(x) = 0, \qquad 0\neq x\in\mathbb{R},$$

as a corollary of the principle of dominated convergence (cf. [401], Section 31). ■

Theorem 3.4 applies to right hand sides of the form

$$\omega(x) = \begin{cases} +e^{-xT^{-1}}Q_+h, & 0<x<\infty, \\ -e^{-xT^{-1}}Q_-h, & -\infty<x<0, \end{cases}$$

where $h\in D(T)$. For this case one finds $(T\omega)' + \omega \equiv 0$.

4. Reduction of dimensionality

In the previous section we have established the equivalence of certain boundary value problems to a vector-valued convolution equation. Although the convolution equation is solved over the given Hilbert space H, that is to say, its solution is represented as an element of $L_\infty(H)_0^\tau$, there seems to be a considerable excess of dimensionality in the space on which the convolution equation is defined. An example of this was already seen in radiative transfer with isotropic scattering, Eqs. (1.1)–(1.2), wherein the convolution equation (1.3) on H may be reduced to the scalar convolution equation (1.5). In this section we will indicate how the dimensionality of such problems may be reduced, in fact, to the rank of the operator B.

Let us consider a closed subspace $\mathbb{B}$ of the (real or complex) Hilbert space H, which contains Ran B^*. We adopt all notations and conventions of the third section. Let $j:\mathbb{B}\rightarrow H$ and $\pi:H\rightarrow\mathbb{B}$ be the unique operators such that πj is the identity on $\mathbb{B}$

and $j\pi$ is the orthogonal projection of H onto $\mathbb{B}$. From Ran $B^* \subset \mathbb{B}$ one derives immediately the equality

$$Bj\pi = B. \tag{4.1}$$

We also note that $\pi^* = j$.

PROPOSITION 4.1. For $0<\tau\leq\infty$, the convolution equation

$$\psi(x) - \int_0^\tau \mathcal{H}(x-y)B\psi(y)dy = \omega(x), \qquad 0<x<\tau, \tag{4.2}$$

is uniquely solvable on $L_p(H)_0^\tau$ if and only if the convolution equation

$$\chi(x) = \int_0^\tau \pi\mathcal{H}(x-y)Bj\chi(y)dy = \pi\omega(x), \qquad 0<x<\tau, \tag{4.3}$$

is uniquely solvable on $L_p(\mathbb{B})_0^\tau$. Then $\chi(x) = \pi\psi(x)$ and

$$\psi(x) = \omega(x) + \int_0^\tau \mathcal{H}(x-y)Bj\chi(y)dy. \tag{4.4}$$

Proof: Eq. (4.4) is obtained from (4.3) by defining $\chi = \pi\psi$ and utilizing (4.1). Consider the operators $W_1{:}L_p(\mathbb{B})_0^\tau \to L_p(H)_0^\tau$ and $W_2{:}L_p(H)_0^\tau \to L_p(\mathbb{B})_0^\tau$ defined by

$$(W_1\chi)(x) = \int_0^\tau \mathcal{H}(x-y)Bj\chi(y)dy$$

and

$$(W_2\chi)(x) = \pi\psi(x).$$

Then (4.2) and (4.3) can be written as

$$(I - W_1W_2)\psi = \omega,$$

$$(I - W_2W_1)\chi = \hat{\omega},$$

respectively, where $\hat{\omega} = \pi\omega$. We easily check the identities

$$(I - W_2W_1)^{-1} = I + W_2(I - W_1W_2)^{-1}W_1,$$

$$(I - W_1W_2)^{-1} = I + W_1(I - W_2W_1)^{-1}W_2,$$

which establish the equivalence of unique solvability. ■

5. Reflecting boundaries

Hitherto we have only considered integral formulations of boundary value problems which do not involve reflecting boundaries. The result was a convolution equation. This situation is quite different if reflecting boundary conditions are imposed. Let J be a signature operator anticommuting with T and commuting with B, and let R be a bounded reflection operator (see Sections III.4 and V.5). For the sake of simplicity we will take T bounded.

In Chapter III we studied the half space problem

$$(T\psi)'(x) = -A\psi(x), \quad 0<x<\infty, \tag{5.1}$$

$$Q_+\psi(0) = \varphi_+ + RJQ_-\psi(0), \tag{5.2}$$

$$\|\psi(x)\|_H = 0(1) \ (x\to\infty), \tag{5.3}$$

which accounts for reflection at the surface x=0. In Section V.5 we studied the problem in slab geometry

$$(T\psi)'(x) = -A\psi(x), \quad 0<x<\tau, \tag{5.4}$$

$$Q_+\psi(0) = \varphi_+, \tag{5.5}$$

$$Q_-\psi(\tau) = \varphi_- + JRQ_+\psi(\tau), \tag{5.6}$$

sometimes called the "abstract planetary problem." We will now indicate how each one of these may be converted to an equivalent integral equation. However, although the derivation will mimic those leading to the equivalence theorems, the resulting equation will not be of convolution type.

Consider first (5.4)–(5.6). According to Theorem 3.3 with

$$\omega(x) = e^{-xT^{-1}}\varphi_+ + e^{(\tau-x)T^{-1}}[\varphi_- + JRe^{\tau T^{-1}}\varphi_+ + JR\int_0^\tau \mathcal{H}(\tau-y)B\psi(y)dy],$$

the solution $\psi\in L_\infty(H)_0^\tau$ satisfies

$$\psi(x) - \int_0^\tau \mathcal{H}(x-y)B\psi(y)dy = e^{-xT^{-1}}Q_+\psi(0) + e^{(\tau-x)T^{-1}}Q_-\psi(\tau). \tag{5.7}$$

Utilizing (5.5)–(5.6) yields

$$\psi(x) - \int_0^\tau \mathcal{H}(x-y)B\psi(y)dy = e^{-xT^{-1}}\varphi_+ + e^{(\tau-x)T^{-1}}[\varphi_- + JRQ_+\psi(\tau)].$$

If we premultiply by Q_+ and set $x=\tau$, we obtain

$$Q_+\psi(\tau) = \int_0^\tau \mathcal{H}(\tau-y)B\psi(y)dy + e^{-\tau T^{-1}}\varphi_+.$$

Finally, combining the last two equations and rearranging terms leads to the integral equation

$$\psi(x) - \int_0^\tau [\mathcal{H}(x-y) + e^{(\tau-x)T^{-1}}JR\mathcal{H}(\tau-y)]B\psi(y)dy =$$

$$= e^{-xT^{-1}}\varphi_+ + e^{(\tau-x)T^{-1}}[\varphi_- + JRe^{-\tau T^{-1}}\varphi_+], \tag{5.8}$$

which is not of convolution type.

Conversely, if $\psi\in L_\infty(H)_0^\tau$ is a solution of (5.8), we may write (5.8) in the form (3.4), with ω given as above. Note that ω contains the solution function ψ, and satisfies the equation

$$(T\omega)'(x) + \omega(x) \equiv 0, \quad 0<x<\tau.$$

The boundary conditions (5.5)–(5.6) are easily confirmed. Hence ψ is a solution of the boundary value problem (5.4)–(5.6).

The half space problem (5.1)–(5.3) is analyzed in a precisely analogous fashion. In this case $\psi\in L_\infty(H)_0^\infty$ satisfies

$$\psi(x) - \int_0^\infty \mathcal{H}(x-y)B\psi(y)dy = e^{-xT^{-1}}[\varphi_+ + RJQ_-\psi(0)],$$

which leads to the integral equation

$$\psi(x) - \int_0^\infty [\mathcal{H}(x-y) + e^{-xT^{-1}}RJ\mathcal{H}(-y)]B\psi(y)dy = e^{-xT^{-1}}\phi_+, \tag{5.9}$$

again not of convolution type.

We remark that in both cases the argument may be modified to deal with unbounded T, with the reflection operator satisfying suitable domain requirements. We note also that the dimensionality of (5.8) and (5.9) can be lowered just as with the convolution equations in Section 4.

The bilateral slab problem

$$(T\psi)'(x) = -A\psi(x), \quad 0<x<\tau, \tag{5.10}$$

$$Q_+\psi(0) = \varphi_+ + RJQ_-\psi(0), \tag{5.11}$$

$$Q_-\psi(\tau) = \varphi_- + RJQ_+\psi(\tau), \tag{5.12}$$

formulated here with bounded T, though this restriction is easily removed, can be written in the form (5.3). Using (5.11) and (5.12), one obtains the equation

$$\mathfrak{T}_R(Q_+\psi(\tau)+Q_-\psi(0)) = e^{-\tau|T|^{-1}}\varphi + \int_0^\tau [\mathcal{H}(\tau-y)+\mathcal{H}(-y)]B\psi(y)dy, \tag{5.13}$$

where $\varphi = \varphi_+ + \varphi_-$ and $\mathfrak{T}_R$ is defined by

$$\mathfrak{T}_R h = h - e^{\mp\tau T^{-1}}RJh. \tag{5.14}$$

If $\mathfrak{T}_R$ is invertible on H, we obtain

$$\psi(x) - \int_0^\tau [\mathcal{H}(x-y)+\{e^{-xT^{-1}}Q_+ + e^{(\tau-x)T^{-1}}Q_-\}\mathfrak{T}_R^{-1}\{\mathcal{H}(\tau-y)+\mathcal{H}(-y)\}]B\psi(y)dy =$$

$$= \{e^{-xT^{-1}}Q_+ + e^{(\tau-x)T^{-1}}Q_-\}\mathfrak{T}_R^{-1}e^{-\tau|T|^{-1}}\varphi. \tag{5.15}$$

Conversely, by rearranging (5.15) and applying Theorem 3.3 we easily recover Eqs.

(5.10)–(5.12). The basic prerequisite to the equivalence of (5.15) to Eqs. (5.10)–(5.12) is the invertibility of $\mathbb{T}_R$ on D(T). This is satisfied if $TR = \hat{R}T$ with R and $\hat{R}$ leading to invertible $\mathbb{T}_R$ and $\hat{\mathbb{T}}_R$ on H. For example, if $R = \alpha I$ with $0<\alpha\leq 1$ and T is bounded, one easily estimates

$$\max \{\|I-\mathbb{T}_R\|_H, \|I-\hat{\mathbb{T}}_R\|_H\} \leq e^{-\tau \|T\|^{-1}} \|R\| < 1, \tag{5.16}$$

and invertibility is guaranteed. For unbounded T, one should require $\alpha \in (0,1)$.

6. Generalizations

We have developed an equivalence theory for boundary value problems and integral equations in a Hilbert space setting. The operator T appearing in these boundary value problems has been assumed to be self adjoint. However, there exist kinetic models for which the selfadjointness assumption is too restrictive.

One obvious generalization is to assume that T is a normal operator on a complex Hilbert space H. Such a generalization arises in a natural way from an abstract time dependent kinetic equation, namely

$$\frac{\partial \phi}{\partial t}(x,t) + \frac{\partial}{\partial x}(\hat{T}\phi)(x,t) = -A\phi(x,t), \quad 0<x<\tau, \quad t\geq 0.$$

Separation of variables, $\phi(x,t) = e^{i\omega t}\psi(x)$ for ω complex, and division by $1+i\omega$ lead to the abstract kinetic equation

$$(1+i\omega)^{-1}(\hat{T}\psi)'(x) = -[I - (1+i\omega)^{-1}B]\psi(x), \tag{6.1}$$

where we might define $T = (1+i\omega)^{-1}\hat{T}$ and note that the right hand side still represents a compact perturbation of the identity.

After some variable transformations one arrives at an equation of the form

$$\varsigma\frac{\partial \phi}{\partial x}(x,\varsigma) + \phi(x,\varsigma) = \iint_M w(\hat{\varsigma})\phi(x,\hat{\varsigma})d\hat{\xi}\,d\hat{\eta}, \quad \varsigma \in M, \tag{6.2}$$

where $\varsigma=\xi+i\eta$ (ξ,η real) belongs to a complex region M which intersects the imaginary axis at $\varsigma=0$ only. Problems of this sort related to velocity dependent neutron transport were first treated by Cercignani [78, 79], using the theory of generalized analytic functions developed by Vekua [373]. A similar treatment is contained in Section 8.1 of [211].

Equations (6.1) and (6.2) lead to abstract kinetic equations of the form

$$(T\psi)'(x) = -A\psi(x), \quad 0<x<\tau,$$

where T is an injective (bounded or unbounded) normal operator on H and A is an arbitrary compact perturbation of the identity. We shall consider a somewhat more restrictive assumption on T, satisfied in both of the examples above, namely the spectral inclusion

$$\sigma(T) \subset \Delta_\delta^+ \cup \Delta_\delta^- \cup \{0\} \tag{6.3}$$

for some $0\leq\delta<\frac{1}{2}\pi$, where the sets $\Delta_\delta^\pm$ are defined by $\Delta_\delta^+ = \{t\in\mathbb{C} : |\arg t|\leq\delta\}$ and $\Delta_\delta^- = \{t\in\mathbb{C} : |\pi-\arg t|\leq\delta\}$. For Eq. (6.1) we can, in fact, choose $\frac{1}{2}\pi>\delta> |\arctan \omega|$. We then find the norm estimates, for $0<\alpha<1$,

$$\| |T|^\alpha \mathcal{H}(x)\| \leq \sup\{|t|^{\alpha-1}|e^{-x/t}| : t\in\Delta_\delta^\pm\} = O(|x|^{\alpha-1}) \ (x\to 0),$$

$$\| |T|^{-1-\alpha} \mathcal{H}(x)\| \leq \sup\{|t|^{-2-\alpha}|e^{-x/t}| : t\in\Delta_\delta^\pm\} = O(|x|^{-2-\alpha}) \ (x\to\pm\infty),$$

where ± corresponds to the sign of x. The important step in deriving these is the estimates $|t| \leq (\cos\delta)^{-1}|\mathrm{Re}\ t|$ and $|e^{-x/t}| \leq e^{-|x|\cos\delta/|t|}$. Using (6.3) we may also extend Lemma 3.2 to injective normal T. Mimicking the proofs while exploiting Lemmas 3.1 and 3.2, we may show that the equivalence theorems 3.3 and 3.4 extend to situations where T is an injective normal operator satisfying condition (6.3).

Another type of generalization of the equivalence theorems is to a Banach space setting. Here we are facing immense problems arising from the absence of a Spectral Theorem for T. Still a Banach space generalization is warranted because of the physical nature of L_1-solutions of neutron transport and radiative transfer equations. In the remainder of this section, we shall give a general discussion of how a Banach space theory of equivalence may be constructed.

In formulating such a theory one needs proper definitions for $Q_\pm$, the propagator function $\mathcal{H}(x)$ and the semigroups associated with T. Let us assume a closed densely defined injective operator T on a (real or complex) Banach space H, whose spectrum satisfies (6.3) for some $0\le\delta<\frac{1}{2}\pi$. We assume a decomposition of H as the direct sum $H = H_+ \oplus H_-$ of two closed subspaces $H_\pm$, which satisfy the following hypotheses:

(A.1) $H_\pm \cap D(T)$ is dense in $H_\pm$ and $T[H_\pm\cap D(T)]\subset H_\pm$.

(A.2) The spectrum of the restriction $T_\pm$ of T to $H_\pm$ is the set $\sigma(T)\cap\Delta_\delta^\pm$, which is contained in the right/left half plane.

(A.3) The operators $\mp T^{-1}$ generate (bounded) analytic semigroups $\{U_\pm(x)\}_{x\ge 0}$ on $H_\pm$.

(A.4) One has $\lim_{x\to\infty} U_\pm(x) = 0$ in the strong operator topology.

(A.1) and (A.2) solve the problem of defining $Q_\pm$. Indeed, one takes $Q_\pm$ as the projection of H onto $H_\pm$ along $H_\mp$. The hypothesis (A.3) solves a number of problems. The propagator function we define by

$$\mathcal{H}(x)h = \begin{cases} -\frac{d}{dx}U_+(x)h, & 0<x<\infty, \\ +\frac{d}{dx}U_-(-x)h, & -\infty<x<0. \end{cases}$$

Further, we pick $0<\gamma<\frac{1}{2}\pi$, called the "opening angle", such that both semigroups are analytic in the cone $S_\gamma = \{z\in\mathbb{C} : |\arg z|\le\gamma\}$ and strongly continuous on its closure. Following the treatment of Krasnoselskii et al. ([224], Chapters 13 and 14), we obtain

$$\|T^{-n}U_\pm(x)Q_\pm\| \le \frac{c(\gamma)\; n!}{(\sin\gamma)^n\; x^n} \tag{6.4}$$

for n=0,1,2,..., and $0\le x<\infty$ (cf. [224], Eq. (13.64) with $\omega=0$). We may define "negative" fractional powers of the generator $(\pm T_\pm)^{-1}$ using the procedure of Section 14.2 of [224]. In terms of these we then define

$$|T|^\alpha h = [(-T_+)^{-1}]^{-\alpha}Q_+h + [(T_-)^{-1}]^{-\alpha}Q_-h, \quad \alpha\ge 0.$$

According to Theorem 14.2 of [224] we have the estimates

$$\|[(\pm T_{\pm})^{-1}]^{-\alpha}h\| \leq k(\alpha,\beta)\|[(\pm T_{\pm})^{-1}]^{-\beta}h\|^{\alpha/\beta}\|h\|^{1-\alpha/\beta},$$

where $h\in D([(\pm T_{\pm})^{-1}]^{-\beta})\subset \mathrm{Ran}Q_{\pm}$ and $0<\alpha<\beta$. Using the previous equation we obtain

$$\|\,|T|^{\alpha}h\| \leq k(\alpha,\beta)\|\,|T|^{\beta}h\|^{\alpha/\beta}\|h\|^{1-\alpha/\beta}, \tag{6.5}$$

where $h\in D(|T|^{\beta})$ and $0<\alpha<\beta$. Inequalities (6.4) and (6.5) imply

$$\|\,|T|^{-n}\mathcal{H}(x)\| \leq \frac{c(\gamma)}{(\sin\gamma)^{n+1}}\frac{\Gamma(n)}{|x|^{n+1}}, \quad n\geq 0, \tag{6.6}$$

where we have exploited the convexity of log $\Gamma(z)$ for $z\in[0,\infty)$ (cf. [114]; differentiate Eq. 1.7(6)). The estimates in the proof of Lemma 3.1 now follow immediately.

Finally, condition (A.4) leads to Lemma 3.2. Repeating the proofs of Section 3, one finds that the equivalence theorems, Theorem 3.3 and Theorem 3.4, extend to Banach space settings if T satisfies assumptions (A.1) to (A.4), and, in addition, T and B satisfy the regularity condition (3.1).

We believe that an equivalence theory as indicated in this section can be the starting point for a Banach space treatment of stationary boundary value problems in abstract kinetic theory. In fact, in Chapters VII and VIII we shall generalize most of our Hilbert space results to the Banach space setting. One expects that there is a complete generalization of reduction of dimension (Section 4) to Banach spaces, by requiring $\mathbb{B} \supset \mathbb{D}$ where $\mathbb{D}\oplus\mathrm{Ker}\ B = H$, whence the imbedding $j:\mathbb{B}\to H$ and the projection $\pi:H\to\mathbb{B}$ satisfy (4.1). The inclusion of reflective boundary conditions (Section 5) would at first seem to be restricted by the need for the semigroups $\{U_{\pm}(x)\}_{x\geq 0}$ to be contractive, in order to obtain the invertibility of the operator $\mathbb{T}_R$ in (5.14). However, one may renorm the Banach space and recover a contraction semigroup, thus obtaining the invertibility of $\mathbb{T}_R$ for $R=\alpha I$ under the same conditions as for self adjoint T.

Chapter VII

SEMIGROUP FACTORIZATION AND RECONSTRUCTION

1. Convolution operators on the half line

In this chapter we shall continue our study of the theory of convolution equations and its applications to abstract kinetic equations. In the first section we will outline the classical method for solving Wiener–Hopf equations on a half line. This will reduce the half space problem to a factorization problem. In the second section we shall study the connection between the semigroups developed in Chapters II and III and the solution of convolution equations corresponding to abstract kinetic equations. In this way we will obtain an alternative way of defining these projections and semigroups. In the following section we will begin a study of explicit representations of the Wiener–Hopf factors of the symbol, which is important in the derivation of representations for the solutions of the half space and the finite slab problem, to be discussed in the next chapter. In the fourth section, we present some recent results on the treatment of nonregular collision operators. Finally, in the last section, we outline the extension of the previous theory to a Banach space setting.

Since its origin in the work of Wiener and Hopf [392], the theory of convolution equations on a half line has been studied extensively. Important contributions were made by Gohberg and Krein [141]. Their finite dimensional results were extended to infinite dimension by Feldman [117]. The corresponding factorization theory in an infinite dimensional setting was developed by Gohberg and Leiterer [138, 148, 149, 150].

Recall the vector valued Wiener–Hopf equation

$$\psi(x) - \int_0^\infty k(x-y)\psi(y)dy = \omega(x), \qquad 0<x<\infty, \tag{1.1}$$

on the (real or complex) Banach space X, introduced in Section VI.2. Assuming $k \in L_1(L(X))_{-\infty}^{\infty}$ we strive for solutions $\psi \in L_p(X)_0^\infty$ for given $\omega \in L_p(X)_0^\infty$. On defining $\omega(x) = 0$ and $\psi(x) = \int_0^\infty k(x-y)\psi(y)dy$ for $-\infty<x<0$, the equation itself can be extended to the full real line. Let us define the Laplace transforms

$$\hat{\psi}_{\pm}(\lambda) = \pm\int_0^{\pm\infty} e^{x/\lambda}\psi(x)dx, \tag{1.2a}$$

$$\hat{\omega}_{+}(\lambda) = \int_0^{\infty} e^{x/\lambda}\omega(x)dx \tag{1.2b}$$

and

$$W(\lambda) = I - \int_{-\infty}^{\infty} e^{x/\lambda}k(x)dx. \tag{1.3}$$

The operator function in (1.3) is continuous on the extended imaginary line (with $W(0) = I$) and is called the **symbol** of Eq. (1.1). This equation now transforms into the equation

$$W(\lambda)\hat{\psi}_{+}(\lambda) + \hat{\psi}_{-}(\lambda) = \hat{\omega}_{+}(\lambda), \quad \text{Re } \lambda=0. \tag{1.4}$$

Let us consider the case $p=1$. Then $\hat{\psi}_{+}$ and $\hat{\omega}_{+}$ are continuous on the extended imaginary line and extend continuously to analytic functions on the open left half plane. The same applies to $\hat{\psi}_{-}$ with regard to the right half plane.

The factorization properties of $W(\lambda)$ play a crucial role in solving Eq. (1.1), as we will see shortly. By a **right Wiener–Hopf factorization** of the operator function W we mean a factorization of the form

$$W(\lambda) = W_{-}(\lambda)[P_0 + \sum_{i=1}^{n} \left(\frac{1+\lambda}{1-\lambda}\right)^{\kappa_i} P_i]W_{+}(\lambda) \tag{1.5}$$

for Re $\lambda=0$, where

(i) $W_{\pm}$ satisfies

$$W_{\pm}(\lambda) = I \pm \int_0^{\pm\infty} e^{x/\lambda}k_{\pm}(x)dx$$

for $k_{+} \in L_1(L(X))_0^{\infty}$ and $k_{-} \in L_1(L(X))_{-\infty}^{0}$;

(ii) $W_{\pm}(\lambda)$ is invertible for all λ in the extended closed left/right half plane;

(iii) $P_1,\dots,P_n$ are mutually disjoint one-dimensional projections;

(iv) $P_0 + P_1 + \dots + P_n = I$.

The integers $\kappa_1,\ldots,\kappa_n$ are uniquely determined by the function W (see [138]) and are called the **right indices** of W. The middle factor $D(\lambda)$ we call the **diagonal function**. Using Theorem VI 2.3 and property (ii) we obtain the existence of functions $\ell_+ \in L_1(L(X))_0^\infty$ and $\ell_- \in L_1(L(X))_{-\infty}^0$ such that

$$\text{(v)} \quad W_\pm(\lambda)^{-1} = I \pm \int_0^{\pm\infty} e^{x/\lambda} \ell_\pm(x)dx.$$

The same theorem implies that $\ell_\pm(x)$ is almost everywhere compact operator valued if $k_\pm(x)$ is. If all right indices vanish, i.e., if $D(\lambda)\equiv I$, then (1.5) has the form $W(\lambda) = W_-(\lambda)W_+(\lambda)$ for $\text{Re}\ \lambda=0$. Such a factorization is called **right canonical**. Assuming $W_\pm(0) = I$ one can determine the factors $W_\pm$ uniquely. Alternatively one could consider a **left Wiener-Hopf factorization** of the form

$$W(\lambda) = W_+(\lambda)[P_0 + \sum_{j=1}^m \left(\frac{1+\lambda}{1-\lambda}\right)^{\rho_j} P_j]W_-(\lambda)$$

for $\text{Re}\ \lambda=0$, which is related to the half line convolution equation

$$\psi(x) - \int_{-\infty}^0 k(x-y)\psi(y)dy = \omega(x), \quad -\infty<x<0. \tag{1.6}$$

Here $W_\pm(\lambda)$ and $P_0,P_1,\ldots,P_m$ are chosen as in the previous conditions (i) to (iv) and $\rho_1,\ldots,\rho_m$ are the **left indices**, which can differ in number and value from the right indices. If these indices all vanish, one has a **left canonical factorization**.

Let us introduce the notation

$$(\mathcal{L}_\pm\psi)(x) = \pm \int_0^{\pm\infty} k(x-y)\psi(y)dy, \quad 0\le\pm x<\infty.$$

Then (1.1) may be written $(I-\mathcal{L}_+)\psi = \omega$, and (1.6) may be written $(I-\mathcal{L}_-)\psi = \omega$.

We now proceed to a solution of (1.1). If we substitute (1.5) in (1.4), we obtain the vector Riemann-Hilbert problem

$$D(\lambda)W_+(\lambda)\hat{\psi}_+(\lambda) + W_-(\lambda)^{-1}\hat{\psi}_-(\lambda) = W_-(\lambda)^{-1}\hat{\omega}_+(\lambda) \tag{1.7}$$

for $\text{Re}\ \lambda=0$. With a bit of algebra, we may derive

$$W_-(\lambda)^{-1}\hat{\omega}_+(\lambda) = \hat{f}_+(\lambda) + \hat{f}_-(\lambda)$$

for Re $\lambda=0$ and functions

$$\hat{f}_+(\lambda) = \int_0^\infty e^{x/\lambda}[\omega(x) + \int_x^\infty \ell_-(x-y)\omega(y)dy]\, dx,$$

$$\hat{f}_-(\lambda) = \int_{-\infty}^0 e^{x/\lambda}[\int_0^\infty \ell_-(x-y)\omega(y)dy]\, dx.$$

On premultiplying (1.7) by P_0 we obtain

$$P_0W_+(\lambda)\hat{\psi}_+(\lambda) - P_0\hat{f}_+(\lambda) = P_0\hat{f}_-(\lambda) - P_0W_-(\lambda)^{-1}\hat{\psi}_-(\lambda).$$

The left hand side is continuous on the closed left half plane, analytic on the open left half plane and zero for $\lambda=0$. The right hand side is continuous on the closed right half plane, analytic on the open right half plane and zero at $\lambda=0$. The Schwarz reflection principle then implies that both sides represent an entire vector function that is bounded and vanishes at $\lambda=0$. Using Liouville's theorem, both sides must vanish identically, which proves

$$P_0W_+(\lambda)\hat{\psi}_+(\lambda) = P_0\hat{f}_+(\lambda),$$

$$P_0W_-(\lambda)^{-1}\hat{\psi}_-(\lambda) = P_0\hat{f}_-(\lambda).$$

Similarly, premultiplication of (1.7) by P_i with $\kappa_i>0$ and the Liouville argument yield

$$P_iW_+(\lambda)\hat{\psi}_+(\lambda) = (\frac{1-\lambda}{1+\lambda})^{\kappa_i}P_i\hat{f}_+(\lambda),$$

$$P_iW_-(\lambda)^{-1}\hat{\psi}_-(\lambda) = P_i\hat{f}_-(\lambda),$$

provided $P_i\hat{f}_+(\lambda)$ has a zero at $\lambda=-1$ of order at least κ_i. If $\hat{f}_+$ does not have such a zero, the Riemann–Hilbert problem obtained from (1.7) by premultiplication by P_i does not have solutions at all. For $\kappa_i<0$ a premultiplication by P_i and the above Liouville argument give

$$P_iW_+(\lambda)\hat{\psi}_+(\lambda) = (\frac{1-\lambda}{1+\lambda})^{\kappa_i}P_i\hat{f}_+(\lambda) + (1-\lambda)^{\kappa_i}\phi_i(\lambda)h_i$$

$$P_iW_-(\lambda)^{-1}\hat{\psi}_-(\lambda) = P_i\hat{f}_-(\lambda) - \phi(\lambda)(1-\lambda)^{\kappa_i}h_i,$$

where ϕ_i is an arbitrary scalar polynomial of degree equal or less than (κ_i-1) and h_i is a fixed vector in Ran P_i. As we shall see, the arbitrariness in ϕ_i will lead to nonuniqueness in the solution of the convolution equation.

It is now possible to separate the terms of (1.7) into those analytic in the left half plane and those analytic in the right half plane. Again using a Liouville argument we obtain

$$\hat{\psi}_+(\lambda) = W_+(\lambda)^{-1}[D(\lambda)^{-1}\hat{f}_+(\lambda) + \sum_{\kappa_i<0} (1-\lambda)^{\kappa_i}\phi_i(\lambda)h_i], \tag{1.8a}$$

$$\hat{\psi}_-(\lambda) = W_-(\lambda)[\hat{f}_-(\lambda) - \sum_{\kappa_i<0} (1-\lambda)^{\kappa_i}\phi_i(\lambda)h_i], \tag{1.8b}$$

which represents the Laplace transform of the solution of the convolution equation (1.1), provided a set of d independent linear constraints on ω is satisfied, where d is the sum of the positive indices κ_i. There are n independent arbitrary constants in this solution, where $(-n)$ is the sum of the negative indices κ_i. The same results can be found in $L_p(X)_0^\infty$.

The next theorem summarizes our findings.

THEOREM 1.1. Let the symbol W have a right Wiener–Hopf factorization with indices $\kappa_1,\dots,\kappa_n$. Then on all spaces $L_p(X)_0^\infty$ $(1\le p\le\infty)$ and $C(X)_0^\infty$ the operator $I - \mathcal{L}_+$ is a Fredholm operator with characteristics

$$\dim \operatorname{Ker} (I-\mathcal{L}_+) = -\sum_{\kappa_i<0} \kappa_i ,$$

$$\operatorname{codim} \operatorname{Ran} (I-\mathcal{L}_+) = +\sum_{\kappa_i>0} \kappa_i .$$

The convolution equation (1.1) has a unique solution for all $\omega \in L_p(X)_0^\infty$ if and only if $\kappa_i=0$ for all i.

Theorem 1.1 implies that W has a right canonical factorization if and only if $I-\mathcal{L}_+$ is invertible on some (and hence all) spaces $L_p(X)_0^\infty$ or $C(X)_0^\infty$. From (1.8) and property (v) one finds

$$[(I - \mathcal{L}_+)^{-1}\omega](x) = \omega(x) + \int_0^\infty \gamma(x,y)\omega(y)dy, \qquad 0\leq x<\infty, \tag{1.9}$$

where the **resolvent kernel** $\gamma(x,y)$ is given by

$$\gamma(x,y) = \begin{cases} \ell_-(x-y) + \int_0^x \ell_+(x-z)\ell_-(z-y)dz, & y>x\geq 0 \\ \ell_+(x-y) + \int_0^y \ell_+(x-z)\ell_-(z-y)dz, & x>y\geq 0. \end{cases}$$

Indeed, if the symbol has a canonical factorization, we obtain for the Laplace transform of the solution to the Riemann–Hilbert problem (1.4)

$$\hat{\psi}_+(\lambda) = W_+(\lambda)^{-1}\hat{f}_+(\lambda)$$

and

$$\hat{\psi}_-(\lambda) = W_-(\lambda)\hat{f}_-(\lambda).$$

These formulas are special cases of (1.8). Exploiting the expressions for $\hat{f}_\pm(\lambda)$ obtained before, part (v) of the definition of a Wiener–Hopf factorization and the fact that the product of Laplace transforms of two functions is the Laplace transform of the convolution product, we find the above expression for the resolvent kernel after some algebra.

In an analogous way one proves the following theorem.

THEOREM 1.2. Let the symbol W have a left Wiener–Hopf factorization with indices $\rho_1,\ldots,\rho_m$. Then on all spaces $L_p(X)^0_{-\infty}$ $(1\leq p\leq\infty)$ and $C(X)^0_{-\infty}$, the operator $I-\mathcal{L}_-$ is a Fredholm operator with characteristics

$$\dim \operatorname{Ker} (I - \mathcal{L}_-) = \sum_{\rho_j>0} \rho_j,$$

$$\operatorname{codim} \operatorname{Ran} (I - \mathcal{L}_-) = - \sum_{\rho_j<0} \rho_j.$$

The convolution equation (1.6) has a unique solution for all $\omega\in L_p(X)^0_{-\infty}$ if and only if $\rho_j=0$ for all j.

The above method for solving Wiener–Hopf equations depends critically on the existence of Wiener–Hopf factorizations of its symbol. In the remainder of this section we will be concerned with sufficient conditions for the existence of Wiener–Hopf and canonical factorizations. Algorithms for the explicit construction of the factors are deferred to later sections. The first result presented is due to Gohberg and Leiterer ([150], Theorems 4.3 or 4.4).

THEOREM 1.3. If $W(\lambda)$ is invertible for all extended imaginary λ and has the form (1.3), where $k\in L_1(L(X))_{-\infty}^{\infty}$, and almost every $k(x)$ is uniformly approximable by operators of finite rank, then W has a right and a left Wiener–Hopf factorization. Moreover,

$$\sum_{i=1}^{n} \kappa_i = \sum_{j=1}^{m} \rho_i. \tag{1.10}$$

The last identity seems to be known, but a proof is difficult to find in the literature. However, if we naturally identify $L_p(X)_{-\infty}^{0}\oplus L_p(X)_0^{\infty} \simeq L_p(X)_{-\infty}^{\infty}$ and define $\mathcal{L}$ by

$$(\mathcal{L}\psi)(x) = \int_{-\infty}^{\infty} k(x-y)\psi(y)dy,$$

then

$$[(\mathcal{L} - (\mathcal{L}_-\oplus \mathcal{L}_+))\psi](x) = \begin{cases} \int_{-\infty}^{0} k(x-y)\psi(y)dy, & 0<x<\infty, \\ \int_{0}^{\infty} k(x-y)\psi(y)dy, & -\infty<x<0. \end{cases}$$

By the compactness of $k(x)$ for almost every x and the data $k\in L_1(L(X))_{-\infty}^{\infty}$, we have the compactness of the difference between $I-\mathcal{L}$ and the direct sum of $I-\mathcal{L}_-$ and $I-\mathcal{L}_+$. But Theorem VI 2.3 implies the invertibility of $I-\mathcal{L}$. Thus the sum of the Fredholm indices of $I-\mathcal{L}_-$ and $I-\mathcal{L}_+$ vanishes. Equation (1.10) is immediate from Theorems 1.1 and 1.2.

The next result is also due to Gohberg and Leiterer [149].

THEOREM 1.4. Let X be a Hilbert space, and let

$$\sup_{\mathrm{Re}\,\lambda=0} \|W(\lambda) - I\|_{L(X)} < 1$$

for functions W of the form (1.3). Then W has a right and a left canonical factorization.

This result is incorrect in a general Banach space setting (see [247] for counterexamples).

2. Semigroup reconstruction

In the study of existence and uniqueness properties of differential equations, one frequently obtains semigroups as operators which map the initial data into the solution. In this section we will, in fact, achieve the same result for abstract kinetic equations in integral form. In what follows we shall assume that T is a (possibly unbounded) injective self adjoint operator and B a (not necessarily self adjoint) compact operator, both defined on the (real or complex) Hilbert space H. We shall also assume the existence of $0<\alpha<1$ such that

$$\mathrm{Ran}\ B \subset \mathrm{Ran}\ |T|^{\alpha} \cap D(|T|^{1+\alpha}). \tag{2.1}$$

In Sections 4 and 5 we will substantially relax these hypotheses.

We define $Q_\pm$ as the orthogonal projections of H onto the maximal T-positive/negative T-invariant subspaces and the propagator function $\mathcal{H}(x)$ by

$$\mathcal{H}(x) = \begin{cases} +\ T^{-1} e^{-xT^{-1}} Q_+, & 0<x<\infty, \\ -\ T^{-1} e^{-xT^{-1}} Q_-, & -\infty<x<0 . \end{cases}$$

With these assumptions the full equivalence theory of Section VI.3 will be available to us. Throughout we put $A = I-B$.

If B is self adjoint and A strictly positive, the operator $T^{-1}A$ is similar to a self adjoint operator (see Chapter II) and suitable invariant subspaces are readily found on which either $-T^{-1}A$ or $+T^{-1}A$ generates a (bounded) semigroup. Such semigroups are of the utmost importance in representing solutions of various abstract boundary value problems (cf. Chapters II to V). We strive for the development of a theory for non-self adjoint B, but for such general cases it is not clear how to obtain suitable invariant subspaces of $T^{-1}A$ and corresponding semigroups. However, we shall draw on one of our few leads, namely the integral form of the boundary value problem (cf. [364, 366]).

Considering the identity

$$(T-\lambda)^{-1}(T-\lambda A) - I = -\lambda(\lambda-T)^{-1}B, \quad \lambda \notin \sigma(T),$$

we see that outside $\sigma(T)$ the function $W(\lambda) = (T-\lambda)^{-1}(T-\lambda A)$ is a compact perturbation of the identity. If we now define the spectrum of W as

$$\Sigma(W) = \{\lambda \notin \sigma(T) : W(\lambda) \text{ is not invertible}\},$$

then a well-known stability result (see, for instance, [137, 152, 286, 319, 338] for this result at various stages of its development) implies that $\Sigma(W)$ is a discrete set in the resolvent set of T such that, at each $\lambda_0 \in \Sigma(W)$,

$$W(\lambda)^{-1} = \sum_{n=-p}^{\infty} (\lambda-\lambda_0)^n W_n, \quad 0<|\lambda-\lambda_0|<\varepsilon(\lambda_0),$$

for certain operators $W_{-1},\dots,W_{-p}$ of finite rank and some Fredholm operator W_0. This in turn implies that the spectrum of $T^{-1}A$ outside $\sigma(T^{-1})$ exists solely of isolated eigenvalues of finite geometric multiplicity.

We can, in fact, rewrite W as

$$W(\lambda) = I - \int_{-\infty}^{\infty} e^{x/\lambda} \mathcal{H}(x)B \, dx, \quad \text{Re } \lambda=0.$$

Thus W turns out to be the symbol of a convolution equation on H with kernel $k(x)=\mathcal{H}(x)B$. This kernel is compact operator valued and belongs to $L_1(L(H))_{-\infty}^{\infty}$ (cf. Lemma VI 3.1). The theory of the previous section will therefore apply if $W(\lambda)$ is invertible for all extended imaginary λ (See Theorem 1.3; note that B is uniformly approximable by operators of finite rank, since it is a compact operator on a Hilbert space). In a natural way we arrive at a distinction between the **regular case**, where $T^{-1}A$ does not have purely imaginary or zero eigenvalues, and the **singular case**, where $T^{-1}A$ has at least one imaginary or zero eigenvalue. In the regular case $W(\lambda)$ is invertible for all extended imaginary λ; this is not true in the singular case.

Let us first consider the regular case. Then A is invertible and $A^{-1}T$ does not have imaginary eigenvalues. Consider the full line convolution equation

$$\psi(x) - \int_{-\infty}^{\infty} \mathcal{H}(x-y)B\psi(y)dy = \omega(x), \qquad -\infty<x<\infty. \tag{2.2}$$

According to Theorem VI 2.3 this equation is uniquely solvable in $L_p(H)_{-\infty}^{\infty}$ for every $\omega \in L_p(H)_{-\infty}^{\infty}$ $(1\leq p\leq\infty)$, and its unique solution is given by

$$\psi(x) = \omega(x) + \int_{-\infty}^{\infty} \ell(x-y)\omega(y)dy, \qquad -\infty<x<\infty, \tag{2.3}$$

for certain compact operator valued functions $\ell \in L_1(L(H))_{-\infty}^{\infty}$. Given $h \in H$, define

$$\omega_h(x) = \begin{cases} +\ e^{-xT^{-1}}Q_+h\,, & 0<x<\infty\,, \\ -\ e^{-xT^{-1}}Q_-h\,, & -0<x<0\,. \end{cases}$$

Then ω_h is bounded and continuous except for a jump at $x=0$ of size

$$\omega_h(0^+) - \omega_h(0^-) = Q_+h + Q_-h = h,$$

and the unique solution ψ_h of Eq. (2.2) with right hand side $\omega=\omega_h$ is bounded and continuous except for a jump at $x=0$ of size

$$\psi_h(0^+) - \psi_h(0^-) = \omega_h(0^+) - \omega_h(0^-) = h. \tag{2.4}$$

We now define the linear operators

$$V_+(x)h = \psi_h(x), \tag{2.5a}$$

$$V_-(x)h = -\psi_h(-x), \tag{2.5b}$$

where for $x=0$ we may write

$$P_\pm h = V_\pm(0)h = \pm\psi_h(0^\pm). \tag{2.6}$$

We may derive the following properties of these operators.

LEMMA 2.1. Suppose $T^{-1}A$ does not have zero or imaginary eigenvalues. Then the following statements hold true:

(i) For all $0 \le x < \infty$ the operator $V_\pm(x)$ is bounded.

(ii) The operator $V_\pm(x)$ depends continuously on $0 \le x < \infty$ in the strong operator topology and $\|V_\pm(x)\| \le M < \infty$ for some constant M and all $0 \le x < \infty$.

(iii) The differences $V_\pm(x) - \exp\{-xT^{-1}\}Q_\pm$, for $0 \le x < \infty$, and $P_\pm - Q_\pm$ are compact operators.

(iv) We have the semigroup properties $V_\pm(x)V_\pm(y) = V_\pm(x+y)$ and $V_\pm(x)V_\mp(y) = V_\mp(x)V_\pm(y) = 0$, for $0 \le x,y < \infty$.

(v) The operators $P_\pm$ are complementary projections.

(vi) The operators $T^{-1}A|_{\mathrm{Ran}\,P_\pm}$ have their spectra in the sets $\{\lambda \in \mathbb{C} : \pm\mathrm{Re}\lambda > 0\} \cup \{0\}$.

(vii) The semigroups $\{V_\pm(x)|_{\mathrm{Ran}\,P_\pm}\}$ are generated by $\mp T^{-1}A|_{\mathrm{Ran}\,P_\pm}$.

Proof: The boundedness statement (i) follows from the estimates

$$\|V_\pm(x)h\| \le \|\psi_h\|_{L_\infty(H)_{-\infty}^{\infty}} \le \|(I-\mathcal{L})^{-1}\| \; \|\omega_h\|_{L_\infty(H)_{-\infty}^{\infty}} \le c\|h\|,$$

where $\mathcal{L}$ is the full line convolution operator, and for (ii) it is only necessary to notice that ψ_h is bounded continuous except for a possible jump at $x=0$. For (iii) we observe that

$$V_+(x) - \exp\{-xT^{-1}\}Q_+ = \int_0^\infty \ell(x-y)\exp\{-yT^{-1}\}Q_+\,dy$$

is compact, since $\ell \in L_1(L(H))_{-\infty}^{\infty}$ is compact operator valued and the operator function $\exp\{-yT^{-1}\}Q_+ \in L_\infty(L(H))_0^\infty$.

To demonstrate the semigroup property, choose $0 \le y < \infty$ and define

$$\psi(x) = \begin{cases} \psi_h(x+y), & 0 \le x < \infty, \\ 0, & -\infty < x < 0. \end{cases}$$

Then ψ satisfies Eq. (2.2) with some right hand side ω. Let us compute ω. For $x<0$ we get

$$\omega(x) = -\int_0^\infty \mathcal{H}(x-z)B\psi_h(z+y)dz = -\int_y^\infty \mathcal{H}(x+y-w)B\psi_h(w)dw =$$

$$= -e^{-xT^{-1}}Q_-\int_{-\infty}^\infty \mathcal{H}(y-w)B\psi_h(w)dw = -e^{-xT^{-1}}Q_-V_+(y)h = \omega_{V_+(y)h}(x),$$

whereas $x>0$ leads to

$$\omega(x) = \psi_h(x+y) - \int_y^\infty \mathcal{H}(x+y-w)B\psi_h(w)dw =$$

$$= \{\psi_h(x+y) - \int_{-\infty}^\infty \mathcal{H}(x+y-z)B\psi_h(z)dz\} + \int_{-\infty}^y \mathcal{H}(x+y-z)B\psi_h(z)dz =$$

$$= \omega_h(x+y) - e^{-xT^{-1}}Q_+\int_{-\infty}^\infty \mathcal{H}(y-z)B\psi_h(z)dz = e^{-xT^{-1}}Q_+V_+(y)h =$$

$$= \omega_{V_+(y)h}(x).$$

This means that $\psi=\psi_{V_+(y)h}$. Hence, for $0\le x,y<\infty$, $V_+(x)V_+(y) = V_+(x+y)$ and $V_-(x)V_+(y) = 0$. Note that the semigroup property implies that $P_\pm$ are disjoint projections (let $x=y=0$). This, along with Eqs. (2.4) and (2.6), gives (v).

For $0<y<\infty$ and $h\in D(T)\cap \text{Ran } P_\pm$ we have, in view of Theorem VI 3.4,

$$T\frac{V_\pm(y)-I}{y}h = \mp A\frac{1}{y}\int_0^y \psi_h(\pm z)dz \to \mp A\psi_h(0^\pm)$$

as $y\to 0$. Thus if $G_\pm$ denote the infinitesimal generators of the semigroups $\{V_\pm(x)|_{\text{Ran } P_\pm}\}_{x\ge 0}$, then $A^{-1}TG_\pm P_\pm h = \mp P_\pm h$ on $D(G_\pm)$, and therefore

$$G_\pm \subset \mp T^{-1}A|_{\text{Ran } P_\pm}. \tag{2.7}$$

Conversely, if $\lambda\ne 0$ is imaginary and $h\in D(T)$, then a simple application of the identity $W(\lambda) = (T-\lambda)^{-1}(T-\lambda A)$ gives

$$\int_{-\infty}^\infty e^{x/\lambda}\psi_h(x)dx = W(\lambda)^{-1}\int_{-\infty}^\infty e^{x/\lambda}\omega_h(x)dx = \lambda A^{-1}T(\lambda-A^{-1}T)^{-1}h.$$

The linear operator which maps h into $T^{-1}A\int_{-\infty}^\infty e^{x/\lambda}\psi_h(x)dx$ is bounded for all imaginary $\lambda\ne 0$. If $h\in \text{Ran } P_\pm$, the integration can be restricted to the half line $\{x\in\mathbb{R} : \pm x\ge 0\}$ and the expression extends analytically to the open left/right half plane. Thus the restriction of $T^{-1}A$ to Ran $P_\pm$ has its spectrum in $\{\lambda\in\mathbb{C}$:

$\pm\text{Re}\lambda>0$ or $\lambda=0\}$. The operator $\mp G_{\pm}$ must, by virtue of the boundedness of the generated semigroup, have its spectrum in the same set. (Note that $G_{\pm}$ is a restriction of $\mp T^{-1}A$). As it is also true that, for $\lambda\neq 0$ in closed left/right half plane,

$$\lambda(\lambda\pm G_{\pm})^{-1}h = \lambda(\lambda-A^{-1}T)^{-1}|_{\text{Ran }P_{\pm}}h,$$

one must have the equality sign in (2.7). We have therefore proved (vi) and (vii). ■

As a result of the foregoing, we shall use the notation $V_{\pm}(x) = \exp\{\mp xT^{-1}A\}P_{\pm}$ for $0\leq x<\infty$. We have then equivalence proofs such as Theorems 3.3 and 3.4 of Chapter VI.

We have defined projections $P_{\pm}$ and corresponding semigroups. We shall now show that these operators may also be obtained using the solutions of the half space problem and that, in fact, they coincide with the corresponding notions encountered in Chapter II.

THEOREM 2.2. Suppose $T^{-1}A$ has no zero or imaginary eigenvalues. Then every function $\psi:(0,\infty)\rightarrow D(T)$ such that $T\psi$ is strongly differentiable and satisfies the equations

$$(T\psi)'(x) = -A\psi(x), \quad 0<x<\infty, \tag{2.8}$$

$$Q_{+}\psi(0) = \varphi_{+}, \tag{2.9a}$$

$$\|\psi(x)\|_{H} = O(1) \quad (x\rightarrow\infty), \tag{2.9b}$$

has the form

$$\psi(x) = \exp\{-xT^{-1}A\}P_{+}h, \quad 0\leq x<\infty, \tag{2.10}$$

where $Q_{+}P_{+}h = \varphi_{+}$ for a vector $h\in D(T)$.

Proof: Equations (2.8) and (2.9) are equivalent to the Wiener-Hopf equation

$$\psi(x) - \int_{0}^{\infty}\mathcal{H}(x-y)B\psi(y)dy = \exp\{-xT^{-1}\}\varphi_{+}, \quad 0<x<\infty, \tag{2.11}$$

in $L_\infty(H)_0^\infty$ and any such solution is continuous on $[0,\infty)$ (cf. Theorem VI 3.4). Simultaneously we consider the equation

$$\psi(x) - \int_{-\infty}^{0} \mathcal{H}(x-y)B\psi(y)dy = \exp\{-xT^{-1}\}\varphi_-, \quad -\infty<x<0, \tag{2.12}$$

where $\varphi_- \in \text{Ran } Q_-$. Let $H_\pm$ be the subspace of H consisting of all initial values $\psi(0^\pm)$ of solutions of Eq. (2.11) (resp. (2.12)), where $\varphi_\pm \in \text{Ran } Q_\pm$. Then we may use the method suggested by Lemma 2.1 to prove that

$$V_\pm(x)\psi(0^\pm) = \psi(\pm x), \quad 0\le x<\infty,$$

is a bounded strongly continuous semigroup on $H_\pm$, whose generator is the restriction of $\pm T^{-1}A$ to its invariant subspace $H_\pm$. Hence, $H_\pm \subset \text{Ran } P_\pm$.

Let us prove that $H_\pm = \text{Ran } P_\pm$. As above, we define

$$(\mathcal{L}_\pm\psi)(x) = \pm\int_0^{\pm\infty} \mathcal{H}(x-y)B\psi(y)dy, \quad 0\le\pm x<\infty.$$

Then the equivalence theorem VI 3.4 and its analog on the half line $(-\infty,0)$ imply that any solution of Eq. (2.11) (resp. (2.12)) satisfies $Q_\pm\psi(0^\pm) = \varphi_\pm$. Hence,

$$\text{Ker}(I-\mathcal{L}_\pm) = \{\psi(x)=\exp\{-xT^{-1}A\}h_\pm : h_\pm \in H_\pm \cap \text{Ran } Q_\mp\}. \tag{2.13}$$

The range of $I-\mathcal{L}_\pm$ can be described to some extent by observing that the function $\psi(x) = \exp\{-xT^{-1}A\}h_\pm$ $(0\le\pm x<\infty)$ for $h_\pm \in H_\pm$ is mapped by the operator $I-\mathcal{L}_\pm$ into the right hand side of Eq. (2.11) (resp. (2.12)), where $Q_\pm h_\pm = \varphi_\pm$. Therefore, $\varphi_\pm = h_\pm - Q_+ h_\pm \in H_\pm + \text{Ran } Q_\mp$, whence

$$\text{Ran}(I-\mathcal{L}_\pm) \supset \{\omega(x)=\exp\{-xT^{-1}\}\varphi_\pm : \varphi_\pm \in H_\pm + \text{Ran } Q_\mp\}. \tag{2.14}$$

We thus conclude that

$$\dim \text{Ker } (I-\mathcal{L}_\pm) = \dim [H_\pm \cap \text{Ran } Q_\mp], \tag{2.15}$$

$$\text{codim Ran } (I-\mathcal{L}_\pm) \le \text{codim } [H_\pm + \text{Ran } Q_\mp]. \tag{2.16}$$

The finiteness of the right hand side of (2.16) will be proved shortly.

Next observe that the operator $V = Q_+P_+ + Q_-P_-$ has the following three properties:

(a) $\operatorname{Ker} V = [\operatorname{Ran} P_+ \cap \operatorname{Ran} Q_-] \oplus [\operatorname{Ran} P_- \cap \operatorname{Ran} Q_+]$,

(b) $\operatorname{Ran} V = [\operatorname{Ran} P_+ + \operatorname{Ran} Q_-] \cap [\operatorname{Ran} P_- + \operatorname{Ran} Q_+]$,

(c) $I-V = (Q_+ - Q_+P_+) + (Q_- - Q_-P_-) = (Q_- - Q_+)(P_+ - P_-)$, which implies that $I-V$ is a compact operator (cf. (iii) of Lemma 2.1).

Therefore, dim Ker V = codim Ran V, or

$$\dim [\operatorname{Ran} P_+ \cap \operatorname{Ran} Q_-] + \dim [\operatorname{Ran} P_- \cap \operatorname{Ran} Q_+] =$$

$$= \operatorname{codim} [\operatorname{Ran} P_+ + \operatorname{Ran} Q_-] + \operatorname{codim} [\operatorname{Ran} P_- + \operatorname{Ran} Q_+]. \tag{2.17}$$

In view of the inclusion $H_\pm \subset \operatorname{Ran} P_\pm$ we also have

$$\dim [H_\pm \cap \operatorname{Ran} Q_\mp] \le \dim [\operatorname{Ran} P_\pm \cap \operatorname{Ran} Q_\mp], \tag{2.18}$$

$$\operatorname{codim} [H_\pm + \operatorname{Ran} Q_\mp] \ge \operatorname{codim} [\operatorname{Ran} P_\pm + \operatorname{Ran} Q_\mp]. \tag{2.19}$$

Finally, combining Theorems 1.1 and 1.2 with Eq. (1.9) we find

$$\dim \operatorname{Ker} (I-\mathcal{L}_+) + \dim \operatorname{Ker} (I-\mathcal{L}_-) =$$

$$= \operatorname{codim} \operatorname{Ran} (I-\mathcal{L}_+) + \operatorname{codim} \operatorname{Ran}(I-\mathcal{L}_-). \tag{2.20}$$

We now consider Eqs. (2.15) to (2.20) and conclude that all inequalities among these are, in fact, equalities. Moreover,

$$\dim [H_\pm \cap \operatorname{Ran} Q_\mp] = \dim [\operatorname{Ran} P_\pm \cap \operatorname{Ran} Q_\mp],$$

$$\operatorname{codim} [H_\pm + \operatorname{Ran} Q_\mp] = \operatorname{codim} [\operatorname{Ran} P_\pm + \operatorname{Ran} Q_\mp].$$

The latter two identities and the inclusion $H_\pm \subset \operatorname{Ran} P_\pm$ in turn imply

$$H_\pm \cap \operatorname{Ran} Q_\mp = \operatorname{Ran} P_\pm \cap \operatorname{Ran} Q_\mp,$$

$$H_{\pm} + \text{Ran } Q_{\mp} = \text{Ran } P_{\pm} + \text{Ran } Q_{\mp}.$$

These, together with the inclusion $H_{\pm} \subset \text{Ran } P_{\pm}$, again imply $H_{\pm} = \text{Ran } P_{\pm}$, which we sought to derive. Theorem 2.2 now follows immediately. ∎

Before paying attention to the singular case we observe that the identity $A(T^{-1}A) = (AT^{-1})A$ and the invertibility of A imply the existence of bounded operators defined by

$$A \exp\{\mp xT^{-1}A\}P_{\pm} = \exp\{\mp xAT^{-1}\}\hat{P}_{\pm}A. \tag{2.21}$$

The generators of the corresponding semigroups satisfy a similar intertwining property. We may use Lemma 2.1 (iii) and the compactness of $B = I-A$ to show that $\exp\{\mp xAT^{-1}\}\hat{P}_{\pm} - \exp\{\mp xT^{-1}\}Q_{\pm}$, $0 \le x < \infty$, is compact as well. Furthermore, Eq. (2.21) implies that $\exp\{\mp xT^{-1}A\}P_{\pm}$ leaves invariant D(T), while

$$T\exp\{\mp xT^{-1}A\}P_{\pm}h = \exp\{\mp xAT^{-1}\}P_{\pm}Th, \quad h \in D(T). \tag{2.22}$$

But $\exp\{\mp xT^{-1}\}Q_{\pm}$ also leaves invariant D(T) and satisfies

$$T\exp\{\mp xT^{-1}\}Q_{\pm}h = \exp\{\mp xT^{-1}\}Q_{\pm}Th, \quad h \in D(T).$$

Defining $\hat{V} = Q_{+}\hat{P}_{+} + Q_{-}\hat{P}_{-}$, we find $V[D(T)] \subset D(T)$ and

$$TVh = \hat{V}Th, \quad h \in D(T). \tag{2.23}$$

From the latter identity and the compactness of $I-V$ and $I-\hat{V}$ we obtain the following proposition, which will play an important role later in the derivation of explicit Wiener-Hopf factors.

PROPOSITION 2.3. We have

$$\dim \text{Ker } V = \dim \text{Ker } \hat{V},$$

$$\text{codim Ran } V = \text{codim Ran } \hat{V}.$$

Consequently, V is invertible if and only if $\hat{V}$ is invertible.

Let us consider the singular case. In this case $T^{-1}A$ has imaginary eigenvalues. These form an at most countable sequence with limit points at $\lambda = \infty$ and $\lambda = 0$. Given a linear operator S and $\lambda \in \mathbb{C}$, put

$$Z_\lambda(S) = \bigcup_{n=1}^{\infty} \text{Ker } (\lambda - S)^n = \bigcup_{n=1}^{\infty} \{h \in D(S^n) : (S-\lambda)^n h = 0\}$$

(cf. Chapter III). We shall make the following simplifying assumptions:

(I) The operator $T^{-1}A$ has at most finitely many imaginary eigenvalues.

(II) For each of these, $Z_\lambda(T^{-1}A)$ has finite dimension.

(III) The subspaces $Z_\lambda(T^{-1}A)$, $\text{Re}\lambda=0$, are contained in D(T).

(IV) The finite dimensional subspace $Z_0 = \bigoplus_\lambda Z_\lambda(T^{-1}A)$, where the direct sum is taken over eigenvalues λ with Re $\lambda=0$, has a closed direct complement Z_1 which is invariant under $T^{-1}A$.

If we now define $\hat{Z}_0 = T[Z_0]$ and $\hat{Z}_1 = A[Z_1]$, we obtain $A[Z_0] \subset \hat{Z}_0$, $T[Z_1] = \hat{Z}_1$ and $\hat{Z}_0 \oplus \hat{Z}_1 = H$. In fact, if $h \in \hat{Z}_0 \cap \hat{Z}_1$, write $h = Tk$ for $k \in Z_0$ and $h = A\ell$ for $\ell \in Z_1$. Then $\ell \in D(T^{-1}A)$ and $k = T^{-1}A\ell \in Z_1$. Together with $k \in Z_0$ we get $k=0$, and therefore $h=0$. The decomposition of H then is a simple consequence of the fact that, in view of Ker $A \subset Z_0$,

$$\dim \hat{Z}_0 = \dim T[Z_0] = \dim Z_0 = \text{codim } Z_1 = \text{codim } A[Z_1] = \text{codim } \hat{Z}_1.$$

We remark that for bounded T all four conditions are automatically satisfied.

In order to construct analogs of the projections $P_\pm$ and the associated semigroups, we first prove a lemma which is related to similar results in Chapter III.

LEMMA 2.4. Let us denote by P_0 and P_1 the complementary projections with ranges Z_0 and Z_1, respectively. Suppose $\beta: Z_0 \to Z_0$ and put $A_\beta = T\beta^{-1}P_0 + AP_1$ and $B_\beta = I - A_\beta$. If β has no imaginary eigenvalues, then A_β is invertible, B_β is compact, and

$$A_\beta^{-1}T = \beta \oplus (T^{-1}A|_{Z_1})^{-1} \tag{2.24}$$

does not have imaginary eigenvalues. Furthermore, there exists $0<\alpha<1$ such that

$$\operatorname{Ran} B_\beta \subset \operatorname{Ran}|T|^\alpha \cap D(|T|^{1+\alpha}), \tag{2.25}$$

provided

$$\operatorname{Ran} B \subset \operatorname{Ran}|T|^\alpha \cap D(|T|^{m+1+\alpha})$$

for $m = \max\limits_{\operatorname{Re}\lambda=0} \min\,\{k\geq 0 : Z_\lambda(T^{-1}A) = \operatorname{Ker}\,(T^{-1}A-\lambda)^k\}$.

Proof: Recall that $A[Z_1]=\hat{Z}_1$ and $T[Z_0] = \hat{Z}_0$. Then the invertibility of A_β is clear and Eq. (2.24) holds true. Also, $B_\beta = B+(A-A_\beta) = B+AP_0 = P_0+BP_1$ is compact and satisfies (2.25) if, for some $0<\alpha<1$,

$$\operatorname{Ran} P_0 \subset \operatorname{Ran}|T|^\alpha \cap D(|T|^{1+\alpha}).$$

However, if for some imaginary λ one has $(T^{-1}A-\lambda)h_k = h_{k-1}$, $1\leq k\leq n$, and $h_0 = 0$, then

$$h_0 = Bh_0 + \lambda Th_0, \tag{2.26a}$$
$$h_1 = Bh_1 + \lambda Th_1 + Th_0, \tag{2.26b}$$
$$\cdots\cdots\cdots\cdots\cdots\cdots$$
$$h_n = Bh_n + \lambda Th_n + Th_{n-1}. \tag{2.26c}$$

Since $\operatorname{Ran} B \subset \operatorname{Ran}|T|^\alpha$, one has $\{h_0,\ldots,h_n\} \subset \operatorname{Ran}|T|^\alpha$, which in turn implies that $\operatorname{Ran} P_0\subset\operatorname{Ran}|T|^\alpha$. Next, we rewrite Eqs. (2.26) as

$$h_0 = (I-\lambda T)^{-1}Bh_0,$$
$$h_1 = (i-\lambda T)^{-1}Bh_1 + (I-\lambda T)^{-1}Th_0,$$
$$\cdots\cdots\cdots\cdots\cdots\cdots$$
$$h_n = (I-\lambda T)^{-1}Bh_n + (I-\lambda T)^{-1}Th_{n-1}.$$

Since $\operatorname{Ran} B \subset D(|T|^{n+2+\alpha})$ for some $0<\alpha<1$ and $(I-\lambda T)^{-1}$ leaves $D(|T|^{k+\alpha})$ invariant for all imaginary λ and $k=1,2,\ldots,n+2$, we successively find $h_0\in D(|T|^{n+1+\alpha})$, $h_1\in D(|T|^{n+\alpha})$, ..., $h_n\in D(|T|^{1+\alpha})$, whence $\operatorname{Ran} P_0 \subset D(|T|^{1+\alpha})$. ∎

The operators T and B_β satisfy the general assumptions of this section and the regular case applies to them. We may then define (β-dependent) projections $P_\pm$ and $\hat{P}_\pm$ and semigroups $\exp\{+xT^{-1}A_\beta\}P_\pm$ and $\exp\{-xA_\beta T^{-1}\}\hat{P}_\pm$ and derive Theorem 2.2 and Proposition 2.3, where A_β plays the role of A. Let us define

$$\exp\{\mp xT^{-1}A\}P_{1,\pm} = \exp\{\mp xT^{-1}A_\beta\}P_\pm P_1,$$

$$\exp\{\mp xAT^{-1}\}\hat{P}_{1,\pm} = \exp\{\mp xA_\beta T^{-1}\}\hat{P}_\pm \hat{P}_1,$$

where $\hat{P}_1 = A_\beta P_1 A_\beta^{-1}$. From Eq. (2.24) it follows that these semigroups do not depend on β and that $\text{Ran } P_{1,\pm} = (\text{Ran } P_\pm) \cap Z_1$ is $T^{-1}A$-invariant. Furthermore, the exponential difference operators $\exp\{\mp xT^{-1}A\}P_{1,\pm} - \exp\{\mp xT^{-1}\}Q_\pm$ and $\exp\{\mp xAT^{-1}\}\hat{P}_{1,\pm} - \exp\{\mp xT^{-1}\}Q_\pm$ are compact. We summarize this with the following theorem.

THEOREM 2.5. Under the conditions (I)–(IV) there is a decomposition of the Hilbert space H as a direct sum of the three $T^{-1}A$-invariant subspaces $\text{Ran } P_{1,+}$, $\text{Ran } P_{1,-}$ and $\text{Ran } P_0$ with the following properties:

(i) The restriction of $\mp T^{-1}A$ to $\text{Ran } P_{1,\pm}$ generates a bounded analytic semigroup.

(ii) $\text{Ran } P_0$ has finite dimension and contains Ker A, while the restriction of $T^{-1}A$ to this subspace has purely imaginary spectrum.

We have developed the basic ingredients for a later treatment of boundary value problems with non-self adjoint A. In Section 4 we shall do the same for normal T and for a Banach space setting (cf. Section VI.5). The notions developed parallel those of Chapters II to V. We remark that all of these semigroups are, in fact, analytic.

3. Factorization of the symbol

The classical factorization method for solving Wiener-Hopf equations on a half line has been discussed in Section 1. Applying this method to the integral form (2.11) of abstract kinetic equations, one arrives at the problem of factorizing its symbol, which has the form

$$W(\lambda) = (T-\lambda)^{-1}(T-\lambda A) = A - T(\lambda-T)^{-1}B, \quad \text{Re } \lambda=0. \tag{3.1}$$

For strictly positive operators A and bounded self adjoint T a formal expression for the right canonical Wiener–Hopf factorization has been found in [25, 359, 360]. It was presented there as an application of a factorization principle for "transfer functions" of the form

$$\mathbb{W}(\lambda) = \mathbb{D} + \mathbb{C}(\lambda-\mathbb{A})^{-1}\mathbb{B}, \tag{3.2}$$

which in its most general form is due to Bart et al. [25, 26] For more general situations it is not straightforward to apply this principle to the symbol (3.1). A canonical factorization of the symbol might not exist, as is the situation, for instance, in the singular case of the previous section. Still, for many such cases factorizations (of non–Wiener–Hopf type) have been constructed (see [366]).

In this section we shall obtain a canonical factorization of the symbol in terms of the albedo operator and prove the equivalence of the existence of a canonical factorization to a certain decomposition of the Hilbert space H. Let us state the factorization principle first (see [25, 26]). The proof is by direct computation.

THEOREM 3.1. On Banach spaces X and Y, suppose $\mathbb{D}:Y\to Y$ is invertible, and $\mathbb{A}:X\to X$, $\mathbb{B}:Y\to X$, $\mathbb{C}:X\to Y$ are bounded. Write $\mathbb{D}=\mathbb{D}_1\mathbb{D}_2$ and $\mathbb{A}^\times=\mathbb{A}-\mathbb{B}\mathbb{D}^{-1}\mathbb{C}$. If $\mathfrak{M}\subset X$ and $\mathfrak{M}^\times\subset X$ are closed invariant subspaces of $\mathbb{A}$ and $\mathbb{A}^\times$, respectively, and the decomposition $\mathfrak{M} \oplus \mathfrak{M}^\times = X$ holds true, then

$$\mathbb{W}(\lambda) = [\mathbb{D}_1 + \mathbb{C}(\lambda-\mathbb{A})^{-1}(I-\mathbb{E}_+)\mathbb{B}\mathbb{D}_2^{-1}][\mathbb{D}_2 + \mathbb{D}_1^{-1}\mathbb{C}\mathbb{E}_+(\lambda-\mathbb{A})^{-1}\mathbb{B}],$$

where

$$[\mathbb{D}_1 + \mathbb{C}(\lambda-\mathbb{A})^{-1}(I-\mathbb{E}_+)\mathbb{B}\mathbb{D}_2^{-1}]^{-1} = \mathbb{D}_1^{-1} - \mathbb{D}_1^{-1}\mathbb{C}(I-\mathbb{E}_+)(\lambda-\mathbb{A}^\times)^{-1}\mathbb{B}\mathbb{D}^{-1},$$

$$[\mathbb{D}_2 + \mathbb{D}_1^{-1}\mathbb{C}\mathbb{E}_+(\lambda-\mathbb{A})^{-1}\mathbb{B}]^{-1} = \mathbb{D}_2^{-1} - \mathbb{D}^{-1}\mathbb{C}(\lambda-\mathbb{A}^\times)^{-1}\mathbb{E}_+\mathbb{B}\mathbb{D}_2^{-1}.$$

Here $\mathbb{E}_+$ is the projection of X onto $\mathfrak{M}^\times$ along $\mathfrak{M}$.

The earliest applications of this theorem to the symbol (3.1) (where $\mathbb{A}=T$, $\mathbb{B}=B$, $\mathbb{C}=-T$ and $\mathbb{D}=A$) considered a product $\mathbb{D}=\mathbb{D}_1\mathbb{D}_2$ with $\mathbb{D}_1=I$ and $\mathbb{D}_2=\mathbb{D}=A$ (cf. [25, 359, 360]). In spite of the fact that $\mathbb{W}(0)=I$, the factors thus obtained did not have this property. Because of the need to study the generalization to singular cases (considered in this section) and to construct the so-called generalized H-functions (considered in the next chapter), we shall make a different choice of $\mathbb{D}_1$ and $\mathbb{D}_2$. A somewhat different generalization, already discussed in Section VI.4, will be of special interest if B has finite rank. Let $\mathbb{B}$ be a closed subspace of H which contains Ran B^*, and let $\pi:H\to\mathbb{B}$ and $j:\mathbb{B}\to H$ be operators such that $j\pi$ is the identity on $\mathbb{B}$ and πj the orthogonal projection of H onto $\mathbb{B}$. Then $Bj\pi=B$ (cf. Section VI.4). We define the **dispersion function** by

$$\Lambda(\lambda) = \pi W(\lambda)j = \pi Aj - \pi T(\lambda-T)^{-1}Bj, \quad \lambda\notin\sigma(T). \tag{3.3}$$

We note that $\Lambda(\lambda)$ is a transfer function of type (3.2), where $\mathbb{A}=T$, $\mathbb{B}=Bj$, $\mathbb{C}=-\pi T$ and $\mathbb{D}=\pi Aj$. If A is invertible, then $\mathbb{D}$ is invertible also and $\mathbb{D}^{-1}=\pi A^{-1}j$. We thus obtain

$$\mathbb{A}^{x} = \mathbb{A}-\mathbb{B}\mathbb{D}^{-1}\mathbb{C} = A^{-1}T. \tag{3.4}$$

We begin with a technical lemma.

LEMMA 3.2. Let T and A satisfy the general assumptions of Section 2, and suppose that $T^{-1}A$ does not have zero or imaginary eigenvalues. Then in the strong operator topology we have

$$\lim_{\lambda\to0,\,\mathrm{Re}\,\lambda\le0} T(T-\lambda)^{-1}Q_+ = Q_+,$$

$$\lim_{\lambda\to\infty,\,\mathrm{Re}\,\lambda\le0} T(T-\lambda)^{-1}Q_+ = 0,$$

$$\lim_{\lambda\to0,\,\mathrm{Re}\,\lambda\le0} A^{-1}T(A^{-1}T-\lambda)^{-1}P_+ = P_+,$$

$$\lim_{\lambda\to\infty,\,\mathrm{Re}\,\lambda\le0} A^{-1}T(A^{-1}T-\lambda)^{-1}P_+ = 0.$$

Similar statements hold true if Q_+ and P_+ are replaced by Q_- and P_-.

Proof: Denoting by $\sigma(.)$ the resolution of the identity of T, we obtain

$$T(T-\lambda)^{-1}Q_+h = \int_0^\infty t(t-\lambda)^{-1}\sigma(dt)h, \quad h\in H.$$

As $|t(t-\lambda)^{-1}|\leq 1$ for $\mathrm{Re}\lambda\leq 0$ and $t>0$, and $\sigma(.)$ is a bounded measure, the first pair of identities follow using the principle of dominated convergence. Next observe that

$$A^{-1}T(A^{-1}T-\lambda)^{-1}h = W(\lambda)^{-1}T(T-\lambda)^{-1}h, \quad h\in H.$$

Thus

$$\lim_{\lambda\to 0,\,\mathrm{Re}\ \lambda=0} A^{-1}T(A^{-1}T-\lambda)^{-1}h = h,$$

$$\lim_{\lambda\to 0,\,\mathrm{Re}\ \lambda=0} A^{-1}T(A^{-1}T-\lambda)^{-1}h = 0.$$

These limits also hold true if h is replaced by P_+h. Using the maximum modulus principle we immediately have the second pair of formulas. ■

THEOREM 3.3. Suppose that $T^{-1}A$ does not have zero or imaginary eigenvalues. Then the following five statements are equivalent:

(i) The convolution equation (2.11) is uniquely solvable in $L_p(H)_0^\infty$ for every $\omega\in L_p(H)_0^\infty$.

(ii) For every $\varphi_+\in Q_+[D(T)]$ the boundary value problem (2.8)–(2.9) is uniquely solvable.

(iii) $\mathrm{Ran}\ P_+ \oplus \mathrm{Ran}\ Q_- = H$.

(iv) $\mathrm{Ran}\ \hat{P}_+ \oplus \mathrm{Ran}\ Q_- = H$.

(v) The symbol (3.1) has a right canonical factorization.

If one of these statements is true and E_+ (resp. $\hat{E}_+$) is the projection of H onto $\mathrm{Ran}\ P_+$ (resp. $\mathrm{Ran}\ \hat{P}_+$) along $\mathrm{Ran}\ Q_-$, then a right canonical factorization of the dispersion function (3.3) is given by

$$\Lambda(\lambda)^{-1} = H_\ell^+(-\lambda)H_r^+(\lambda), \quad \mathrm{Re}\ \lambda=0, \tag{3.5}$$

where

$$H_\ell^+(-\lambda) = I - \lambda\pi(T-\lambda A)^{-1}\hat{E}_+Bj, \tag{3.6}$$

$$H_r^+(\lambda) = I - \lambda\pi(I-E_+)(T-\lambda A)^{-1}Bj. \tag{3.7}$$

The inverses of these factors are given by

$$H_\ell^+(-\lambda)^{-1} = I - \lambda\pi E_+(\lambda-T)^{-1}Bj, \tag{3.8}$$

$$H_r^+(\lambda)^{-1} = I - \lambda\pi(\lambda-T)^{-1}(I-\hat{E}_+)Bj. \tag{3.9}$$

Proof: The equivalence of (i) and (v) follows from Theorem 1.1, while Theorem VI 3.3 implies that (ii) follows from (i). Using Eqs. (2.15) and (2.16) (the latter with equality sign) and the conclusion $H_+ = \text{Ran } P_+$ of the proof of Theorem 2.2, one establishes the equivalence of (ii) and (iii). From the proof of Theorem 2.2 and the inclusion $\text{Ker}(I-\mathcal{L}_+)\subset D(T)$, we easily find

$$T[\text{Ran } P_+ \cap \text{Ran } Q_-] \subset \text{Ran } \hat{P}_+ \cap \text{Ran } Q_-.$$

In the latter inclusion one has, in fact, equality (see Proposition 2.3). Hence, (iii) and (iv) are equivalent. We shall next prove that (iii) and (iv) imply (v), actually by explicit construction.

Observe that $\mathfrak{M} = \text{Ran } Q_-$ is T-invariant, $\mathfrak{M}^\times = \text{Ran } P_+$ is $A^{-1}T$-invariant and $\mathfrak{M} \oplus \mathfrak{M}^\times = H$ (by assumption (iii)). Applying Theorem 3.1 for $\mathbb{A}=T$, $\mathbb{B}=Bj$, $\mathbb{C}=-\pi T$ and $\mathbb{D}=\pi Aj$ with arbitrary splitting $\mathbb{D}=\mathbb{D}_1\mathbb{D}_2$, we obtain the factorization

$$\Lambda(\lambda) = [\mathbb{D}_1 + \pi T(T-\lambda)^{-1}(I-E_+)Bj\mathbb{D}_2^{-1}][\mathbb{D}_2 + \mathbb{D}_1^{-1}\pi TE_+(T-\lambda)^{-1}Bj],$$

where

$$[\mathbb{D}_1 + \pi T(T-\lambda)^{-1}(I-E_+)Bj\mathbb{D}_2^{-1}]^{-1} = \mathbb{D}_1^{-1} - \mathbb{D}_1^{-1}\pi T(I-E_+)(T-\lambda A)^{-1}Bj,$$

$$[\mathbb{D}_2 + \mathbb{D}_1^{-1}\pi TE_+(T-\lambda)^{-1}Bj]^{-1} = \mathbb{D}_2^{-1} - \pi(T-\lambda A)^{-1}TE_+Bj\mathbb{D}_2^{-1}.$$

Since for some $0<\varphi<\frac{1}{2}\pi$ one has

$$\sigma(T|_{\mathrm{Ker}E_+}) \subset \{0\} \cup \{\lambda : |\pi-\arg\lambda| \le \varphi\}, \tag{3.10}$$

$$\sigma(A^{-1}T|_{\mathrm{Ran}E_+}) \subset \{0\} \cup \{\lambda : |\arg\lambda| \le \varphi\}, \tag{3.11}$$

the left factor and its inverse extend analytically to the right half plane, while the right factor and its inverse extend analytically to the left half plane. All four factors have continuous extensions to the closures of these half planes, except possibly at $\lambda=\infty$ (for unbounded T) and $\lambda=0$.

We note that all limits in the formulation of Lemma 3.2 are valid uniformly on compact subsets of H. Thus, exploiting the compactness of B, the factors of $\Lambda(\lambda)$ and their inverses all have uniform limits for $\lambda\to 0$ and $\lambda\to\infty$ approaching from the appropriate half plane. Hence, we have, indeed, found a right canonical Wiener–Hopf factorization of $\Lambda(\lambda)$.

The next objective is to adjust $\mathbb{D}_1$ and $\mathbb{D}_2$ so that the factors will coincide with I as $\lambda\to 0$ (from the appropriate half plane). We get

$$\mathbb{D}_1 + \pi(I-E_+)Bj\mathbb{D}_2^{-1} = I = [\mathbb{D}_1^{-1} - \mathbb{D}_1^{-1}\pi(I-\hat{E}_+)Bj]^{-1}, \tag{3.12}$$

$$\mathbb{D}_2 + \mathbb{D}_1^{-1}\pi\hat{E}_+Bj = I = [\mathbb{D}_2^{-1} - \pi E_+Bj\mathbb{D}_2^{-1}]^{-1}. \tag{3.13}$$

Here we have used $E_+[D(T)] \subset D(T)$ and $TE_+ = \hat{E}_+T$ on D(T), which follows from Eq. (2.11) (yielding $\psi(x)-\omega(x)\in D(T)$) and Eq. (2.23). Substituting $\mathbb{D}_1\mathbb{D}_2 = \pi Aj$, one obtains

$$\mathbb{D}_1 = \pi[A + \hat{E}_+B]j, \qquad \mathbb{D}_1^{-1} = \pi[A^{-1} - E_+BA^{-1}]j,$$

$$\mathbb{D}_2 = \pi[I - E_+B]j, \qquad \mathbb{D}_2^{-1} = \pi[I + A^{-1}\hat{E}_+B]j.$$

Utilizing these, one may simplify the expressions for the factors of $\Lambda(\lambda)$ and their inverses. The result is, in fact, obtained in the following manner. First observe

$$H_r^+(\lambda)^{-1} = \mathbb{D}_1 + \pi T(T-\lambda)^{-1}(I-E_+)Bj\mathbb{D}_2^{-1} =$$

$$= \pi[A + \hat{E}_+B + T(T-\lambda)^{-1}(I-E_+)B + T(T-\lambda)^{-1}(I-E_+)BA^{-1}\hat{E}_+B]j. \tag{3.14}$$

Writing

$$(I-E_+)BA^{-1}\hat{E}_+ =$$

$$= (I-E_+)A^{-1}\hat{E}_+ + (I-E_+)(I-\hat{E}_+) - (I-E_+) = (I-\hat{E}_+) - (I-E_+), \tag{3.15}$$

we apply a simplification, yielding

$$H_r^+(\lambda)^{-1} = \pi[I - (I-\hat{E}_+)B + T(T-\lambda)^{-1}(I-\hat{E}_+)B]j,$$

which implies (3.9). Similarly,

$$H_\ell^+(-\lambda)^{-1} = \mathbb{D}_2 + \mathbb{D}_1^{-1}\pi TE_+(T-\lambda)^{-1}Bj =$$

$$= \pi[I - E_+B + A^{-1}TE_+(T-\lambda)^{-1}B - E_+BA^{-1}TE_+(T-\lambda)^{-1}B]j.$$

Writing $TE_+ = \hat{E}_+T$ (noting that Ran $B \subset D(|T|^{1+\alpha}) \subset D(T)$) and using (3.15), we obtain

$$H_\ell^+(-\lambda)^{-1} = \pi[I - E_+B + E_+T(T-\lambda)^{-1}B]j,$$

which simplifies to (3.8). Next,

$$H_\ell^+(-\lambda) = \mathbb{D}_2^{-1} - \pi(T-\lambda A)^{-1}TE_+Bj\mathbb{D}_2^{-1} =$$

$$= \pi[I + A^{-1}\hat{E}_+B - (T-\lambda A)^{-1}TE_+B - (T-\lambda A)^{-1}TE_+BA^{-1}\hat{E}_+B]j.$$

Again using (3.15) we obtain

$$H_\ell^+(-\lambda) = \pi[I + A^{-1}\hat{E}_+B - (T-\lambda A)^{-1}TA^{-1}\hat{E}_+B]j,$$

which gives (3.6). Finally,

$$H_r^+(\lambda) = \mathbb{D}_1^{-1} - \mathbb{D}_1^{-1}\pi T(I-E_+)(T-\lambda A)^{-1}Bj =$$

$$= \pi[A^{-1} - E_+BA^{-1} - A^{-1}T(I-E_+)(T-\lambda A)^{-1}B + E_+BA^{-1}T(I-E_+)(T-\lambda A)^{-1}B]j.$$

With the help of the intertwining relation $T\hat{E}_+=E_+T$ and (3.15), this simplifies to

$$H_r^+(\lambda) = \pi[I + (I-E_+)A^{-1}B - (I-E_+)A^{-1}T(T-\lambda A)^{-1}B]j,$$

which yields (3.7). ∎

We provide next three sufficient conditions for the existence of right and left Wiener-Hopf factorizations. The sufficiency of the first condition follows easily from the results of Chapter II. We will prove the sufficiency of the second and third conditions. The second condition was studied in the context of multigroup neutron transport with isotropic scattering by Bowden, Sancaktar and Zweifel [57].

THEOREM 3.4. The boundary value problem (2.8)–(2.9) is uniquely solvable and the dispersion function $\Lambda(z)$ has a right and a left canonical factorization, provided one of the following three conditions is satisfied:

(i) A is a strictly positive operator;
(ii) B has norm less than one;
(iii) A is invertible and $C = A^{-1}B$ has norm less than one.

Proof: Part (i) follows from the equivalence theorem VI 3.3 and the results of Section II.2. To prove parts (ii) and (iii), we write the dispersion function in the form $\Lambda(z) = I-z\pi(z-T)^{-1}Bj$ for Re $z=0$, with π and j of unit norm. Thus,

$$\|I-\Lambda(z)\| \le \sup_{\mathrm{Re}\ z=0} \|z(z-T)^{-1}\| \ \|B\| \le \|B\|.$$

Hence, if $\|B\|<1$, then $\Lambda(z)$ satisfies the hypothesis of Theorem 1.4, and so it has a right and a left canonical factorization.

If A is invertible, then πAj is invertible (with inverse $\pi A^{-1}j$) and

$$\Lambda(z)(\pi Aj)^{-1} = \pi[I+T(T-z)^{-1}C]j$$

for Re $z=0$. Therefore,

$$\|I-\Lambda(z)(\pi Aj)^{-1}\| \le \sup_{\mathrm{Re}\ z=0} \|T(T-z)^{-1}\| \ \|C\| \le \|C\|,$$

and a similar application of Theorem IX 1.4 completes the proof. ∎

On inspecting Eqs. (3.6) to (3.9), one is surprised at the conspicuous absence of A^{-1}. It is therefore reasonable to seek a generalization to singular cases. Thus, under the assumptions of the previous section let us consider the singular case, possibly with non-invertible A. In order to have a suitable analogue of statement (iii) of Theorem 3.3, we assume the existence of a $T^{-1}A$-invariant subspace N_+ of Z_0 such that

$$\text{Ran } P_{1,+} \oplus N_+ \oplus \text{Ran } Q_- = H. \tag{3.16}$$

Defining $\hat{N}_+ = T[N_+]$, we must also have

$$\text{Ran } \hat{P}_{1,+} \oplus \hat{N}_+ \oplus \text{Ran } Q_- = H. \tag{3.17}$$

Let E_+(resp. $\hat{E}_+$) be the projection of H onto Ran $P_{1,+} \oplus N_+$ (resp. Ran $\hat{P}_{1,+} \oplus \hat{N}_+$) along Ran Q_-. The right hand sides of Eqs. (3.8) and (3.9) are analytic on one half plane and continuous up to the extended imaginary line. The right hand sides of (3.6) and (3.7) are analytic on one half plane too, but their behavior on approaching the imaginary line is more involved. (We shall describe it shortly.) Clear-cut computation (as in [366]) yields that (i) the right hand side of (3.6) is the inverse of the right hand side of (3.8); (ii) the right hand side of (3.7) is the inverse of the right hand side of (3.9); and (iii) $\Lambda(\lambda) = H_r^+(\lambda)^{-1}H_\ell^+(-\lambda)$, $\text{Re}\lambda = 0$. Thus we obtain a generalization of the factorization part of Theorem 3.3 to singular cases.

Let us investigate the behavior of the right hand sides of (3.6) and (3.7) as λ approaches the imaginary axis. We have to distinguish between three cases:

(a) Let λ_0 be a non-zero imaginary eigenvalue and let

$$(T-\lambda A)^{-1} = \sum_{n=-p}^{\infty} (\lambda-\lambda_0)^n S_n$$

be the Laurent series of $(T-\lambda A)^{-1}$ in a deleted neighborhood of λ_0. Then $H_\ell^+(-\lambda)$ and $H_r^+(\lambda)$ have poles at λ_0 with the Laurent series expansions

$$H_\ell^+(-\lambda) = \sum_{n=-p}^{\infty} (\lambda-\lambda_0)^n \{\delta_{n0}I-\pi(\lambda_0 S_n+S_{n-1})\hat{E}_+ Bj\},$$

$$H_r^+(\lambda) = \sum_{n=-p}^{\infty} (\lambda-\lambda_0)^n\{\delta_{n0}I-\pi(I-E_+)(\lambda_0 S_n+S_{n-1})Bj\},$$

where we read $S_{-p-1}=0$.

(b) Near $\lambda=0$ we have $H_\ell(0^+) = H_r(0^+) = I$ as one approaches $\lambda=0$ from the closed right half plane.

(c) Near $\lambda=\infty$ the behavior of $H_\ell^+(-\lambda)$ and $H_r^+(\lambda)$ is described by the Laurent series

$$(A-\lambda T)^{-1}\hat{P}_0 = \sum_{n=-p}^{\infty} \lambda^n T_n: \text{Ran } \hat{P}_0 \rightarrow \text{Ran } P_0$$

in a deleted neighborhood of $\lambda=0$. We then have

$$H_\ell^+(-\lambda) = I - \lambda\pi(T-\lambda A)^{-1}\hat{P}_{1,+}\hat{E}_+Bj + \sum_{n=-p}^{\infty} \lambda^{-n}\pi T_n\hat{P}_0\hat{E}_+Bj, \tag{3.18}$$

$$H_r^+(\lambda) = I - \lambda\pi(I-E_+)(T-\lambda A)^{-1}\hat{P}_{1,-}Bj + \sum_{n=-p}^{\infty} \lambda^{-n}\pi(I-E_+)T_n\hat{P}_0Bj, \tag{3.19}$$

where $(T-\lambda A)^{-1}\hat{P}_{1,\pm}$ is an operator with values in Ran $P_{1,\pm}$. The asymptotic expansions on approaching $\lambda=\infty$ from the appropriate directions are then given by

$$H_r^+(-\lambda) \cong [I+\pi A^{-1}\hat{P}_{1,+}\hat{E}_+Bj] + \sum_{n=-p}^{0} \lambda^{-n}\pi T_n\hat{P}_0\hat{E}_+Bj, \tag{3.20}$$

$$H_r^+(\lambda) \cong [I+\pi(I-E_+)A^{-1}\hat{P}_{1,-}Bj] + \sum_{n=-p}^{0} \lambda^{-n}\pi(I-E_+)T_n\hat{P}_0Bj, \tag{3.21}$$

where $A^{-1}\hat{P}_{1,\pm}$ is an operator with values in Ran $P_{1,\pm}$. The difference of $H_\ell^+(-\lambda)$ (resp. $H_r^+(\lambda)$) and its asymptotic expansion vanishes as $\lambda\rightarrow\infty$ from the left (resp. right) half plane.

We have thus accomplished a generalization of Eqs. (3.6) and (3.9) to singular cases. We summarize this in the following theorem.

THEOREM 3.5. Let (3.16) be satisfied, and let E_+ and $\hat{E}_+$ be the projections along Ran Q_- onto Ran $P_{1,+}\oplus N_+$ and Ran $\hat{P}_{1,+}\oplus\hat{N}_+$, respectively. Then the symbol $W(\lambda)$ has the factorization (3.5) with the factors given by (3.6)–(3.9). At $\lambda=\infty$ the factors have the asymptotic expansions (3.18)–(3.21).

4. Construction in a Banach space setting

The first three sections were written for a Hilbert space setting, where T is injective and self adjoint. As was pointed out, for many applications one should allow that T is either a normal Hilbert space operator or a suitable Banach space operator. Such generalizations were discussed in Section VI.6, where an extension of the equivalence theorems was outlined. In the present section we shall also generalize the semigroup reconstruction and factorization of Sections 1 to 3.

Let T be an operator on a Banach space H which satisfies the assumptions (A.1) to (A.4) of Section VI.6. We assume that B is uniformly approximable by operators of finite rank and satisfies the regularity condition

$$\exists\alpha>0: \ \text{Ran } B \subset \text{Ran } |T|^{\alpha} \cap D(|T|^{1+\alpha}).$$

Then B is a compact operator. (Note that not every compact operator is uniformly approximable by an operator of finite rank; see [112, 307] for counterexamples. However, on Hilbert spaces and L_p-spaces ($1\le p<\infty$), such counterexamples do not exist.) By Eq. (2.1) we define the symbol of a convolution equation on H with kernel $k(x) = \mathcal{H}(x)B$. As $\mathcal{H}(x)B$ is norm-integrable over $\mathbb{R}$ (cf. Lemma VI 3.1) and k(x) is almost everywhere uniformly approximable by operators of finite rank, Theorems 1.1 to 1.3 apply, provided $W(\lambda)$ is invertible for all extended imaginary λ. The latter is equivalent to requiring that $T^{-1}A$ does not have zero or imaginary eigenvalues. As in the Hilbert space setting we may again distinguish between regular and singular cases.

In the regular case we can repeat the construction of Section 2. We use the unique solution of Eq. (2.3) to construct bounded semigroups $\{V_{\pm}(x)\}_{x\ge0}$ and bounded complementary projections $P_{\pm}$. We prove the properties (i) to (vii) in the statement of Lemma 2.1, establish Theorem 2.2 and Proposition 2.3, and derive Eq. (2.23). In the singular case we again make the assumptions (I) to (IV), which are automatically fulfilled for bounded T, derive Lemma 2.4 and Theorem 2.5 and construct the relevant semigroups and projections. All proofs carry over in a natural way.

Section 3 can be developed in the general Banach space setting by a repetition of those arguments. There are only two significant changes in the theory. The first change, involving the operator T, arises in the first half of Lemma 3.2. Here we should prove the identities

$$\lim_{\lambda\to 0,\,\mathrm{Re}\,\lambda\le 0} T(T-\lambda)^{-1}Q_+ = Q_+, \tag{4.1a}$$

$$\lim_{\lambda\to\infty,\,\mathrm{Re}\,\lambda\le 0} T(T-\lambda)^{-1}Q_+ = 0, \tag{4.1b}$$

(in the strong operator topology) by arguments which do not involve the Spectral Theorem for self adjoint operators. Indeed, using Eq. (13.53) of [224] (for $A = (-T|_{\mathrm{Ran}\,Q_+})^{-1}$, $\sigma_0=0$, $\alpha=\gamma$) one finds, for some $\varepsilon>0$,

$$\|T(T-\lambda)^{-1}Q_+\| \le c(\gamma)(1+|\lambda|)^{-1}, \qquad |\pi-\arg\lambda| < \tfrac{1}{2}\pi+\varepsilon.$$

Hence, (4.1b) holds true. Also, for all $k=Th$ we get

$$\lim_{\lambda\to 0,\,\mathrm{Re}\lambda\le 0} \lambda(\lambda-T)^{-1}Q_+k = \lim_{\lambda\to 0,\,\mathrm{Re}\lambda\le 0} -\lambda T(T-\lambda)^{-1}Q_+h = 0.$$

Since $\|\lambda(\lambda-T)^{-1}Q_+\| \le \|Q_+\| + c(\gamma)$ for $\mathrm{Re}\lambda\le 0$ and Ran T is dense in H, one obtains (4.1a). The second change involves Theorem 3.4. Obviously, one cannot generalize part (i). It is also impossible to generalize parts (ii) and (iii) in a straightforward way, since they are based on a Hilbert space result, namely Theorem 1.4.

Finally, in order to define and factorize the dispersion function we have to construct $\pi:H\to\mathbb{B}$ and $j:\mathbb{B}\to H$ such that πj is the identity on $\mathbb{B}$, $j\pi$ a projection of H with range $\mathbb{B}$ and $Bj\pi=B$. The latter is satisfied if $\mathrm{Ker}\, j\pi\subset\mathrm{Ker}\, B$. In this manner we may generalize the factorization results of Section 3.

We now have available a Banach space theory of equivalence, semigroup reconstruction and factorization. In many transport theory applications one is accustomed, for reasons of mathematical convenience, to develop the mathematical theory in an L_2-setting, while the L_1 setting is often the physically relevant one. Therefore, we would like to have a procedure for transferring results from the L_2- to the L_1-setting. For this reason we consider two Banach spaces H and $\mathcal{H}$, where H is densely and continuously imbedded in $\mathcal{H}$. (Note, in transferring results, say from $L_2(\mathbb{R})$ to $L_1(\mathbb{R})$, as is typical in gas dynamics, one may apply the subsequent theory in two steps, going through the intersection space $L_1\cap L_2$ with norm $\|\cdot\|_1+\|\cdot\|_2$.) On H we have operators T and B satisfying the general assumptions of Section VI.6; on

$\mathcal{H}$ we have operators $\mathcal{T}$ and $\mathcal{B}$ satisfying the same assumptions. Further, we assume that T and B are the restrictions to H of $\mathcal{T}$ and $\mathcal{B}$, respectively. We put A=I−B and $\mathcal{A}=I-\mathcal{B}$.

Let us write the symbols

$$W(\lambda) = I - \lambda(\lambda-T)^{-1}B,$$

$$\mathcal{W}(\lambda) = I - \lambda(\lambda-\mathcal{T})^{-1}\mathcal{B},$$

for $\mathrm{Re}\lambda = 0$. Then $W(\lambda)$ is the restriction of $\mathcal{W}(\lambda)$ to H.

LEMMA 4.1. Let H and $\mathcal{H}$ be Banach spaces such that H is densely and continuously embedded in $\mathcal{H}$. Let $F\in L(H)$ and $\mathcal{F}\in L(\mathcal{H})$ be Fredholm operators such that F is the restriction of $\mathcal{F}$ to H. Then

$$\{h\in H : Fh=0\} = \{h\in\mathcal{H} : \mathcal{F}h=0\}; \tag{4.2}$$

$$\overline{\{Fh : h\in H\}}^{\mathcal{H}} = \{\mathcal{F}h : h\in\mathcal{H}\}. \tag{4.3}$$

Hence, F and $\mathcal{F}$ have the same Fredholm index, while

$$\dim \mathrm{Ker}\, F = \dim \mathrm{Ker}\, \mathcal{F}, \tag{4.4}$$

$$\mathrm{codim}\, \mathrm{Ran}\, F = \mathrm{codim}\, \mathrm{Ran}\, \mathcal{F}. \tag{4.5}$$

Proof: Certainly,

$$\mathrm{Ker}\, F = \{h\in H : Fh=0\} \subset \{h\in\mathcal{H} : \mathcal{F}h=0\} = \mathrm{Ker}\, \mathcal{F}$$

and

$$\mathrm{Ran}\, F = \{Fh : h\in H\} \subset \{\mathcal{F}h : h\in\mathcal{H}\} = \mathrm{Ran}\, \mathcal{F}.$$

Since H is dense in $\mathcal{H}$ and $\mathcal{F}\in L(\mathcal{H})$, the closure of Ran F in $\mathcal{H}$ will be the closed subspace Ran $\mathcal{F}$, which settles (4.3) and (4.5). By replacing H and $\mathcal{H}$, and F and $\mathcal{F}$ by Banach adjoints and on applying (4.5) to these adjoints, one finds (4.4). Finally, using Ker $F \subset$ Ker $\mathcal{F}$ one immediately has (4.2). ■

Let us apply the lemma to $W(\lambda)$ and $\mathcal{W}(\lambda)$.

PROPOSITION 4.2. Let T,B: $H \to H$ and $\mathcal{T},\mathcal{B}$: $\mathcal{H} \to \mathcal{H}$ be as before. Then

(i) $T^{-1}A$ and $\mathcal{T}^{-1}\mathcal{A}$ have the same zero or imaginary eigenvalues;

(ii) $Z_\lambda(T^{-1}A) = Z_\lambda(\mathcal{T}^{-1}\mathcal{A})$ for all zero or imaginary λ;

(iii) Assumptions (I) to (IV) of Section 2 are fulfilled for T and B if and only if they are fulfilled for $\mathcal{T}$ and $\mathcal{B}$.

Proof: Observe that $W(\lambda)$ and $\mathcal{W}(\lambda)$ are Fredholm operators for all zero or imaginary λ (including $\lambda=\infty$). Thus Ker $W(\lambda)$ = Ker $\mathcal{W}(\lambda) \subset H$, and therefore $T^{-1}A$ and $\mathcal{T}^{-1}\mathcal{A}$ have the same zero or imaginary eigenvalues with the same multiplicities. If $h_0 \in Z_\lambda(\mathcal{T}^{-1}\mathcal{A})$ for some imaginary λ, then there exists a sequence $\{h_1,\ldots,h_n\} \subset \mathcal{H}$ such that

$$(\mathcal{A}-\lambda\mathcal{T})h_0 = Th_1, \quad (\mathcal{A}-\lambda\mathcal{T})h_1 = \mathcal{T}h_2, \quad \ldots, \quad (\mathcal{A}-\lambda\mathcal{T})h_n = 0.$$

This implies

$$\mathcal{W}(1/\lambda)h_0 = (I-\lambda\mathcal{T})^{-1}\mathcal{T}h_1, \quad \mathcal{W}(1/\lambda)h_1 = (I-\lambda\mathcal{T})^{-1}\mathcal{T}h_2, \quad \ldots, \quad \mathcal{W}(1/\lambda)h_n = 0.$$

Working backwards and applying Lemma 4.1 one finds eventually that $h_0 \in H$, which settles (ii).

Obviously, assumptions (I), (II) and (III) of Section 2 are fulfilled for T and B if and only if they are fulfilled for $\mathcal{T}$ and $\mathcal{B}$. Furthermore,

$$Z_0 = \bigoplus_{\mathrm{Re}\,\lambda=0} Z_\lambda(T^{-1}A) = \bigoplus_{\mathrm{Re}\,\lambda=0} Z_\lambda(\mathcal{T}^{-1}\mathcal{A}) \subset H.$$

Suppose (IV) applies to $\mathcal{T}$ and $\mathcal{B}$. Then Z_0 has a closed complement $\mathcal{Z}_1$ in $\mathcal{H}$, which is invariant under $\mathcal{T}^{-1}\mathcal{A}$. Put $Z_1=\mathcal{Z}_1 \cap H$. Then Z_1 is closed in H and invariant under $T^{-1}A$, while $Z_0 \cap Z_1=\{0\}$ and Z_0+Z_1 is dense in H. Since dim $Z_0<\infty$ and Z_1 is closed in H, also Z_0+Z_1 is closed in H and therefore Z_1 is a closed complement of Z_0 in H.

Conversely, suppose (IV) applies to T and B. Then Z_0 has a closed complement

Z_1 in H, which is invariant under $T^{-1}A$. If we set $\mathcal{Z}_1=\overline{Z_1}^{\mathcal{H}}$, then $\mathcal{Z}_1$ is closed in H and invariant under $\mathcal{T}^{-1}\mathcal{A}$, while $Z_0+\mathcal{Z}_1$ is dense in $\mathcal{H}$. Since Z_0 has finite dimension and $\mathcal{Z}_1$ is closed in $\mathcal{H}$, we must have $Z_0+\mathcal{Z}_1=H$. However, repeatedly using (4.5), one finds that the codimensions of Z_1 in H and $\mathcal{Z}_1$ in $\mathcal{H}$ coincide. This in turn implies that

$$\dim \frac{Z_0}{Z_0\cap\mathcal{Z}_1} = \dim \frac{Z_0+\mathcal{Z}_1}{\mathcal{Z}_1} = \dim \frac{\mathcal{H}}{\mathcal{Z}_1} = \dim \frac{H}{Z_1} = \dim Z_0,$$

whence $Z_0\cap\mathcal{Z}_1=\{0\}$. Hence, $\mathcal{Z}_1$ is a closed complement of Z_0 in $\mathcal{H}$. ∎

Proposition 4.2 permits us to apply the construction of Lemma 2.3 with the same choice of β for T, B and $\mathcal{T}$, $\mathcal{B}$. Below, $\mathcal{Q}_+$ denotes the maximal projection associated with $\mathcal{T}$.

THEOREM 4.3. Let $T,B:H\to H$ and $\mathcal{T},\mathcal{B}:\mathcal{H}\to\mathcal{H}$ be as before. Then the boundary value problem (2.8)–(2.9) in H is uniquely solvable for all $\varphi_+\in Q_+[H]$, if and only if the boundary value problem

$$(\mathcal{T}\psi)'(x) = -\mathcal{A}\psi(x), \quad 0<x<\infty,$$

$$\mathcal{Q}_+\psi(0) = \overset{*}{\varphi}_+,$$

$$\|\psi(x)\| = O(1) \ (x\to\infty)$$

in $\mathcal{H}$ is uniquely solvable for all $\overset{*}{\varphi}_+\in\mathcal{Q}_+[\mathcal{H}]$.

Proof: The problems are equivalent to the Wiener–Hopf equations

$$\psi(x) - \int_0^\infty \mathcal{H}(x-y)B\psi(y)dy = \exp\{-xT^{-1}\}\varphi_+, \quad 0<x<\infty,$$

and

$$\psi(x) - \int_0^\infty \overset{*}{\mathcal{H}}(x-y)\mathcal{B}\psi(y)dy = \exp\{-x\mathcal{T}^{-1}\}\overset{*}{\varphi}_+, \quad 0<x<\infty,$$

respectively, where $\overset{*}{\mathcal{H}}(x)$ is the propagator function associated with $\mathcal{T}$. In the regular case (which then is regular for both T, B and $\mathcal{T}$, $\mathcal{B}$), both convolution equations are Fredholm operator equations on $L_\infty(H)_0^\infty$ and $L_\infty(\mathcal{H})_0^\infty$, respectively. The theorem then is immediate from Lemma 3.1. In the singular case (for T, B as well as $\mathcal{T}$, $\mathcal{B}$) one performs a reduction to corresponding regular cases with one and the same β. ∎

5. Nonregularity of the collision operator

In the analyses of both the integral form and the integrodifferential form of the abstract kinetic equation, the regularity condition (2.1) has played a crucial role. As a result, one might be led to conjecture that it is a necessary condition for the existence theorems and for the equivalence theorems posed in this and the previous chapters. In fact, this is not the case, as we shall demonstrate in this section. Such a generalization is important to treat kinetic models with collision kernels which may violate the regularity condition, for example, Boltzmann models in gas dynamics [218] and radiative transfer and neutron transport equations with arbitrary L_1-scattering kernels (cf. Sections IX.1 – IX.4). Ideally, one would like to remove the regularity condition for all collision operators of the form A=I−B with B compact. Although the results presented here are for B trace class, the abstract approach we shall outline appears to give hints on how one might obtain a generalization to B compact.

We will start the section by introducing notions connected with the tensor product of Banach spaces [175, 176, 307]. Then we will investigate the invertibility of the full line convolution operator by considering it as an element of an algebra of Wiener type. We will utilize the notion of a bisemigroup (see Section IV.4), which is a pair of complementary strongly continuous semigroups acting to the left and right, corresponding to the left/right half space problems which have been the theme of the stationary theory in this monograph, and will obtain the desired existence proofs by proving a theorem on the perturbation of bisemigroups for "weakly" integrable kernels. We refer to [127,129] for additional details.

Let X and Y be Banach spaces and X^*, Y^* the dual Banach spaces. By $X\otimes Y$ we denote the algebraic tensor product consisting of all finite linear combinations $\sum x_i\otimes y_i$ with $x_i\in X$, $y_i\in Y$. A norm α on $X\otimes Y$ is called a **reasonable cross norm** if $\alpha(x\otimes y)\leq \|x\|_X\ \|y\|_Y$ for all $x\in X$, $y\in Y$, and if $x^*\otimes y^*$ is a functional on the normed space $(X\otimes Y,\alpha)$ for all $x^*\in X^*$, $y^*\in Y^*$ with functional norm less than or equal to $\|x^*\|_{X^*}\ \|y^*\|_{Y^*}$. A reasonable cross norm α on $X\otimes Y$ is called **uniform** if $\alpha((A\otimes B)v)\leq$

$\|A\| \|B\| \alpha(v)$ for all $A \in L(X)$, $B \in L(Y)$ and $v \in X \otimes Y$. If X and Y are Banach algebras (even C* algebras), in general a cross norm will not be a Banach algebra norm. However, if $\mathfrak{X} \subset L(X)$ and $\mathfrak{Y} \subset L(Y)$ are Banach subalgebras, then a uniform norm α on $X \otimes Y$ induces a cross norm $\bar{\alpha}$ on $\mathfrak{X} \otimes \mathfrak{Y}$ via

$$\bar{\alpha}(C) = \sup_{v \in X \otimes Y} \frac{\alpha(Cv)}{\alpha(v)}\}, \qquad C \in \mathfrak{X} \otimes \mathfrak{Y},$$

which is a Banach algebra norm.

We will be interested in the **least reasonable cross norm** $\|\cdot\|_\varepsilon$, called also the **ε-tensor product**, **λ-tensor product** or **injective tensor product norm**, and the **greatest reasonable cross norm** $\|\cdot\|_\pi$, called also the **π-tensor product**, **γ-tensor product** or **projective tensor product norm**. By definition

$$\|u\|_\varepsilon = \sup \{ |(x^* \otimes y^*)(u)| : \|x^*\|, \|y^*\| \leq 1\} \tag{5.1}$$

$$\|u\|_\pi = \inf \{ \sum_{i=1}^n \|x_i\| \, \|y_i\| : u = \sum_{i=1}^n x_i \otimes y_i \}. \tag{5.2}$$

The **injective tensor product** $X \otimes_\varepsilon Y$ is the completion of $X \otimes Y$ with respect to $\|\cdot\|_\varepsilon$, and similarly the completion with respect to $\|\cdot\|_\pi$ is called the **projective tensor product**, denoted $X \otimes_\pi Y$. For $u \in X \otimes Y$ and any reasonable cross norm α one has $\|u\|_\varepsilon \leq \alpha(u) \leq \|u\|_\pi$. It is straightforward to show that the cross norms ε and π are uniform, which implies that $\bar{\varepsilon}$ and $\bar{\pi}$ are Banach algebra norms.

Consider a function $f: \mathbb{R} \to X^*$ from the real line to a dual Banach space. We will say that f is weak-* integrable (with respect to Lebesgue measure on $\mathbb{R}$) if the scalar valued function $\langle f(\cdot), x\rangle$ is in $L_1(\mathbb{R})$ for all $x \in X$.

LEMMA 5.1. Let X be a Banach space and $f: \mathbb{R} \to X^*$ be weak-* integrable. Then the map $X \to L_1(\mathbb{R})$ given by $x \to \langle f(\cdot), x\rangle$ is a bounded operator.

Proof: We wish to apply the Closed Graph Theorem. Suppose $\{x_n\}_{n=1}^\infty$ converges to x in X and $\langle f(\cdot), x_n\rangle$ converges to some function $\varphi(\cdot)$ in $L_1(\mathbb{R})$. Choose a subsequence $x_{n(i)}$ so that $\langle f(t), x_{n(i)}\rangle \to \varphi(t)$ pointwise a.e. Obviously, for almost every $t \in \mathbb{R}$ we have $\langle f(t), x_{n(i)}\rangle \to \langle f(t), x\rangle$. Hence $\varphi(\cdot) = \langle f(\cdot), x\rangle$. ∎

From this lemma we may conclude that there is an element x^* in X^* such that $\langle x^*, x\rangle = \int \langle f(t), x\rangle dt$ for all $x \in X$. This element x^* will be denoted $\int f(t)dt$ and is called the Gelfand integral of f. We also have that if f is weak-* integrable then

$$\|f\| = \sup \{\|<f(\cdot),x>\|_1 : x \in X, \|x\| \leq 1\} \tag{5.3}$$

is a finite number. We will call it the weak-* L_1 norm or the Gelfand norm of f. Thus the space $w^*L_1(\mathbb{R},X^*)$ of all weak-* integrable functions from $\mathbb{R}$ to X^* is a normed space with the Gelfand norm. Clearly we can identify the algebraic tensor product $L_1(\mathbb{R})\otimes X^*$ with a subspace of $w^*L_1(\mathbb{R},X^*)$. Also the Gelfand norm (5.5) coincides with the ε-norm. We may see then that the closure of the weak-* integrable functions is precisely $L_1(\mathbb{R})\otimes_\varepsilon X^*$. Unlike the case of weak or weak-* integrable functions, the space $L_1(X)_{-\infty}^{\infty}$ of Bochner integrable functions is complete. It coincides with the projective tensor product $L_1(\mathbb{R})\otimes_\pi X$ and the π-norm is the Bochner norm. On the other hand, an application of the uniform boundedness principle shows that in the spaces of bounded or bounded continuous functions, weak and strong notions coincide: $wL_\infty(\mathbb{R},X)=L_\infty(X)_{-\infty}^{\infty}$ and $wC(\mathbb{R},X)=C(X)_{-\infty}^{\infty}$.

Consider a Hilbert space H. Recall that every trace class operator C on H can be written in the form $C=\sum_{i=1}^{\infty}\alpha_i e_i(\cdot,f_i)$, where $\{e_i\}$ and $\{f_i\}$ are orthonormal systems of vectors in H and α_i are positive numbers, the singular values of C, which are summable. If A is any bounded operator, $\mathrm{tr}(AC) = \sum_{i=1}^{\infty}\alpha_i(Ae_i,\hat{e}_i)$. We denote the trace norm on $K_1(H)$, the trace class operators, by $\|\cdot\|_{(1)}$, to differentiate it from the $\|\cdot\|_1$ norm on $L_1(\mathbb{R})$, and have

$$\|C\|_{(1)} = \mathrm{tr}\,|C| = \sum_{i=1}^{\infty}\alpha_i.$$

The finite rank operators on H can be identified with $H\otimes H$. It is easy to see that the ε-norm is the usual operator norm, so $H\otimes_\varepsilon H = K(H)$, the compact operators on H, and the π-norm is the trace norm, so $H\otimes_\pi H = K_1(H)$. Moreover, viewing them as Banach spaces we have $K_1(H) = (K(H))^*$ and $(K_1(H))^* = L(H)$, the bounded operators on H (cf. [142]). For future reference we note that every element of the tensor product $L_1(\mathbb{R})\otimes L(H)$ induces a bounded operator on $L_\infty(\mathbb{R})\otimes_\varepsilon H$ by

$$\Big[\sum_{i=1}^{n} f_i(t)\otimes V_i\Big]\Big(\sum_{j=1}^{m} g_j\otimes w_j\Big) = \sum_{i=1}^{n}\Big(\sum_{j=1}^{m}(f_i * g_j(t))\otimes V_i w_j\Big)$$

for $f_i\in L_1(\mathbb{R})$, $g_j\in L_\infty(\mathbb{R})$, $V_i\in L(H)$ and $w_j\in H$.

Let T be an injective self adjoint operator on H with resolution of the identity σ, and define the propagator function $\mathcal{H}(x)h = \int_{\sigma(S)}E(t,x)\sigma(dt)h$ for all $h\in H$, where $E(x,t) = |t|^{-1}\exp\{-x/t\}$ for $xt>0$ and $E(x,t) = 0$ for $xt<0$. We wish to construct the projections $P_{\pm}$ and the corresponding semigroups from the solutions of the full line convolution equation

$$\psi(x) - \int_{-\infty}^{\infty} \mathcal{H}(x-y)B\psi(y)dy = \omega(x). \tag{5.4}$$

LEMMA 5.2. Let B be a trace class operator on H. Then the convolution operator

$$(\mathcal{L}\psi)(x) = \int_{-\infty}^{\infty} \mathcal{H}(x-y)B\psi(y)dy$$

is a bounded operator from $L_{\infty}(H)_{-\infty}^{\infty}$ into $C(H)_{-\infty}^{\infty}$.

Proof: Let us first consider the case when $Bh = \langle h,g\rangle f$ for certain $f,g \in H$ of unit norm, and consider an arbitrary vector $e \in H$ of unit norm. Then for all $\psi \in L_{\infty}(H)_{-\infty}^{\infty}$ we have

$$\langle(\mathcal{L}\psi)(x),e\rangle = \int_{-\infty}^{\infty} \langle\mathcal{H}(x-y)f,e\rangle\langle\psi(y),g\rangle dy.$$

In a straightforward way we obtain the estimate

$$\|\langle(\mathcal{L}\psi)(\cdot),e\rangle\|_{\infty} \le \|\langle\mathcal{H}(\cdot)f,e\rangle\|_{1}\|\langle\psi(\cdot),g\rangle\|_{\infty}.$$

Taking the supremum on both sides with e ranging over the vectors of unit norm and using the previous lemma, we get

$$\|\mathcal{L}\psi\|_{\infty} \le \|f\|\,\|g\|\,\|\psi\|_{\infty} = \|\psi\|_{\infty}.$$

Next, let us suppose that $Bh = \Sigma_{i=1}^{\infty}\gamma_i\langle h,g_i\rangle f_i$, where $\{f_i\}_{i=1}^{\infty}$ and $\{g_i\}_{i=1}^{\infty}$ are orthonormal sets in H and $\{\gamma_i\}_{i=1}^{\infty}$ is a nonincreasing sequence in ℓ_1 of nonnegative numbers. Then

$$\|\mathcal{L}\psi\|_{\infty} \le \sum_{i=1}^{\infty}\gamma_i\|f_i\|\,\|g_i\|\,\|\psi\|_{\infty} = \|B\|_{(1)}\|\psi\|_{\infty},$$

where $\|B\|_{(1)}$ is the trace norm of B.

In order to prove that $\mathcal{L}$ is a bounded operator as a map from $L_{\infty}(H)_{-\infty}^{\infty}$ into

$C(H)_{-\infty}^{\infty}$, we approximate B in the trace norm by a sequence of trace class operators $\{B_n\}_{n=1}^{\infty}$ satisfying the regularity condition (2.1). (Such an approximation is constructed in the proof of Theorem 5.5.) Since the regularity condition is sufficient to guarantee that $\mathcal{L}$ maps $L_{\infty}(H)_{-\infty}^{\infty}$ into $C(H)_{-\infty}^{\infty}$, we may conclude that $\mathcal{L}$ has the same property for any trace class operator B. ■

We will prove shortly that the operator $I-\mathcal{L}$ is invertible if and only if $T^{-1}A$ does not have zero or purely imaginary eigenvalues. This will be accomplished using the generalization by Bochner and Phillips to noncommutative Banach algebras of the theory of Fourier integrals and maximal ideals by Wiener and Gelfand. We shall give a brief summary of this theory.

Let Z and $\mathcal{F}$ be Banach algebras with unit. Assume that Z is commutative. By $Z\otimes\mathcal{F}$ we denote the algebraic tensor product consisting of finite sums $\sum z_i f_i$. Evidently this is an algebra. Because both Z and $\mathcal{F}$ have a unit, we may assume that Z and $\mathcal{F}$ are inside $Z\otimes\mathcal{F}$, and, moreover, Z will be in the center of $Z\otimes\mathcal{F}$. Next suppose that $\mathcal{A}$ is a Banach algebra with unit e, such that $Z\otimes\mathcal{F}$ is dense in $\mathcal{A}$ and the norms on Z and $\mathcal{F}$ coincide with the norms induced on Z and $\mathcal{F}$ as subspaces of $\mathcal{A}$. Every multiplicative functional $\varphi:Z\to\mathbb{C}$ induces an algebra homomorphism $\Phi:Z\otimes\mathcal{F}\to\mathcal{F}$ via $\Phi(\sum z_i f_i) = \sum\varphi(z_i)f_i$. Following [149] we call the algebra $\mathcal{A}$ a $Z\otimes\mathcal{F}$-algebra if all the induced homomorphisms Φ are bounded operators from $Z\otimes\mathcal{F}$ with the $\mathcal{A}$-norm to $\mathcal{F}$. Therefore we can extend each induced homomorphism to a Banach algebra homomorphism $\Phi:\mathcal{A}\to\mathcal{F}$.

Wiener [391] proved the following important lemma, which was crucial in the proof of his Tauberian theorems.

LEMMA 5.3. (Wiener) Let f be a function from the unit circle $\mathbf{T}$ to the complex numbers $\mathbb{C}$ and suppose f(z) is invertible for every $z\in\mathbf{T}$, i.e., $f:\mathbf{T}\to\mathbb{C}\setminus\{0\}$. Then if f has an absolutely convergent Fourier expansion, so does $1/f$.

The above lemma was one of the first testing grounds for the maximal ideal theory for commutative Banach algebras developed by Gelfand [132, 134]. Gelfand's proof was extremely short and elegant compared with the proof of Wiener. Bochner and Phillips obtained a substantial generalization of Wiener's Lemma ([45], also [4, 149]). Roughly speaking they changed the scalars from the field of complex numbers to a noncommutative Banach algebra by tensoring the commutative Wiener algebra with the noncommutative algebra.

Theorem 5.4. (Bochner–Phillips) Let $\mathcal{A}$ be a $Z\otimes\mathcal{F}$-algebra. An element $a\in\mathcal{A}$ has a left, right or two-sided inverse in $\mathcal{A}$ if, for each induced homomorphism Φ, the element $\Phi(a)\in\mathcal{F}$ has a left, right or two-sided inverse, respectively.

Let us define a norm $\|u\|_\alpha$ on $\mathbb{B}_0 = L_1(\mathbb{R})\otimes L(H)$ to be the operator norm of $u\in\mathbb{B}_0$ viewed as a bounded operator on $L_\infty(\mathbb{R})\otimes_\varepsilon H$, and consider a multiplication on $\mathbb{B}$ induced by convolution in $L_1(\mathbb{R})$ and operator multiplication in $L(H)$. One sees readily that $\|\cdot\|_\alpha$ is a reasonable cross norm on $\mathbb{B} = L_1(\mathbb{R})\otimes_\alpha L(H)$, and, even more, is a Banach algebra norm. In fact, the submultiplicativity and the bound $\|u\|_\alpha\le\|u\|_\pi$ for $u\in\mathbb{B}_0$ are evident. Suppose that $\psi\in(L_1(\mathbb{R}))^*=L_\infty(\mathbb{R})$ and $\varphi\in(L(H))^*$ with $\|\psi\|_\infty=\|\varphi\|=1$. Since $L_1(\mathbb{R})$ contains an approximate unit $\{u_k\}_{k=1}^\infty$ satisfying $u_k\in L_\infty(\mathbb{R})\cap L_1(\mathbb{R})$, $\|u_k\|_1=1$ and $\lim_k\|f*u_k-f\|_1=0$, one has, by Fubini's Theorem,

$$\psi(f*u_k) = \int_{-\infty}^{\infty}\psi(t)\int_{-\infty}^{\infty}f(s)u_k(t-s)dsdt = \int_{-\infty}^{\infty}u_k(-t)\int_{-\infty}^{\infty}f(s)\psi(-(t-s))dsdt =$$

$$= \hat{u}_k(f*\hat{\psi}),$$

where $\hat{u}_k(t)=u_k(-t)$ and $\hat{\psi}(t)=\psi(-t)$. Therefore, for $\sum_{i=1}^n f_i\otimes V_i\in L_1(\mathbb{R})\otimes L(H)$, we obtain

$$|(\psi\otimes\varphi)(\sum_{i=1}^n f_i\otimes V_i)| \le \sup\{\|\sum_{i=1}^n (f_iV_ih)\| : h\in H,\ \|h\|=1\} \le$$

$$\le \lim_k \sup\{\|\sum_{i=1}^n \hat{u}_k(f_i*\hat{\psi})V_ih\| : h\in H,\ \|h\|=1\} \le \|\sum_{i=1}^n f_i\otimes V_i\|_\alpha,$$

and $\|\cdot\|_\alpha$ is a reasonable cross norm.

Writing I for identity in $L(H)$ and $\mathbb{C}\oplus L_1(\mathbb{R})$ for $L_1(\mathbb{R})$ with an identity adjoined, we will define $Z = (\mathbb{C}\oplus L_1(\mathbb{R}))\otimes I$ and $\mathbb{A}$ the completion of $(\mathbb{C}\oplus L_1(\mathbb{R}))\otimes L(H)$ with respect to the norm

$$\|\sum_{i=1}^n(\beta_i+f_i)\otimes V_i\|_{\mathbb{A}} = \|\sum_{i=1}^n\beta_iV_i\|_{L(H)} + \|\sum_{i=1}^n f_i\otimes V_i\|_\alpha.$$

Then $\|\cdot\|_{\mathbb{A}}$ is a Banach algebra norm and $\mathbb{A}$ is a $Z\otimes L(H)$-algebra. Indeed, the multiplicative functionals on Z are precisely $\Phi_\lambda(\beta\oplus f(\cdot)I)=\hat{f}(\lambda)$ for $\lambda\in\mathbb{R}$ and $\Phi_\infty(\beta\oplus f(\cdot)I)=\beta$, where ^ represents the Fourier transform. Thus, we may compute (for $\lambda\in\mathbb{R}$)

$$\|\Phi_\lambda(\sum_{i=1}^{n} f_iV_i)\| = \sup \{\| \sum_{i=1}^{n} (e^{\lambda t}f_i(t)*e^{-\lambda t})V_ih\| : h\epsilon H, \|h\|=1\} \leq$$

$$\leq \sup \{\| \sum_{i=1}^{n} (f_i(t)*g(t))V_ih\| : h\epsilon H, g\epsilon L_\infty(\mathbb{R}), \|h\|=\|g\|_\infty=1\} \leq$$

$$\leq \| \sum_{i=1}^{n} f_iV_i\|_\alpha.$$

According to the Bochner-Phillips Theorem 5.4, it is necessary and sufficient for the unique solvability of Eq. (5.4) on $L_\infty(H)_{-\infty}^{\infty}$ that the values of its symbol be invertible elements of the Banach algebra $\mathbb{A}$. For the latter to be true it is necessary and sufficient that the symbol

$$W(\lambda) = I - \int_{-\infty}^{\infty} e^{x/\lambda}\mathcal{H}(x)Bdx$$

of Eq. (5.4) be invertible for all extended imaginary λ. Since $W(\lambda) = (T-\lambda)^{-1}(T-\lambda A)$, it is necessary and sufficient for unique solvability that $T^{-1}A$ does not have zero or purely imaginary eigenvalues.

We recall now the notion of a strongly continuous bisemigroup E(t) on a Hilbert space H, and maintain the notation of Section IV.4. In particular, the bounded projections

$$\Pi_\pm = \operatorname*{s-lim}_{\pm t \downarrow 0}(\pm E(t))$$

are the separating projectors of the bisemigroup, and we will denote by S the generator of E(t). We will also write E(t;S) and exp{-tS} for the bisemigroup generated by S. In the case that S generates a decaying semigroup one observes that $\Pi_\pm$ are the maximal positive/negative spectral projectors for the operator S, i.e., $\sigma(S\Pi_\pm) \subset \{\lambda \epsilon \mathbb{C} : \pm\mathrm{Re}\lambda \leq 0\}$.

For an angle $0<\theta\leq\pi/2$ we denote sectors about the real axis by $\Gamma_{\theta\pm}$ with $\Gamma_{\theta\pm}=\{z\epsilon\mathbb{C} : |\arg(\pm z)|<\theta\}$ and $\Gamma_\Gamma=\Gamma_{\theta+}\cup\Gamma_{\theta-}$. Let us assume that S is a spectral operator of scalar type (see Section XIII.2; also [105], [109] vol. III) on a Hilbert space H with spectral measure $d\Sigma(\lambda)$, i.e.,

$$S = \int_{\sigma(S)} \lambda d\Sigma(\lambda).$$

Assume also that $\sigma(S)\subset\overline{\Gamma}_{\pi/2-\theta_1}$ for some $0<\theta_1<\pi/2$ and that zero is either in the resolvent set or in the continuous spectrum of S. It is immediate to check that S is a generator of a strongly decaying analytic bisemigroup of angle at least θ_1, with separating projectors given by the spectral projectors $\Pi_\pm=\Sigma(\sigma(S)\cap\{\pm\mathrm{Re}z\geq 0\})$. If S^{-1} is a bounded operator, the bisemigroup is exponentially decaying.

We wish to establish sufficient conditions so that a perturbation $S^\times=SA$ will still generate a bisemigroup. Besides the assumptions on S made above, suppose also that $\sigma(S^\times)$ is contained in a sector about the real axis and that $B = I-A$ is trace class with Ker $A = \{0\}$.

THEOREM 5.5. If S is a spectral operator of scalar type with $\sigma(S)$ and $\sigma(SA)$ contained in a sector about the real axis and zero in the resolvent set or continuous spectrum of S, and if $B = I-A$ is trace class with Ker $A = \{0\}$, then $S^\times$ generates an analytic bisemigroup $E^\times(t)$ with separating projectors $\Pi^\times_\pm$. For any $t\in\mathbb{R}\setminus\{0\}$, $E(t)-E^\times(t)$ and $\Pi_\pm-\Pi^\times_\pm$ are compact. The bisemigroup $E^\times(t)$ is strongly decaying. If $\sigma(S)$ has a gap at zero (i.e., S^{-1} is bounded), then $E^\times(t)$ is exponentially decaying.

Proof: Consider the operator valued function $\Bbbk(t)=SE(t)B$ defined for $t\in\mathbb{R}\setminus\{0\}$. We claim that $\Bbbk\in L_1(\mathbb{R})\otimes_\alpha L(H)$. Indeed, let us write

$$\Bbbk(t)h = \begin{cases} \int_0^\infty \lambda e^{-t\lambda} d\Sigma(\lambda)Bh, & t>0, \\ -\int_{-\infty}^0 \lambda e^{-t\lambda} d\Sigma(\lambda)Bh, & t<0, \end{cases}$$

let $B = \Sigma_{i=1}^\infty s_i f_i(\cdot,e_i)$, with $\{e_i\}_{i=1}^\infty$ and $\{f_i\}_{i=1}^\infty$ orthonormal families and $\{s_i\}_{i=1}^\infty$ the singular numbers, and denote the trace norm $\|B\|_{(1)}=\Sigma_{i=1}^\infty s_i$. For $g\in L_\infty(H)_{-\infty}^\infty$, $\|g\|_1=1$ and $h\in H$, we have

$$|(Bg(t-\tau),\Sigma(\lambda)h)| \leq \sum_{i=1}^\infty \sum_{j=1}^4 \tfrac14 s_i |(h_{ij},\Sigma(\lambda)h_{ij})|,$$

where $h_{ij}=h+(-1)^{j/2}e_i$ for $j=1,2,3,4$. Therefore, by Fatou's Lemma and Fubini's Theorem,

$$\int_0^\infty d\tau\int_0^\infty \lambda e^{-\tau\lambda}|d(Bg(t-\tau),\Sigma(\lambda)h)| \leq \sum_{i=1}^\infty\sum_{j=1}^4 \tfrac14 s_i\int_0^\infty d\tau\int_0^\infty \lambda e^{-\tau\lambda} d\|\Sigma(\lambda)h_{ij}\|^2 =$$

$$= \sum_{i=1}^{\infty} \sum_{j=1}^{4} \tfrac{1}{4} s_i \int_0^{\infty} d\|\Sigma(\lambda)h_{ij}\|^2. \tag{5.5}$$

Similarly, we obtain

$$-\int_{-\infty}^{0} d\tau \int_{-\infty}^{0} \lambda e^{-\tau\lambda} \,|\, d(Bg(t-\tau),\Sigma(\lambda)h) \,|\, \leq \sum_{i=1}^{\infty} \sum_{j=1}^{4} \tfrac{1}{4} s_i \int_{-\infty}^{0} d\|\Sigma(\lambda)h_{ij}\|^2. \tag{5.6}$$

By (5.5) and (5.6), we get

$$\|\mathbb{k}*g\|_{\varepsilon} \leq \sup_{t\in\mathbb{R}} \{-\int_{-\infty}^{0} d\tau \int_{-\infty}^{0} \lambda e^{-\tau\lambda} \,|\, d(Bg(t-\tau),\Sigma(\lambda)h) \,|\, +$$

$$+ \int_{0}^{\infty} d\tau \int_{0}^{\infty} \lambda e^{-\tau\lambda} \,|\, d(Bg(t-\tau),\Sigma(\lambda)h) \,|\, : h\in H,\ \|h\|=1\} \leq$$

$$\leq 4 \sum_{i=1}^{\infty} s_i = 4\|B\|_{(1)}.$$

Hence, $\|\mathbb{k}\|_{\alpha} \leq 4\|B\|_{(1)}$.

Now consider $\mathbb{k}_n(t) = SE(t)B_n$, where $B_n = \sum_{i=1}^{n} s_i f_i(\cdot, e_i)$ is a finite rank approximate of B. By the above argument, we have also

$$\|(\mathbb{k}_n - \mathbb{k})*g\|_{\varepsilon} \leq 4\|B_n - B\|_{(1)}$$

for each $g\in L_\infty(H)_{-\infty}^{\infty}$, $\|g\|_{\varepsilon}=1$. Since $\mathcal{H}(\cdot)=SE(\cdot)\in L_1(\mathbb{R})\otimes_{\varepsilon} L(H)$ by an easy calculation (cf. [118]), there exists $\{u_j(t)\}_{j=1}^{\infty} \subset L_1(\mathbb{R})\otimes L(H)$ such that $\|\mathcal{H}(\cdot)-u_j(\cdot)\|_{\varepsilon}\to 0$ as $j\to\infty$. Define $\mathbb{k}_{j,n}(t)=u_j B_n$. Then

$$\|(u_j B_n)*g\|_{\varepsilon} \leq \sup_{t\in\mathbb{R}} \{\sup_{h\in H} \sum_{i=1}^{n} s_i \int_{-\infty}^{\infty} d\tau \,|\, (g(t-\tau),e_i)(u_j(\tau)f_i,h) \,|\, / \|h\|\} \leq$$

$$\leq \sup_{t\in\mathbb{R}} \sum_{i=1}^{n} s_i \|u_j\|_{\varepsilon} = \|B_n\|_{(1)} \|u_j\|_{\varepsilon},$$

and so $\mathbb{k}_{j,n} \in L_1(\mathbb{R})\otimes_{\alpha} L(H)$. A similar estimate gives

$$\|(\mathbb{k}_{j,n} - \mathcal{H}(\cdot)B_n)*g\|_{\varepsilon} \leq \sup_{h\in H} \{ \sum_{i=1}^{n} s_i \|((u_j - \mathcal{H})f_i, h)\|_1 / \|h\|\}.$$

Therefore, $\mathbb{k}_{j,n}$ approaches $\mathbb{k}_n$ as $j\to\infty$ in the topology of $\|\cdot\|_{\alpha}$, whence $\mathbb{k}\in L_1(\mathbb{R})\otimes_{\alpha} L(H)$.

Next, let us consider the convolution operator,

$$(\mathcal{L}\psi)(t) = \int_{-\infty}^{\infty} \Bbbk(t-s)\psi(s)ds,$$

n $L_\infty(H)_{-\infty}^{\infty}$. From the assumptions on B it follows that the symbol $W(\lambda) = -S(S-\lambda)^{-1}B = (S-\lambda)^{-1}(S^x-\lambda)$ has a bounded inverse on the extended imaginary xis. We can apply the Bochner–Phillips Theorem to conclude that the operator $I-\mathcal{L}$ invertible with inverse $I+\mathcal{L}^x$, where

$$(\mathcal{L}^x\psi)(t) = \int_{-\infty}^{\infty} \Bbbk^x(t-s)\psi(s)ds,$$

nd $\Bbbk^x \in L_1(\mathbb{R})\otimes_\alpha L(H)$.

We claim that $E^x(t)h=(I+\mathcal{L}^x)E(t)h$, for $h\in H$ and $t\in\mathbb{R}\setminus\{0\}$, is the bisemigroup enerated by S^x. First let us check that $E^x(t)$ defined above is a bisemigroup. Fix ome $s>0$ and $h\in H$ and define $\psi(t)=E^x(t+s)h$ if $t>0$ and $\psi(t)=0$ if $t<0$. When >0 we have

$$(I-\mathcal{L})\psi(t) = E^x(t+s)h - \int_s^{\infty} \Bbbk(t+s-r)E^x(r)hdr =$$

$$= (I-\mathcal{L})E^x(t+s)h + \int_{-\infty}^{s} \Bbbk(t+s-r)E^x(r)hdr,$$

ince for $r<s$ we have $t+s-r>0$ and $\Bbbk(t+s-r)=E(t)\Bbbk(s-r)$, while for $r>s$ we have $)=E(t)\Bbbk(s-r)$. We can rewrite the above as

$$(I-\mathcal{L})\psi(t) = E(t+s)h + E(t)\mathcal{L}E^x(s)h = E(t)(E(s) + \mathcal{L}E^x(s))h = E(t)E^x(s)h.$$

Therefore, $\psi(t)=E^x(t)E^x(s)h$ for $t>0$. When $t<0$ we have

$$(I-\mathcal{L})\psi(t) = E(t)\int_{-\infty}^{\infty} \Bbbk(s-r)E^x(r)hdr = E(t)(E(s) + \mathcal{L}E^x(s))h = E(t)E^x(s)h.$$

Combining the two cases and recalling the definition of $\psi(t)$, we get $E^x(t)E^x(s)=E^x(t+s)$ for $t,s>0$ and $E^x(t)E^x(s)=0$ for $t<0$, $s>0$.

It is easy to see that $E^x(t)$ is strongly continuous and bounded. To check that $\Pi_+^x+\Pi_-^x=I$, note that convolutions are smoothing, hence the jump of $E^x(t)$ at $t=0$ is equal to the jump of $E(t)$ at $t=0$, i.e.,

$$(\Pi_+^x+\Pi_-^x)h = E^x(+0)h - E^x(-0)h = E(+0)h - E(-0)h = (\Pi_++\Pi_-)h = h.$$

Therefore $E^x(t)$ is a bounded strongly continuous bisemigroup.

By assumption, $\sigma(S^x)\subset\bar{\Gamma}_{\pi/2-\theta_2}$ for some $0<\theta_2<\pi/2$. Taking a double sided Laplace transform, we get immediately

$$(S^x-\lambda)^{-1}h = \int_{-\infty}^{\infty} e^{\lambda t}E^x(t)h\,dt, \quad \mathrm{Re}\lambda=0, \quad \lambda\neq 0, \quad h\in H.$$

Hence S^x is the generator of $E^x(t)$ and $\Pi_\pm^x$ are positive/negative spectral projectors for S^x.

If $\varphi\in[0,\theta)$, where $\theta=\min\{\theta_1,\theta_2\}$ then all the above applies as well for $e^{i\varphi}S$ and $e^{i\varphi}S^x$, i.e., $e^{i\varphi}S^x$ is the generator of a bounded bisemigroup, so S^x generates a bounded analytic bisemigroup of angle at least θ (cf. [213], Theorem IX 1.23).

We will show next the compactness of $E(t)-E(t)^x$, and consequently of $\Pi_\pm^x-\Pi_\pm$. Set $\Phi_n(t)= |t|^{1/n}\chi_{[-n,n]}(t)$, where $\chi_{[a,b]}$ is the characteristic function of the interval $[a,b]$. Denote $B_n=\Phi_n(S)B$. Obviously $SE(\cdot)B_n$ is a Bochner integrable function. On the other hand, using the functional calculus for S and the dominated convergence theorem, we get that $\|(\Phi_n(S)-I)h\|\to 0$ as $n\to\infty$ for every $h\in H$, hence $\|(\Phi_n(S)-I)C\|\to 0$ as $n\to\infty$ for every compact operator C. Now, B can be written as $B=CD$, where C is compact self adjoint and D is trace class. Indeed, denote by σ the sum of the singular numbers s_i and by σ_n the partial sums. Choose a strictly increasing sequence of natural numbers $\{k(r)\}$ such that $\sigma-\sigma_{k(r)}\leq 3^{-r}\sigma$ for $r=1,2,\ldots$. Setting $c_i=2^{-r}$ and $d_i=2^r s_i$ for $k(r-1)<i\leq k(r)$, the desired operators are $C=\sum c_i f_i(\cdot,f_i)$ and $D=\sum d_i f_i(\cdot,e_i)$. Note that $\|D\|_{(1)} = \sum d_i \leq 3\sigma = 3\|B\|_{(1)}$. Then we will have that $\|(\Phi_n(S)-I)B\|_{(1)} \leq \|D\|_{(1)}\|(\Phi_n(S)-I)C\| \to 0$ as $n\to\infty$. From before we have

$$\left\|\int_{-\infty}^{\infty} k(t-s)E^x(s)ds\right\| \leq \mathrm{const.}\|B\|_{(1)}.$$

Let $E_n^x(t)$ denote the bisemigroup generated by $S(I-B_n)$. We have then that $E(t)-E_n^x(t)$ is compact for any n by Bochner integrability. From this and the above considerations we may conclude that $E(t)-E^x(t)$ is compact.

Because $D((S^x)^{-1})=D(S^{-1})$ is dense, zero is either in the resolvent set or in the continuous spectrum of S^x, and hence E^x is strongly decaying. If the operator S has a gap at zero, then it is immediate that $E(t)$, and hence $k(t)$, is exponentially decaying. It then follows [134] that $k^x(t)$ will be exponentially decaying, implying the exponential decay of $E^x(t)$. ∎

In effect, Theorem 5.5 provides for the unique solvability of the boundary value problem (2.8)–(2.9). Let us assume that the operator T is injective and self adjoint on the Hilbert space H, and that A is accretive with $B = I-A$ finite rank and Ker $A = \mathrm{Ker}(A+A^*) = 0$. With $S=T^{-1}$ the separating projectors $\Pi_\pm^x$ corresponding to the analytic bisemigroup generated by $S^x=T^{-1}A$ are precisely the complementary projections $P_\pm$ associated with the transport operator $K=T^{-1}A$. The unique solvability of the boundary value problem is equivalent to the invertibility of the operator $V=Q_+P_++Q_-P_-$. The existence of the albedo operator $E=V^{-1}$ follows from the accretiveness assumption by the argument presented Section IV.4 for nonsymmetric collision operators.

COROLLARY 5.6. For every $\varphi_+ \in Q_+D(T)$ the boundary value problem (2.8)–(2.9) has a unique solution $\psi(x)=\exp(-xT^{-1}A)E\varphi_+$, which is decaying at infinity and is square integrable in x. If T is a bounded operator, then the solutions are exponentially decaying.

These results may be extended to collision operators with nontrivial kernel satisfying Ker A = Ker(Re A) by a decomposition of the type introduced in Chapter III. For details see [129].

Chapter VIII

ALBEDO OPERATORS, H-EQUATIONS AND REPRESENTATION OF SOLUTIONS

1. Albedo operators and H-equations: the regular case

In the previous chapters we have defined the albedo operator, which specifies the full boundary value of the solution of a half space problem in terms of partial range boundary data. In this section we shall construct, under the general assumptions of Section VII.2, the albedo operator in terms of certain special functions. These functions generalize the H-functions, which were first extensively studied by Chandrasekhar [89].

The method we are about to describe was introduced in transport theory by Burniston, Mullikin and Siewert [58] for two group neutron transport with isotropic scattering. It was applied to one speed neutron transport with degenerate anisotropic scattering by Mullikin [274] and generalized to a large class of multigroup type models by Kelley [216]. An abstract approach presented by van der Mee [366] will be the basis of the present and the next section.

The first part of the construction can be carried out for general convolution equations on a half line (see [139, 141]). It was only in a rather late stage of the development of convolution equations theory that Krein [227] observed the connection with H-functions. For certain scalar Wiener-Hopf equations, this connection was already thoroughly analyzed by Busbridge [61].

Let us consider a (real or complex) Banach space X, and let $k \in L_1(L(X))_{-\infty}^{\infty}$. According to Theorem VII 1.1, if the symbol

$$W(\lambda) = I - \int_{-\infty}^{\infty} e^{x/\lambda} k(x)dx$$

for Re $\lambda=0$ has a right canonical factorization, then for every $\omega \in L_p(X)_0^\infty$ $(1 \leq p \leq \infty)$ and for every $\omega \in C(X)_0^\infty$ the Wiener-Hopf equation

$$\psi(x) - \int_0^\infty k(x-y)\psi(y)dy = \omega(x), \quad 0\le x<\infty, \tag{1.1}$$

has a unique solution $\psi\in L_p(X)_0^\infty$, or correspondingly $\psi\in C(X)_0^\infty$, which is given by

$$\psi(x) = \omega(x) + \int_0^\infty \gamma(x,y)\omega(y)dy, \quad 0\le x<\infty. \tag{1.2}$$

PROPOSITION 1.1. The resolvent kernel $\gamma(x,y)$ satisfies

$$\int_0^\infty dy \int_0^\infty dz\ e^{y/\mu}e^{-z/\nu}\{\delta(y-z) + \gamma(y,z)\} = \frac{\mu\nu}{\mu-\nu} H_\ell(-\mu)H_r(\nu), \tag{1.3}$$

where $\mathrm{Re}\mu\le 0$, $\mathrm{Re}\nu\ge 0$, $\delta(\cdot)$ is the usual δ-function, and

$$W(\lambda)^{-1} = H_\ell(-\lambda)H_r(\lambda)$$

for Re $\lambda=0$ is a right canonical factorization of $W(\lambda)^{-1}$.

Proof: Let us solve Eq. (1.1) with right hand side $\omega(x)=e^{-x/\nu}h$, where $\mathrm{Re}\nu>0$ and $h\in X$. Then its solution ψ satisfies

$$\int_0^\infty e^{y/\mu}\psi(y)dy = \int_0^\infty e^{y/\mu}\{e^{-y/\nu} + \int_0^\infty e^{-z/\nu}\gamma(y,z)dz\}h\ dy,$$

which is the left hand side of Eq. (1.3) acting on the vector $h\in X$. Using the Wiener-Hopf method explained in Section VII.1 and the factorization of $W(\lambda)$, we obtain the Riemann-Hilbert problem

$$H_\ell(-\mu)^{-1}\int_0^\infty e^{y/\mu}\psi(y)dy + H_r(\mu)\int_{-\infty}^0 e^{y/\mu}\psi(y)dy = H_r(\mu)\int_0^\infty e^{y(1/\mu-1/\nu)}h dy.$$

The right hand side we simplify to $\mu\nu(\mu-\nu)^{-1}H_r(\mu)h$. As the solution we find

$$\int_0^\infty e^{y/\mu}\psi(y)dy = \frac{\mu\nu}{\mu-\nu}H_\ell(-\mu)H_r(\nu)h,$$

$$\int_{-\infty}^0 e^{y/\mu}\psi(y)dy = \mu\nu H_r(\mu)^{-1}\,\frac{H_r(\mu)-H_r(\nu)}{\mu-\nu}\,h,$$

which implies (1.3). ■

In the next theorem we give a simple application. Let T be an injective self adjoint operator and B a compact operator on a Hilbert space H satisfying the regularity condition

$$\exists\alpha>0:\ \mathrm{Ran}\,B \subset \mathrm{Ran}\,|T|^{\alpha} \cap D(|T|^{1+\alpha}). \tag{1.4}$$

Put $A = I-B$, and suppose that $T^{-1}A$ does not have zero or imaginary eigenvalues (the regular case). Choose a closed subspace $\mathbb{B} \supset \mathrm{Ran}B^*$, and let $\pi: H \to \mathbb{B}$ and $j: \mathbb{B} \to H$ be operators such that πj is the identity on $\mathbb{B}$ and $j\pi$ the orthogonal projection of H onto $\mathbb{B}$. As in Chapter VII we define the dispersion function $\Lambda(\lambda)$ by

$$\Lambda(\lambda) = I - \int_{-\infty}^{\infty} e^{x/\lambda} \pi\, \mathcal{H}(x) Bj\,dx = \pi Aj + \pi T(T-\lambda)^{-1}Bj. \tag{1.5}$$

By Theorem VII 3.2, the dispersion function has a right canonical factorization $\Lambda(\lambda)^{-1} = H_\ell^+(-\lambda)H_r^+(\lambda)$ for Re $\lambda=0$, and the boundary value problem

$$(T\psi)'(x) = -A\psi(x), \quad 0<x<\infty, \tag{1.6}$$

$$Q_+\psi(0) = \varphi_+, \tag{1.7a}$$

$$\|\psi(x)\| = O(1) \ (x\to\infty) \tag{1.7b}$$

has a unique solution in H for every $\varphi_+ \in Q_+[D(T)]$.

THEOREM 1.2. Let $\sigma(\cdot)$ be the resolution of the identity of T, and $\psi(x)$ the unique solution of (1.6)–(1.7). Then the boundary value $\psi(0) = E_+\varphi_+$ is given by

$$E_+\varphi_+ = \varphi_+ + \int_{-\infty}^{0}\int_{0}^{\infty} \sigma(d\mu)\, \frac{\nu}{\nu-\mu}\, BjH_\ell^+(-\mu)H_r^+(\nu)\pi\sigma(d\nu)\varphi_+, \tag{1.8}$$

where E_+ is the projection of H onto Ran P_+ along Ran Q_-. Here $\sigma(d\mu)$ is integrated over $(-\infty,0)$ and $\sigma(d\nu)$ over $(0,\infty)$.

Proof: According to Theorem VI 3.4 problem (1.6)–(1.7) is equivalent to the Wiener–Hopf equation

$$\psi(x) - \int_0^\infty \mathcal{H}(x-y)B\psi(y)dy = e^{-xT^{-1}}\varphi_+, \quad 0\le x<\infty.$$

Using $Bj\pi=B$ and putting $\chi(y) = \pi\psi(y)$, we obtain

$$\psi(x) = e^{-xT^{-1}}\varphi_+ + \int_0^\infty \mathcal{H}(x-y)Bj\chi(y)dy,$$

where

$$\chi(x) - \int_0^\infty \pi\mathcal{H}(x-y)Bj\chi(y)dy = \pi e^{-xT^{-1}}\varphi_+, \quad 0\le x<\infty.$$

The latter Wiener-Hopf equation is uniquely solvable (on an appropriate space of H-valued functions), because its symbol $\Lambda(\lambda)$ has a right canonical factorization. In fact, with the help of the representations

$$\mathcal{H}(-y) = -\int_{-\infty}^0 \frac{1}{\mu} e^{y/\mu}\sigma(d\mu) \tag{1.9a}$$

and

$$e^{-xT^{-1}}Q_+ = \int_0^\infty e^{-x/\nu}\sigma(d\nu), \tag{1.9b}$$

one obtains

$$E_+\varphi_+ = \psi(0) = \varphi_+ - \int_0^\infty dy \int_{-\infty}^0 \sigma(d\mu)\frac{1}{\mu} e^{y/\mu}Bj\chi(y).$$

Applying Theorem 1.1, this gives

$$E_+\varphi_+ = \varphi_+ - \int_0^\infty dy \int_{-\infty}^0 \sigma(d\mu)\frac{1}{\mu} e^{y/\mu}Bj \times$$

$$\times [\pi\int_0^\infty e^{-y/\nu}\sigma(d\nu)\varphi_+ + \int_0^\infty dz \int_0^\infty \gamma(y,z)\pi e^{-z/\nu}\sigma(d\nu)\varphi_+].$$

Changing the order of integration (allowed because of Fubini's theorem applied to the multiple integrals obtained by computing $(E_+\varphi_+,h)$) we obtain

$$E_+\varphi_+ = \varphi_+ - \int_{-\infty}^0 \int_0^\infty \sigma(d\mu) \times$$

$$\times \frac{1}{\mu} \mathrm{Bj}\Big[\int_0^\infty dy \int_0^\infty dz\; e^{y/\mu} e^{-z/\nu} \{\delta(y-z)+\gamma(y,z)\}\Big]\pi\sigma(d\nu)\varphi_+,$$

where $\sigma(d\mu)$ is integrated over $(-\infty,0)$ and $\sigma(d\nu)$ over $(0,\infty)$. One may now use (1.3) to obtain the boundary value (1.8). ■

In Section 4 we shall give several applications of formula (1.8). We notice that

$$\hat{E}_+\varphi_+ = \varphi_+ + \int_{-\infty}^0 \int_0^\infty \sigma(d\mu)\, \frac{\mu}{\nu-\mu}\, \mathrm{Bj}H_\ell^+(-\mu)H_r^+(\nu)\pi\sigma(d\nu)\varphi_+$$

is a bounded operator on H which leaves invariant the domain of T and satisfies $T\hat{E}_+h = \hat{E}_+Th$ for $h \in D(T)$. Evidently, $\hat{E}_+$ is the projection of H onto Ran $\hat{P}_+$ along Ran Q_-.

Let us derive the analogs of Chandrasekhar's H-equations. For this purpose, we normalize our factors by requiring $H_\ell^+(0^+)=H_r^+(0^+)=I$.

THEOREM 1.3. We have the coupled non-linear integral equations

$$H_\ell^+(z)^{-1} = I - z\int_0^\infty (z+t)^{-1} H_r^+(t)\pi\sigma(dt)\mathrm{Bj}, \tag{1.10}$$

$$H_r^+(z)^{-1} = I - z\int_0^\infty (z+t)^{-1} \pi\sigma(-dt)\mathrm{Bj}H_\ell^+(t). \tag{1.11}$$

These functions are the only bounded strongly measurable functions from $(0,\infty)$ into L(H) satisfying Eqs. (1.10) and (1.11) whose inverses extend to functions analytic on the open right half plane and continuous on its closure.

Proof: Premultiply (1.3) (with μ,ν replaced by $-t,z$) by $t^{-1}\pi\sigma(-dt)\mathrm{Bj}$ and integrate over $(0,\infty)$. Then

$$\int_0^\infty \frac{z}{z+t}\pi\sigma(-dt)\mathrm{Bj}H_\ell^+(t)H_r^+(z) =$$

$$= \int_0^\infty \int_0^\infty dx \int_0^\infty dy\; e^{-x/t} e^{-y/z} \frac{1}{t}\pi\sigma(-dt)\mathrm{Bj}\{\delta(x-y)+\gamma(x,y)\}.$$

Performing the t-integration at the right, one finds

$$\int_0^\infty \frac{z}{z+t}\pi\sigma(-dt)\mathrm{Bj}H_\ell^+(t)H_r^+(z) =$$

$$= \int_0^\infty dx \int_0^\infty dy \ \pi \mathcal{H}(-x)Bj\{\delta(x-y)+\gamma(x,y)\}e^{-y/z}.$$

In terms of the solution $\Psi_z(x)$ of the Wiener-Hopf equation

$$\Psi_z(x) - \int_0^\infty \mathcal{H}(x-y)B\Psi_z(y)dy = e^{-y/z}I,$$

one gets

$$\int_0^\infty \frac{z}{z+t} \pi\sigma(-dt)BjH_\ell^+(t)H_r^+(z) = \int_0^\infty \pi\mathcal{H}(-x)Bj\Psi_z(x)dx,$$

which equals $\Psi_z(0)-I$. But also (cf. Section VII.1),

$$\Psi_z(0) = \int_0^\infty e^{-x/z}\{\delta(x)+\gamma(0,x)\}dx = I + \int_{-\infty}^0 e^{x/z}\ell_-(x)dx = H_r^+(z),$$

which establishes (1.11).

On the other hand, we may postmultiply (1.3) (with μ,ν replaced by $-z,t$) by $t^{-1}\pi\sigma(dt)Bj$, integrate over $(0,\infty)$ and perform the t-integration. We find

$$\int_0^\infty \frac{z}{z+t} H_\ell^+(z)H_r^+(t)\pi\sigma(dt)Bj = \int_0^\infty dy \int_0^\infty dx\{\delta(x-y)+\gamma(x,y)\}e^{-x/z}\pi\mathcal{H}(y)Bj.$$

The expression between square brackets is the solution of the equation

$$\Phi_z(y) - \int_0^\infty \Phi_z(x)\pi\mathcal{H}(x-y)Bjdx = e^{-y/z}I$$

(cf. [141]; for infinite dimension see [117]). Hence,

$$\int_0^\infty \frac{z}{z+t} H_r^+(z)H_\ell^+(t)\pi\sigma(dt)Bj = \int_0^\infty \Phi_z(y)\pi\mathcal{H}(y)Bjdy = \Phi_z(0) - I.$$

However (cf. Section VII.1),

$$\Phi_z(0) = \int_0^\infty e^{-x/z}\{\delta(x)+\gamma(x,0)\}dx = I + \int_0^\infty e^{-x/z}\ell_+(x)dx = H_\ell^+(z),$$

which establishes (1.10).

In order to establish the uniqueness part of this theorem, let H_ℓ^+ and H_r^+ be two bounded strongly measurable functions from $(0,\infty)$ into H that satisfy Eqs. (1.10) and (1.11). Obviously, the right hand sides of these equations are analytic on the open right half plane and continuous on the closed right half plane (with ∞ included), while $H_\ell^+(0^+)=H_r^+(0^+)=I$. Equations (1.10) and (1.11) imply that their right hand sides are invertible for all but finitely many Re $z\geq 0$ (see after the proof of Theorem 2.2 for details). Then, writing $\Gamma(z) \equiv H_r^+(z)^{-1}H_\ell^+(-z)^{-1}$, we have, for all but finitely many Re $z=0$,

$$\Gamma(z) = I - z\int_0^\infty (z-t)^{-1}H_r^+(t)\pi\sigma(dt)Bj - z\int_0^\infty (z+t)^{-1}\pi\sigma(-dt)BjH_\ell^+(t) +$$

$$+ z^2\int_0^\infty\int_0^\infty (z+t)^{-1}(z-u)^{-1}\pi\sigma(-dt)BjH_\ell^+(t)H_r^+(u)\pi\sigma(du)Bj.$$

Writing

$$z(z+t)^{-1}(z-u)^{-1} = \{t(t+z)^{-1}-u(u-z)^{-1}\}(t+u)^{-1},$$

one finds

$$\Gamma(z) = I - z\int_0^\infty (z+t)^{-1}\pi\sigma(-dt)BjH_\ell^+(t) - z\int_0^\infty (z-u)^{-1}H_r^+(u)\pi\sigma(du)Bj +$$

$$+ z\int_0^\infty\int_0^\infty (z+t)^{-1}\pi\sigma(-dt)BjH_\ell^+(t)t(u+t)^{-1}H_r^+(u)\pi\sigma(du)Bj +$$

$$+ z\int_0^\infty\int_0^\infty u(t+u)^{-1}\pi\sigma(-dt)BjH_\ell^+(t)(z-u)^{-1}H_r^+(u)\pi\sigma(du)Bj.$$

By analytic continuation Eqs. (1.10) and (1.11) hold true for imaginary z. Then,

$$H_r^+(z)^{-1}H_\ell^+(-z)^{-1} = I - z\int_0^\infty (z+t)^{-1}\pi\sigma(-dt)Bj - z\int_0^\infty (z-u)^{-1}\pi\sigma(du)Bj = \Lambda(z),$$

since $\Lambda(z)=\pi\{I-z(z-T)^{-1}B\}j$. Hence, $H_r^+(z)$ and $H_\ell^+(-z)$ are factors appearing in a right canonical factorization of $\Lambda(z)^{-1}$ and which satisfy $H_r^+(0^+)=H_\ell^+(0^+)=I$. Thus they are unique. ∎

Finally, let us discuss a possible Banach space generalization. Since Theorems 1.1 and 1.2 rely on the resolution of the identity of T, it would seem that one must assume, in addition to the hypotheses of Section VI.6, that T is a scalar-type spectral operator. This means that T should admit the representation $T = \int t\sigma(dt)$, where $\sigma(\cdot)$ is a suitable (non-orthogonal) projection valued measure. For an account of the theory of spectral operators we refer to the monographs of Dunford and Schwartz

[109] and Dowson [105]. Similar remarks on Banach space generalizations can be made with regard to Sections 2 and 3 of this chapter.

2. Albedo operators and H-equations: the singular case

In the previous section we derived a formula for the albedo operator appearing in the solution of half space problems. We restricted ourselves solely to the regular case, where $T^{-1}A$ does not have zero or imaginary eigenvalues. Here we shall drop this assumption and adopt the hypotheses (I) to (IV) of Section VII.2. Generally, $E_+=EQ_+$ is replaced by the projection of H onto Ran $P_{1,+}\oplus N_+$ along Ran Q_-, where $N_+\subset Z_0$ for

$$Z_0 = \bigoplus_{\mathrm{Re}\,\lambda=0} Z_\lambda(T^{-1}A).$$

Assuming the decomposition

$$\mathrm{Ran}\ P_{1,+} \oplus N_+ \oplus \mathrm{Ran}\ Q_- = H, \tag{2.1}$$

we seek to construct the projection E_+.

There is one obvious strategy to compute E_+. Let us take a complement N_- of N_+ in Z_0, and an operator β on Z_0 without zero or imaginary eigenvalues, which is reduced by the decomposition $N_+\oplus N_- = Z_0$ and which satisfies $\sigma(\beta\,|\,N_\pm)\subset\{\pm\lambda\in\mathbb{C} : \mathrm{Re}\ \lambda>0\}$. We then replace A by $A_\beta = T\beta^{-1}P_0 + AP_1$, where P_0 and P_1 are the complementary projections with ranges Z_0 and Z_1, respectively (see Lemma VII 2.3). Then E_+ is the projection obtained from the albedo operator for the regular half space problem with T and A replaced by T and A_β, and Theorems 1.2 and 1.3 provide E_+ and generalized H-equations.

The above procedure for finding E_+ utilizes, of course, a modified dispersion function

$$\Lambda_\beta(\lambda) = I - \lambda\pi(\lambda-T)^{-1}B_\beta j$$

for Re $\lambda=0$ and $B_\beta = I-A_\beta$. The usual applications in transport theory, however, are based on the (unmodified) dispersion function

$$\Lambda(\lambda) = I - \lambda\pi(\lambda-T)^{-1}Bj.$$

We would like to study the abstract problem, utilizing the latter dispersion function also. In this respect we are thwarted by a considerable lack of theory of Wiener–Hopf equations. Existing theory is developed for the case when the symbol of the equation (i.e., $\Lambda(\lambda)$) has a Wiener–Hopf factorization. In the singular case, this is no longer satisfied. Nevertheless, the equations derived in Section 1 make sense for singular cases, if H_ℓ^+ and H_r^+ are suitably defined. We shall give an ad hoc proof of Theorems 1.2 and 1.3, by using the factorization results of Section VII.3 for singular cases.

THEOREM 2.1. Assume the decomposition (2.1), and let E_+ be the projection of H onto Ran $P_{1,+}\oplus N_+$ along Ran Q_-. Then E_+ is given by Eq. (1.8), provided H_ℓ^+ and H_r^+ are the factors in the factorization of $W(\lambda)^{-1}$ given by

$$H_\ell^+(-\lambda) = I - \lambda\pi(T-\lambda A)^{-1}\hat{E}_+Bj, \tag{2.2}$$

$$H_r^+(\lambda) = I - \lambda\pi(I-E_+)(T-\lambda A)^{-1}Bj. \tag{2.3}$$

Here $\hat{E}_+$ is the unique bounded projection satisfying $TE_+ = \hat{E}_+T$ on D(T).

Proof: First we compute, for $\varphi_+ \in Q_+[D(T)]$ and $\mu>0$,

$$\int_0^\infty \frac{\nu}{\nu-\mu}H_r^+(\nu)\pi\sigma(d\nu) = \pi T(T-\mu)^{-1}Q_+ - \int_0^\infty \frac{\nu}{\nu-\mu}\pi(I-E_+)(T-\nu A)^{-1}(\nu B)\sigma(d\nu),$$

where $\nu(I-E_+)(T-\nu A)^{-1}$ has an analytic continuation to the right half plane. Using $\nu B = (\nu-T)+(T-\nu A)$, one finds

$$\int_0^\infty \frac{\nu}{\nu-\mu}H_r^+(\nu)\pi\sigma(d\nu) = \pi T(T-\mu)^{-1}Q_+ - \pi(I-E_+)T(T-\mu)^{-1}Q_+ =$$

$$= \pi E_+T(T-\mu)^{-1}Q_+, \tag{2.4}$$

and hence

$$\int_{-\infty}^{0}\int_{0}^{\infty}\sigma(d\mu)\,\frac{\nu}{\nu-\mu}BjH_{\ell}^{+}(-\mu)H_{r}^{+}(\nu)\pi\sigma(d\nu) =$$
$$= \int_{-\infty}^{0}\sigma(d\mu)BjH_{\ell}^{+}(-\mu)\pi E_{+}T(T-\mu)^{-1}Q_{+}.$$

Substituting (2.2) we obtain

$$\int_{-\infty}^{0}\int_{0}^{\infty}\sigma(d\mu)\,\frac{\nu}{\nu-\mu}BjH_{\ell}^{+}(-\mu)H_{r}^{+}(\nu)\pi\sigma(d\nu) =$$
$$= \int_{-\infty}^{0}\sigma(d\mu)BE_{+}T(T-\mu)^{-1}Q_{+} - \int_{-\infty}^{0}\sigma(d\mu)(\mu B)(T-\mu A)^{-1}\hat{E}_{+}BE_{+}T(T-\mu)^{-1}Q_{+}.$$

Noticing $\mu(T-\mu A)^{-1}\hat{E}_{+}$ has an analytic continuation to the right half plane, we obtain

$$\int_{-\infty}^{0}\int_{0}^{\infty}\sigma(d\mu)\frac{\nu}{\nu-\mu}BjH_{\ell}^{+}(-\mu)H_{r}^{+}(\nu)\pi\sigma(d\nu) = \int_{-\infty}^{0}\sigma(d\mu)(I-\hat{E}_{+})BE_{+}T(T-\mu)^{-1}Q_{+}$$

with the help of (2.4). Next, it should be observed that $B=I-A$ and $A[\text{Ran } E_{+}]=\text{Ran } \hat{E}_{+}$. Thus, the right hand side allows the following simplification:

$$\int_{-\infty}^{0}\sigma(d\mu)(I-\hat{E}_{+})BE_{+}T(T-\mu)^{-1}Q_{+} = \int_{-\infty}^{0}\sigma(d\mu)(I-\hat{E}_{+})E_{+}T(T-\mu)^{-1}Q_{+} =$$
$$= \int_{-\infty}^{0}\sigma(d\mu)(E_{+}-\hat{E}_{+})T(T-\mu)^{-1}Q_{+} =$$
$$= Q_{-}(E_{+}-\hat{E}_{+}) + \int_{-\infty}^{0}\sigma(d\mu)(\hat{E}_{+}-E_{+})\mu(\mu-T)^{-1}Q_{+}.$$

Further simplification is obtained by using the identity

$$\mu(\mu-T)^{-1}(\hat{E}_{+}-E_{+})\mu(\mu-T)^{-1} = \mu(\mu-T)^{-1}\hat{E}_{+} - E_{+}\mu(\mu-T)^{-1},$$

which follows directly from the intertwining relation $TE_{+}=\hat{E}_{+}T$ on $D(T)$. We get

$$\int_{-\infty}^{0}\int_{0}^{\infty}\sigma(d\mu)\frac{\nu}{\nu-\mu}BjH_{\ell}^{+}(-\mu)H_{r}^{+}(\nu)\pi\sigma(d\nu) = Q_{-}(E_{+}-\hat{E}_{+}) + Q_{-}\hat{E}_{+} = Q_{-}E_{+},$$

which finally establishes Eq. (1.8). ∎

Next we shall derive Eqs. (1.10)–(1.11).

THEOREM 2.2. Assume the decomposition (2.1), and let E_{+} be the projection of H onto Ran $P_{1,+} \oplus N_+$ along Ran Q_-. Then the functions defined by Eqs. (2.2) and (2.3) satisfy the generalizations (1.10) and (1.11) of Chandrasekhar's H-equations.

Proof: Let us consider the right hand side of Eq. (1.10), where Re $z \geq 0$. Inserting (2.2) one finds

$$I - z\int_0^\infty (z+t)^{-1} H_r^+(t)\pi\sigma(dt)Bj =$$

$$= I - z\pi(z+T)^{-1}Q_+Bj + z\int_0^\infty (z+t)^{-1}\pi(I-E_+)(T-tA)^{-1}(tB)\sigma(dt)Bj,$$

where $t(I-E_+)(T-tA)^{-1}$ extends to an analytic function in the right half plane. Using Eqs. (2.4) and (VII 3.8), the expression simplifies to

$$I - z\int_0^\infty (z+t)^{-1} H_r^+(t)\pi\sigma(dt)Bj = I - z\pi E_+(z+T)^{-1}Bj = H_\ell^+(z)^{-1}.$$

Next, substituting (2.3) in the right hand side of Eq. (1.11), one obtains

$$I - z\int_0^\infty (z+t)^{-1}\pi\sigma(-dt)BjH_\ell^+(t) = I - z\int_{-\infty}^0 (z-t)^{-1}\pi\sigma(dt)BjH_\ell^+(-t) =$$

$$= I - z\pi(z-T)^{-1}Q_-Bj + z\int_{-\infty}^0 (z-t)^{-1}\pi\sigma(dt)(tB)(T-tA)^{-1}\hat{E}_+Bj,$$

where $t(T-tA)^{-1}E_+$ has an analytic continuation to the left half plane. With the help of Eqs. (2.4) and (VII 3.9), we obtain the simple expression

$$I - z\int_0^\infty (z+t)^{-1}\pi\sigma(-dt)BjH_\ell^+(t) = I - z\pi(z-T)^{-1}(I-\hat{E}_+)Bj = H_r^+(z)^{-1},$$

which completes the proof of the theorem. ■

The uniqueness problem for the solution of the generalized H-equations (1.10) and (1.11) is far more complicated for the singular case than it is for the regular case. The main reason is that though $H_\ell^+(z)^{-1}$ and $H_r^+(z)^{-1}$ must be analytic in and continuous up to the boundary of the right half plane, this no longer holds for $H_\ell^+(z)$ and $H_r^+(z)$ themselves. Starting from bounded strongly measurable solutions H_ℓ^+, $H_r^+:(0,\infty)\to L(H)$, the right hand sides of Eqs. (1.10) and (1.11) must be analytic in and continuous up the boundary of the right half plane. Because of the condition that B is a compact operator satisfying (1.4), the right hand sides are compact perturbations of the identity and approach the identity uniformly as $z\to 0$ from the right half plane. Then, according to the analytic version of the Fredholm alternative, the right hand sides are invertible for all but finitely many Re $z\geq 0$. (Observe that the continuity extends to ∞, when approached from the right half plane.) For all but finitely many imaginary z we repeat the last part of the proof of Theorem 1.3 and obtain $H_r^+(z)^{-1}H_\ell^+(-z)^{-1} = \Lambda(z)$ for Re $z=0$. Note that on requiring analyticity of $H_r^+(z)$ and $H_\ell^+(z)$ in the open right half plane, these functions may still have poles at imaginary points of non-invertibility of $\Lambda(z)$ and may still be $O(z^n)$ for some $n\geq 0$ on approaching ∞ from the right half plane. It is not at all clear how physical considerations should lead to uniqueness. Uniqueness might be obtained, for example, by specifying the principal parts at the imaginary points of non-invertibility of $\Lambda(z)$, by specifying operator polynomials $P_\ell(z)$ and $P_r(z)$ such that

$$\lim_{z\to\infty,\, \mathrm{Re}\, z\geq 0} \|H_\ell^+(z)-P_\ell(z)\| = 0,$$

$$\lim_{z\to\infty,\, \mathrm{Re}\, z\geq 0} \|H_r^+(z)-P_r(z)\| = 0,$$

and by requiring analyticity in the open right half plane. For these principal parts and polynomials we refer to statements (a) and (c) at the end of Section VII.3.

3. Reflection and transmission operators and X-and Y-equations

In this section convolution equations theory on finite intervals is applied to the integral version,

$$\psi(x) - \int_0^\tau \mathcal{H}(x-y)B\psi(y)dy = \omega(x), \qquad 0<x<\tau, \tag{3.1}$$

of the abstract finite slab problem (see Sections V.1 and V.2). Unfortunately, convolution equations theory on finite intervals is less advanced than the corresponding half line theory of Wiener-Hopf equations, and most of the results have been given for finite dimensional spaces. We are able, therefore, to give proofs for some of the results only for finite dimensional Banach spaces. We shall rely on Section VI.2.

The following result has been proved in a finite dimensional context by Gohberg and Heinig ([140], Lemma 1.1). Our proof, which applies to the infinite dimensional problem, will be different.

PROPOSITION 3.1. Let $k \in L_1(L(X))_{-\tau}^{\tau}$ be compact operator-valued. Then the operator

$$(\mathcal{L}\psi)(x) = \int_0^\tau k(x-y)\psi(y)dy$$

is compact on the spaces $L_p(X)_0^\tau$ $(1 \le p \le \infty)$ and $C(X)_0^\tau$.

Proof: Since the measurable step functions are dense in $L_1(L(X))_{-\tau}^{\tau}$, and on both of the above function spaces the norm of $\mathcal{L}$ is bounded above by the L_1-norm of $k(x)$, it suffices to consider measurable step functions only. We can, in fact, get a further simplification by assuming the existence of a measurable subset E of $(-\tau,\tau)$ and a compact operator K in L(X) such that

$$k(x) = \begin{cases} K , & x \in E \\ 0 , & x \notin E . \end{cases}$$

Let $\mathcal{U}$ be a bounded subset of one of the function spaces $L_p(X)_0^\tau$ or $C(X)_0^\tau$ on which we seek to prove the compactness of $\mathcal{L}$. Then, for all $\psi \in \mathcal{U}$,

$$(\mathcal{L}\psi)(x) = K \int_E \psi(x-y)dy$$

is continuously dependent on $x \in [0,\tau]$ (cf. Proposition VI 2.2). Notice that for every $x \in [0,\tau]$ the set $\{\mathcal{L}\psi(x) : \psi \in \mathcal{U}\}$ is relatively compact in X, because K is a compact operator on X and $\{\int_0^\tau \psi(x-y)dy : \psi \in \mathcal{U}\}$ is bounded in X. Furthermore, the family of functions $\{\mathcal{L}\psi : \psi \in \mathcal{U}\}$ is equicontinuous on $[0,\tau]$. As a consequence of Ascoli's theorem ([385], Theorem 14.24), the set $\{\mathcal{L}\psi : \psi \in \mathcal{U}\}$ is relatively compact in $C(X)_0^\tau$. Because $C(X)_0^\tau$ is continuously embedded in $L_p(X)_0^\tau$, this set is also relatively compact in $L_p(X)_0^\tau$ $(1 \le p \le \infty)$. Hence, $\mathcal{L}$ is a compact operator on both of the above function spaces. ∎

The above proposition implies that the convolution equation

$$\psi(x) - \int_0^\tau k(x-y)\psi(y)dy = \omega(x), \quad 0<x<\tau, \tag{3.2}$$

where $k \in L_1(L(X))_{-\tau}^{\tau}$ is compact operator-valued, has the characteristics of a Fredholm integral equation. Moreover, one immediately derives from Lemma VII 4.1 that the number dim Ker$(I-\mathcal{L})$ = codim Ran$(I-\mathcal{L})$ is the same for all solution spaces $(1 \leq p \leq \infty)$ and $C(X)_0^\tau$.

The next theorem gives representations of the solutions of Eq. (3.2). It was proved by Gohberg and Heinig [140] in a finite dimensional setting, and generalizes the scalar result of Gohberg and Semençul [151]. The theorem has been known in radiative transfer for scalar even kernels (see [61, 340]).

THEOREM 3.2. Assume that Eq. (3.2) is uniquely solvable. Then the unique solution has the form

$$\psi(x) = \omega(x) + \int_0^\tau \gamma(x,y)\omega(y)dy, \quad 0<x<\tau, \tag{3.3}$$

where the resolvent kernel $\gamma(x,y)$ is given by either of the expressions

$$\gamma(x,y) = \ell_+(x-y) + \int_0^y [\ell_+(x-z)\ell_-(z-y) - m_-(x-z-\tau)m_+(z-y+\tau)]dz, \quad x>y,$$

$$\gamma(x,y) = \ell_-(x-y) + \int_0^x [\ell_+(x-z)\ell_-(z-y) - m_-(x-z-\tau)m_+(z-y+\tau)]dz, \quad x<y,$$

or

$$\gamma(x,y) = m_+(x-y) + \int_x^\tau [m_-(x-z)m_+(z-y) - \ell_+(x-z+\tau)\ell_-(z-y-\tau)]dz, \quad x>y,$$

$$\gamma(x,y) = m_-(x-y) + \int_y^\tau [m_-(x-z)m_+(z-y) - \ell_+(x-z+\tau)\ell_-(z-y-\tau)]dz, \quad x<y.$$

The operator functions $\ell_\pm$ and $m_\pm$ are the unique solutions of the equations

$$\ell_+(x) - \int_0^\tau k(x-y)\ell_+(y)dy = k(x),$$

$$\ell_-(-x) - \int_0^\tau \ell_-(-y)k(y-x)dy = k(-x),$$

$$m_+(x) - \int_0^\tau m_+(y)k(x-y)dy = k(x),$$

$$m_-(-x) - \int_0^\tau k(y-x)m_-(-y)dy = k(-x),$$

for $0<x<\tau$.

Proof: The proof can be given by direct substitution of one of the expressions for $\gamma(x,y)$ in Eq. (3.3), changing the order of integration in the double integrals obtained and applying the integral equations for $\ell_\pm$ and $m_\pm$ together with Eq. (3.2). Although the proof in [140] is given for a special type of kernel k and the result obtained there is extended to general k by approximation, such approximation generally cannot be repeated in an infinite dimensional setting, and a direct computation is necessary. ■

From the two expressions for $\gamma(x,y)$ we get the useful identities

$$\ell_+(x) = \gamma(x,0), \quad \ell_-(-y) = \gamma(0,y), \tag{3.4}$$

$$m_+(\tau-y) = \gamma(\tau,y), \quad m_-(x-\tau) = \gamma(x,\tau). \tag{3.5}$$

The next result goes back to Ambarzumian [7] for radiative transfer with isotropic scattering (see also [89, 340, 342]). For a large class of scalar even kernels the result was found by Busbridge [61]. In a finite dimensional setting it was misstated in [140], but correctly proved by Dym and Gohberg [110]. A new proof of the scalar result, with a generalization to integrodifferential equations of convolution type, is due to Sakhnovich [323]. As the presentation in [110] is quite inaccessible to our present purpose, we prefer to give a full proof, which, in fact, repairs the incorrect proof in [140].

THEOREM 3.3. Assume that Eq. (3.2) is uniquely solvable. Then

$$\int_0^\tau dy \int_0^\tau dz\ e^{y/\mu}e^{-z/\nu}\{\delta(y-z)+\gamma(y,z)\} = \frac{\mu\nu}{\mu-\nu}\{X_\ell^+(-\mu)X_r^+(\nu)-Y_\ell^-(-\mu)Y_r^-(\nu)\},$$

$$\int_0^\tau dy \int_0^\tau dz\ e^{-(\tau-y)/\mu}e^{-z/\nu}\{\delta(y-z)+\gamma(y,z)\} = \frac{\mu\nu}{\mu-\nu}\{Y_\ell^+(\mu)X_r^+(\nu)-X_\ell^-(\mu)Y_r^-(\nu)\},$$

$$\int_0^\tau dy \int_0^\tau dz\ e^{-(\tau-y)/\mu}e^{(\tau-z)/\nu}\{\delta(y-z)+\gamma(y,z)\} =$$

$$= \frac{\mu\nu}{\mu-\nu}\{Y_\ell^+(\mu)Y_r^+(-\nu)-X_\ell^-(\mu)X_r^-(-\nu)\},$$

$$\int_0^\tau dy \int_0^\tau dz\ e^{y/\mu} e^{(\tau-z)/\nu} \{\delta(y-z)+\gamma(y,z)\} =$$

$$= \frac{\mu\nu}{\mu-\nu}\{X_\ell^+(-\mu)Y_r^+(-\nu)-Y_\ell^-(-\mu)X_r^-(-\nu)\},$$

where

$$X_\ell^+(\mu) = I + \int_0^\tau e^{-y/\mu}\ell_+(y)dy, \qquad Y_\ell^+(\mu) = e^{-\tau/\mu} + \int_0^\tau e^{-(\tau-y)/\mu}\ell_+(y)dy,$$

$$X_r^+(\mu) = I + \int_0^\tau e^{-y/\mu}\ell_-(-y)dy, \qquad Y_r^+(\mu) = e^{-\tau/\mu} + \int_0^\tau e^{-(\tau-y)/\mu}\ell_-(-y)dy,$$

$$X_\ell^-(\mu) = I + \int_0^\tau e^{-y/\mu}m_-(-y)dy, \qquad Y_\ell^-(\mu) = e^{-\tau/\mu} + \int_0^\tau e^{-(\tau-y)/\mu}m_-(-y)dy,$$

$$X_r^-(\mu) = I + \int_0^\tau e^{-y/\mu}m_+(y)dy, \qquad Y_r^-(\mu) = e^{-\tau/\mu} + \int_0^\tau e^{-(\tau-y)/\mu}m_+(y)dy.$$

Proof: We shall make use of one of the ancillary results of [140], namely that the vector function

$$\varsigma(x,y) = \lim_{\epsilon\to 0} (\gamma(x,y) - \gamma(x-\epsilon,y-\epsilon))/\epsilon$$

satisfies the convolution equation

$$\varsigma(x,y) - \int_0^\tau k(x-z)\varsigma(z,y)dz = k(x)\ell_-(-y) - k(x-\tau)m_+(\tau-y),$$

and at the same time satisfies the equality

$$\varsigma(x,y) = \ell_+(x)\ell_-(-y) - m_-(x-\tau)m_+(\tau-y). \tag{3.6}$$

Starting out as in the proof of the corollary of [140], we have for $\mathrm{Re}\mu<0$ and $\mathrm{Re}\nu>0$ the identity

$$\frac{1-\exp[-\varepsilon(-\mu^{-1}+\nu^{-1})]}{\varepsilon}\int_0^\tau dy\int_0^\tau dz\ e^{y/\mu}e^{-z/\nu}\{\delta(y-z)+\gamma(y,z)\} =$$

$$= \frac{1}{\varepsilon}\int_0^\tau dy\int_0^\tau dz\ e^{y/\mu}e^{-z/\nu}\{\delta(y-z)+\gamma(y,z)\} +$$

$$+ \frac{1}{\varepsilon}\int_{\varepsilon}^{\tau+\varepsilon} dy \int_{\varepsilon}^{\tau+\varepsilon} dz\ e^{y/\mu} e^{-z/\nu} \{-\delta(y-z) - \gamma(y-\varepsilon, z-\varepsilon)\}$$

$$= \frac{1}{\varepsilon}\int_{\varepsilon}^{\tau} dy \int_{\varepsilon}^{\tau} dz\ e^{y/\mu} e^{-z/\nu} \{\gamma(y,z) - \gamma(y-\varepsilon, z-\varepsilon)\} + \frac{1}{\varepsilon}\left(\int_{0}^{\varepsilon} - \int_{\tau}^{\tau+\varepsilon}\right) e^{y/\mu} e^{-y/\nu} dy +$$

$$+ \frac{1}{\varepsilon}\left(\int_{0}^{\tau} dy \int_{0}^{\varepsilon} dz + \int_{0}^{\varepsilon} dy \int_{0}^{\tau} dz - \int_{0}^{\varepsilon} dy \int_{0}^{\varepsilon} dz\right) e^{y/\mu} e^{-z/\nu} \gamma(y,z) +$$

$$+ \frac{1}{\varepsilon}\left(\int_{\tau}^{\tau+\varepsilon} dy \int_{\tau}^{\tau+\varepsilon} dz - \int_{\varepsilon}^{\tau+\varepsilon} dy \int_{\tau}^{\tau+\varepsilon} dz - \int_{\tau}^{\tau+\varepsilon} dy \int_{\varepsilon}^{\tau+\varepsilon} dz\right) e^{y/\mu} e^{-z/\nu} \times$$

$$\times\ \gamma(y-\varepsilon, z-\varepsilon),$$

where the z-integration has been performed in all terms involving $\delta(y-z)$. Thus,

$$\frac{\mu-\nu}{\mu\nu} \int_{0}^{\tau} dy \int_{0}^{\tau} dz\ e^{y/\mu} e^{-z/\nu} \{\delta(y-z) + \gamma(y,z)\} =$$

$$= \int_{0}^{\tau} dy \int_{0}^{\tau} dz\ e^{y/\mu} e^{-z/\nu} \varsigma(y,z) + \left(1 - \exp\left[\tau\left(\frac{1}{\mu} - \frac{1}{\nu}\right)\right]\right) + \int_{0}^{\tau} e^{x/\mu} \gamma(x,0) dx +$$

$$+ \int_{0}^{\tau} e^{-y/\nu} \gamma(0,y) dy - \int_{0}^{\tau} e^{x/\mu} e^{-\tau/\nu} \gamma(x,\tau) dx - \int_{0}^{\tau} e^{\tau/\mu} e^{-y/\nu} \gamma(\tau,y) dy.$$

Utilizing (3.6) and the identities (3.4) and (3.5), one finds

$$\int_{0}^{\tau} dy \int_{0}^{\tau} dz\ e^{y/\mu} e^{-z/\nu} \{\delta(y-z) + \gamma(y,z)\} = \frac{\mu\nu}{\mu-\nu} \{X_{\ell}^{+}(-\mu) X_{r}^{+}(\nu) - Y_{\ell}^{-}(-\mu) Y_{r}^{-}(\nu)\},$$

which establishes the first part of the proof. The other parts can be proved analogously. ∎

We now apply the previous results to the convolution equation

$$\psi(x) - \int_{0}^{\tau} \mathcal{H}(x-y) B\psi(y) dy = \omega(x), \quad 0<x<\tau, \tag{3.7}$$

where

$$\omega(x) = [e^{-xT^{-1}} Q_{+} + e^{(\tau-x)T^{-1}} Q_{-}]h, \quad 0 \le x < \tau.$$

The convolution equation (3.1) then is equivalent to the boundary value problem

$$(T\psi)'(x) = -(I-B)\psi(x), \quad 0<x<\tau, \tag{3.8}$$

$$\lim_{x\downarrow 0}\|Q_+\psi(0)-Q_+h\| = 0, \tag{3.9a}$$

$$\lim_{x\uparrow \tau}\|Q_-\psi(\tau)-Q_-h\| = 0, \tag{3.9b}$$

provided $h\in D(T)$ (see Theorem VI 3.3). Let us choose a closed subspace $\mathbb{B}$ containing Ran B^* and operators $j:\mathbb{B}\to H$ and $\pi:H\to\mathbb{B}$ such that πj is the identity on $\mathbb{B}$ and $j\pi$ is the orthogonal projection of H onto $\mathbb{B}$. Then the solution has the form

$$\psi(x) = \omega(x) + \int_0^\tau \mathcal{H}(x-y)Bj\chi(y)dy, \quad 0<x<\tau, \tag{3.10}$$

where $\chi(x) = \pi\psi(x)$ satisfies the convolution equation

$$\chi(x) - \int_0^\tau \pi\mathcal{H}(x-y)Bj\chi(y)dy = \pi\omega(x), \quad 0<x<\tau.$$

Whenever Eq. (3.1) is uniquely solvable, we can, as in Chapter V, define reflection operators $R_{\pm\tau}$ and transmission operators $T_{\pm\tau}$ by the equations

$$\psi(0) = (R_{+\tau} + T_{-\tau})h, \tag{3.11a}$$

$$\psi(\tau) = (R_{-\tau} + T_{+\tau})h, \tag{3.11b}$$

where

$$R_{\pm\tau}Q_{\mp} = T_{\pm\tau}Q_{\mp} = 0. \tag{3.12}$$

These equations define $R_{\pm\tau}$ and $T_{\pm\tau}$ uniquely. Applying the previous theorem to the kernel $k(x) = \pi\mathcal{H}(x)Bj$ and adopting the notation previously introduced in this section, we obtain

THEOREM 3.4. Let Eq. (3.1) be uniquely solvable. Then

$$R_{+\tau} = Q_+ + \int_{-\infty}^0\int_0^\infty \frac{\nu}{\nu-\mu}\,\sigma(d\mu)Bj[X_\ell^+(-\mu)X_r^+(\nu)-Y_\ell^-(-\mu)Y_r^-(\nu)]\pi\sigma(d\nu),$$

$$T_{+\tau} = e^{-\tau T^{-1}}Q_{+} - \int_0^\infty\int_0^\infty \frac{\nu}{\nu-\mu}\,\sigma(d\mu)Bj[Y_\ell^+(\mu)X_r^+(\nu)-X_\ell^-(\mu)Y_r^-(\nu)]\pi\sigma(d\nu),$$

$$R_{-\tau} = Q_{-} - \int_0^\infty\int_{-\infty}^0 \frac{\nu}{\nu-\mu}\,\sigma(d\mu)Bj[Y_\ell^+(\mu)Y_r^+(-\nu)-X_\ell^-(\mu)X_r^-(-\nu)]\pi\sigma(d\nu),$$

$$T_{-\tau} = e^{\tau T^{-1}}Q_{-} + \int_{-\infty}^0\int_{-\infty}^0 \frac{\nu}{\nu-\mu}\times$$

$$\times\,\sigma(d\mu)Bj[X_\ell^+(-\mu)Y_r^+(-\nu)-Y_\ell^-(-\mu)X_r^-(-\nu)]\pi\sigma(d\nu).$$

All these operators are bounded.

Proof: Let us write $\psi(0)$ and $\psi(\tau)$ in the form (3.10), where

$$\chi(y) = \pi\omega(y) + \int_0^\tau \gamma(y,z)\pi\omega(z)dz.$$

Subsequently, use (3.7) and the spectral representations (1.9) in order to reduce the proof to a simple application of the previous result. ■

COROLLARY 3.5. The reflection and transmission operators are invariant on D(T), and may also be represented as

$$R_{+\tau} = Q_{+} + \int_0^\tau \Gamma(0,y)e^{-yT^{-1}}Q_{+}dy,$$

$$T_{+\tau} = e^{-\tau T^{-1}}Q_{+} + \int_0^\tau \Gamma(\tau,y)e^{-yT^{-1}}Q_{+}dy,$$

$$R_{-\tau} = Q_{-} + \int_0^\tau \Gamma(\tau,y)e^{(\tau-y)T^{-1}}Q_{-}dy,$$

$$T_{-\tau} = e^{\tau T^{-1}}Q_{-} + \int_0^\tau \Gamma(0,y)e^{(\tau-y)T^{-1}}Q_{-}dy,$$

and as

$$R_{+\tau} = [P_{+} + e^{\tau T^{-1}A}P_{-} + P_0]V_\tau^{-1}Q_{+},$$

$$T_{+\tau} = [P_{-} + e^{-\tau T^{-1}A}P_{+} + \sum_{k=0}^{N-1}\frac{(-\tau)^k}{k!}(T^{-1}A)^k P_0]V_\tau^{-1}Q_{+},$$

$$R_{-\tau} = [P_{-} + e^{-\tau T^{-1}A}P_{+} + \sum_{k=0}^{N-1}\frac{(-\tau)^k}{k!}(T^{-1}A)^k P_0]V_\tau^{-1}Q_{-},$$

$$T_{-\tau} = [P_+ + e^{\tau T^{-1}A}P_- + P_0]V_\tau^{-1} Q_-,$$

where $\Gamma(x,y)$ is the resolvent kernel of Eq. (3.1) and N is the maximum length of Jordan chains in $Z_0(T^{-1}A)$.

Proof: The first representation is immediate, since the reflection and transmission operators are defined by (3.11) and (3.12) and the solution of (3.1) can be expressed in the right hand side using the resolvent kernel. Using the compactness and Bochner integrability of $\Gamma(0,y)$ and $\Gamma(\tau,y)$ (cf. Section VII.1) one may prove that $R_{\pm\tau}-Q_\pm$ and $T_{\pm\tau}-\exp\{\mp\tau T^{-1}\}Q_\pm$ are compact operators.

The second representation of the reflection and transmission operators is obtained by using the projections and semigroups introduced in Section VII.2. Premultiplying Eqs. (3.8) and (3.9) by P_+, P_- and P_0 and solving the corresponding two boundary value problems, we find

$$\psi(x) = [e^{-xT^{-1}A}P_+ + e^{(\tau-x)T^{-1}A}P_- + \sum_{k=0}^{N-1} \frac{(-x)^k}{k!}(T^{-1}A)^kP_0]\varphi,$$

where $(T^{-1}A)^NP_0=0$ and $\varphi\in H$. (Compare with Section V.2.) Substituting this expression in Eqs. (3.9) we obtain $V_\tau\varphi=h$, where

$$V_\tau = Q_+[P_+ + e^{\tau T^{-1}A}P_- + P_0] + Q_-[P_- + e^{-\tau T^{-1}A}P_+ + \sum_{k=0}^{N-1}\frac{(-\tau)^k}{k!}(T^{-1}A)^kP_0].$$

The unique solvability of Eqs. (3.8) and (3.9) and the compactness of $I-V_\tau$ (implied by the compactness of $P_\pm-Q_\pm$ and P_0) lead to the invertibility of V_τ. The defining equations (3.11) and (3.12) for the reflection and transmission operators now imply the claimed representation.

Clearly, V_τ leaves invariant D(T) and $TV_\tau = \hat{V}_\tau T$ on D(T) for $\hat{V}_\tau$ a suitable invertible operator. We may then conclude that $R_{\pm\tau}$ and $T_{\pm\tau}$ leave invariant D(T), while

$$TR_{\pm\tau} = \hat{R}_{\pm\tau}T, \tag{3.13a}$$

$$TT_{\pm\tau} = \hat{T}_{\pm\tau}T, \tag{3.13b}$$

on D(T) for $\hat{R}_{\pm\tau}$ and $\hat{T}_{\pm\tau}$ suitable bounded operators. We note that representations for the latter operators are easily found by replacing $\nu/(\nu-\mu)$ by

$\mu/(\nu-\mu)$ in the formulas of Theorem 3.4, as a straightforward application of (3.13). ■

Next we shall derive nonlinear integral equations relating the functions $X_\ell^\pm$, $Y_\ell^\pm$, $X_r^\pm$ and $Y_r^\pm$. For radiative transfer with isotropic scattering such equations were found by Ambarzumian [7] and extended to a larger class of radiative transfer problems by Chandrasekhar [89]. For scalar even kernels they were studied by Busbridge [61]. A full proof of their solvability for all subcritical ($c<1$) and critical ($c=1$) one speed transport problems is due to van der Mee [363], who also studied such scalar equations in general (see [361]).

THEOREM 3.6. The functions $X_\ell^\pm$, $Y_\ell^\pm$, $X_r^\pm$ and $Y_r^\pm$ satisfy the nonlinear integral equations

$$X_\ell^+(z) = I + z\int_0^\infty \frac{1}{t+z}\{X_\ell^+(z)X_r^+(t)-Y_\ell^-(z)Y_r^-(t)\}\pi\sigma(dt)Bj,$$

$$Y_\ell^+(z) = e^{-\tau/z} - z\int_0^\infty \frac{1}{t-z}\{Y_\ell^+(z)X_r^+(t)-X_\ell^-(z)Y_r^-(t)\}\pi\sigma(dt)Bj,$$

$$X_r^+(z) = I + z\int_0^\infty \frac{1}{t+z}\pi\sigma(-dt)Bj\{X_\ell^+(t)X_r^+(z)-Y_\ell^-(t)Y_r^-(z)\},$$

$$Y_r^+(z) = e^{-\tau/z} - z\int_0^\infty \frac{1}{t-z}\pi\sigma(-dt)Bj\{X_\ell^+(t)Y_r^+(z)-Y_\ell^-(t)X_r^-(z)\},$$

$$X_\ell^-(z) = I - z\int_0^\infty \frac{1}{t+z}\{X_\ell^-(z)X_r^-(t)-Y_\ell^+(z)Y_r^+(t)\}\pi\sigma(-dt)Bj,$$

$$Y_\ell^-(z) = e^{-\tau/z} - z\int_0^\infty \frac{1}{t-z}\{Y_\ell^-(z)X_r^-(t)-X_\ell^+(z)Y_r^+(t)\}\pi\sigma(-dt)Bj,$$

$$X_r^-(z) = I - z\int_0^\infty \frac{1}{t+z}\pi\sigma(dt)Bj\{X_\ell^-(t)X_r^-(z)-Y_\ell^+(t)Y_r^+(z)\},$$

$$Y_r^-(z) = e^{-\tau/z} - z\int_0^\infty \frac{1}{t-z}\pi\sigma(dt)Bj\{X_\ell^-(t)Y_r^-(z)-Y_\ell^+(t)X_r^+(z)\}.$$

Proof: All eight formulas can be proved in the same manner as Eqs. (1.10) and (1.11). In fact, the proof of Theorem 1.3 contains a procedure to derive (1.11) and a procedure to derive (1.10). If we now denote the right hand sides of the four equations appearing in the statement of Theorem 3.3 by I, II, III and IV, respectively, and replace μ, ν by a suitable choice among $\pm t$ and $\pm z$, the procedure to derive (1.10) or (1.11) will reproduce the above equations for $X_\ell^\pm$, $Y_\ell^\pm$, $X_r^\pm$ and $Y_r^\pm$. We have to employ Eqs. (3.4) and (3.5), the definitions of $X_\ell^\pm$, $Y_\ell^\pm$, $X_r^\pm$, and $Y_r^\pm$ in the statement of Theorem 3.3, and the following convolution equations:

$$k(-y) + \int_0^\tau k(-x)\gamma(x,y)dx = \gamma(0,y), \tag{3.14}$$

$$k(x) + \int_0^\tau \gamma(x,y)k(y)dy = \gamma(x,0), \tag{3.15}$$

$$k(\tau-y) + \int_0^\tau k(\tau-x)\gamma(x,y)dy = \gamma(\tau,y), \tag{3.16}$$

$$k(x-\tau) + \int_0^\tau \gamma(x,y)k(y-\tau)dy = \gamma(x,\tau), \tag{3.17}$$

where $k(w) = \pi\mathcal{H}(w)Bj$. Equations (3.14) to (3.17) are easily derived and can, in fact, be found in [140].

We exhibit in Table 1 the substitutions and auxiliary equations to be used in deriving each of the integral equations. This concludes the proof. ■

Unknown	**Procedure**	**Starting Eqn.**	**Substitution**	**Auxiliary Eqn.**
X_ℓ^+	(1.10)	I	$\mu=-z$, $\nu=t$	(3.15)
Y_ℓ^+	(1.10)	II	$\mu=z$, $\nu=t$	(3.15)
X_r^+	(1.11)	I	$\mu=-t$, $\nu=z$	(3.14)
Y_r^+	(1.11)	IV	$\mu=-t$, $\nu=-z$	(3.14)
X_ℓ^-	(1.11)	III	$\mu=z$, $\nu=-t$	(3.17)
Y_ℓ^-	(1.10)	IV	$\mu=-z$, $\nu=-t$	(3.17)
X_r^-	(1.10)	III	$\mu=t$, $\nu=-z$	(3.16)
Y_r^-	(1.11)	II	$\mu=t$, $\nu=z$	(3.16)

TABLE 1. Derivation of Integral Equations for X- and Y-Functions

We finish this section by observing the well-known symmetry relations

$$Y_\ell^\pm(\mu) = e^{-\tau/\mu} X_\ell^\pm(-\mu),$$

$$Y_r^\pm(\mu) = e^{-\tau/\mu} X_r^\pm(-\mu),$$

which hold for the analytic continuations of the functions to the Riemann sphere with zero removed. All these functions are, in fact, entire in $1/\mu$ of order at most τ. They satisfy

$$\lim_{\mu\to 0,\,\mathrm{Re}\,\mu\geq 0} X_\ell^\pm(\mu) = I,$$

$$\lim_{\mu\to 0,\,\mathrm{Re}\,\mu\geq 0} X_r^\pm(\mu) = I,$$

$$\lim_{\mu\to 0,\,\mathrm{Re}\,\mu\geq 0} \|Y_\ell^\pm(\mu)-e^{-\tau/\mu}I\| = 0,$$

$$\lim_{\mu\to 0,\,\mathrm{Re}\,\mu\geq 0} \|Y_r^\pm(\mu)-e^{-\tau/\mu}I\| = 0.$$

4. Linear H–equations, uniqueness properties and constraints

In the first three sections we have expressed albedo, reflection and transmission operators in terms of generalized H–, X– and Y–functions, which satisfy nonlinear integral equations. In the present section we study uniqueness properties of solutions of H–equations. Since nonlinear integral equations usually are inconvenient for this purpose, it is preferable to reduce them to non–homogeneous linear equations. Then uniqueness can be studied by determining the number of linearly independent solutions of the corresponding homogeneous linear equation. For the H–equations of Section 1 (the regular case) one can simply use the Wiener–Hopf factorization of the dispersion function to derive such linear equations. For the corresponding singular case the procedure is more involved, but its implementation is still possible. Once linear equations have been obtained one may impose constraints to single out the solution of interest. This was first done by Mullikin [272] for certain scalar X– and Y–equations. Another advantage of linear equations is their amenability to accurate numerical methods, such as the collocation method (called F_N–method) introduced by Siewert and Benoist [331]. In the present section we shall restrict ourselves to H–equations.

Let us start with the regular case where $T^{-1}A$ does not have imaginary eigenvalues, and exploit the right canonical Wiener–Hopf factorization

$$\Lambda(z)^{-1} = H_\ell^+(-z)H_r^+(z), \quad \text{Re}\,z=0. \tag{4.1}$$

Substituting this into the nonlinear integral equations (1.10) and (1.11) we get

$$H_r^+(-z)\Lambda(-z) = I - z\int_0^\infty (z+t)^{-1}H_r^+(t)\pi\sigma(dt)Bj, \tag{4.2}$$

$$\Lambda(z)H_\ell^+(-z) = I - z\int_0^\infty (z+t)^{-1}\pi\sigma(-dt)BjH_\ell^+(t). \tag{4.3}$$

The advantage is that we obtain linear equations, in which H_ℓ^+ and H_r^+ are uncoupled. The disadvantage is that it is not entirely clear how to formulate Eqs. (4.2) and (4.3) for $z\in\sigma(T)\cap[0,\infty)$ and $z\in\sigma(-T)\cap[0,\infty)$, respectively. Indeed, in the formula (1.8) for the albedo operator we need $H_r^+(z)$ and $H_\ell^+(z)$ for these values of z, and we need to be able to define $\Lambda(z)$ for $z\in\sigma(T)$. For this reason we assume that T is bounded, $\sigma(T)$ has finitely many connected components, T is absolutely continuous with Radon–Nikodym derivative $\Delta(\mu)$, and $\pi\Delta(\mu)Bj$ is uniformly Hölder continuous on $\sigma(T)$, except for finitely many jump discontinuities. Let Σ denote the set of points of $\sigma(T)$ which are neither boundary points nor imbedded eigenvalues of $\sigma(T)$. Then Σ is an open subset of the real line. As the dispersion function is given by

$$\Lambda(z) = I - z\int_{\sigma(T)} (z-\mu)^{-1}\pi\Delta(\mu)Bj\, d\mu, \quad z\notin\sigma(T),$$

we find

$$\lim_{\varepsilon\downarrow 0} \Lambda(t\pm i\varepsilon) = I - tP\int_{\sigma(T)} \frac{\pi\Delta(\mu)Bj}{t-\mu}d\mu \pm i\pi t\{\pi\Delta(t)Bj\}, \quad t\in\Sigma,$$

(see [276] for the finite dimensional theory). In fact, the limit is uniformly Hölder continuous on compact subintervals of Σ. The symbol P indicates a principal value integral is to be taken. As an abbreviated notation we shall write

$$\lambda(t) = \lim_{\varepsilon\downarrow 0} \tfrac{1}{2}\{\Lambda(t+i\varepsilon) + \Lambda(t-i\varepsilon)\} = I - tP\int_{\sigma(T)} \frac{\pi\Delta(\mu)Bj}{t-\mu}\, d\mu, \quad t\in\Sigma.$$

An obvious replacement for Eqs. (4.2) and (4.3) now is the pair of equations

$$H_r^+(t)\lambda(t) = I - tP\int_{\Sigma_+} (t-\mu)^{-1} H_r^+(\mu)\pi\Delta(\mu)Bj\,d\mu, \tag{4.4}$$

$$\lambda(-t)H_\ell^+(t) = I - tP\int_{\Sigma_-} (t-\mu)^{-1} \pi\Delta(-\mu)BjH_\ell^+(\mu)\,d\mu, \tag{4.5}$$

where $\Sigma_\pm = \sigma(\pm T)\cap(\pm\Sigma)\cap(0,\infty)$ and $t\in\Sigma_\pm$. These singular integral equations cannot be solved using the standard methods presented in the monographs of Muskhelishvili [276], Gohberg and Krupnik [143] and Prössdorf [311] (possibly with infinite dimensional generalization to rank B $= \infty$), because $\lambda(\pm t)$ usually is not bounded on $\Sigma_\pm$. We shall instead follow procedures inspired by the work of Mullikin [271, 272]. We shall again restrict ourselves to H-equations only.

THEOREM 4.1. In addition to the general hypotheses of this section, let the following assumptions be fulfilled:

(i) For $t\in\Sigma\cap(0,\infty)$ (resp. $t\in\Sigma\cap(-\infty,0)$) the limiting values

$$\lim_{\varepsilon\downarrow 0} \Lambda(t\pm i\varepsilon) = \lambda(t) \pm i\pi t\{\pi\Delta(t)Bj\}$$

are invertible operators.

(ii) For every $t\in(\sigma(T)\setminus\Sigma)\cap[0,\infty)$ (resp. $t\in(\sigma(T)\setminus\Sigma)\cap(-\infty,0]$) there is a neighborhood U_t of t such that $\|\Lambda(z)^{-1}\|$ is bounded on $U_t\setminus\mathbb{R}$.

Then for regular cases, where $T^{-1}A$ does not have imaginary eigenvalues, the nonlinear H-equations (1.10) and (1.11) as well as the linear H-equations (4.4) and (4.5) have a unique bounded uniformly Hölder continuous solution satisfying the constraints

$$\delta_{k0}I + t_0 P\int_{\Sigma_+} (\mu-t_0)^{-(k+1)} H_r^+(\mu)\pi\Lambda(\mu)Bj\,d\mu = 0, \quad k=0,1,\ldots,p-1,$$

and

$$\delta_{k0}I + t_0 P\int_{\Sigma_-} (\mu-t_0)^{-(k+1)} \pi\Delta(-\mu)BjH_\ell^+(\mu)\,d\mu = 0, \quad k=0,1,\ldots,p-1,$$

respectively. This unique solution is analytic in and continuous up to the boundary of the right half plane and has invertible values there. If $T^{-1}A$ is not regular, then more than one solution satisfying the above constraints exists, but all of them are analytic and invertible on the open right half plane and meromorphic at all λ for which $1/\lambda$ is an imaginary eigenvalue of $T^{-1}A$. Uniqueness is obtained if the principal parts of $H_r^+(1/z)$ (resp. $H_\ell^+(1/z)$) at the imaginary eigenvalues of $T^{-1}A$ are specified.

Proof: Let us replace Eqs. (4.2) and (4.3) by their continuations

$$H_r^+(z)\Lambda(z) = I - z\int_{\Sigma_+} (z-\mu)^{-1}H_r^+(\mu)\pi\Delta(\mu)Bjd\mu, \tag{4.6}$$

$$\Lambda(-z)H_\ell^+(z) = I - z\int_{\Sigma_-} (z-\mu)^{-1}\pi\Delta(-\mu)BjH_\ell^+(\mu)d\mu, \tag{4.7}$$

where $z\notin\sigma(T)\cap[0,\infty)$ (resp. $z\notin\sigma(-T)\cap[0,\infty)$). Assuming uniform Hölder continuity of H_r^+ and H_ℓ^+ on $\sigma(T)\cap[0,\infty)$ and $\sigma(-T)\cap[0,\infty)$, we stipulate H_r^+ and H_ℓ^+ for $z\notin\sigma(\pm T)\cap[0,\infty)$ by

$$H_r^+(z) = \{I - z\int_{\Sigma_+} (z-\mu)^{-1}H_r^+(\mu)\pi\Delta(\mu)Bjd\mu\}\Lambda(z)^{-1}, \tag{4.8}$$

$$H_\ell^+(z) = \Lambda(-z)^{-1}\{I - z\int_{\Sigma_-} (z-\mu)^{-1}\pi\Delta(-\mu)BjH_\ell^+(\mu)d\mu\}. \tag{4.9}$$

Putting $z=t\pm i\varepsilon$ for $t\in\Sigma_+$ (resp. $t\in\Sigma_-$) and taking the limit as $\varepsilon\downarrow 0$, we get

$$H_r^+(t) = \{I - t\mathcal{P}\int_{\Sigma_+} (t-\mu)^{-1}H_r^+(\mu)\pi\Delta(\mu)Bjd\mu \pm i\pi tH_r^+(t)(\pi\Delta(t)Bj)\} \times$$
$$\times [\lambda(t)\pm i\pi t(\pi\Delta(t)Bj)]^{-1},$$

where we have used condition (i). Writing

$$\lim_{\varepsilon\downarrow 0} H_r^+(t\pm i\varepsilon) = k(t) \pm i\ell(t),$$

we obtain

$$[k(t)\ \ell(t)] \begin{bmatrix} \lambda(t) & \pi t(\pi\Delta(t)Bj) \\ -\pi t(\pi\Delta(t)Bj) & \lambda(t) \end{bmatrix} =$$

$$= \begin{bmatrix} I - tP\int_\Sigma (t-\mu)^{-1} H_r^+(\mu)\pi\Delta(\mu)Bj\,d\mu \\ \pi t H_r^+(t)(\pi\Delta(t)Bj) \end{bmatrix},$$

where the right hand side is a row vector and the matrix is the invertible operator

$$\tfrac{1}{2}\begin{bmatrix} I & iI \\ -iI & I \end{bmatrix} \begin{bmatrix} \lambda(t)+i\pi t(\pi\Delta(t)Bj) & 0 \\ 0 & \lambda(t)-i\pi t(\pi\Delta(t)Bj) \end{bmatrix} \begin{bmatrix} I & -iI \\ iI & I \end{bmatrix}.$$

The above row vector can, in fact, be written as $[H_r^+(t) \quad \lambda(t)H_r^+(t)\pi t(\pi\Delta(t)Bj)]$ (see Eq. (4.4)), and therefore $k(t)=H_r^+(t)$ and $\ell(t)=0$ is a solution. Condition (i) and the above representation of the 2×2 matrix operator imply that this solution is unique. Hence,

$$\lim_{\varepsilon\downarrow 0} H_r^+(t\pm i\varepsilon) = H_r^+(t).$$

Similarly, we find

$$\lim_{\varepsilon\downarrow 0} H_\ell^+(t\pm i\varepsilon) = H_\ell^+(t).$$

Thus H_r^+ and H_ℓ^+ have analytic continuations to the right half plane, except possibly for singularities in the set $(\sigma(T)\setminus\Sigma_\pm)\cap[0,\infty)$ and for poles at the zeros of $\Lambda(z)$. The Laurent series of $H_r^+(z)\Lambda(z)$ and $\Lambda(-z)H_\ell^+(z)$ at a zero t_0 of order p in the open right half plane are given by

$$H_r^+(z)\Lambda(z) = I + \sum_{k=0}^{\infty} t_0(z-t_0)^k\, P\int_{\Sigma_+} (\mu-t_0)^{-(k+1)} H_r^+(\mu)\pi\Delta(\mu)Bj\,d\mu.$$

As a consequence of the constraints, the right hand sides have zeros at t_0 of order at least p. Therefore, H_ℓ^+ and H_r^+ are analytic at t_0. We have proved so far that H_ℓ^+ and H_r^+ are analytic on the open right half plane, except possibly at points in the set $(\sigma(T)\setminus\Sigma_\pm)\cap[0,\infty)$, with continuous boundary values at any imaginary z which is not a zero of $\Lambda(z)$ (i.e., such that $1/z$ is not an eigenvalue of $T^{-1}A$).

Let us consider the analytical behavior of H_ℓ^+ and H_r^+ at the finitely many points of the set $(\sigma(T)\setminus\Sigma_+)\cap(0,\infty)$. Since H_r^+ and H_ℓ^+ are uniformly Hölder continuous on $\sigma(\pm T)\cap[0,\infty)$, these points cannot be poles. So they are either essential singularities or points of analyticity. Let us rewrite Eqs. (4.6) and (4.7) as follows:

$$H_r^+(z)[I - z\int_\Sigma \frac{\pi\Delta(\mu)Bj}{z-\mu}d\mu]\Lambda(z)^{-1} =$$

$$= [I + z\int_\Sigma \frac{H_r(\mu)-H_r(z)}{\mu-z}\pi\Delta(\mu)Bjd\mu]\Lambda(z)^{-1}, \tag{4.10}$$

$$\Lambda(-z)^{-1}[I - z\int_\Sigma \frac{\pi\Delta(-\mu)Bj}{z-\mu}d\mu]H_\ell^+(z) =$$

$$= \Lambda(-z)^{-1}[I + z\int_\Sigma \frac{H_\ell(\mu)-H_\ell(z)}{\mu-z}\pi\Delta(-\mu)Bjd\mu]. \tag{4.11}$$

According to condition (ii), for every $t\in(\sigma(T)\setminus\Sigma)\cap(0,\infty)$ (resp. $t\in(\sigma(T)\setminus\Sigma)\cap(-\infty,0)$) there is a neighborhood U_t of t and an $\varepsilon>0$ such that $\Lambda(z)^{-1}$ is bounded on $U_t\setminus\mathbb{R}$. Thus the right hand sides of the above equations are bounded on $\pm U_t$. However, $[I-\int_{\Sigma_\pm}\frac{\pi\Delta(\pm\mu)Bj}{z-\mu}d\mu]^{-1}$ is a meromorphic operator function on the open right half plane with continuous extension to the closed right half plane. Thus, both of the functions H_r^+ and H_ℓ^+ must be meromorphic on all of the open right half plane, which establishes their analyticity at the above points.

Postmultiplying Eq. (4.10) by $\Lambda(z)$ and premultiplying Eq. (4.11) by $\Lambda(-z)$, we get

$$H_r^+(z)[I - z\int_{\Sigma_-} \frac{\pi\Delta(\mu)Bj}{z-\mu}d\mu] = I + z\int_{\Sigma_+} \frac{H_r(\mu)-H_r(z)}{\mu-z}\pi\Delta(\mu)Bjd\mu,$$

$$[I - z\int_{\Sigma_+} \frac{\pi\Delta(-\mu)Bj}{z-\mu}d\mu]H_\ell^+(z) = I + z\int_{\Sigma_-} \frac{H_\ell(\mu)-H_\ell(z)}{\mu-z}\pi\Delta(-\mu)Bjd\mu.$$

Note that $I-z\int_\Sigma(z-\mu)^{-1}\pi\Delta(\pm\mu)Bjd\mu$ tends to I for $z\to0$ along the right half plane. Further, notice that, by the uniform Hölder continuity of H_r^+ and H_ℓ^+, the right hand side tends to I for $z\to0$ along the right half plane. Hence,

$$\lim_{z\to\infty,\, \mathrm{Re}\, z\geq0} \|H_\ell^+(z)-I\| = 0,$$

$$\lim_{z\to\infty,\, \mathrm{Re}\, z\geq 0} \|H_r^+(z)-I\| = 0.$$

The behavior of $H_r^+(z)$ and $H_\ell^+(z)$ as z approaches the finite imaginary line from the right is most easily seen from Eqs. (4.8) and (4.9). If $z_0\neq 0$ is not an imaginary zero of $\Lambda(z)$, then obviously $H_\ell^+(z)$ and $H_r^+(z)$ have analytic continuations at $z=z_0$. If $z_0\neq 0$ is an imaginary zero of $\Lambda(z)$ (or, alternatively, if $1/z_0$ is an imaginary eigenvalue of $T^{-1}A$), the functions $H_\ell^+(z)$ and $H_r^+(z)$ are meromorphic at $z=z_0$ and one should specify the principal parts of their Laurent series in order to single out H_ℓ^+ and H_r^+ among the solutions of Eqs. (4.4) and (4.5). This is most conveniently done by imposing suitable constraints on H_ℓ^+ and H_r^+.

For the behavior of $H_\ell^+(z)$ and $H_r^+(z)$ at ∞ we use the boundedness of T and repeat the argument for the finite imaginary zeros of $\Lambda(z)$. ■

This result has many applications in radiative transfer and neutron physics. Most research in this area, especially the earlier work, has been obtained for scalar dispersion functions of the form

$$\Lambda(z) = 1 - z\int_{-1}^{1} (z-\mu)^{-1}\psi(\mu)d\mu, \quad z\notin[-1,1],$$

where $\psi(\mu)\geq 0$ and $c=2\int_0^1 \psi(\mu)d\mu\leq 1$. Following the work of Crum [95] on the H–equation (also treated in [89]), Chandrasekhar [89] constructed solutions of the X– and Y–equations (some of them for $c<1$, all of them for $c=1$), assuming one (physical) solution. Busbridge [61] first established the existence of H–, X– and Y–functions and determined all solutions of the H–equation (for the latter, also [63]). Mullikin [271] constructed all solutions of the X– and Y–equations and isolated the physically relevant solution using linear constraints under the above assumptions [272]. Van der Mee [363] extended these results to anisotropic one–speed neutron transport with $c\leq 1$.

We have proved Theorem 4.1 under the major restriction that T is bounded. Note that, on dropping the boundedness of T, one worsens the behavior of $H_\ell^+(z)$ and $H_r^+(z)$ for $z\to\infty$, while the remaining part of the proof still goes through. Conditions (i) and (ii) are technical conditions, which imply that the symbol of the associated singular integral equation has invertible values. For one speed neutron transport with degenerate anisotropic scattering these conditions were proved to be satisfied by Garcia and Siewert [130]. In multigroup neutron transport one can easily construct models in which these conditions are violated. In many applications it has been usual to assume these conditions explicitly or tacitly, for mathematical convenience rather than physical content. Indeed, for a variety of problems, it is not known whether or when conditions (i) and (ii) are fulfilled.

5. Adding method

In this section we shall represent solutions in the interior of the spatial domain by exploiting identities existing between reflection and transmission operators for a medium of thickness τ and reflection and transmission operators for the constituent media of thicknesses τ_1 and τ_2, with $\tau=\tau_1+\tau_2$.

Let us consider a self adjoint injective operator T on the Hilbert space H, and a collision operator A = I−B which is a compact perturbation of the identity satisfying the range condition (1.4). We shall assume that the finite slab problem

$$(T\psi)'(x) = -A\psi(x), \quad 0<x<\tau, \tag{5.1}$$

$$Q_+\psi(0) = Q_+\varphi, \tag{5.2a}$$

$$Q_-\psi(\tau) = Q_-\varphi, \tag{5.2b}$$

is uniquely solvable for every $\varphi\in D(T)\subset H$. As we have seen in Sections V.4 and Section 3 of this chapter, we may write the solution at x=0 and at $x=\tau$ in the form

$$\psi(0) = (R_{+\tau}+T_{-\tau})\varphi,$$

$$\psi(\tau) = (T_{+\tau}+R_{-\tau})\varphi,$$

where $R_{\pm\tau}Q_{\mp}=T_{\pm\tau}Q_{\mp}=0$ for reflection and transmission operators $R_{\pm\tau}$ and $T_{\pm\tau}$. It has been shown in Section 3 that the equivalence of (5.1)–(5.2) to a convolution equation with compact operator kernel implies that the operators $R_{\pm\tau}$ − $Q_{\pm}$ and $T_{\pm\tau}$ − $\exp\{\mp\tau T^{-1}\}Q_{\pm}$ are compact.

Consider Eqs. (5.1)–(5.2) for three values of τ, namely, τ_1,τ_2 and $\tau=\tau_1+\tau_2$. The adding method gives a procedure for computing the reflection and transmission operators $R_{\pm\tau}$ and $T_{\pm\tau}$ from the operators $R_{\pm\tau_1}$, $T_{\pm\tau_1}$, $R_{\pm\tau_2}$, and $T_{\pm\tau_2}$. Thus the reflection and transmission properties of the constituent layers $x\epsilon(0,\tau_1)$ and $x\epsilon(0,\tau_2)$ are used to obtain the reflection and transmission operators of the combined layer $x\epsilon(0,\tau_1+\tau_2)$. In the special case $\tau_1=\tau_2$, where the constituent layers are equal, this method is called the **doubling method**. Note that when applying these methods one assumes implicitly the unique solvability of (5.1)–(5.2) for $\tau=\tau_1$, τ_2 and $\tau_1+\tau_2$.

As we shall see shortly, the adding method requires the computation of an inverse operator, which can be expressed as a Neumann series whose convergence may be proved in general. For many specific models the adding method can be implemented fairly easily on a modest computer for many specific models, by iterating the equivalent integral equation

$$\psi(x) - \int_0^{\tau_0} \mathcal{H}(x-y)B\psi(y)dy = \left[e^{-xT^{-1}}Q_+ + e^{(\tau-x)T^{-1}}Q_-\right]\varphi,$$

for thin layers $\tau_0<<1$ and repeatedly applying adding. As a result, one obtains $R_{\pm\tau}$ and $T_{\pm\tau}$ for $\tau=\tau_0$, $2\tau_0$, $3\tau_0$, ..., and from these operators the solution $\psi(x)$ with $\tau=n\tau_0$ for $x=0$, τ_0, $2\tau_0$, ..., $(n-1)\tau_0$, $n\tau_0$. One thus accomplishes a spatially discrete representation of the solution.

The doubling method was first applied by Peebles and Plesset [302] to the scattering of γ-rays. The earliest systematic application to a large class of radiative transfer problems is due to van de Hulst [356, 357] and was extended to polarized light problems by Hansen [185] and Hovenier [199]. Comprehensive discussions of the method, focused on numerical applications, have been given by Hansen and Travis [186], van de Hulst [357], and de Haan et al. [100] A different approach to the adding method, stemming from the theory of compartimental systems, was expounded by Ribarič [318]. Using positive cones in L_1-spaces, he derived the dependence of the albedo operator for a system on the albedo operators for its constituent subsystems.

Let us derive the **adding equations**. To this purpose we first introduce the **transfer (matrix) operator S_τ**, closely related to the albedo operator of [318], by

$$S_\tau = \begin{bmatrix} S_\tau^{++} & S_\tau^{+-} \\ S_\tau^{-+} & S_\tau^{--} \end{bmatrix},$$

where $S_\tau^{\pm\pm}:Q_\pm[H]\to Q_\pm[H]$, $S_\tau^{\pm\mp}:Q_\mp[H]\to Q_\pm[H]$, and

$$S_\tau\varphi = Q_-\psi(0) + Q_+\psi(\tau). \tag{5.3}$$

Here ψ is the unique solution of Eqs. (5.1) and (5.2). We have the equalities

$$S_\tau^{++}\varphi = T_{+\tau}\varphi, \tag{5.4a}$$

$$S_\tau^{--}\varphi = T_{-\tau}\varphi, \tag{5.4b}$$

$$S_\tau^{-+}\varphi = (R_{+\tau}-Q_+)\varphi, \tag{5.5a}$$

$$S_\tau^{+-}\varphi = (R_{-\tau}-Q_-)\varphi. \tag{5.5b}$$

It is then clear that $S_\tau-\exp\{-\tau|T|^{-1}\}$ is a compact operator.

THEOREM 5.1. Suppose that Eqs. (5.1) and (5.2) are uniquely solvable for two constituent layers with $\tau=\tau_1$ and $\tau=\tau_2$, as well as for the combined layer with $\tau=\tau_1+\tau_2$. Then the operators $(I-S_{\tau_1}^{+-}S_{\tau_2}^{-+})$ and $(I-S_{\tau_2}^{-+}S_{\tau_1}^{+-})$ are invertible, and we have the adding equations

$$S_\tau^{++} = S_{\tau_2}^{++}(I-S_{\tau_1}^{+-}S_{\tau_2}^{-+})^{-1}S_{\tau_1}^{++}, \tag{5.6}$$

$$S_\tau^{+-} = S_{\tau_2}^{+-} + S_{\tau_2}^{++}(I-S_{\tau_1}^{+-}S_{\tau_2}^{-+})^{-1}S_{\tau_1}^{+-}S_{\tau_2}^{--}, \tag{5.7}$$

$$S_\tau^{-+} = S_{\tau_1}^{-+} + S_{\tau_1}^{--}(I-S_{\tau_2}^{-+}S_{\tau_1}^{+-})^{-1}S_{\tau_2}^{-+}S_{\tau_1}^{++}, \tag{5.8}$$

$$S_\tau^{--} = S_{\tau_1}^{--}(I-S_{\tau_2}^{-+}S_{\tau_1}^{+-})^{-1}S_{\tau_2}^{--}. \tag{5.9}$$

Proof: Straightforward application of the definition (5.3) to the constituent layers with $\varphi_\pm=Q_\pm\varphi$ gives the identities

$$Q_+\psi(\tau_1) = S_{\tau_1}^{++}\varphi_+ + S_{\tau_1}^{+-}Q_-\psi(\tau_1), \tag{5.10}$$

$$Q_-\psi(0) = S_{\tau_1}^{-+}\varphi_+ + S_{\tau_1}^{--}Q_-\psi(\tau_1), \tag{5.11}$$

$$Q_+\psi(\tau) = S^{++}_{\tau_2}Q_+\psi(\tau_1) + S^{+-}_{\tau_2}\varphi_-, \tag{5.12}$$

$$Q_-\psi(\tau_1) = S^{-+}_{\tau_2}Q_+\psi(\tau_1) + S^{--}_{\tau_2}\varphi_-, \tag{5.13}$$

whence

$$Q_+\psi(\tau_1) = S^{++}_{\tau_1}\varphi_+ + S^{+-}_{\tau_1}\{S^{-+}_{\tau_2}Q_+\psi(\tau_1) + S^{--}_{\tau_2}\varphi_-\}, \tag{5.14}$$

$$Q_-\psi(\tau_1) = S^{--}_{\tau_2}\varphi_- + S^{-+}_{\tau_2}\{S^{++}_{\tau_1}\varphi_+ + S^{+-}_{\tau_1}Q_-\psi(\tau_1)\}. \tag{5.15}$$

These last equations imply

$$(I-S^{+-}_{\tau_1}S^{-+}_{\tau_2})Q_+\psi(\tau_1) = S^{++}_{\tau_1}\varphi_+ + S^{+-}_{\tau_1}S^{--}_{\tau_2}\varphi_-, \tag{5.16}$$

$$(I-S^{-+}_{\tau_2}S^{+-}_{\tau_1})Q_-\psi(\tau_1) = S^{--}_{\tau_2}\varphi_- + S^{-+}_{\tau_2}S^{++}_{\tau_1}\varphi_+. \tag{5.17}$$

If $(I-S^{+-}_{\tau_1}S^{-+}_{\tau_2})$ and $(I-S^{-+}_{\tau_2}S^{+-}_{\tau_1})$ were not invertible, then $S^{+-}_{\tau_1}$ and $S^{-+}_{\tau_2}$ being compact would imply that Eqs. (5.16) and (5.17) have more than one solution. This would imply in turn that Eqs. (5.1) and (5.2) for the combined layer $\tau=\tau_1+\tau_2$ have more than one solution, thereby contradicting the assumption of unique solvability. Indeed, there would be unique solutions of the finite slab problem on $(0,\tau_1)$ with boundary data $\{0,Q_-\psi(\tau_1)\}$ and the finite slab problem on (τ_1,τ_2) with boundary data $\{Q_+\psi(\tau_1),0\}$. Since $T\psi$ would then be continuously differentiable at $x=\tau_1$, it would yield a solution of Eqs. (5.1)–(5.2) with zero incoming data, whence $Q_\pm\psi(\tau_1)=0$ and the contradiction follows. Therefore, one may eliminate $Q_+\psi(\tau_1)$ and $Q_-\psi(\tau_1)$ from (5.10)–(5.13) using (5.16) and (5.17) and obtain Eqs. (5.6)–(5.9). ■

The nonzero eigenvalues of the compact operators $S^{+-}_{\tau_1}S^{-+}_{\tau_2}$ and $S^{-+}_{\tau_2}S^{+-}_{\tau_1}$ coincide, and their resolvents are related as follows:

$$(I-cS^{+-}_{\tau_1}S^{-+}_{\tau_2})^{-1} = I + cS^{+-}_{\tau_1}(I-cS^{-+}_{\tau_2}S^{+-}_{\tau_1})^{-1}S^{-+}_{\tau_2},$$

$$(I-cS^{-+}_{\tau_2}S^{+-}_{\tau_1})^{-1} = I + cS^{-+}_{\tau_2}(I-cS^{+-}_{\tau_1}S^{-+}_{\tau_2})^{-1}S^{+-}_{\tau_1}.$$

We also have the identities

$$(R_{+\tau_1}+R_{-\tau_1}-I)(R_{+\tau_2}+R_{-\tau_2}-I) = S^{+-}_{\tau_1}S^{-+}_{\tau_2} \oplus S^{-+}_{\tau_1}S^{+-}_{\tau_2}, \tag{5.18}$$

$$(R_{+\tau_2}+R_{-\tau_2}-I)(R_{+\tau_1}+R_{-\tau_1}-I) = S^{+-}_{\tau_2}S^{-+}_{\tau_1} \oplus S^{-+}_{\tau_2}S^{+-}_{\tau_1}, \tag{5.19}$$

as one easily computes.

In numerical applications the inverse operators appearing in (5.6)–(5.9) are usually computed by expansion as a Neumann series. In doing so, Eqs. (5.6)–(5.9) are cast in the form

$$S^{++}_{\tau} = \sum_{n=0}^{\infty} S^{++}_{\tau_2}(S^{+-}_{\tau_1}S^{-+}_{\tau_2})^n S^{++}_{\tau_1}, \tag{5.20}$$

$$S^{+-}_{\tau} = S^{+-}_{\tau_2} + \sum_{n=0}^{\infty} S^{++}_{\tau_2}(S^{+-}_{\tau_1}S^{-+}_{\tau_2})^n S^{+-}_{\tau_1}S^{--}_{\tau_2}, \tag{5.21}$$

$$S^{-+}_{\tau} = S^{-+}_{\tau_1} + \sum_{n=0}^{\infty} S^{--}_{\tau_1}(S^{-+}_{\tau_2}S^{+-}_{\tau_1})^n S^{-+}_{\tau_2}S^{++}_{\tau_1}, \tag{5.22}$$

$$S^{--}_{\tau} = \sum_{n=0}^{\infty} S^{--}_{\tau_1}(S^{-+}_{\tau_2}S^{+-}_{\tau_1})^n S^{--}_{\tau_2}. \tag{5.23}$$

When applied to radiative transfer in planetary atmospheres (cf. Sections IX.1 and IX.2), these expansions are easily interpreted physically. For a layer with (optical) thickness σ and incident fluxes φ_+ and φ_-, $S^{++}_{\sigma}\varphi$ and $S^{--}_{\sigma}\varphi$ describe the light directly transmitted from the top ($x=0$) to the bottom ($x=\sigma$) and the light directly transmitted from the bottom ($x=\sigma$) to the top ($x=0$), respectively. The incident light reflected at the top is represented by $S^{-+}_{\sigma}\varphi$ and the incident light reflected at the bottom by $S^{+-}_{\sigma}\varphi$. Then the expansion (5.20) can be read in the following fashion. The light transmitted from $x=0$ through both layers to the opposite side at $x=\tau=\tau_1+\tau_2$ is the sum, for $n=0,1,2,\ldots$, of the light transmitted through the first layer, subsequently undergoing n pairs of events consisting of reflection by the top of the second layer and back reflection by the bottom of the first layer, and finally transmitted through the second layer. The other expansions allow similar physical interpretations. We may therefore call these expansions multiple interface reflection expansions. The adding equations are generally derived using multiple scattering arguments, without resorting to a convergence proof of these expansions, although numerical convergence has been studied in some detail (cf. [100, 357]). We shall provide a general convergence proof, drawing on ideas used for the polarized light transfer problem [368].

THEOREM 5.2. Let H be a Banach lattice, and suppose the operators F(T) and B=I−A are positive, in the lattice sense, for all nonnegative real continuous functions F. If $0 < \tau_1+\tau_2 < \tau_c$, where τ_c is the critical slab width,

$$\tau_c = \inf\{\tau \in (0,\infty) : (5.1)-(5.2) \text{ are not uniquely solvable}\},$$

then the multiple interface reflection expansions (5.20)−(5.23) converge in the norm topology.

Proof: If $\tau=\tau_1+\tau_2 \in (0,\tau_c)$, then clearly the operators S_{τ_1}, S_{τ_2} and S_τ are well-defined. Since they depend analytically on τ_1, τ_2 and τ (cf. Section VIII.3), the compact operators $S^{+-}_{\tau_1}S^{-+}_{\tau_2}$ and $S^{-+}_{\tau_2}S^{+-}_{\tau_1}$ depend analytically on τ_1 and τ_2 in the region $D_c=\{(\tau_1,\tau_2) : 0<\tau_1+\tau_2<\tau_c\}$. Then each eigenvalue $\lambda_k(\tau_1,\tau_2)$ is separately analytic for $(\tau_1,\tau_2)\in D_c$, except for algebraic branch points (cf. Theorem VIII.8 of [213]).

The nonnegativity of solutions of (5.1)−(5.2) for nonnegative φ and (5.3) imply that the operators $S^{+-}_{\tau_1}S^{-+}_{\tau_2}$ and $S^{-+}_{\tau_2}S^{+-}_{\tau_2}$ are positive in the lattice sense if $(\tau_1,\tau_2)\in D_c$. Therefore, the spectral radius $r(\tau_1,\tau_2)$ is an eigenvalue of both operators and is separately continuous for $(\tau_1,\tau_2)\in D_c$ (cf. Theorem I 4.2). Since $r(\tau_1,\tau_2)$ vanishes as $\tau_1\to 0$ with τ_2 fixed, or as $\tau_2\to 0$ with τ_1 fixed, and since $r(\tau_1,\tau_2)=1$ would contradict the invertibility of $(I-S^{+-}_{\tau_1}S^{-+}_{\tau_2})$, we must have $r(\tau_1,\tau_2)<1$ for every $(\tau_1,\tau_2)\in D_c$. Hence, the expansions (5.20) and (5.23) converge in the norm topology if $\tau_1+\tau_2\in(0,\tau_c)$. ∎

THEOREM 5.3. Suppose H is a Hilbert space and A is positive self adjoint. If $0<\tau_1,\tau_2<\infty$, then the multiple interface reflection expansions (5.20)−(5.23) converge in the norm topology.

Proof: Observe that the operator S_τ is well defined (cf. Section V.1). Next, observe that $(R_{+\sigma}+R_{-\sigma}-I)$ is a self adjoint strict contraction on H_T, i.e., on the completion of D(T) with respect to the inner product

$$(h,k)_T = (|T|h,k).$$

By virtue of (5.18) and (5.19), $S_{\tau_1}^{+-}S_{\tau_2}^{-+}$ and $S_{\tau_2}^{-+}S_{\tau_1}^{+-}$ are strict contractions on $Q_+[H_T]$ and Q_-H_T, respectively, and therefore their (coinciding) spectral radii are strictly less than one. To prove that this statement remains true on $Q_\pm[H]$, we consider

$$\hat{S}_\tau = \begin{bmatrix} \hat{S}_\tau^{++} & \hat{S}_\tau^{+-} \\ \hat{S}_\tau^{-+} & \hat{S}_\tau^{--} \end{bmatrix},$$

where $\hat{S}_\tau^{\pm\pm}:Q_\pm[H]\to Q_\pm[H]$ and $\hat{S}_\tau^{\pm\mp}:Q_\mp[H]\to Q_\pm[H]$. These operators satisfy $\hat{S}_\tau - \exp\{-\tau|T|^{-1}\}$ compact and have the intertwining properties

$$S_\tau[D(T)] \subset D(T), \tag{5.24a}$$

$$TS_\tau h = \hat{S}_\tau Th, \qquad h\in D(T). \tag{5.24b}$$

As a result of (5.24) and their compactness we have

$$\sigma(S_{\tau_1}^{+-}S_{\tau_2}^{-+}) = \sigma(S_{\tau_2}^{-+}S_{\tau_1}^{+-}) = \sigma(\hat{S}_{\tau_1}^{+-}\hat{S}_{\tau_2}^{-+}) = \sigma(\hat{S}_{\tau_2}^{-+}\hat{S}_{\tau_1}^{+-}).$$

Finally, using Proposition II 3.1 applied to all their integer powers, one may show that $S_{\tau_1}^{+-}S_{\tau_2}^{-+}$ and $S_{\tau_2}^{-+}S_{\tau_1}^{+-}$ have spectral radii on $Q_+[H]$ and $Q_-[H]$ that are strictly less than one. ■

Chapter IX

APPLICATIONS OF THE STATIONARY THEORY

1. Radiative transfer without polarization

A good approximation to the mathematical description of radiative transfer processes in planetary atmospheres is to consider the atmosphere as a medium that is plane parallel and invariant under arbitrary translations in horizontal directions. Although one excludes in this way processes such as zodiacal light where the light is incident at small angle with the planetary surface, it permits the study of the most important radiative phenomena. As position coordinate in the vertical direction one frequently employs the **optical depth**, τ, which is defined as minus the antiderivative of the extinction coefficient $\kappa_e(z)$ as a function of vertical position z: $\tau=\int_z^\infty \kappa_e(\hat{z})d\hat{z}$. At the top of the atmosphere we set $\tau=0$, while τ coincides with the **optical thickness** b at the bottom, i.e., at the planetary surface. When considering a stationary radiative transfer problem, sunlight is thought to be incident at the top $\tau=0$, and "reflection" (more precisely, a combination of genuine reflection and true absorption) of light takes place at the bottom $\tau=b$. Sometimes one also accounts for radiation emitted by the surface and by sources within the atmosphere. In any case, the physical quantities of interest are the specific intensity and the state of polarization of the light at any position and in any direction, as a function of incident radiation at the top (and sometimes at the bottom), reflection properties of the bottom and internal sources. In this section we shall only discuss radiative transfer models in which polarization of light is neglected. The quantity of primary interest then is the (specific) intensity. In the next section we shall discuss more general models that account for polarization phenomena.

Let us consider a (vertically) homogeneous atmosphere, i.e. an atmosphere in which single scattering events do not depend on vertical position. On single scattering a fraction of the light, the **albedo of single scattering** $a \in (0,1]$, undergoes true scattering. As a function of the **scattering angle** $\theta \in [0,\pi]$ between the incident and the scattered beam of light, with $\theta=0$ representing forward and $\theta=\pi$ representing backward scattering, the fraction of light scattered is described by a probability distribution, the **phase function** $a_1(\theta)$, which is nonnegative and measurable and satisfies the normalization condition

$$\tfrac{1}{2}\int_{-1}^{1} a_1(\theta)d(\cos\theta) = 1. \tag{1.1}$$

If one uses polar coordinates $\nu \in [0,\pi]$ and $\varphi \in [0,2\pi)$ to indicate direction and writes $u=-\cos\nu$, with $u>0$ in the direction of increasing optical depth τ, one may derive the equation of radiative transfer

$$u\frac{\partial}{\partial\tau}I(\tau,u,\varphi) + I(\tau,u,\varphi) = \frac{a}{4\pi}\int_{-1}^{1}\int_{0}^{2\pi} a_1(\theta)I(\tau,\hat{u},\hat{\varphi})d\hat{\varphi}d\hat{u} + S(\tau,u,\varphi), \tag{1.2}$$

where $0<\tau<b$, with boundary conditions

$$I(0,u,\varphi) = D(u,\varphi) \tag{1.3}$$

$$I(b,-u,\varphi) = \frac{1}{\pi}\int_{0}^{1}\int_{0}^{2\pi} \hat{u}R_g(u,\hat{u},\varphi-\hat{\varphi})I(b,\hat{u},\hat{\varphi})d\hat{\varphi}d\hat{u} + D(-u,\varphi) \tag{1.4}$$

for $u>0$. In these equations the various quantities have the following meaning:

(i) $I(\tau,u,\varphi)$ is the **specific intensity** as a function of the optical thickness τ and the direction (u,φ). Similarly, $S(\tau,u,\varphi)$ accounts for internal sources, and $D(u,\varphi)$ and $D(-u,\varphi)$, with $u>0$, is the **incident radiation** at top and bottom, respectively.

(ii) The scattering angle θ is related to the direction $(\hat{u},\hat{\varphi})$ of the incident light and the direction (u,φ) of the singly scattered light, by

$$\cos\theta = \cos\nu\cos\nu' + \sin\nu\sin\nu'\cos(\varphi-\hat{\varphi}), \tag{1.5}$$

where $u=-\cos\nu$ and $\hat{u}=-\cos\nu'$.

(iii) $R_g(u,\hat{u},\varphi-\hat{\varphi})$ is the **reflection function** of the ground. It is nonnegative, always satisfies the property of **reciprocity symmetry**

$$R_g(u,\hat{u},\varphi-\hat{\varphi}) = R_g(\hat{u},u,\hat{\varphi}-\varphi), \tag{1.6}$$

and in most applications **mirror symmetry**

$$R_g(u,\hat{u},\varphi-\hat{\varphi}) = R_g(u,\hat{u},\hat{\varphi}-\varphi), \tag{1.7}$$

and obeys the energy bound

$$0 \le \frac{1}{\pi}\int_0^1\int_0^{2\pi} \hat{u}R_g(u,\hat{u},\varphi-\hat{\varphi})d(\varphi-\hat{\varphi})d\hat{u} \le 1. \tag{1.8}$$

The integral expression in (1.8), which is bounded by 1, is called the **plane albedo** of the planetary surface. From the above it is clear that the corresponding total radiative fluxes, obtained by integrating the specific intensities with respect to direction using the measure $|u|\,d\varphi du$, are finite.

In order to obtain more explicit information about the solutions of the equation of radiative transfer, one usually resorts to Fourier decomposition. We first write the expansion

$$I(\tau,u,\varphi) = I^0(\tau,u) + 2\sum_{m=1}^{\infty} I^m(\tau,u)\cos m\varphi, \tag{1.9}$$

and analogously for $S(\tau,u,\varphi)$, $D(u,\varphi)$ and $D(-u,\varphi)$. For the reflection function we have the expansion

$$R_g(u,\hat{u},\varphi-\hat{\varphi}) = R_g^0(u,\hat{u}) + 2\sum_{m=1}^{\infty} R_g^m(u,\hat{u})\cos m(\varphi-\hat{\varphi}). \tag{1.10}$$

As a result we obtain the "component" equations of radiative transfer

$$u\frac{\partial}{\partial\tau}I^m(\tau,u) + I^m(\tau,u) = \tfrac{1}{2}a\int_{-1}^{1} p^m(u,\hat{u})I^m(\tau,\hat{u})d\hat{u} + S^m(\tau,u), \tag{1.11}$$

where $0<\tau<b$, with boundary conditions

$$I^m(0,u) = D^m(u), \quad u>0, \tag{1.12}$$

$$I^m(b,-u) = 2\int_0^1 \hat{u}R_g^m(u,\hat{u})I^m(b,\hat{u})d\hat{u} + D^m(-u), \quad u>0. \tag{1.13}$$

Here we have defined

$$p^m(u,\hat{u}) = \frac{1}{2\pi}\int_0^{2\pi} a_1(u\hat{u}+(1-u^2)^{1/2}(1-\hat{u}^2)^{1/2}\cos(\varphi-\hat{\varphi}))\cos m(\varphi-\hat{\varphi})d\hat{\varphi} \tag{1.14}$$

and note the formula for the expansion coefficient

$$R_g^m(u,\hat{u}) = \frac{1}{2\pi}\int_0^{2\pi} R_g(u,\hat{u},\varphi-\hat{\varphi})\cos m(\varphi-\hat{\varphi})d\hat{\varphi}. \tag{1.15}$$

For $m=0$ the functions $I^m(\tau,u)$, $S^m(\tau,u)$, $D^m(u)$ and $D^m(-u)$ are nonnegative, while this need no longer be the case for $m\geq 1$. The energy bound (1.8) now takes the form

$$0 \leq 2\int_0^1 \hat{u}R_g^0(u,\hat{u})d\hat{u} \leq 1. \tag{1.16}$$

In the case of an atmosphere of infinite optical thickness ($b=\infty$), the basic boundary value problem is less complicated to formulate, since the incoming sunlight can never reach the planetary surface and therefore reflection by a bottom surface is out of the question. Instead one imposes a growth condition on the solution as $\tau\to\infty$. Thus we endow the equation of transfer (1.2), where $0<\tau<b=\infty$, with the boundary conditions

$$I(0,u,\varphi) = D_+(u,\varphi), \quad u>0, \tag{1.17}$$

$$\limsup_{\tau\to\infty} \|I(\tau,\cdot,\cdot)\| < \infty. \tag{1.18}$$

with the sense of the norm to be specified. Analogously the Fourier component equation (1.11), where $0<\tau<b=\infty$, is endowed with the boundary conditions

$$I^m(0,u) = D_+^m(u), \quad u>0, \tag{1.19}$$

$$\limsup_{\tau\to\infty} \|I^m(\tau,\cdot)\| < \infty. \tag{1.20}$$

Another problem in semi-infinite geometry originates from the study of stellar atmospheres. In 1921 E.A. Milne [265] proposed a model for the transfer of light in a stellar atmosphere, which consisted of computing the outgoing radiation of the star from the flux coming from its interior. True absorption and internal sources were assumed absent, i.e. a=1. Generalizing beyond Milne's originally isotropic scattering problem to general anisotropic scattering, we obtain the boundary value problem

$$u\frac{\partial}{\partial\tau}I(\tau,u,\varphi) + I(\tau,u,\varphi) = \frac{1}{4\pi}\int_{-1}^{1}\int_{0}^{2\pi} a_1(\theta)I(\tau,\hat{u},\hat{\varphi})d\hat{\varphi}d\hat{u}, \tag{1.21}$$

$$I(0,u,\varphi) = 0, \quad u>0, \tag{1.22}$$

$$\lim_{\tau\to\infty}\int_{-1}^{1}\int_{0}^{2\pi}\hat{u}I(\tau,\hat{u},\hat{\varphi})d\hat{\varphi}d\hat{u} = -F/4\pi, \tag{1.23}$$

where $0<\tau<b=\infty$ and F is a given positive constant, the radiative flux at large optical depth.

So far we have restricted ourselves to a brief description of the physical background and the mathematical formulation of radiative transfer problems. For more detailed information we refer to the classical monograph of Chandrasekhar [89], which still is relevant to present day radiative transfer, as well as to the more recent monographs of Sobolev [340, 342], which mainly deal with invariant imbedding techniques, and van de Hulst [357], which contains a thorough discussion of numerical work.

Let us now introduce the functional formulation which will enable us to apply the existence and uniqueness theory of the previous chapters. Let Ω denote the unit sphere in $\mathbb{R}^3$, and let $\omega\equiv(u,\varphi)\in\Omega$. Here the Cartesian coordinates of ω are $(1-u^2)^{1/2}\cos\varphi$, $(1-u^2)^{1/2}\sin\varphi$ and u. By $L_p(\Omega)$ we denote the (real or complex) Banach space of measurable functions on Ω endowed with the L_p-norm

$$\|h\|_p = \left[\int_{-1}^{1}\int_{0}^{2\pi}|h(u,\varphi)|^p d\varphi du\right]^{1/p}, \quad 1\le p<\infty, \tag{1.24a}$$

or

$$\|h\|_\infty = \operatorname*{ess\,sup}_{(u,\varphi)\in\Omega}|h(u,\varphi)|. \tag{1.24b}$$

Then $L_2(\Omega)$ is a Hilbert space with inner product

$$(h,k) = \int_{-1}^{1}\int_{0}^{2\pi} h(u,\varphi)\overline{k(u,\varphi)}\,d\varphi du. \tag{1.25}$$

On $L_p(\Omega)$ we define the bounded linear operators

$$(Th)(u,\varphi) = uh(u,\varphi),$$

$$(Q_\pm h)(u,\varphi) = \begin{cases} h(u,\varphi), & \pm u>0, \\ 0, & \pm u<0, \end{cases}$$

$$(Ah)(u,\varphi) = h(u,\varphi) - a(Bh)(u,\varphi),$$

$$(Bh)(u,\varphi) = \frac{1}{4\pi}\int_{-1}^{1}\int_{0}^{2\pi} a_1(\theta)h(\hat{u},\hat{\varphi})\,d\hat{\varphi}d\hat{u},$$

$$(Rh)(u,\varphi) = \frac{1}{\pi}\int_{-1}^{1}\int_{0}^{2\pi} \hat{u}R_g(u,\hat{u},\varphi-\hat{\varphi})h(\hat{u},\pi-\hat{\varphi})\,d\hat{\varphi},$$

$$(Jh)(u,\varphi) = h(-u,\pi-\varphi),$$

and the vectors

$$I(\tau)(u,\varphi) = I(\tau,u,\varphi),$$

$$S(\tau)(u,\varphi) = S(\tau,u,\varphi).$$

We then obtain the boundary value problem

$$(TI)'(\tau) = -AI(\tau) + S(\tau), \quad 0<\tau<b, \tag{1.26}$$

$$Q_+I(0) = Q_+D, \tag{1.27}$$

$$Q_-I(b) = RJQ_+I(b) + Q_-D. \tag{1.28a}$$

on $L_p(\Omega)$ if $b<\infty$. For the half space problem $b=\infty$, (1.28a) would be supplanted by

$$\limsup_{\tau\to\infty} \|I(\tau)\| < \infty. \tag{1.28b}$$

On $L_p[-1,1]$ we obtain boundary value problems of exactly the same form (apart from upper indices m), if we define the bounded linear operators

$$(T^m h)(u) = uh(u),$$

$$(Q_\pm^m h)(u) = \begin{cases} h(u), & \pm u>0, \\ 0, & \pm u<0, \end{cases}$$

$$(A^m h)(u) = h(u) - a(B^m h)(u),$$

$$(B^m h)(u) = \tfrac{1}{2}\int_{-1}^{1} p^m(u,\hat{u})h(\hat{u})d\hat{u},$$

$$(R^m h)(u) = 2\int_{-1}^{1} R_g^m(u,\hat{u})h(\hat{u})d\hat{u},$$

$$(J^m h)(u) = h(-u),$$

and obtain analogues of (1.26)–(1.28) for $I^m(\tau)$ with $S^m(\tau)$ as source term. Note that, in writing $A=I-aB$, we have departed somewhat from the convention of the previous chapters, in which we denoted $A=I-B$. It is convenient in the next several sections to distinguish the albedo of single scattering a in this fashion.

It is clear that on the Hilbert space $L_2(\Omega)$ the operators T, A and B are self adjoint, that T has spectrum $[-1,1]$ which is absolutely continuous, and that Q_+ and Q_- are the orthogonal projections of $L_2(\Omega)$ onto the maximal T-positive and T-negative T-invariant subspaces. On $L_p(\Omega)$, $1\le p<\infty$, the operator T is a scalar-type spectral operator (cf. [105, 109]) and Q_+ and Q_- are complementary projections onto the maximal subspaces which are T-invariant and on which T has restrictions with nonnegative and nonpositive spectrum, respectively. On $L_p(\Omega)$, $1\le p\le\infty$, J is an isometry having the identity operator as its square; we shall call such an operator an **inversion symmetry**. It satisfies the identities

$$JT = -TJ, \tag{1.29a}$$
$$JQ_\pm = Q_\mp J, \tag{1.29b}$$
$$JA = AJ, \tag{1.29c}$$
$$JB = BJ, \tag{1.29d}$$
$$JR = RJ, \tag{1.29e}$$

where the last identity follows from the reciprocity symmetry (1.6). If the mirror symmetry (1.7) is satisfied, $|T|R$ will be self adjoint on $L_2(\Omega)$.

The following result is due to Vladimirov ([377], Appendix XII.8). A new proof is given for the form of the eigenfunctions, suggested by E.J.G. Thomas.

LEMMA 1.1. The operator B is compact on all of the spaces $L_p(\Omega)$ $(1\le p<\infty)$. A complete set of eigenfunctions consists of the spherical harmonics $Y_\ell^m(\nu,\varphi)$ satisfying

$$BY_\ell^m = (2\ell+1)^{-1}\beta_\ell Y_\ell^m, \quad \ell=0,1,2,..., \quad m=-\ell,-\ell+1,...,\ell, \tag{1.30}$$

where

$$\beta_\ell = (\ell+½)\int_{-1}^{1} a_1(\theta)P_\ell(\cos\theta)d(\cos\theta) \tag{1.31}$$

with $P_\ell(\cos\theta)$ the usual Legendre polynomial of degree ℓ.

Proof: If $\mathcal{R}$ is a pure rotation in three dimensional space, then B commutes with the invertible isometry $(U_\mathcal{R}h)(\omega) = h(\mathcal{R}(\omega))$, $\omega\in\Omega$. The operators $U_\mathcal{R}$ form a compact group of invertible isometries on $L_p(\Omega)$ which is (group theoretically) isomorphic and (topologically) homeomorphic to the unitary group SO(3) of pure rotations on $\mathbb{R}^3$. Let

$$Y_\ell^m(\nu,\varphi) = (-1)^m\left[\frac{2\ell+1}{4\pi}\,\frac{(\ell-m)!}{(\ell+m)!}\right]^{½} e^{im\varphi}P_\ell^m(\cos\nu)$$

be the spherical harmonics, where

$$P_\ell^m(u) = \frac{1}{2^\ell(\ell!)}(1-u^2)^{m/2}\left(\frac{d}{du}\right)^{\ell+m}(u^2-1)^\ell.$$

Then the $(2\ell+1)$ dimensional subspace of $L_p(\Omega)$ spanned by the vectors Y_ℓ^m, where $m=-\ell,-\ell+1,...,\ell$, is an invariant subspace for all of the operators $U_\mathcal{R}$. Moreover, the matrix representations of the restrictions of $U_\mathcal{R}$ to this subspace with respect to the basis $\{Y_\ell^m : -\ell\le m\le\ell\}$ form a unitary group, which is an irreducible representation of SO(3) of dimension $2\ell+1$ (see [393]). It is then clear that B leaves invariant the subspace spanned by Y_ℓ^m, $m=-\ell,-\ell+1,...,\ell$, and that the restriction of B to this subspace has a matrix with respect to the basis $\{Y_\ell^m : -\ell\le m\le\ell\}$ which commutes with all of the matrices from the above irreducible unitary representation. As a consequence of Schur's lemma (cf. [393]), we have

$$BY_\ell^m = (2\ell+1)^{-1}\beta_\ell Y_\ell^m, \quad m=-\ell,\ldots,\ell, \quad \ell=0,1,2,\ldots,$$

for certain constants β_ℓ that do not depend on m. However, for m=0 the spherical harmonic Y_ℓ^m is proportional to the Legendre polynomial $P_\ell(u)=(-1)^\ell P_\ell(\cos\nu)$ and therefore

$$BP_\ell = (2\ell+1)^{-1}\beta_\ell P_\ell,$$

which implies

$$\beta_\ell = (\ell+\tfrac{1}{2})\int_{-1}^{1} a_1(\theta)P_\ell(\cos\theta)d(\cos\theta).$$

Next we deal with the compactness statement of the theorem. If a_1 is a continuous function of $\cos\theta$ on $[-1,1]$, then the estimate

$$|(Bh)(\omega_1)-(Bh)(\omega_2)| \le \frac{1}{4\pi}\int_\Omega |a_1(\omega_1\cdot\hat\omega)-a_1(\omega_2\cdot\hat\omega)| \cdot |h(\hat\omega)|\, d\hat\omega$$

in combination with the uniform continuity of the function $(\omega,\hat\omega)\to a_1(\omega\cdot\hat\omega)$ from $\Omega\times\Omega$ into $\mathbb{C}$ gives the boundedness of B as an operator from $L_1(\Omega)$ into the Banach space $C(\Omega)$ of continuous functions $h:\Omega\to\mathbb{C}$ with supremum norm; we then have

$$\|B\|_{L_1\to C} \le \frac{1}{4\pi}\max\,\{|a_1(\theta)| \,:\, -1\le\cos\theta\le 1\}.$$

However, by the Weierstrass approximation theorem we can approximate a_1 by polynomials uniformly in $\cos\theta$ on $[-1,1]$. Since a_1 is a polynomial and therefore a finite linear combination of Legendre polynomials, the operator B has finite rank. Thus B is compact as an operator from $L_1(\Omega)$ into $C(\Omega)$. Since $C(\Omega)\subset L_p(\Omega)\subset L_1(\Omega)$, $1\le p<\infty$, in the sense of continuous imbeddings, the operator B is compact on all of the spaces $L_p(\Omega)$ whenever a_1 is continuous on $[-1,1]$. However, the continuous functions on $[-1,1]$ are dense in $L_1[-1,1]$. Moreover, we easily prove that $\|Bh\|_\infty\le M\|h\|_\infty$ and $\|Bh\|_1\le M\|h\|_1$, where $M=\tfrac{1}{2}\int_{-1}^{1}|a_1(\theta)|\,d(\cos\theta)=1$ (cf. (1.1)). Hence, by interpolation, $\|Bh\|_p\le M\|h\|_p$, $1\le p\le\infty$. Using these bounds, the density of the continuous functions in $L_1[-1,1]$ and the compactness of B on $L_p(\Omega)$ if a_1 is continuous, we obtain the compactness of B on $L_p(\Omega)$, $1\le p<\infty$. ■

As a consequence of the nonnegativity of $a_1(\theta)$ and the normalization condition (1.1), we easily find $\beta_0=1$ and $-(2\ell+1)<\beta_\ell<(2\ell+1)$ for $\ell\geq 1$. If we consider the real spaces $L_p(\Omega)$, $1\leq p\leq\infty$, and $C(\Omega)$ as Banach lattices, then the operators B, $|T|$, Q_+, Q_- and R are positive operators, provided R is bounded on the space under consideration. (As a result of (1.8), R is bounded on $L_\infty(\Omega)$.) Let us consider the operators

$$(e^{\mp xT^{-1}}Q_\pm h)(u,\varphi) = \begin{cases} e^{\mp x/u}h(u,\varphi), & u>0, \\ 0, & u<0, \end{cases} \tag{1.32}$$

where $0\leq x<\infty$, and define

$$(\mathcal{H}(x)h)(u,\varphi) = \begin{cases} |u|^{-1}e^{-x/u}h(u,\varphi), & xu>0, \\ 0, & xu<0, \end{cases} \tag{1.33}$$

where $0\neq x\in\mathbb{R}$. Then both of the families of operators (1.32) are analytic semigroups, which are defined on $Q_\pm[L_p(\Omega)]$ and have $-\mathcal{H}(\pm x)$ restricted to $Q_\pm[L_p(\Omega)]$ as their strong derivatives. It is easily seen that the operators defined by (1.32) and (1.33) are positive (in the lattice sense) and that $\mathcal{H}(x)$ is related to T as the propagator function introduced in Section VI.3. Similar notions can be introduced with respect to T^m by restriction.

LEMMA 1.2. Let $a_1\in L_r[-1,1]$ for some $r>1$. Then $B:L_p(\Omega)\to L_{pr}(\Omega)$ is bounded.

This lemma, which is a straightforward consequence of Theorem 1(2.X) of [206], follows directly by application of Hölder estimates, and leads to the following useful result.

LEMMA 1.3. Let $a_1\in L_r[-1,1]$ for some $r>1$. Then

$$B[L_p(\Omega)] \subset |T|^\alpha[L_p(\Omega)], \qquad 0<\alpha<(r-1)/pr. \tag{1.34}$$

Proof: By virtue of Lemma 1.2, B is a bounded operator from $L_p(\Omega)$ into $L_{pr}(\Omega)$. As one easily shows,

$$\| |T|^{-\alpha}h\|_p^p = \int_\Omega |u|^{-\alpha p} |h(\omega)|^p d\omega \leq$$

$$\leq \left[\int_\Omega |u|^{-\alpha pr/(r-1)} d\omega\right]^{(r-1)/r} \left[\int_\Omega |h(\omega)|^{pr} d\omega\right]^{1/r} =$$

$$= \left[\frac{4\pi}{1-(\alpha pr/(r-1))}\right]^{(r-1)/r} \|h\|_{pr}^p.$$

Therefore, if $0<\alpha<(r-1)/pr$, the operator $|T|^{-\alpha}$ is bounded from $L_{pr}(\Omega)$ into $L_p(\Omega)$. Hence, in this case $|T|^{-\alpha}B$ is bounded on $L_p(\Omega)$ and (1.34) holds true. ■

As a result we retrieve Feldman's observation (cf. [116]) that

$$\int_{-\infty}^{\infty} \|\mathcal{H}(\tau)B\|_{L_p(\Omega)} d\tau < \infty, \quad 1\leq p<\infty, \tag{1.35}$$

if $a_1 \in L_r[-1,1]$ for some $r>1$.

We shall now state a number of results on the existence and uniqueness of solutions of the radiative transfer problems outlined above. To the extent that they are formulated in an L_2-setting, they are applications of the theory presented in Chapters II, III and V. In this case we have to distinguish between phase functions $a_1 \in L_r[-1,1]$ with $r>1$ where all results can be stated in $L_2(\Omega)$, and phase functions $a_1 \in L_1[-1,1]$ where a larger Hilbert space is introduced because of a breakdown of the proof in $L_2(\Omega)$. For phase functions $a_1 \in L_r[-1,1]$ with $r>1$, we may apply the theory of Sections VI.6 and VII.4 and obtain existence and uniqueness results in $L_p(\Omega)$, $1\leq p<\infty$. It is not known whether these results, both the ones in $L_2(\Omega)$ and the ones in $L_p(\Omega)$, extend to $a_1 \in L_1[-1,1]$ in general.

We shall first consider the elementary problems (1.2)-(1.17)-(1.18) and (1.21)-(1.23). We put $\Omega_+=\{\omega\equiv(u,\varphi)\in\Omega : u\geq 0\}$.

THEOREM 1.4. Let $1\leq p<\infty$, $0<a\leq 1$ and $a_1 \in L_r[-1,1]$ with $r>1$. Let $S(\tau,u,\varphi)$ satisfy the Hölder continuity condition

$$\left[\int_{-1}^{1}\int_0^{2\pi} |S(\tau_1,u,\varphi)-S(\tau_2,u,\varphi)|^p d\varphi du\right]^{1/p} \leq M|\tau_1-\tau_2|^\gamma \tag{1.36}$$

for some $0<\gamma<1$ and all $0\leq\tau_1,\tau_2<\infty$, as well as the growth condition

$$\int_1^\infty \tau\left[\int_{-1}^1 |S(\tau;u,\varphi)|^p d\varphi du\right]^{1/p} d\tau < \infty. \tag{1.37}$$

Then for every $D_+ \in L_p(\Omega_+)$ there exists a unique solution $I(\tau,u,\varphi)$ of the equation of transfer (1.2) for $\tau \in (0,\infty)$ with boundary conditions (1.17) and (1.18). More precisely, there exists a unique continuous vector function $I:[0,\infty)\to L_p(\Omega)$ such that TI is differentiable for $\tau \in (0,\infty)$ in the strong topology of $L_p(\Omega)$ and the following equations are satisfied:

$$(TI)'(\tau) = -AI(\tau) + S(\tau), \quad 0<\tau<\infty, \tag{1.38a}$$

$$Q_+ I(0) = D_+, \tag{1.38b}$$

$$\limsup_{\tau\to\infty} \|I(\tau)\|_p < \infty. \tag{1.38c}$$

THEOREM 1.5. Let $1\le p<\infty$, and let $a_1 \in L_r[-1,1]$ with $r>1$. Then for every F there exists a unique solution $I(\tau,u,\varphi)$ of the Milne problem (1.21)–(1.23). More precisely, there exists a unique continuous vector function $I:[0,\infty)\to L_p(\Omega)$ such that TI is differentiable for $\tau \in (0,\infty)$ in the strong topology of $L_p(\Omega)$ and the following equations are satisfied:

$$(TI)'(\tau) = -AI(\tau), \quad 0<\tau<\infty, \tag{1.39a}$$

$$Q_+ I(0) = 0, \tag{1.39b}$$

$$\lim_{\tau\to\infty} \int_{-1}^1 \int_0^{2\pi} \hat{u} I(\tau,\hat{u},\hat{\varphi}) d\hat{\varphi} d\hat{u} = -F/4\pi. \tag{1.39c}$$

The proofs of these results follow directly from an analysis of the kernel of A. Since the eigenvalues of B are the numbers $\beta_\ell/(2\ell+1)$, $\ell=0,1,2,\ldots$, which belong to $(-1,1)$ for $\ell\ge 1$, while $a\le 1$, the operator A will be positive self adjoint on $L_2(\Omega)$. In fact, since A+aB is the identity, A is strictly positive self adjoint on $L_2(\Omega)$ for $0<a<1$, and has the one dimensional kernel of constant functions if $a=1$. However, we have $(TP_0,P_0)=0$ in the inner product of $L_2(\Omega)$ while Ker A=span$\{P_0\}$. Thus for $p=2$ Theorems 1.3 and 1.4 follow as applications of Theorems II 2.7 and III 2.2. The extension of these results to all $1\le p<\infty$ can be proved using Sections VI.6 and VII.4. If one only assumes $a_1 \in L_1[-1,1]$, analogs of the above theorems can be proved on the Hilbert space of functions on Ω which are square integrable with respect to the weighted measure $|u|d\varphi du$. That is to say, if $H=L_2(\Omega)$, then this Hilbert space is the space H_T introduced in Section II.3 and these more general results are applications of Sections II.3 and III.2.

Next, let us consider the boundary value problem (1.26)–(1.28). If the surface reflection operator R is bounded on $L_p(\Omega)$, we can state an existence and uniqueness result in $L_p(\Omega)$, $1\le p<\infty$. Since it is easily seen that Eqs. (1.6)–(1.8) imply

$$\left|\int_\Omega u(Rh)(\omega)\overline{h}(\omega)d\omega\right| \le \int_\Omega u|h(\omega)|^2 d\omega,$$

we have, as an application of the results in Section V.4, the following theorem.

THEOREM 1.6. Let $1\le p<\infty$, and let $a_1\in L_r[-1,1]$ with $r>1$. Let R be bounded on $L_2(\Omega)$ and $L_p(\Omega)$, and let $S(\tau,u,\varphi)$ satisfy the Hölder continuity condition

$$\left[\int_{-1}^{1}\int_0^{2\pi}|S(\tau_1,u,\varphi)-S(\tau_2,u,\varphi)|^p d\varphi du\right]^{1/p} \le M|\tau_1-\tau_2|^\gamma$$

for some $0<\gamma<1$ and all $0\le\tau_1,\tau_2\le b$. Then for every $D\in L_p(\Omega)$ there exists a unique solution of the boundary value problem (1.26)–(1.28).

If $a_1\in L_r[-1,1]$ with $r>1$, we can prove quite easily that the solutions $I(\tau,u,\varphi)$ of the above problems are nonnegative if the internal source term $S(\tau,u,\varphi)$ and the incident radiation $D(u,\varphi)$ are nonnegative. Let us illustrate this fact for Eqs. (1.2)–(1.17)–(1.18). As a consequence of (1.35), we may write the problem in the integral form

$$I(\tau) - a\int_0^\infty \mathcal{H}(\tau-\hat{\tau})BI(\hat{\tau})d\hat{\tau} = e^{-\tau T^{-1}}Q_+D_+, \quad 0<\tau<\infty, \tag{1.40}$$

where we have used the definitions (1.32) and (1.33). Consider the Banach space $L_q(L_p(\Omega))_0^\infty$ of strongly measurable L_q-functions $I:(0,\infty)\to L_p(\Omega)$, $1\le p<\infty$, $1\le q\le\infty$. Then we may write Eq. (1.40) in the form

$$I - a\mathcal{L}_\infty I = \omega, \tag{1.41}$$

with $\mathcal{L}_\infty: L_q(L_p(\Omega))_0^\infty \to L_q(L_p(\Omega))_0^\infty$. It is clear that $1-a\mathcal{L}_\infty$ is invertible on the spaces $L_q(L_p(\Omega))_0^\infty$ for $0<a<1$. Moreover, $\mathcal{L}_\infty$ is a positive operator on $L_q(L_p(\Omega))_0^\infty$ in the sense that it leaves invariant the reproducing and normal cone of nonnegative functions in this space. As a consequence of Theorem I 4.2, its spectral radius must belong to the spectrum, and therefore $r(L_\infty)\leq 1$. We may then iterate Eq. (1.41) for $0<a<1$ and conclude that the unique solution is nonnegative. For $a=1$ the reasoning necessary to prove nonnegativity of the solution is more involved and will be omitted. For the boundary value problem (1.26)-(1.28) we can prove nonnegativity of the solution for $0<a\leq 1$ using a similar iteration argument.

Existence and uniqueness of solutions for the equation of radiative transfer has been proved for many special cases in various degrees of mathematical preciseness, sometimes for the integrodifferential form and sometimes for the integral form of the equation. For this reason it is not easy to give a fair account of the history of these results. Therefore we shall restrict ourselves to some major developments.

For isotropic scattering ($a_1(\theta)\equiv 1$) the existence and uniqueness issue for the equation in integral form was settled by Hopf [196] and Busbridge [61]. In integrodifferential form we mention the later developments by Case [68] and van Kampen [372]. Their singular eigenfunction approach was made rigorous by the work of Larsen and Habetler [239, 241], and Hangelbroek and Lekkerkerker [180, 181, 184].

Using the integral form of the equation of radiative transfer, the asymptotics of solutions were studied in great detail by Maslennikov (for $a_1\in L_2[-1,1]$, cf. [259]) and Feldman (for $a_1\in L_r[-1,1]$ with $r>1$, cf. [116]; also for general $a_1\in L_1[-1,1]$, cf. [118]). Relying on cone preservation methods Nelson [281] showed subcriticality to imply unique solvability. In integrodifferential form half range completeness and orthogonality results have been given by many authors, although aspects of these works have frequently been heuristic (see [70, 233, 263] for references). Using the abstract formulation of the previous chapters, existence and uniqueness of solutions were proved by Beals [32] in the weighted Hilbert space $L_2(\Omega,|u|\,du\,d\varphi)$. As to existence and uniqueness results in $L_p(\Omega)$, we mention the work of Hangelbroek ($0<a<1$, $a_1(\theta)$ polynomial, $p=2$, cf. [182]) and van der Mee ($0<a\leq 1$, $a_1\in L_r[-1,1]$ with $r>1$, $1\leq p<\infty$, cf. [359, 360]). Existence and uniqueness theorems that account for reflection by the planetary surface have been given by Greenberg and van der Mee (same assumptions, $p=2$, cf. [164]; $1\leq p<\infty$, appearing as a special case of the results for polarized light transfer).

2. Radiative transfer with polarization

In this section we shall analyze in detail the existence and uniqueness properties of the solution of an equation of transfer for polarized light. Contrary to the unpolarized light case, the solution is not a scalar function, the (specific) intensity $I(\tau,u,\varphi)$, but a four-vector $\mathbf{I}(\tau,u,\varphi)$ which specifies the intensity and state of polarization of the light. As a result the theory is considerably more involved than the theory presented in Section 1.

The intensity and state of polarization of a beam of light can be characterized completely by using four polarization parameters, I, Q, U and V, which are the components of the **Stokes vector** $\mathbf{I}=\{I,Q,U,V\}$, after G. Stokes [344] who first introduced these parameters in 1852. According to the (equivalent) conventions of Chandrasekhar [89] and van de Hulst [358], the Stokes parameters I, Q, U and V are specified as follows:

(i) I is the **specific intensity** of the beam, which is nonnegative.

(ii) $P=(Q^2+U^2+V^2)^{1/2}/I$ is the **degree of polarization** of the light. If $P=0$, we have natural (i.e., completely unpolarized) light. For $P=1$ we have completely polarized light. For $0<P<1$ the light is partially polarized. As a result we have

$$0 \leq (Q^2+U^2+V^2)^{1/2} \leq I. \tag{2.1}$$

Every Stokes vector $\mathbf{I}=\{I,Q,U,V\}$ can be written as the sum of a beam of natural light with Stokes vector $\{I-(Q^2+U^2+V^2)^{1/2},0,0,0\}$ and a beam of completely polarized light with Stokes vector $\{(Q^2+U^2+V^2)^{1/2},Q,U,V\}$.

(iii) $P_\ell=(Q^2+U^2)^{1/2}/I$ denotes the fraction of the light that is linearly polarized, while $P_c=|V|/I$ denotes the fraction of the light that is circularly polarized. Linear polarization occurs if $V=0$, circular polarization if $Q=U=0$, and elliptic polarization if both $(Q^2+U^2)^{1/2}$ and V are non-zero.

(iv) Right-handed polarization (when looking in the direction of propagation) occurs if $V>0$, while left-handed polarization occurs if $V<0$.

(v) If $(Q^2+U^2)^{1/2}>0$, i.e. if the light is neither natural nor circularly polarized, the orientation of the major semiaxis of the polarization ellipse in space is specified by the fraction U/Q.

For further details we refer the reader to the textbooks of Chandrasekhar [89] and Van de Hulst [358] and the article of Hovenier and van der Mee [201].

The equation of transfer of polarized light was first formulated by Chandrasekhar [89] for Rayleigh scattering and later by Kuščer and Ribarič [234] in general, and has the form

$$u\frac{\partial}{\partial\tau}\mathbf{I}(\tau,u,\varphi) + \mathbf{I}(\tau,u,\varphi) =$$
$$= \frac{a}{4\pi}\int_{-1}^{1}\int_{0}^{2\pi}\mathbf{Z}(u,\hat{u},\varphi-\hat{\varphi})\mathbf{I}(\tau,\hat{u},\hat{\varphi})d\hat{\varphi}d\hat{u} + \mathbf{S}(\tau,u,\varphi), \tag{2.2}$$

where $0<\tau<b$. The **phase matrix** $\mathbf{Z}(u,\hat{u},\varphi-\hat{\varphi})$ is given by the product

$$\mathbf{Z}(u,\hat{u},\varphi-\hat{\varphi}) = \mathbf{L}(\pi-\sigma_2)\mathbf{F}(\theta)\mathbf{L}(-\sigma_1) \tag{2.3}$$

of two rotation matrices of the type

$$\mathbf{L}(\alpha) = \begin{bmatrix} 1 & 0 & 0 & 0 \\ 0 & \cos 2\alpha & \sin 2\alpha & 0 \\ 0 & -\sin 2\alpha & \cos 2\alpha & 0 \\ 0 & 0 & 0 & 1 \end{bmatrix}$$

and the **scattering matrix**

$$\mathbf{F}(\theta) = \begin{bmatrix} a_1(\theta) & b_1(\theta) & 0 & 0 \\ b_1(\theta) & a_2(\theta) & 0 & 0 \\ 0 & 0 & a_3(\theta) & b_2(\theta) \\ 0 & 0 & -b_2(\theta) & a_4(\theta) \end{bmatrix} \tag{2.4}$$

The quantities $u=-\cos\nu$ and $\hat{u}=-\cos\nu'$, $0\le\nu,\nu',\theta<\pi$, on the one hand and the angles φ, $\hat{\varphi}$, σ_1 and σ_2 on the other hand are related by the equations

$$\cos\theta = \cos\nu\ \cos\nu' + \sin\nu\ \sin\nu'\ \cos(\hat{\varphi}-\varphi), \tag{2.5a}$$

$$\cos\sigma_1 = \frac{\cos\nu - \cos\nu'\ \cos\theta}{\sin\nu'\ \sin\theta}, \tag{2.5b}$$

$$\cos\sigma_2 = \frac{\cos\nu' - \cos\nu \; \cos\theta}{\sin\nu \; \sin\theta}, \tag{2.5c}$$

where $\sin\sigma_1$ and $\sin\sigma_2$ have the same sign as $\sin(\hat{\varphi}-\varphi)$. It turns out (cf. [201]) that for $0<\hat{\varphi}-\varphi<\pi$ (resp. $\pi<\hat{\varphi}-\varphi<2\pi$) the quantities $\hat{\varphi}-\varphi$, σ_1 and σ_2 (resp. $\varphi-\hat{\varphi}$, $-\sigma_1$ and $-\sigma_2$) form the angles and θ, ν and ν' (in both cases) the opposite sides of a spherical triangle. On inspection the phase matrix satisfies the following symmetry relations [197]:

(i) Reciprocity symmetry:

$$\mathbf{Z}(-\hat{u},-u,\hat{\varphi}-\varphi) = \mathbf{P}\tilde{\mathbf{Z}}(u,\hat{u},\varphi-\hat{\varphi})\mathbf{P} \tag{2.6}$$

(ii) Symmetry with respect to the equatorial plane:

$$\mathbf{Z}(-u,-\hat{u},\hat{\varphi}-\varphi) = \mathbf{Z}(u,\hat{u},\varphi-\hat{\varphi}) \tag{2.7}$$

(iii) Reflection symmetry with respect to the meridian plane of incidence:

$$\mathbf{Z}(u,\hat{u},\hat{\varphi}-\varphi) = \mathbf{D}\mathbf{Z}(u,\hat{u},\varphi-\hat{\varphi})\mathbf{D}, \tag{2.8}$$

where $\mathbf{P}=\mathrm{diag}(1,1,-1,1)$ and $\mathbf{D}=\mathrm{diag}(1,1,-1,-1)$. As a consequence we have

$$\tilde{\mathbf{Z}}(\hat{u},u,\hat{\varphi}-\varphi) = \mathbf{Q}\mathbf{Z}(u,\hat{u},\varphi-\hat{\varphi})\mathbf{Q}, \tag{2.9}$$

where $\mathbf{Q}=\mathrm{diag}(1,1,1,-1)$ and tilde above a matrix denotes transposition. For physical reasons one demands that the scattering matrix and the phase matrix leave invariant the vectors $\mathbf{I}=\{I,Q,U,V\}$ satisfying (2.1). We also have the normalization condition (1.1).

We may now write down the complete analogs of the boundary value problems of Section 1. On the ground reflection matrix we have to impose the symmetry relations [197, 198] of reciprocity symmetry,

$$\mathbf{R}_g(u,\hat{u},\varphi-\hat{\varphi}) = \mathbf{P}\tilde{\mathbf{R}}_g(\hat{u},u,\hat{\varphi}-\varphi)\mathbf{P}, \tag{2.10}$$

and mirror symmetry,

$$\mathbf{R}_g(u,\hat{u},\varphi-\hat{\varphi}) = \mathbf{D}\mathbf{R}_g(u,\hat{u},\hat{\varphi}-\varphi)\mathbf{D}, \tag{2.11}$$

and the energy conservation law that the plane albedo does not exceed unity,

$$0 \leq \frac{1}{\pi}\int_0^1\int_0^{2\pi} \hat{u}[\mathbf{R}_g(u,\hat{u},\varphi-\hat{\varphi})]_{11}\,d(\varphi-\hat{\varphi})d\hat{u} \leq 1, \tag{2.12}$$

where $[\mathbf{R}]_{ij}$ denotes the (i,j)-element of a matrix $\mathbf{R}$. For finite optical layers we get the analog of problem (1.2)–(1.4), which is Eq. (1.2) with boundary conditions

$$\mathbf{I}(0,u,\varphi) = \mathbf{D}(u,\varphi), \quad u>0, \tag{2.13}$$

$$\mathbf{I}(b,-u,\varphi) = \frac{1}{\pi}\int_0^1\int_0^{2\pi} \hat{u}\mathbf{R}_g(u,\hat{u},\varphi-\hat{\varphi})\,\mathbf{I}(b,\hat{u},\hat{\varphi})d\hat{\varphi}d\hat{u} + \mathbf{D}(-u,\varphi), \quad u>0. \tag{2.14}$$

For semi-infinite media we again consider two boundary value problems: (i) the equation of transfer (2.2) with boundary conditions

$$\mathbf{I}(0,u,\varphi) = \mathbf{D}_+(u,\varphi), \quad u>0, \tag{2.15}$$

$$\lim_{\tau\to\infty} \sup \|\mathbf{I}(\tau,\cdot,\cdot)\| < \infty; \tag{2.16}$$

and (ii) the Milne problem

$$u\frac{\partial}{\partial\tau}\mathbf{I}(\tau,u,\varphi) + \mathbf{I}(\tau,u,\varphi) = (1/4\pi)\int_{-1}^1\int_0^{2\pi} \mathbf{Z}(u,\hat{u},\varphi-\hat{\varphi})\,\mathbf{I}(\tau,\hat{u},\hat{\varphi})d\hat{\varphi}d\hat{u}, \tag{2.17}$$

$$\mathbf{I}(0,u,\varphi) = 0, \quad u>0, \tag{2.18}$$

$$\lim_{\tau\to\infty} \int_{-1}^1\int_0^{2\pi} \hat{u}\mathbf{I}(\tau,\hat{u},\hat{\varphi})d\hat{\varphi}d\hat{u} = -(F/4\pi)\ \{1,0,0,0\}. \tag{2.19}$$

On writing

$$\mathbf{I}(\tau,u,\varphi) = \mathbf{I}^{co}(\tau,u) + 2\sum_{j=1}^{\infty}[\mathbf{I}^{cj}(\tau,u)\cos j\varphi + \mathbf{I}^{sj}(\tau,u)\sin j\varphi],$$

$$\mathbf{Z}(u,\hat{u},\varphi-\hat{\varphi}) = \mathbf{Z}^{co}(u,\hat{u}) + 2\sum_{j=1}^{\infty}[\mathbf{Z}^{cj}(u,\hat{u})\cos\{j(\varphi-\hat{\varphi})\} + \mathbf{Z}^{sj}(u,\hat{u})\sin(j(\varphi-\hat{\varphi})\}],$$

$$\mathbf{R}_g(u,\hat{u},\varphi-\hat{\varphi}) = \mathbf{R}_g^{co}(u,\hat{u}) + 2\sum_{j=1}^{\infty}[\mathbf{R}_g^{cj}(u,\hat{u})\cos\{j(\varphi-\hat{\varphi})\} + \mathbf{R}_g^{sj}(u,\hat{u})\sin\{j(\varphi-\hat{\varphi})\}],$$

and analogous expansions for $\mathbf{S}(\tau,u,\varphi)$, $\mathbf{D}(u,\varphi)$ and $\mathbf{D}_+(u,\varphi)$, we accomplish a Fourier decomposition of the above boundary value problems into a problem for $\mathbf{I}^{co}(\tau,u)$ and problems coupling $\mathbf{I}^{cj}(\tau,u)$ and $\mathbf{I}^{sj}(\tau,u)$ for $j\geq1$. As a result of (2.8) and (2.11) we have the symmetry relations

$$\mathbf{Z}^{co}(u,\hat{u}) = \mathbf{D}\mathbf{Z}^{co}(u,\hat{u})\mathbf{D},$$

$$\mathbf{Z}^{cj}(u,\hat{u}) = \mathbf{D}\mathbf{Z}^{cj}(u,u)\mathbf{D},$$

$$\mathbf{Z}^{sj}(u,\hat{u}) = -\mathbf{D}\mathbf{Z}^{sj}(u,\hat{u})\mathbf{D},$$

and

$$\mathbf{R}_g^{co}(u,\hat{u}) = \mathbf{D}\mathbf{R}_g^{co}(u,\hat{u})\mathbf{D},$$

$$\mathbf{R}_g^{cj}(u,\hat{u}) = \mathbf{D}\mathbf{R}_g^{cj}(u,\hat{u})\mathbf{D},$$

$$\mathbf{R}_g^{sj}(u,\hat{u}) = -\mathbf{D}\mathbf{R}_g^{sj}(u,\hat{u})\mathbf{D}.$$

If we now put

$$\mathbf{W}^j(u,\hat{u}) = \mathbf{Z}^{cj}(u,\hat{u}) - \mathbf{D}\mathbf{Z}^{sj}(u,\hat{u}) = \mathbf{Z}^{cj}(u,\hat{u}) + \mathbf{Z}^{sj}(u,\hat{u})\mathbf{D},$$

we obtain the pair of component equations

$$u\frac{\partial}{\partial\tau}\mathbf{Y}^j(\tau,u) = -\mathbf{Y}^j(\tau,u) + \tfrac{1}{2}a\int_{-1}^{1}\mathbf{W}^j(u,\hat{u})\mathbf{Y}^j(\tau,\hat{u})d\hat{u} \tag{2.20}$$

and

$$u\frac{\partial}{\partial\tau}\mathbf{X}^j(\tau,u) = -\mathbf{X}^j(\tau,u) + \tfrac{1}{2}a\int_{-1}^{1}\mathbf{W}^j(u,\hat{u})\mathbf{X}^j(\tau,\hat{u})d\hat{u}, \tag{2.21}$$

where

$$\mathbf{Y}^j(\tau,u) = \tfrac{1}{2}(1+\mathbf{D})\mathbf{I}^{cj}(\tau,u) + \tfrac{1}{2}(1-\mathbf{D})\mathbf{I}^{sj}(\tau,u)$$

and

$$\mathbf{X}^j(\tau,u) = \tfrac{1}{2}(1-\mathbf{D})\mathbf{I}^{cj}(\tau,u) - \tfrac{1}{2}(1+\mathbf{D})\mathbf{I}^{sj}(\tau,u).$$

If we write $\mathbf{Z}^{co}(u,\hat{u})$, $\mathbf{Z}^{cj}(u,\hat{u})$ and $\mathbf{Z}^{sj}(u,\hat{u})$ in block matrix form with square blocks of matrix order 2, we have, as a result of the above symmetry relations,

$$\mathbf{Z}^{co} = \begin{bmatrix} \mathbf{Z}^{cos} & 0 \\ 0 & \mathbf{Z}^{coa} \end{bmatrix},$$

$$\mathbf{Z}^{cj} = \begin{bmatrix} \mathbf{Z}^{cjs} & 0 \\ 0 & \mathbf{Z}^{cja} \end{bmatrix},$$

$$\mathbf{Z}^{sj} = \begin{bmatrix} 0 & \mathbf{Z}^{sjs} \\ \mathbf{Z}^{sja} & 0 \end{bmatrix},$$

(s=symmetric, a=antisymmetric), and similarly for $\mathbf{R}_g^{co}(u,\hat{u})$, $\mathbf{R}_g^{cj}(u,\hat{u})$ and $\mathbf{R}_g^{sj}(u,\hat{u})$. If we now write $\mathbf{I}=\{I,Q,U,V\}$ for each indexed four-vector $\mathbf{I}$, we get

$$\mathbf{Y}^0(\tau,u) = \{I^{co}(\tau,u),\ Q^{co}(\tau,u),\ 0,\ 0\},$$

$$\mathbf{X}^0(\tau,u) = \{0,\ 0,\ U^{co}(\tau,u),\ V^{co}(\tau,u)\},$$

$$\mathbf{W}^0(u,\hat{u}) = \begin{bmatrix} \mathbf{Z}^{cos}(u,\hat{u}) & 0 \\ 0 & \mathbf{Z}^{coa}(u,\hat{u}) \end{bmatrix}, \tag{2.22}$$

and, for $j \geq 1$,

$$\mathbf{Y}^j(\tau,u) = \{I^{cj}(\tau,u),\ Q^{cj}(\tau,u),\ U^{sj}(\tau,u),\ V^{sj}(\tau,u)\},$$

$$\mathbf{X}^j(\tau,u) = \{I^{sj}(\tau,u),\ Q^{sj}(\tau,u),\ -U^{sj}(\tau,u),\ -V^{sj}(\tau,u)\},$$

$$\mathbf{W}^j(u,\hat{u}) = \begin{bmatrix} \mathbf{Z}^{cjs}(u,\hat{u}) & -\mathbf{Z}^{sjs}(u,\hat{u}) \\ \mathbf{Z}^{sja}(u,\hat{u}) & \mathbf{Z}^{cja}(u,\hat{u}) \end{bmatrix}. \tag{2.23}$$

The ground reflection matrix corresponding to Eqs. (2.20) and (2.21) is constructed in the same way as the phase matrix $\mathbf{W}^j(u,\hat{u})$ of the component equation.

For j=0 the equations (2.20) and (2.21) are equations with two-vector functions as solutions. Their kernels $\mathbf{Z}^{cos}(u,\hat{u})$ and $\mathbf{Z}^{coa}(u,\hat{u})$ satisfy the symmetry relations (cf. (2.6), (2.7), (2.22))

$$\mathbf{Z}^{cos}(u,\hat{u}) = \tilde{\mathbf{Z}}^{cos}(\hat{u},u) = \mathbf{Z}^{cos}(-u,-\hat{u}) \tag{2.24}$$

and

$$\mathbf{Z}^{coa}(u,\hat{u}) = \mathbf{E}\tilde{\mathbf{Z}}^{coa}(\hat{u},u)\mathbf{E} = \mathbf{Z}^{coa}(-u,-\hat{u}), \tag{2.25}$$

where $\mathbf{E}$=diag (1,−1). For j≥1 we have the symmetry relations (cf (2.6), (2.7), (2.23))

$$\mathbf{W}^j(u,\hat{u}) = \mathbf{Q}\tilde{\mathbf{W}}^j(\hat{u},u)\mathbf{Q}, \tag{2.26a}$$

$$\mathbf{W}^j(u,\hat{u}) = \mathbf{D}\mathbf{W}^j(-u,-\hat{u})\mathbf{D}, \tag{2.26b}$$

where j≥1. Following the work of Kuščer and Ribarič [234] yielding the j=0 case and complex component equations, the complete decomposition in terms of real component equations was derived by Siewert [329]. Here we have essentially followed the treatment by Hovenier and van der Mee [201]. Explicit representations of $\mathbf{Z}^{cos}(u,\hat{u})$, $\mathbf{Z}^{coa}(u,\hat{u})$ and $\mathbf{W}^j(u,\hat{u})$ can be found in [201, 329].

For the case when, as a function of $\cos\theta$, $a_1 \in L_r[-1,1]$ for some r>1, a complete existence and uniqueness theory for the boundary value problems (2.2)−(2.13)−(2.14), (2.2)−(2.15)−(2.16) and (2.2)−(2.17)−(2.18)−(2.19) has been presented by van der Mee [365]. His analysis was based in part on the observation by Germogenova and Konovalov [136] that on the Banach space $L_p^{(4)}(\Omega)$ of L_p−functions $\mathbf{I}:\Omega\to\mathbb{R}^4$, where Ω is the unit sphere in three-dimensional space, the subset

$$\{\mathbf{I} = \{I,Q,U,V\} : I(\tau)\geq\{Q(\tau)^2+U(\tau)^2+V(\tau)^2\}^{1/2} \text{ a.e.}\}$$

is a (reproducing and normal) cone. This observation paved the way to the application of cone preservation methods to the integral form of the boundary value problem (cf. Section I.4). Earlier the application of the same technique to the "eigenvalue equation"

$$(1-ku)\Phi_k(u,\varphi) = (a/4\pi)\int_{-1}^{1}\int_{0}^{2\pi} \mathbf{Z}(u,\hat{u},\varphi-\hat{\varphi})\ \Phi_k(\hat{u},\hat{\varphi})d\hat{\varphi}d\hat{u}, \tag{2.27}$$

as implemented by Germogenova and Konovalov [136] and Kuzmina [235, 236], led to valuable information on the position of eigenvalues of the characteristic equation (2.27).

In this section we shall restrict ourselves to the component equation with phase matrix $\mathbf{Z}^{\cos}(u,\hat{u})$, which has the form

$$u\frac{\partial}{\partial\tau}\mathbf{I}^{\cos}(u,\hat{u}) + \mathbf{I}^{\cos}(u,\hat{u}) = \tfrac{1}{2}a\int_{-1}^{1}\mathbf{Z}^{\cos}(u,\hat{u})\,\mathbf{I}^{\cos}(\tau,\hat{u})d\hat{u} + \mathbf{S}^{\cos}(\tau,u). \tag{2.28}$$

On the Banach space $L_p^{(2)}[-1,1]$ of measurable functions $\mathbf{I}^{\cos}:[-1,1]\to\mathbb{C}^2$ with L_p-norm

$$\|\mathbf{I}^{\cos}\|_p = \Big[\int_{-1}^{1}\{|I^{\cos}(u)|^p + |Q^{\cos}(u)|^p\}du\Big]^{1/p}, \quad 1\le p<\infty,$$

and

$$\|\mathbf{I}^{\cos}\|_\infty = \max\{\sup_{-1\le u\le 1}|I^{\cos}(u)|,\ \sup_{-1\le u\le 1}|Q^{\cos}(u)|\},$$

we define the bounded linear operators

$$(T\mathbf{I}^{\cos})(u) = u\mathbf{I}^{\cos}(u),$$

$$(Q_\pm\mathbf{I}^{\cos})(u) = \begin{cases} \mathbf{I}^{\cos}(u), & \pm u>0, \\ 0, & \pm u<0, \end{cases}$$

$$(A\mathbf{I}^{\cos})(u) = \mathbf{I}^{\cos}(u) - a(B\mathbf{I}^{\cos})(u),$$

$$(B\mathbf{I}^{\cos})(u) = \tfrac{1}{2}\int_{-1}^{1}\mathbf{Z}^{\cos}(u,\hat{u})\,\mathbf{I}^{\cos}(\hat{u})d\hat{u},$$

$$(R\mathbf{I}^{\cos})(u) = 2\int_{-1}^{1}\hat{u}\mathbf{R}_g^{\cos}(u,\hat{u})\,\mathbf{I}^{\cos}(\hat{u})d\hat{u},$$

$$(J\mathbf{I}^{\cos})(u) = \mathbf{I}^{\cos}(-u),$$

and the vectors

$$\mathbf{I}^{\cos}(\tau)(u) = \mathbf{I}^{\cos}(\tau,u),$$

$$\mathbf{S}^{\cos}(\tau)(u) = \mathbf{S}^{\cos}(\tau,u).$$

As a result we obtain vector-valued differential equations with boundary conditions for each of the radiative transfer problems.

The operator B can be proved compact on all spaces $L_p^{(4)}(\Omega)$, but a complete description of its eigenvalues and eigenfunctions is more difficult to give. The analyses of Germogenova and Konovalov ([136], where $a_1 \epsilon L_1[-1,1]$), and Kuzmina ([235, 236], where $a_1 \epsilon L_2[-1,1]$), gave results in terms of equations involving complex polarization parameters and the results we need must be derived from theirs. Siewert [329] and Hovenier and van der Mee [201] provided, for scattering matrices $\mathbf{F}(\theta)$ for which $a_1(\theta)$, $(1 \pm \cos\theta)^{-2}(a_2(\theta) \pm a_3(\theta))$, $a_4(\theta)$, $(\sin\theta)^{-1}b_1(\theta)$ and $(\sin\theta)^{-1}b_2(\theta)$ are polynomials in $\cos\theta$, analytical expressions for $\mathbf{Z}^{\cos}(u,\hat{u})$, $\mathbf{Z}^{coa}(u,\hat{u})$, and $\mathbf{W}^j(u,\hat{u})$ for $j \geq 1$, but they did not relate them to the spectral structure of the integral operator in the equation of transfer. All of the expansions appearing in these expressions make use of the generalized spherical functions of Gelfand and Shapiro ([135], also [133]; the reader should be aware of the sudden convention changes and many printing errors in these works), which were first applied to polarized light problems by Kuščer and Ribarič [234].

Let us consider the Legendre polynomials and associated Legendre functions

$$P_\ell(u) = \frac{1}{2^\ell(\ell!)}\left(\frac{d}{du}\right)^\ell (u^2-1)^\ell,$$

$$P_\ell^j(u) = (1-u^2)^{j/2}\left(\frac{d}{du}\right)^j P_\ell(u),$$

where $\ell \geq j \geq 0$, and the special functions (cf. [329])

$$R_\ell^j(u) =$$

$$= \frac{1}{2^{j+1}} \frac{(\ell+j)!}{[(\ell-2)!(\ell+2)!]^{1/2}} (1-u^2)^{j/2}\left[\frac{1+u}{1-u} P_{\ell-j}^{j-2,\,j+2}(u) + \frac{1-u}{1+u} P_{\ell-j}^{j+2,\,j-2}(u)\right],$$

$$T_\ell^j(u) =$$

$$= \frac{1}{2^{j+1}} \frac{(\ell+j)!}{[(\ell-2)!(\ell+2)!]^{1/2}} (1-u^2)^{j/2}\left[\frac{1+u}{1-u} P_{\ell-j}^{j-2,\,j+2}(u) - \frac{1-u}{1+u} P_{\ell-j}^{j+2,\,j-2}(u)\right],$$

where $\ell \geq j \geq 2$ and $P_s^{\alpha\beta}(u)$ is a Jacobi polynomial (see [347] for its definition and major properties). These functions satisfy the orthogonality and normalization conditions

$$\int_{-1}^{1} P_\ell(u)P_r(u)du = \frac{2}{2\ell+1}\delta_{\ell r},$$

$$\int_{-1}^{1} P_\ell^j(u)P_r^j(u)du = \frac{2}{2\ell+1}\frac{(\ell+j)!}{(\ell-j)!}\delta_{\ell r},$$

$$\int_{-1}^{1} [R_\ell^j(u)R_r^j(u) + T_\ell^j(u)T_r^j(u)]du = \frac{2}{2\ell+1}\frac{(\ell+j)!}{(\ell-j)!}\delta_{\ell r},$$

$$\int_{-1}^{1} [R_\ell^j(u)T_r^j(u) + T_\ell^j(u)R_r^j(u)]du = 0,$$

as one easily computes from Eqs. (182), (183), (A12), (A13), (A21) and (A22) of [201]. One may also introduce the special functions (cf. [329, 330])

$$R_\ell(u) = \tfrac{1}{4}\left[\frac{(\ell-2)!(\ell+2)!}{\ell!}\right]^{1/2}(1-u^2)P_{\ell-2}^{2,2}(u), \quad \ell \geq 2,$$

where

$$\int_{-1}^{1} R_\ell(u)R_r(u)du = \frac{2}{2\ell+1}\delta_{\ell r}.$$

THEOREM 2.1. Let $a_1 \in L_1[-1,1]$. The operator B has the separated form

$$(BI^{\cos})(u) =$$

$$= \tfrac{1}{2}\int_{-1}^{1}\sum_{\ell=0}^{\infty}\begin{bmatrix} P_\ell(u) & 0 \\ 0 & R_\ell(u)\end{bmatrix}\begin{bmatrix} \beta_\ell & \gamma_\ell \\ \gamma_\ell & \alpha_\ell\end{bmatrix}\begin{bmatrix} P_\ell(\hat{u}) & 0 \\ 0 & R_\ell(\hat{u})\end{bmatrix} I^{\cos}(\hat{u})d\hat{u}, \quad (2.29)$$

and both this and the analogous operators B_L, where the summation is from $\ell=0$ to $\ell=L$, are compact operators satisfying $\lim_{L\to\infty} \|B-B_L\|=0$ on $L_2^{(2)}[-1,1]$. Here the operators B_L are finite rank approximations of B. The expansion coefficients α_ℓ, β_ℓ and γ_ℓ are given by

$$\alpha_0 = \alpha_1 = 0,$$

$$\alpha_\ell=(\ell+\tfrac{1}{2})\left[\frac{(\ell-2)!}{(\ell+2)!}\right]^{\frac{1}{2}}\int_{-1}^{1}\{a_2(\theta)R_\ell^2(\cos\theta)+a_3(\theta)T_\ell^2(\cos\theta)\}d(\cos\theta),$$

$$\beta_0 = 1,$$

$$\beta_\ell=(\ell+\tfrac{1}{2})\int_{-1}^{1}a_1(\theta)P_\ell(\cos\theta)d(\cos\theta),$$

$$\gamma_0 = \gamma_1 = 0,$$

$$\gamma_\ell = (\ell+\tfrac{1}{2})\left[\frac{(\ell-2)!}{(\ell+2)!}\right]^{\frac{1}{2}}\int_{-1}^{1}b_1(\theta)P_\ell^2(\cos\theta)d(\cos\theta).$$

The operator B is self adjoint on $L_2^{(2)}[-1,1]$, $A=1-aB$ has its spectrum on the interval $[1-a,1+a]$ and $1-a$ is a simple eigenvalue of A.

Proof: Let us expand the elements of the scattering matrix $\mathbf{F}(\theta)$ as follows:

$$a_1(\theta) = \sum_{\ell=0}^{\infty}\beta_\ell P_\ell(\cos\theta), \tag{2.30}$$

$$a_2(\theta) = \sum_{\ell=2}^{\infty}\left[\frac{(\ell-2)!}{(\ell+2)!}\right]^{\frac{1}{2}}\{\alpha_\ell R_\ell^2(\cos\theta)+\varsigma_\ell T_\ell^2(\cos\theta)\}, \tag{2.31}$$

$$a_3(\theta) = \sum_{\ell=2}^{\infty}\left[\frac{(\ell-2)!}{(\ell+2)!}\right]^{\frac{1}{2}}\{\varsigma_\ell R_\ell^2(\cos\theta)+\alpha_\ell T_\ell^2(\cos\theta)\}, \tag{2.32}$$

$$a_4(\theta) = \sum_{\ell=0}^{\infty}\delta_\ell P_\ell(\cos\theta), \tag{2.33}$$

$$b_1(\theta) = \sum_{\ell=2}^{\infty}\left[\frac{(\ell-2)!}{(\ell+2)!}\right]^{\frac{1}{2}}\gamma_\ell P_\ell^2(\cos\theta), \tag{2.34}$$

$$b_2(\theta) = -\sum_{\ell=2}^{\infty}\left[\frac{(\ell-2)!}{(\ell+2)!}\right]^{\frac{1}{2}}\epsilon_\ell P_\ell^2(\cos\theta), \tag{2.35}$$

where it is assumed that $a_1\in L_2[-1,1]$ as a function of $\cos\theta$. Since $\mathbf{F}(\theta)$ leaves invariant the set of four-vectors satisfying (2.1), a simple application of this rule to the vectors $\{1,\pm1,0,0\}$, $\{1,0,1,0\}$ and $\{1,0,0,1\}$ gives the inequalities (cf. [125, 201, 234])

$$|b_1(\theta)| \le \tfrac{1}{2}\{a_1(\theta)+a_2(\theta)\} \le a_1(\theta)$$

and

$$b_1(\theta)^2 + b_2(\theta)^2 + a_k(\theta)^2 \le a_1(\theta)^2, \qquad k=3,4.$$

As a result, all of the elements of $\mathbf{F}(\theta)$ are L_2-functions of $\cos\theta$ if this is the case for $a_1(\theta)$. The expansions (2.30)–(2.35) then follow as a simple consequence, in view of the various orthogonality and completeness properties. We easily find the formulas for α_ℓ, β_ℓ, and γ_ℓ, with $\beta_0=1$ resulting from (1.1), while

$$\delta_\ell = (\ell+½)\int_{-1}^{1} a_4(\theta)P_\ell(\cos\theta)d(\cos\theta),$$

$$\varepsilon_0 = \varepsilon_1 = 0,$$

$$\varepsilon_\ell = -(\ell+½)\left[\frac{(\ell-2)!}{(\ell+2)!}\right]^{½}\int_{-1}^{1} b_2(\theta)P_\ell^2(\cos\theta)d(\cos\theta)$$

and

$$\varsigma_0 = \varsigma_1 = 0,$$

$$\varsigma_\ell = -(\ell+½)\left[\frac{(\ell-2)!}{(\ell+2)!}\right]^{½}\int_{-1}^{1} \{a_3(\theta)R_\ell^2(\cos\theta)+a_2(\theta)T_\ell^2(\cos\theta)\}d(\cos\theta).$$

Using the above expansions and the addition formula for generalized spherical functions (cf. [133, 135]; unambiguous formula in [201], Eq. (A24)) we may derive

$$\mathbf{Z}^{\cos}(u,\hat{u}) = \sum_{\ell=0}^{\infty} \begin{bmatrix} P_\ell(u) & 0 \\ 0 & R_\ell(u) \end{bmatrix} \begin{bmatrix} \beta_\ell & \gamma_\ell \\ \gamma_\ell & \alpha_\ell \end{bmatrix} \begin{bmatrix} P_\ell(\hat{u}) & 0 \\ 0 & R_\ell(\hat{u}) \end{bmatrix}, \tag{2.36}$$

$$\mathbf{Z}^{\text{coa}}(u,\hat{u}) = \sum_{\ell=0}^{\infty} \begin{bmatrix} P_\ell(u) & 0 \\ 0 & R_\ell(u) \end{bmatrix} \begin{bmatrix} \varsigma_\ell & -\varepsilon_\ell \\ \varepsilon_\ell & \delta_\ell \end{bmatrix} \begin{bmatrix} P_\ell(\hat{u}) & 0 \\ 0 & R_\ell(\hat{u}) \end{bmatrix}, \tag{2.37}$$

and, for $j\geq1$,

$$\mathbf{W}^j(u,\hat{u}) = \sum_{\ell=j}^{\infty} \frac{(\ell-j)!}{(\ell+j)!} \begin{bmatrix} P_\ell^j(u) & 0 & 0 & 0 \\ 0 & R_\ell^j(u) & -T_\ell^j(u) & 0 \\ 0 & -T_\ell^j(u) & R_\ell^j(u) & 0 \\ 0 & 0 & 0 & P_\ell^j(u) \end{bmatrix} \times$$

$$\times\begin{bmatrix}\beta_\ell & \gamma_\ell & 0 & 0\\ \gamma_\ell & \alpha_\ell & 0 & 0\\ 0 & 0 & \varsigma_\ell & -\xi_\ell\\ 0 & 0 & \xi_\ell & \delta_\ell\end{bmatrix}\begin{bmatrix}P_\ell^j(\hat{u}) & 0 & 0 & 0\\ 0 & R_\ell^j(\hat{u}) & -T_\ell^j(\hat{u}) & 0\\ 0 & -T_\ell^j(\hat{u}) & R_\ell^j(\hat{u}) & 0\\ 0 & 0 & 0 & P_\ell^j(\hat{u})\end{bmatrix}. \tag{2.38}$$

As a result we obtain (2.29) and the analogous formulas for the other components of the equation of transfer. If one only assumes $a_1 \epsilon L_1[-1,1]$ as a function of $\cos\theta$, we may replace the elements of the scattering matrix $\mathbf{F}(\theta)$ by the approximates a_1^L, a_2^L, a_3^L, a_4^L, b_1^L and b_2^L, which are given by (2.30)–(2.35) with the summation running up to $\ell=L$. For these approximates one then derives (2.36)–(2.38) with the summation up to $\ell=L$. Now let P_L be the projection of $L_p^{(4)}[-1,1]$, $1\le p<\infty$, onto the linear span of the vectors $(P_\ell,0)$ and $(0,R_\ell)$ with $\ell\le L$ along the closed linear span of the vectors $(P_\ell,0)$ and $(0,R_\ell)$ with $\ell\ge L+1$. Then the projections P_L converge strongly to the identity on $L_2^{(2)}[-1,1]$, and $B_L=P_LB$ is the operator given by (2.29) with summation up to $\ell=L$. Since, as we shall see shortly, B is a compact operator, we have $\lim_{L\to\infty}\|B-B_L\|=0$ on $L_2^{(2)}[-1,1]$.

It remains to establish the compactness of B and the positive selfadjointness of A. It is easily seen, using Lemma 1.1, that on the Banach space $L_p^{(4)}(\Omega)$ of measurable functions $\mathbf{I}:\Omega\to\mathbb{C}^4$ for $1\le p<\infty$ the operator

$$(B^{\text{total}}\mathbf{I})(u,\varphi) = \frac{-1}{4\pi}\int_{-1}^{1}\int_0^{2\pi}\mathbf{Z}(u,\hat{u},\varphi-\hat{\varphi})\,\mathbf{I}(\hat{u},\hat{\varphi})\,d\hat{\varphi}\,d\hat{u},$$

with the phase matrix $\mathbf{Z}(u,\hat{u},\varphi-\hat{\varphi})$ satisfying (2.3), is compact. In fact, if we write B^{total} in matrix form with respect to the decomposition of $L_p^{(4)}(\Omega)$ into four copies of $L_p(\Omega)$, we obtain

$$B^{\text{total}} = \begin{bmatrix}B_{a_1} & D_{12} & D_{13} & 0\\ D_{21} & D_{22} & D_{23} & D_{24}\\ D_{31} & D_{32} & D_{33} & D_{34}\\ 0 & D_{42} & D_{43} & B_{a_4}\end{bmatrix},$$

where B_c, with c being one of the elements of the scattering matrix, is an operator of the type to which Lemma 1.1 relates. Moreover, D_{12}, D_{13}, D_{21}, D_{31}, D_{42}, D_{43}, D_{24},

D_{34}, D_{22}, D_{23}, D_{32} and D_{33} are integral operators, whose respective kernels are given by $b_1(\theta)C_1$, $-b_1(\theta)S_1$, $b_1(\theta)C_2$, $b_1(\theta)S_2$, $-b_2(\theta)S_1$, $-b_2(\theta)C_1$, $-b_2(\theta)S_2$, $b_2(\theta)C_2$, $C_2a_2(\theta)C_1-S_2a_3(\theta)S_1$, $-C_2a_2(\theta)S_1-S_2a_3(\theta)C_1$, $S_2a_2(\theta)C_1+C_2a_3(\theta)S_1$, and $-S_2a_2(\theta)S_1+C_2a_3(\theta)C_1$. Here $C_k=\cos 2\sigma_k$ and $S_k=\sin 2\sigma_k$ for k=1,2, where σ_1 and σ_2 are given by (2.5b) and (2.5c). Thus the compactness of these additional operators is easily obtained using Lemma 1.1. Hence, B^{total} is a compact operator. If we restrict B^{total} to the subspace of $L_p^{(4)}(\Omega)$ consisting of all four-vectors $\mathbf{I}$ = {I,Q,U,V} such that I and Q do not depend on φ and U and V vanish almost everywhere, and identify this subspace with $L_p^{(2)}[-1,1]$ in the natural way, we obtain the operator B, which must then be compact. In a similar way one proves the compactness of the integral operators related to the other component equations.

Following [202], one easily proves that

$$(2\ell+1)A\begin{bmatrix}P_\ell\\0\end{bmatrix} = (2\ell+1-a\beta_\ell)\begin{bmatrix}P_\ell\\0\end{bmatrix} - a\gamma_\ell\begin{bmatrix}0\\R_\ell\end{bmatrix} \tag{2.39}$$

and

$$(2\ell+1)A\begin{bmatrix}0\\R_\ell\end{bmatrix} = -a\gamma_\ell\begin{bmatrix}P_\ell\\0\end{bmatrix} + (2\ell+1-a\alpha_\ell)\begin{bmatrix}0\\R_\ell\end{bmatrix}. \tag{2.40}$$

Since B is self adjoint on $L_2^{(2)}[-1,1]$ (see (2.29)), it suffices to show that all of the matrices

$$\mathbf{M}_\ell^s = \begin{bmatrix}2\ell+1-a\beta_\ell & -a\gamma_\ell\\ -a\gamma_\ell & 2\ell+1-a\alpha_\ell\end{bmatrix}, \quad \ell\geq 0, \quad 0<a\leq 1,$$

have their eigenvalues in the interval $[(2\ell+1)(1-a),\ (2\ell+1)(1+a)]$ with $(2\ell+1)(1-a)$ only occurring as a (simple) eigenvalue if $\ell=0$, in order to prove that A=1−aB has its spectrum in [1−a,1+a] with 1−a as a simple eigenvalue. Indeed, let us suppose that

$$\mathbf{M}_\ell^s\begin{bmatrix}p\\q\end{bmatrix} = (2\ell+1)\lambda a\begin{bmatrix}p\\q\end{bmatrix} \tag{2.41}$$

for some $\lambda<0$, and let c be the minimum of those constants satisfying

$$cP_0(u) + pR_\ell(u) \geq |qR_\ell(u)| \geq 0, \quad u\in[-1,1].$$

We first compute (cf. (2.39)–(2.41))

$$B\left[c\begin{bmatrix}P_0\\0\end{bmatrix}+\begin{bmatrix}pP_\ell\\qR_\ell\end{bmatrix}\right]=c\begin{bmatrix}P_0\\0\end{bmatrix}+\frac{1-\lambda a}{a}\begin{bmatrix}pP_\ell\\qR_\ell\end{bmatrix},$$

which is a vector $\{I^{\cos},Q^{\cos}\}$ satisfying $I^{\cos}\geq|Q^{\cos}|\geq 0$. Hence,

$$cP_0(u)+\frac{1-\lambda a}{a}pP_\ell(u)\geq\left|\frac{1-\lambda a}{a}qR_\ell(u)\right|,$$

which in turn implies

$$\frac{ca}{1-\lambda a}P_0(u)+pP_\ell(u)\geq|qR_\ell(u)|,\qquad u\in[-1,1]. \tag{2.42}$$

As a result of the minimality requirement on c and the nonnegativity of c, we get $a/(1-\lambda a)\geq 1$, which contradicts $\lambda<0$. Hence, all eigenvalues of $\mathbf{M}_\ell^s$ are nonnegative numbers on the interval $[(2\ell+1)(1-a),\ \infty)$. In fact, for $\ell\geq 1$ the equality sign in (2.42) can only hold true for at most finitely many $u\in[-1,1]$, and so for $\ell\geq 1$ we have $a/(1-\lambda a)>1$. For $\ell=0$ the matrix $\mathbf{M}_\ell^s$ reduces to

$$\mathbf{M}_0^s=\begin{bmatrix}1-a & 0\\0 & 1\end{bmatrix},$$

which has a simple eigenvalue at $1-a$ with corresponding eigenvector $\{1,0\}$. We may therefore conclude that $1-a$ is a simple eigenvalue of the positive self adjoint operator $A=1-aB$ with eigenvector $\{1,0\}$. Finally, since B leaves invariant the positive (reproducing and normal) cone of vector functions $\mathbf{I}^{\cos}=\{I^{\cos},Q^{\cos}\}$ on $L_2^{(2)}[-1,1]$ satisfying $I^{\cos}(u)\geq|Q^{\cos}(u)|\geq 0$ for $u\in[-1,1]$, the spectral radius of B belongs to the spectrum of B (see Theorem I 4.2). Since B is self adjoint having 1 as its largest eigenvalue, we have $\sigma(B)\subset[-1,1]$ and $\sigma(A)\subset[1-a,1+a]$. ∎

As a corollary of the above theorem we see that $A=1-aB$ is strictly positive self adjoint for $0<a<1$, and positive self adjoint with simple zero eigenvalue for $a=1$. For $a=1$ the eigenvector $\{1,0\}$ satisfies the identity

$$\left(T\begin{bmatrix}1\\0\end{bmatrix},\begin{bmatrix}1\\0\end{bmatrix}\right)=\int_{-1}^{1}u\{(1)^2+(0)^2\}du=0,$$

whence Ker A is neutral with respect to the indefinite inner product $[\mathbf{h},\mathbf{k}]=(T\mathbf{h},\mathbf{k})$.

Before presenting some existence and uniqueness results for the polarized light equation, let us prove a lemma which represents energy conservation and allows the reflection operators to be taken in H_T without boundedness assumptions. For a vector $\mathbf{I}^{\cos}=\{I^{\cos},Q^{\cos}\}$ we shall write $\|\mathbf{I}^{\cos}\|_p=(|I^{\cos}|^p+|Q^{\cos}|^p)^{1/p}$, where $1\le p<\infty$.

LEMMA 2.2. The operator $|T|R$ is self adjoint on $L_2^{(2)}[-1,1]$, and

$$\left|\int_{-1}^{1} |u|(R\mathbf{I}^{\cos})(u)\cdot\mathbf{I}^{\cos}(u)du\right| \le \int_{-1}^{1} \|\mathbf{I}^{\cos}(u)\|_2^2 du. \tag{2.43}$$

Proof: If we transfer the symmetry relations (2.10) and (2.11) for the ground reflection matrix to the Fourier components appearing in the expansion

$$\mathbf{R}_g(u,\hat{u},\varphi-\hat{\varphi}) = \mathbf{R}_g^{co}(u,\hat{u}) + 2\sum_{j=1}^{\infty}[\mathbf{R}_g^{cj}(u,\hat{u})\cos\{j(\varphi-\hat{\varphi})\}+\mathbf{R}_g^{sj}(u,\hat{u})\sin\{j(\varphi-\hat{\varphi})\}],$$

we obtain the symmetry relations for the component $\mathbf{R}_g^{co}(u,\hat{u})$, i.e.,

$$\mathbf{R}_g^{co}(u,\hat{u}) = \tilde{\mathbf{P}}\tilde{\mathbf{R}}_g^{co}(\hat{u},u)\mathbf{P}, \tag{2.44a}$$

$$\mathbf{R}_g^{co}(u,\hat{u}) = \mathbf{D}\mathbf{R}_g^{co}(u,\hat{u})\mathbf{D}. \tag{2.44b}$$

For the other components we obtain

$$\mathbf{R}_g^{cj}(u,\hat{u}) = \tilde{\mathbf{P}}\tilde{\mathbf{R}}_g^{cj}(\hat{u},u)\mathbf{P}, \tag{2.45a}$$

$$\mathbf{R}_g^{cj}(u,\hat{u}) = \mathbf{D}\mathbf{R}_g^{cj}(u,\hat{u})\mathbf{D}, \tag{2.45b}$$

$$\mathbf{R}_g^{sj}(u,\hat{u}) = -\tilde{\mathbf{P}}\tilde{\mathbf{R}}_g^{sj}(\hat{u},u)\mathbf{P}, \tag{2.46a}$$

$$\mathbf{R}_g^{sj}(u,\hat{u}) = \mathbf{D}\mathbf{R}_g^{sj}(u,\hat{u})\mathbf{D}. \tag{2.46b}$$

From (2.44) we obtain

$$\mathbf{R}_g^{co}(u,\hat{u}) = \begin{bmatrix}\mathbf{R}_g^{cos}(u,\hat{u}) & 0\\ 0 & \mathbf{R}_g^{coa}(u,\hat{u})\end{bmatrix},$$

where

$$\mathbf{R}_g^{cos}(u,\hat{u}) = \tilde{\mathbf{R}}_g^{cos}(\hat{u},u),$$

$$\mathbf{R}_g^{coa}(u,\hat{u}) = \mathbf{E}\tilde{\mathbf{R}}_g^{coa}(\hat{u},u)\mathbf{E},$$

with $\mathbf{E}=\mathrm{diag}(1,-1)$. Hence, by definition, the surface reflection operator R has the property that $|T|R$ is self adjoint on $L_2^{(2)}[-1,1]$.

On defining the unitary operator U on $L_2^{(2)}[-1,1]$ satisfying

$$(U\mathbf{I}^s)(\tau) = \frac{1}{\sqrt{2}}\begin{bmatrix} 1 & -1 \\ 1 & 1 \end{bmatrix} \mathbf{I}^s(\tau),$$

$$(U^{-1}\mathbf{I}^s)(\tau) = \frac{1}{\sqrt{2}}\begin{bmatrix} 1 & 1 \\ -1 & 1 \end{bmatrix} \mathbf{I}^s(\tau),$$

we see that URU^{-1} leaves invariant the cone of $L_2^{(2)}[-1,1]$ consisting of all vectors $\{I_x,I_y\}$ with $I_x\geq 0$ and $I_y\geq 0$, while $\mathbf{R}_g(u,\hat{u})$ is transformed into the kernel

$$\mathbf{R}_g^U(u,\hat{u}) = \tfrac{1}{2}\begin{bmatrix} R_{11}+R_{22}-(R_{12}+R_{21}) & R_{11}-R_{22}+(R_{12}-R_{21}) \\ R_{11}-R_{22}-(R_{12}-R_{21}) & R_{11}+R_{22}+(R_{12}+R_{21}) \end{bmatrix}$$

with $R_{ij}=R_{ij}(u,\hat{u}) = [\mathbf{R}_g(u,\hat{u})]_{ij}$. Defining $\hat{R} = URU^{-1}$ and using the positivity estimates

$$|R_{12}(u,\hat{u})\pm R_{21}(u,\hat{u})| \leq R_{11}(u,\hat{u})\pm R_{22}(u,\hat{u}), \tag{2.47}$$

a straightforward estimate yields

$$\left|\int_0^1 u(\hat{R}\mathbf{I})(u)\cdot\bar{\mathbf{I}}(u)du\right| \leq \int_0^1\int_0^1 u\hat{u}\{(R_{11}(u,\hat{u})+R_{22}(u,\hat{u})) +$$

$$+ |R_{12}(u,\hat{u})+R_{21}(u,\hat{u})| + (R_{11}(u,\hat{u})- R_{22}(u,\hat{u})) + |R_{12}(u,\hat{u})-R_{21}(u,\hat{u})|\} \times$$

$$\times \{(|I(u)|^2+|Q(u)|^2)(|I(\hat{u})|^2+|Q(\hat{u})|^2)\}^{1/2}du d\hat{u}.$$

On using the estimate (2.47), Schwarz's inequality and the dissipativity condition (2.12), we obtain

$$\left|\int_0^1 u(\hat{R}\mathbf{I})(u)\cdot\bar{\mathbf{I}}(u)du\right| \leq \int_0^1 u\|\mathbf{I}(u)\|_2^2 du.$$

Hence, by the unitarity of U, we have (2.43). ∎

We conclude this section with a number of existence and uniqueness results for the component equation (2.28). When formulated in $L_2^{(2)}[-1,1]$ and for phase functions $a_1 \epsilon L_r[-1,1]$ with $r>1$, these results are direct applications of the theory of Chapters II, III and V. When formulated in $L_p^{(2)}[-1,1]$ with $1 \le p < \infty$ and for phase functions $a_1 \epsilon L_r[-1,1]$ with $r>1$, they can be derived from the corresponding results in the L_2-setting using Sections VI.6 and VII.4.

THEOREM 2.3. Let $1 \le p < \infty$, and let $a_1 \epsilon L_r[-1,1]$ with $r>1$. Let $\mathbf{S}^{\cos}(\tau,u)$ satisfy the Hölder continuity condition

$$[\int_{-1}^{1} \| \mathbf{S}^{\cos}(\tau_1,u) - \mathbf{S}^{\cos}(\tau_2,u) \|_p^p du]^{1/p} \le M |\tau_1 - \tau_2|^\gamma$$

for some $0<\gamma<1$, where $0 \le \tau_1, \tau_2 < \infty$, and let the growth condition

$$\int_1^\infty \tau [\int_{-1}^{1} \| \mathbf{S}^{\cos}(\tau,u) \|_p^p du]^{1/p} d\tau < \infty$$

be satisfied. Then for every $\mathbf{D}_+^{\cos} \epsilon L_p^{(2)}[-1,1]$ there exists a unique continuous vector function $\mathbf{I}^{\cos}:[0,\infty) \to L_p^{(2)}[-1,1]$ such that $T\mathbf{I}^{\cos}$ is differentiable for $\tau \epsilon (0,\infty)$ in the strong topology of $L_p^{(2)}[-1,1]$, and which satisfies the component equation of transfer (cf. Eq. (2.28))

$$(T\mathbf{I}^{\cos})'(\tau) = -A\mathbf{I}^{\cos}(\tau) + \mathbf{S}^{\cos}(\tau), \quad 0<\tau<\infty,$$

and the boundary conditions

$$Q_+ \mathbf{I}^{\cos}(0) = \mathbf{D}_+^{\cos},$$

$$\limsup_{\tau \to \infty} \| \mathbf{I}^{\cos}(\tau) \|_p < \infty.$$

THEOREM 2.4. Let $1 \le p < \infty$, and let $a_1 \epsilon L_r[-1,1]$ with $r>1$. Then for every F there exists a unique solution $\mathbf{I}^{\cos}(\tau,u)$ of the Milne problem (2.17)–(2.19). More precisely, there exists a unique continuous vector function $\mathbf{I}^{\cos}:[0,\infty) \to L_p^{(2)}[-1,1]$ such that $T\mathbf{I}^{\cos}$ is differentiable for $\tau \epsilon (0,\infty)$ in the strong topology of $L_p^{(2)}[-1,1]$ and the following equations are satisfied:

$$(T\mathbf{I}^{\cos})'(\tau) = -A\mathbf{I}^{\cos}(\tau), \quad 0<\tau<\infty,$$

$$Q_+ \mathbf{I}^{\cos}(0) = 0,$$

$$\lim_{\tau\to\infty} \int_{-1}^{1} \hat{u}\mathbf{I}^{\cos}(\tau,\hat{u})d\hat{u} = -(F/4\pi)\ \{1,0\}.$$

THEOREM 2.5. Let $a_1 \epsilon L_r[-1,1]$ with $r>1$, and let R be bounded on $L_2^{(2)}[-1,1]$ and $L_p^{(2)}[-1,1]$. Let $\mathbf{S}^{\cos}(\tau,u)$ satisfy the Hölder continuity condition

$$[\int_{-1}^{1} \|\mathbf{S}^{\cos}(\tau_1,u)-\mathbf{S}^{\cos}(\tau_2,u)\|_p^p du]^{1/p} \le M|\tau_1-\tau_2|^{\gamma}$$

for some $0<\gamma<1$, where $0\le\tau_1,\tau_2\le b$. Then for every $\mathbf{D}^{\cos} \epsilon L_p^{(2)}[-1,1]$ there exists a unique solution of the boundary value problem

$$(T\mathbf{I}^{\cos})'(\tau) = -A\mathbf{I}^{\cos}(\tau) + \mathbf{S}^{\cos}(\tau), \quad 0<\tau<b,$$

$$Q_+\mathbf{I}^{\cos}(0) = Q_+\mathbf{D}^{\cos},$$

$$Q_-\mathbf{I}^{\cos}(b) = Q_-\mathbf{D}^{\cos} + JRQ_+\mathbf{I}^{\cos}(b).$$

If $a_1 \epsilon L_r[-1,1]$ with $r>1$, the solution $\mathbf{I}^{\cos}(\tau,u)$ satisfies the condition $\mathrm{I}^{\cos}(\tau,u) \ge |Q^{\cos}(\tau,u)|$, if the internal source term $\mathbf{S}^{\cos}(\tau,u)$ and the incident radiation $\mathbf{D}^{\cos}(u)$ satisfy such a condition also.

3. One speed neutron transport

When neutrons move through a plane parallel reactor medium with constant speed, the stationary transport of neutrons may be modeled by the time independent one speed neutron transport equation. If one also assumes the absorption, scattering and fission processes for a single neutron to be independent of position (i.e., if the medium is assumed homogeneous) and delayed fission is neglected, the equation describing the transport of neutrons is given by (cf. Section I.1)

$$\mu \frac{\partial\psi}{\partial x}(x,\omega) + \psi(x,\omega) = (c/4\pi)\int_{\Omega} p(\omega\cdot\hat{\omega})\psi(x,\hat{\omega})d\hat{\omega} + q(x,\omega). \tag{3.1}$$

Here $x \epsilon (o,\tau)$ is the position variable measured perpendicular to the surface and τ is the thickness of the medium, both measured in units of neutron mean free path. The direction of the neutrons is indicated by points ω on the unit sphere Ω in three dimensional space, where (θ,φ) denote the usual spherical coordinates of $\omega\epsilon\Omega$ and therefore $\omega=(\sin\theta\cos\varphi, \sin\theta\sin\varphi, \cos\theta)$. We have $\theta=0$ in the direction of increasing x and $\theta=\pi$ in the direction of decreasing x, while $\mu=\cos\theta$. As a consequence, if the slab medium is thought vertical, then μ is positive if the neutron has its velocity component to the right. The positive number c is the average number of secondary neutrons per collision. True absorption dominates if $c<1$, there is conservation of the number of secondary neutrons if $c=1$, and multiplication by fission dominates if $c>1$. The **redistribution function** $p(t)$ describes the probability distribution for the scattering of neutrons as a function of the scattering angle (i.e., the angle between the directions of incidence and scattering). Therefore, $p(t)$ is nonnegative and measurable and satisfies the normalization condition $\frac{1}{2}\int_{-1}^{1}p(t)dt = 1$. The unknown function $\psi(x,\omega)$ is the neutron angular density as a function of position x and direction ω, and therefore must be nonnegative and measurable. The function $q(x,\omega)$, which describes the internal neutron sources, must likewise be nonnegative and measurable.

The boundary conditions to be imposed specify the angular densities of incoming neutrons. For a finite slab medium $(0<\tau<\infty)$ we demand that

$$\psi(0,\omega) = \varphi(\omega), \qquad \mu=\cos\theta>0, \tag{3.2a}$$

$$\psi(\tau,\omega) = \varphi(\omega), \qquad \mu=\cos\theta<0, \tag{3.2b}$$

so that one function, $\varphi(\omega)$, describes the angular densities of incoming neutrons at both surfaces. It is clear physically that one should assume $\varphi(\omega)\geq 0$. For a half space medium $(\tau=\infty)$ we demand that

$$\psi(0,\omega) = \varphi_{+}(\omega), \qquad \mu=\cos\theta>0, \tag{3.3a}$$

$$\limsup_{x\to\infty} \int_{\Omega} |\psi(x,\omega)|^{p}d\omega < \infty. \tag{3.3b}$$

Again, one assumes $\varphi_{+}(\omega)\geq 0$.

On applying Fourier decomposition we may write

$$\psi(x,\omega) = \psi_{0}(x,\mu) + 2\sum_{m=1}^{\infty}\psi_{m}(x,\mu)\cos m\varphi \tag{3.4}$$

and similarly for $q(x,\omega)$, $\varphi(\omega)$ and $\varphi_+(\omega)$, and derive the component equations

$$\mu\frac{\partial\psi_m}{\partial x}(x,\mu) + \psi_m(x,\mu) = \tfrac{1}{2}c\int_{-1}^{1} \mathbf{p}_m(\mu,\hat{\mu})\psi_m(x,\hat{\mu})d\hat{\mu} + q_m(x,\mu)$$

with boundary conditions

$$\psi_m(0,\mu) = \varphi_m(\mu), \quad \mu>0,$$

$$\psi_m(\tau,\mu) = \varphi_m(\mu), \quad \mu<0,$$

for finite slabs, and

$$\psi_m(0,\mu) = \varphi_+(\mu), \quad \mu>0,$$

$$\limsup_{x\to\infty} \int_{-1}^{1} |\psi_m(x,\mu)|^p d\mu < \infty$$

for half spaces. The scattering kernel $\mathbf{p}_m(\mu,\hat{\mu})$ is given by

$$\mathbf{p}_m(\mu,\hat{\mu}) = \frac{1}{2\pi}\int_0^{2\pi} \mathbf{p}(\mu\hat{\mu}+(1-\mu^2)^{1/2}(1-\hat{\mu}^2)^{1/2}\cos(\varphi-\hat{\varphi}))\cos m(\varphi-\hat{\varphi})d\hat{\varphi}.$$

For $m=0$ all functions involved are nonnegative, but this need no longer be the case for $m\geq 1$.

Mathematically the one speed neutron transport equation with the above boundary conditions coincides with the equation of transfer of unpolarized light with non-reflective boundary conditions. This identification is realized by identifying position x and optical depth τ, slab thickness τ and optical thickness b, phase function $a_1(\omega\cdot\hat{\omega})$ and redistribution function $\mathbf{p}(\omega\cdot\hat{\omega})$, albedo of single scattering a and number of secondaries per collision c, and (specific) intensity $I(\tau,u,\varphi)$ and angular density $\psi(x,\omega)$, respectively. In radiative transfer problems, however, the albedo of single scattering a is confined to the interval $(0,1]$, whereas in neutron transport the number of secondary neutrons per collision, c, is an arbitrary positive number which may well exceed unity. As a result, the existence and uniqueness theory for the underlying boundary value problems in one speed neutron transport does not differ from the analogous theory in radiative transfer, provided $c\in(0,1]$. For $c>1$ neutron transport displays phenomena that do not have parallels in radiative transfer. Among them one has the existence of a critical value $c_0(\tau)>1$ for the number of secondaries per collision, with non-uniqueness of stationary solutions at $c=c_0$ and non-existence of nonnegative solutions for $c>c_0$. These are non-existence and non-uniqueness results alien to radiative transfer. For the literature of neutron transport theory we refer to the textbooks of Davison [98] and Williams [394], which use Wiener-Hopf type techniques, Case and Zweifel [70], which particularly emphasizes eigenfunction expansion methods, and Duderstadt and Martin [108], which contains a relatively up-to-date review of neutron transport theory. For the underlying nuclear reactor physics we refer to Bell and Glasstone [38] and Zweifel [406].

Let us introduce the functional formulation to be used in this section. As in Section 1, we let $L_p(\Omega)$ be the (real or complex) Banach space of measurable functions on Ω endowed with L_p-norm, and define on $L_p(\Omega)$ the operators

$$(Th)(\omega) = \mu h(\omega),$$

$$(Q_\pm h)(\omega) = \begin{cases} h(\omega), & \pm\mu>0, \\ 0, & \pm\mu<0, \end{cases}$$

$$(Ah)(\omega) = h(\omega) - c(Bh)(\omega),$$

$$(Bh)(\omega) = \frac{1}{4\pi}\int_\Omega p(\omega\cdot\hat{\omega})h(\hat{\omega})d\hat{\omega},$$

as well as the vector-valued functions

$$\psi(x)(\omega) = \psi(x,\omega),$$

$$q(x)(\omega) = q(x,\omega).$$

We then obtain the boundary value problem

$$(T\psi)'(x) = -A\psi(x) + q(x), \quad 0<x<\tau, \tag{3.5a}$$

$$Q_+\psi(0) = Q_+\varphi, \tag{3.5b}$$

$$Q_-\psi(\tau) = Q_-\varphi, \tag{3.5c}$$

for finite slabs of thickness τ, and

$$(T\psi)'(x) = -A\psi(x) + q(x), \qquad 0<x<\infty, \tag{3.6a}$$

$$Q_+\psi(0) = \varphi_+, \tag{3.6b}$$

$$\limsup_{x\to\infty} \|\psi(x)\|_p < \infty, \tag{3.6c}$$

for half spaces. On $L_p[-1,1]$ we may formulate similar boundary value problems if we define the operators

$$(T_m h)(\mu) = \mu h(\mu), \tag{3.7}$$

$$(Q_{\pm,m}h)(\mu) = \begin{cases} h(\mu), & \pm\mu>0, \\ 0, & \pm\mu<0, \end{cases}$$

$$(A_m h)(\mu) = h(\mu) - c(B_m h)(\mu), \tag{3.8}$$

$$(B_m h)(\mu) = \tfrac{1}{2}\int_{-1}^{1} p_m(\mu,\hat{\mu})h(\hat{\mu})d\hat{\mu}, \tag{3.9}$$

and the vector-valued functions

$$\psi_m(x)(\mu) = \psi_m(x,\mu),$$

$$q_m(x)(\mu) = q_m(x,\mu).$$

In one speed neutron transport, reciprocity symmetry leads to the existence of an inversion symmetry J satisfying (1.29a)–(1.29d), namely the operator defined by $(Jh)(\omega)=h(-\omega)$. Similar properties hold for the operator $(J_m h)(\mu)=h(-\mu)$, where now all of the operators bear the subscript m.

On the Hilbert space $L_2(\Omega)$ the operators T, A and B are self adjoint, T has absolutely continuous spectrum $[-1,1]$ and Q_+ and Q_- are the orthogonal projections of $L_2(\Omega)$ onto the maximal T-positive and T-negative T-invariant subspaces. On $L_p(\Omega)$, $1\le p<\infty$, T is a scalar-type spectral operator (cf. [105, 109] for the definitions). As is clear from Lemma 1.1, B is a compact operator on $L_p(\Omega)$, $1\le p<\infty$ with the spherical harmonics as a complete set of eigenfunctions. We denote by $C(\Omega)$ the (real or complex) Banach space of continuous functions $h:\Omega\to\mathbb{C}$ endowed with the supremum norm.

For $r=2$ the next result is due to Maslennikov [259]. Our proof was inspired by the argument of the proof of his Theorem 4(iii).

PROPOSITION 3.1. Let $1\le p<\infty$, and let $p\in L_r[-1,1]$ for some $r>1$. Then for some $m\in\mathbb{N}$ the operator B^m maps $L_p(\Omega)$ into $C(\Omega)$. Also, for every nonzero nonnegative function $h\in L_p(\Omega)$ there exists $n\in\mathbb{N}$ such that

$$0 < \alpha \le (B^n h)(\omega) \le \beta < \infty, \quad \omega\in\Omega. \tag{3.10}$$

Proof: It follows directly from Lemma 1.2 that B maps $L_p(\Omega)$ into $L_{pr}(\Omega)$, and, consequently, that B^{m-1} maps $L_p(\Omega)$ into $L_{pr^{m-1}}(\Omega)$. However, a straightforward application of the Hölder inequality, i.e., the estimate

$$|(Bh)(\omega)| \le \frac{1}{4\pi}\Big[\int_\Omega p(\omega\cdot\hat\omega)^r d\hat\omega\Big]^{1/r}\|h\|_{r/(r-1)},$$

implies that B maps $L_{r/(r-1)}(\Omega)$ into $L_\infty(\Omega)$. Since there is a sequence of continuous functions $p_{(n)}:[-1,1]\to\mathbb{R}$ satisfying

$$\lim_{n\to\infty} \int_{-1}^{1} |p(t)-p_{(n)}(t)|^r dt = 0,$$

and the estimate

$$|(B_{(n)}h)(\omega_1)-(B_{(n)}h)(\omega_2)| \le M_n\|\omega_1-\omega_2\|\,\|h\|_{r/(r-1)}$$

holds true, we have, in fact, $B[L_{r/(r-1)}(\Omega)]\subset C(\Omega)$, and consequently $B^m[L_p(\Omega)]\subset C(\Omega)$ for $pr^{m-1}\ge r/(r-1)$.

If $r>1$, then some iterate has the property that it maps $L_2(\Omega)$ into $C(\Omega)$, i.e., that it is a Hilbert-Schmidt operator. In view of Lemma 1.1, all iterates B^m of B have the form

$$(B^m h)(\omega) = \frac{1}{4\pi}\int_\Omega p^m(\omega\cdot\hat\omega)h(\hat\omega)d\hat\omega,$$

where $p^m(t)\geq 0$ and $\int_{-1}^1 p^m(t)dt=2$. Hence, for some iterate B^m we must have

$$\int_{-1}^1 [p^m(t)]^2 dt = 2\sum_{\ell=0}^{\infty}(2\ell+1)^{-1}[\beta_\ell^m]^2 < \infty,$$

where β_ℓ^m is the expansion coefficient β_ℓ defined for the redistribution function $p^m(t)$. Then, since $P_0(t)\equiv 1$, $1\pm P_\ell(t)\geq 0$, and therefore $\int_{-1}^1 p^m(t)(1\pm P_\ell(t))dt>0$ for $\ell\geq 1$, we have $|\beta_\ell^m|<2\ell+1$ for $\ell\geq 1$, whence

$$\theta = \sup\{(2\ell+1)^{-1}|\beta_\ell^m| : \ell=1,2,3,...\}\in[0,1).$$

For $n\geq 2$ we obtain for the kernel of the iterate B^{mn}:

$$p^{mn}(t) = 1+\varepsilon_n(t),$$

$$\varepsilon_n(t) = \sum_{\ell=0}^{\infty}(2\ell+1)[(2\ell+1)^{-1}\beta_\ell^m]^n P_\ell(t).$$

Now choose $0<\varepsilon<1$, take $L\in\mathbb{N}$ such that

$$\sum_{\ell=L+1}^{\infty}(2\ell+1)^{-1}[\beta_\ell^m]^2 < \tfrac{1}{2}\varepsilon,$$

and take N such that $\theta^N<\frac{1}{2}\varepsilon/L(L+2)$. Then for $n>N$ we have immediately

$$|\varepsilon_n(t)| \leq \sum_{\ell=1}^{L}(2\ell+1)\theta^n + \sum_{\ell=L+1}^{\infty}(2\ell+1)^{n/2}[(2\ell+1)^{-1}\beta_\ell^m]^n\cdot|P_\ell(t)| \leq$$

$$\leq \sum_{\ell=1}^{L}(2\ell+1)\theta^n + \sum_{\ell=L+1}^{\infty}[(2\ell+1)^{-1}(\beta_\ell^m)^2]^{n/2} \leq$$

$$\leq \sum_{\ell=1}^{L}(2\ell+1)\theta^n + \sum_{\ell=L+1}^{\infty}(2\ell+1)^{-1}[\beta_\ell^m]^2 \leq \theta^N\sum_{\ell=1}^{L}(2\ell+1) + \tfrac{1}{2}\varepsilon < \varepsilon,$$

whence $\lim_{n\to\infty} p^{mn}(t)=1$ uniformly in t on $[-1,1]$. Thus some iterate B^{mn} has a kernel $p^{mn}(t)$ which is strictly positive and satisfies $|1-p^{mn}(t)|<\frac{1}{2}\varepsilon$ for every $t\in[-1,1]$. This immediately gives (3.10) and the proof is complete. ■

The above proof implies that, for $p\in L_r[-1,1]$ with $r>1$, B^n converges to the operator $(B_{\text{isotropic}}h)(\omega) = \frac{1}{4\pi}\int_\Omega h(\hat{\omega})d\hat{\omega}$. Physically, this means that the redistribution of neutrons tends to behave isotropically after a large number of collisions.

For $0<c\leq 1$, when the reactor medium is nonmultiplying, the existence and uniqueness theory of the relevant boundary value problems coincides with the existence and uniqueness theory for transfer of unpolarized light, if reflection by the planetary surface is neglected. For these results we refer to Section 1.

In nonmultiplying media neutron transport with isotropic scattering leads in the Hilbert space $H=L_2[-1,1]$ to the uniquely solvable half space problem (3.6). The subscript m=0 in the notation of (3.7)-(3.9) will be dropped for the isotropic operators, $\mathbf{p}_0(\mu,\hat{\mu})=1$, and we define the vector $e(\mu)=1$ and denote by $\sigma(\tau)$ the resolution of the identity associated with the self adjoint operator T:

$$(\sigma(\tau)f)(\mu) = \begin{cases} f(\mu), & \mu\in\tau\cap[-1,1], \\ 0, & \mu\notin\tau\cap[-1,1]. \end{cases}$$

Let us observe that B has the one dimensional range of constant functions. We put $\mathbb{B} = \text{span}\{e\}$, identify $\mathbb{B}$ with $\mathbb{C}$ and define $\pi:H\to\mathbb{B}$ and $j:\mathbb{B}\to H$ by

$$(\pi f)(\mu) = \tfrac{1}{2}\int_{-1}^{1} f(\hat{\mu})d\hat{\mu},$$

$$(j\xi)(\mu) = \xi e,$$

where $\xi\in\mathbb{C}$. The dispersion function can now be expressed as a scalar function on $\mathbb{B}$ by

$$\Lambda(z) = 1 + \tfrac{1}{2}cz\int_{-1}^{1}(t-z)^{-1}dt.$$

The function is analytic on the Riemann sphere cut along $[-1,1]$ and is even and nonzero along the imaginary axis. It allows a unique Wiener-Hopf factorization

$$\Lambda(z)^{-1} = H(-z)H(z)$$

for Re $z=0$, where $H(z)$ is analytic on the Riemann sphere cut along $[-1,0]$, continuous and nonzero on the closed right half plane, analytic at ∞ for $0<c<1$ and having a simple pole at ∞ for $c=1$ (cf. [63] for the original result of Busbridge; also [61]).

Using this factorization we find

$$(E\varphi_+)(\mu) = \varphi_+(\mu), \quad \mu>0,$$

and

$$(E\varphi_+)(\mu) = \tfrac{1}{2}c\int_0^1 \nu(\nu-\mu)^{-1}H(-\mu)H(\nu)\varphi_+(\nu)d\nu, \quad \mu<0.$$

This is a well-known expression, which is already retrievable from the work of Chandrasekhar [89] and has been derived many times since.

Let us give some more details on the way in which the albedo operator is derived from Eq. VIII (1.8). For small intervals $d\mu\subset(0,1)$ and $dt\subset(-1,0)$ we compute

$$[\sigma(dt)BjH_\ell^+(-t)H_r^+(\mu)\pi\sigma(d\mu)\varphi_+](\mu) = \tfrac{1}{2}c\chi_{dt}(\mu)H(-t)H(\mu)\int_{d\mu}\varphi_+(\nu)d\nu,$$

where $\chi_{\mathcal{A}}(\mu)=1$ for $\mu\in\mathcal{A}$ and $\chi_{\mathcal{A}}(\mu)=0$ for $\mu\notin\mathcal{A}$. Finishing the substitution of the above data in VIII (1.8), we obtain for $\mu<0$ the desired representation.

Next we substitute all data in the H-equation VIII (1.10) or VIII (1.11). For either of these one obtains the same equation, namely

$$H(\mu)^{-1} = 1 - \tfrac{1}{2}c\mu\int_0^1 (\mu+t)^{-1}H(t)dt.$$

This is the famous Chandrasekhar H-equation (cf. [89]).

If we consider instead neutron transport with anisotropic scattering, an equation for transport may be obtained by averaging over azimuthal angle. Writing $P_\ell(\mu)$ for the usual Legendre polynomial and assuming $f_0=1$, $0<c\le1$, and

$$\sum_{\ell=0}^{L} f_\ell(2\ell+1)P_\ell(\mu)P_\ell(\hat{\mu}) \ge 0,$$

leads to the uniquely solvable half space problem ($0\le x<\infty$, $-1\le\mu\le1$):

$$\mu\frac{\partial\psi}{\partial x}(x,\mu) + \psi(x,\mu) = \tfrac{1}{2}c\sum_{\ell=0}^{L} f_\ell(2\ell+1)P_\ell(\mu)\int_{-1}^{1} P_\ell(\hat{\mu})\psi(x,\hat{\mu})d\hat{\mu},$$

$$\psi(0,\mu) = \varphi_+(\mu), \quad 0\le\mu\le1,$$

$$\int_{-1}^{1} |\psi(x,\mu)|^2 d\mu = O(1) \ (x\to\infty).$$

We have terminated the expansion in Legendre polynomials in the scattering term above; this is referred to as the **degenerate anisotropic** neutron transport equation. It should be noted that the nonnegativity of the redistribution function is often lost on truncation. Although this does not affect the unique solvability of the above boundary value problems, it was shown by Feldman [119] that one may truncate the expansion and at the same time modify the coefficients in such a way that the truncated redistribution functions are nonnegative, converge to the untruncated redistribution functions in $L_1[-1,1]$, and conserve the wellposedness of the boundary value problems.

Let us study this problem in the Hilbert space $H = L_2[-1,1]$ and introduce the operators

$$(Tf)(\mu) = \mu f(\mu),$$

$$(Bf)(\mu) = \tfrac{1}{2}c \sum_{\ell=0}^{L} f_\ell (2\ell+1) P_\ell(\mu) \int_{-1}^{1} P_\ell(\hat{\mu}) f(\hat{\mu}) d\hat{\mu},$$

$$(Q_\pm f)(\mu) = \begin{cases} f(\mu), & \pm\mu\geq 0, \\ 0, & \pm\mu<0. \end{cases}$$

Let $\mathbb{B}$ be the linear span of the (Legendre) polynomials of degree equal or less than L, let $j:\mathbb{B}\to H$ denote the natural imbedding, and put

$$\pi f = \sum_{\ell=0}^{L} \frac{2\ell+1}{2} (f,P_\ell) P_\ell.$$

The Legendre polynomials are eigenvectors of B with eigenvalues cf_ℓ. The dispersion function, in matrix representation with respect to the basis $\{P_\ell\}_{\ell=0}^{L}$ of $\mathbb{B}$, is now computed to have the form

$$[\Lambda(z)]_{ik} = \delta_{ik} - \tfrac{1}{2}cz \int_{-1}^{1} (z-t)^{-1} (2i+1) f_k P_i(t) P_k(t) dt$$

for $z\notin[-1,1]$. There are unique functions $\mathbf{H}_\ell$ and $\mathbf{H}_r$ which are analytic and continuous up to the boundary of the closed right half plane and assume invertible values there, such that

$$\Lambda(z)^{-1} = \mathbf{H}_\ell(-z)\mathbf{H}_r(z)$$

for Re $z=0$. For $0<c<1$ these functions are analytic at ∞, but for $c=1$ they have a simple pole there. The albedo operator is found to be given by

$$(E\varphi_+)(\mu) = \varphi_+(\mu)$$

for $0\leq\mu\leq1$ and

$$(E\varphi_+)(\mu) = \tfrac{1}{2}c\int_0^1 \frac{\nu}{\nu-\mu}\sum_{i=0}^{L}\sum_{j=0}^{L}(2j+1)f_i[\mathbf{H}_\ell(-\mu)\mathbf{H}_r(\nu)]_{ij}P_i(\mu)P_j(\nu)\varphi_+(\nu)d\nu,$$

for $-1\leq\mu<0$. We also find the coupled H-equations

$$[\mathbf{H}_\ell(z)^{-1}]_{ik} = \delta_{ik} - \tfrac{1}{2}cz\int_0^1 \frac{1}{z+t}\sum_{j=0}^{L}(2j+1)f_k[\mathbf{H}_r(t)]_{ij}P_j(t)P_k(t)dt,$$

$$[\mathbf{H}_r(z)^{-1}]_{ik} = \delta_{ik} - \tfrac{1}{2}cz\int_0^1 \frac{1}{z+t}\sum_{j=0}^{L}(-1)^{i+j}(2i+1)f_j[\mathbf{H}_\ell(t)]_{jk}P_i(t)P_j(t)dt.$$

The above identities were derived by Mullikin [274]. Note that it is also possible to express the albedo operator in scalar H-functions, which appear as the determinants of $\mathbf{H}_\ell$ and $\mathbf{H}_r$, and polynomials which are obtained by solving linear equations whose coefficients involve moments of the scalar H-functions. This was first done in general for $0<c<1$ by Busbridge [61] and improved and extended for $0<c\leq1$ by Pahor [296] and Kuščer and McCormick [233]. For most purposes, the latter results are more expedient than the ones derived above, but their derivation requires more elaborate analysis.

Let us consider next the integral form of the boundary value problem:

$$\psi(x) - c\int_0^\tau \mathcal{H}(x-y)B\psi(y)dy = \varsigma(x), \quad 0<x<\tau\leq\infty.$$

For finite slabs the internal source term $q(x,\omega)$ is taken to satisfy

$$\left[\int_\Omega |q(x,\omega)-q(y,\omega)|^p d\omega\right]^{1/p} \leq M_p|x-y|^\gamma, \quad 0\leq x,y\leq\tau,$$

for some $0<\gamma<1$, when using an L_p-setting, and the right hand side is given by

$$\varsigma(x,\omega) = \begin{cases} e^{-x/\mu}\varphi(\mu) + \int_0^x \mu^{-1} e^{-(x-y)/\mu} q(y,\mu)\,dy, & \mu>0 \\ e^{(\tau-x)/\mu}\varphi(\mu) - \int_x^\tau \mu^{-1} e^{-(x-y)/\mu} q(y,\mu)\,dy, & \mu<0. \end{cases}$$

For half spaces one assumes the condition

$$[\int_\Omega |q(x,\omega)-q(y,\omega)|^p d\omega]^{1/p} \le M_p |x-y|^\gamma, \quad 0\le x,y\le\infty,$$

for some $0<\gamma<1$, as well as the conditions

$$\int_\Omega |q(x,\omega)|^p d\omega = O(1)\ (x\to\infty),$$

$$\int_1^\infty x[\int_\Omega |q(x,\omega)|^p d\omega]^{1/p} dx < \infty,$$

when using an L_p-setting, and the right hand side is given by

$$\varsigma(x,\omega) = \begin{cases} e^{-x/\mu}\varphi_+(\mu) + \int_0^x \mu^{-1} e^{-(x-y)/\mu} q(y,\mu)\,dy, & \mu>0 \\ -\int_x^\infty \mu^{-1} e^{-(x-y)/\mu} q(y,\mu)\,dy, & \mu<0. \end{cases}$$

Details of the equivalence of the boundary value problem and the integral equation, which is valid if $\mathfrak{p}\in L_r[-1,1]$ for some $r>1$, can be found in Chapter VI.

As in Section VI.2 we define, for $1\le q\le\infty$ and $1\le p<\infty$, $L_q(L_p(\Omega))_a^b$ to be the (real or complex) Banach space of all strongly measurable functions $\psi:(a,b)\to L_p(\Omega)$ which are bounded with respect to the L_p-norm and $C(L_p(\Omega))_a^b$ to be the Banach space of all bounded continuous functions from $[a,b]$ into $L_p(\Omega)$, endowed with the supremum norm. One may then show (cf. Section VI.2) that on all of the spaces $L_q(L_p(\Omega))_0^\tau$ and $C(L_p(\Omega))_0^\tau$, $\tau\in(0,\infty]$, the convolution operator

$$(\mathcal{L}_\tau\psi)(x) = \int_0^\tau \mathcal{H}(x-y)B\psi(y)dy$$

is bounded, where the propagator function $\mathcal{H}(x)$ has the form

$$(\mathcal{H}(x)h)(\mu) = \begin{cases} |\mu|^{-1} e^{-x/\mu} h(\omega), & x\mu>0 \\ 0 \quad , & x\mu<0 , \end{cases}$$

Moreover, for $\tau \in (0,\infty)$ the operator $\mathcal{L}_\tau$ is compact. In fact, all of these properties follow using the integrability estimate (1.35).

PROPOSITION 3.2. Let $1\le p<\infty$, $1\le q\le\infty$ and $0<\tau<\infty$, and let $\mathfrak{p}\in L_r[-1,1]$ for some $r>1$. Then, for every nonnegative function $\psi\in L_q(L_p(\Omega))_0^\tau$ (resp. $\psi\in C(L_p(\Omega))_0^\tau$), there exist $n\in\mathbb{N}$ and $\alpha,\beta>0$ such that

$$0 < \alpha \le (\mathcal{L}_\tau^n\psi)(x)(\omega) \le \beta < \infty. \tag{3.11}$$

Proof: By Lemma 1.2, B maps $L_p(\Omega)$ into $L_{pr}(\Omega)$. Moreover, as in the proof of Lemma 1.3, one may show that $\|\mathcal{H}(x)B\| = O(|x|^{\alpha-1})$ $(x\to 0)$, where $0<\alpha<(r-1)/pr$. Again applying Lemma 1.2, but this time to the integral operator with kernel $\|\mathcal{H}(x-y)B\|$ on $L_q(0,\tau)$, τ finite, one gets

$$\mathcal{L}_\tau^m[L_q(L_p(\Omega))_0^\tau] \subset C(C(\Omega))_0^\tau$$

for suitable $m=m(p,q,r)\in\mathbb{N}$.

In order to establish (3.11), take a nonnegative function $h\in C(C(\Omega))_0^\tau$ and identify $C(C(\Omega))_0^\tau$ with the Banach space $C([0,\tau]\times\Omega)$ of continuous functions $\tilde{h}:[0,\tau]\times\Omega\to\mathbb{C}$ endowed with the supremum norm, using $\tilde{h}(x,\omega)=(h(x))(\omega)$ for $(x,\omega)\in[0,\tau]\times\Omega$. Then there exists an open interval $(a,b)\subset(0,\tau)$ such that $h(x)\neq 0$ for $x\in(a,b)$. As a result we have

$$(\mathcal{L}_\tau h)(x) = \int_0^\tau \mathcal{H}(x-y)Bh(y)dy \neq 0, \quad x\in[0,\tau],$$

since otherwise $Bh(y)=0$ and consequently $h(y)=0$ (cf. (3.10)). So we may as well assume that $h(x)\neq 0$ for $x\in[0,\tau]$. Now observe that $\mathcal{L}_\tau^m$ can be naturally identified with the integral operator on $C([0,\tau]\times\Omega)$ with kernel

$$\left(\frac{1}{4\pi}\right)^m \int_\Omega d\omega_1 \cdots \int_\Omega d\omega_{m-1} \int_0^\tau dy_1 \cdots \int_0^\tau dy_{m-1} |\mu\mu_1\cdots\mu_m|^{-1} \psi\left[\frac{x-y_1}{\mu}\right] \cdots \psi\left[\frac{y_{m-1}-y_m}{\mu_{m-1}}\right] \times$$

$$\times e^{-(x-y)/\mu} e^{-(y_1-y_2)/\mu_1} \dots e^{-(y_{m-1}-y_m)/\mu_{m-1}} p(\omega\cdot\omega_1)p(\omega_1\cdot\omega_2)\dots p(\omega_{m-1}\cdot\omega_m),$$

where $\psi(t)=1$ for $t>0$ and $\psi(t)=0$ for $t<0$. Hence, we may choose an interval $(c,d)\subset(0,\tau)$ and a constant $\alpha>0$ satisfying $(B^m h(y_m))(\omega) \geq \alpha > 0$ for $\omega\in\Omega$ and $y_m \in (c,d)$, whence

$$(\mathcal{L}_\tau^m h)(x) \geq \alpha \int_0^\tau dy_1 \dots \int_0^\tau dy_m \,|\mu\mu_1\dots\mu_m|^{-1} \psi\left[\frac{x-y_1}{\mu}\right]\dots\psi\left[\frac{y_{m-1}-y_m}{\mu_{m-1}}\right] \times$$

$$\times\, e^{-(x-y)/\mu} e^{-(y_1-y_2)/\mu_1} \dots e^{-(y_{m-1}-y_m)/\mu_{m-1}} \psi_{(c,d)}(y_m) = \gamma > 0,$$

where $\psi_{(c,d)}(t)=1$ for $t\in(c,d)$ and $\psi_{(c,d)}(t)=0$ for $t\in(c,d)$. On the other hand, if $\delta=\max\{\tilde{h}(x,\omega) : (x,\omega)\in[0,\tau]\times\Omega\}=\|h\|$, then

$$(\mathcal{L}_\tau^m h)(x) \leq \delta \int_0^\tau dy_1 \dots \int_0^\tau dy_m \,|\mu\mu_1\dots\mu_m|^{-1} \psi\left[\frac{x-y_1}{\mu}\right]\dots\psi\left[\frac{y_{m-1}-y_m}{\mu_{m-1}}\right] \times$$

$$\times\, e^{-(x-y_1)/\mu} e^{-(y_1-y_2)/\mu_1} \dots e^{-(y_{m-1}-y_m)/\mu_{m-1}} = \beta < \infty,$$

which completes the proof. ∎

We have shown that, for $a_1 \in L_r[-1,1]$ with $r>1$, the operator $\mathcal{L}_\tau$ is u_0-positive on all spaces $L_q(L_p(\Omega))_0^\tau$ and $C(L_p(\Omega))_0^\tau$, $q\in[1,\infty]$, $p\in[1,\infty)$. Here u_0-positivity is defined as in Section I.4 Since the cones of nonnegative functions on those spaces are reproducing and normal, we may apply Theorems I 4.3 and I 4.4 and obtain the following results.

THEOREM 3.3. Let $a_1\in L_r[-1,1]$ with $r>1$, $1\leq p<\infty$ and $1\leq q\leq\infty$. Then for finite τ there exists a unique critical eigenvalue $c=c(\tau)>0$ such that

$$\psi(x) = c\int_0^\tau \mathcal{H}(x-y)B\psi(y)dy = 0, \quad 0<x<\tau, \tag{3.12}$$

has a nontrivial nonnegative solution in $L_q(L_p(\Omega))_0^\tau$ and $C(L_p(\Omega))_0^\tau$. The corresponding generalized eigenvector space $\bigcup_n \mathrm{Ker}(I-cL_\tau)^n$ is one dimensional, $c(\tau)^{-1}$ is the spectral radius of $\mathcal{L}_\tau$ and $\mathcal{L}_\tau$ does not have other eigenvalues with modulus $c(\tau)^{-1}$.

THEOREM 3.4. Let $a_1 \in L_r[-1,1]$ with $r>1$, $1 \le p < \infty$ and $1 \le q < \infty$. Then for finite τ the inhomogeneous convolution equation

$$\psi(x) - c\int_0^\tau \mathcal{H}(x-y)B\psi(y)dy = \varsigma(x), \quad 0<x<\tau, \tag{3.13}$$

where $\varsigma(x)$ is nontrivial and nonnegative, has the following properties:

(i) For $0<c<c(\tau)$, Eq. (3.13) is uniquely solvable on all of the spaces $L_q(L_p(\Omega))_0^\tau$ and $C(L_p(\Omega))_0^\tau$ and the solution is nonnegative and given by the Neumann series

$$\psi(x) = \sum_{n=1}^{\infty} c^n(\mathcal{L}_\tau^n\varsigma)(x), \quad 0<x<\tau.$$

(ii) For $c \ge c(\tau)$, Eq. (3.13) does not have nonnegative solutions on $L_q(L_p(\Omega))_0^\tau$ and $C(L_p(\Omega))_0^\tau$.

Using an argument of Mullikin [270], based on Theorem I 4.4, and an argument of Borgioli et al. [49], based on the analytic perturbation theory of Kato [213], we can prove that $\tau \to c(\tau)$ is a strictly monotonically decreasing C^∞-function from $(0,\infty)$ onto $(1,\infty)$. As a result, we have the existence of the strictly monotonically decreasing C^∞-function $c \to \tau(c)$ from $(1,\infty)$ onto $(0,\infty)$, which gives the "critical" size, τ, as a function of the number, $c>1$, of secondaries per collision. Let us first give the argument for strict monotonicity. Let $0<\tau_1<\tau_2<\infty$ and let ψ be a nonzero nonnegative solution of Eq. (3.11) with $\tau=\tau_2$. We then write

$$\psi(x) - c(\tau_2)\int_0^{\tau_1} \mathcal{H}(x-y)B\psi(y)dy = c(\tau_2)\int_{\tau_1}^{\tau_2} \mathcal{H}(x-y)B\psi(y)dy, \quad 0<x<\tau_1,$$

where

$$\omega(x) = c(\tau_2)\int_{\tau_1}^{\tau_2} \mathcal{H}(x-y)B\psi(y)dy = 0$$

for some $x \in (\tau_1,\tau_2)$ would imply $B\psi(y) \equiv 0$ for all $y \in (\tau_1,\tau_2)$ and therefore $\psi(y) \equiv 0$ for all $y \in (\tau_1,\tau_2)$ (cf. (3.10)). Because one cannot have $\psi(y)(\omega)=0$ for $(\omega,y) \in \Omega \times (\tau_1,\tau_2)$ if ψ is a nonnegative eigenfunction at the critical eigenvalue (cf. Section I.4; note that $\mathcal{L}_\tau$ is u_0-positive), we have $\omega(x) \ne 0$ almost everywhere. As a consequence of part 3 of Theorem I 3.4 we obtain $c(\tau_2)<c(\tau_1)$. To prove the

continuity of $c(\tau)$, we replace $\mathcal{L}_\tau$ by the operator

$$(\hat{\mathcal{L}}_\tau\varphi)(x) = \tau\int_0^1 \mathcal{H}(\tau(x-y))B\varphi(y)dy$$

on $L_q(L_p(\Omega))_0^1$ or on $C(L_p(\Omega))_0^1$, obtained by similarity. Inasmuch as $\frac{\partial}{\partial\tau}\mathcal{H}(\tau(x-y)) = -(x-y)T^{-1}\mathcal{H}(\tau(x-y))$ and

$$\int_0^1 |x-y|\, \|T^{-1}\mathcal{H}(\tau(x-y))B\| dy \le M\tau^{-2} < \infty$$

with M independent of x and τ, the operator function $\tau \to \hat{\mathcal{L}}_\tau$ is analytic on the open right half plane. But then the compactness of $\mathcal{L}_\tau$ and the simplicity of the eigenvalue $c(\tau)^{-1}$ of $\hat{\mathcal{L}}_\tau$ imply the analyticity of $\tau\to c(\tau)$ on a neighborhood of $(0,\infty)$, which proves $\tau\to c(\tau)$ to be a C^∞-function on $(0,\infty)$. Since obviously the spectral radius of $\mathcal{L}_\infty$ is unity, we must have $\lim_{\tau\to\infty} c(\tau)=1$, while $\lim_{\tau\to 0} c(\tau)=\infty$.

The monotonicity property can be used to derive the following result.

THEOREM 3.5. Let $a_1 \in L_r[-1,1]$ with $r>1$, $1\le p<\infty$ and $1\le q<\infty$. Then for $0<c<1$ the Wiener-Hopf equation

$$\psi(x) - c\int_0^\infty \mathcal{H}(x-y)B\psi(y)dy = \varsigma(x), \qquad 0<x<\infty, \tag{3.14}$$

where $\varsigma(x)$ is nontrivial and nonnegative, is uniquely solvable on all spaces $L_q(L_p(\Omega))_0^\infty$ and $C(L_p(\Omega))_0^\infty$, and the solution is nonnegative and given by the Neumann series

$$\psi(x) = \sum_{n=0}^\infty c^n(\mathcal{L}_\infty^n\varsigma)(x), \qquad 0<x<\infty.$$

For $0<c\le 1$, Theorems 3.4 and 3.5 yield the nonnegativity of the solution of the equation of transfer of unpolarized light, where reflection by the planetary surface is neglected (see also Section 1). For $\varsigma(x)\equiv 0$ and $c=1$, Eq. (3.14) does not have nonzero solutions in $L_q(L_p(\Omega))_0^\infty$ or in $C(L_p(\Omega))_0^\infty$, and therefore the spectral radius $c=1$ of $\mathcal{L}_\infty$ is not an eigenvalue.

We conclude this section with some historical remarks. For the existence and uniqueness theory for $0<c\le 1$ and its history, also relevant to radiative transfer, we refer to Section 1. Many of the prerequisities for the present results on critical eigenvalues and the non-existence of positive solutions for $c>1$ can be found in the monograph of Maslennikov [259]. Some of the ideas of using positivity arguments in order to study criticality go back to Mullikin [270]. For neutron transport in multiplying media, spectral results for the isotropic one speed equation were provided by Ball and Greenberg [21], and the boundary value problem in an abstract setting was studied by Greenberg and van der Mee [163] and Greenberg and Walus [167, 386].

4. Multigroup neutron transport

In the present section we shall analyze the neutron transport equation under the same physical assumptions as in the previous section, but now we shall drop the hypothesis of constant neutron speed. We shall retain the remaining physical assumptions, such as spatial homogeneity. As a result of plane symmetry we have the transport equation

$$v\mu\frac{\partial\psi}{\partial x}(x,v,\omega) + v\sigma(v)\psi(x,v,\omega) = \int_0^\infty \int_\Omega \hat{v}\sigma(v,\hat{v},\omega\cdot\hat{\omega})\psi(x,\hat{v},\hat{\omega})d\hat{\omega}d\hat{v} + q(x,v,\omega),$$

where $x\epsilon(0,\tau)$ is a position variable, μ the direction cosine of propagation, v the speed and $\omega\epsilon\Omega$ the angular direction of the neutron. The angular density $\psi(x,v,\omega)$ now also is a function of the neutron speed v. In the **multigroup approximation** the speed variable v is discretized. The complete speed interval, which usually is assumed bounded below and above, is written as the disjoint union of finitely many, N, subintervals on which the speed, the macroscopic cross section, the average number of secondaries per collision and the collision kernel $\sigma(\hat{v},v,\omega.\hat{\omega})$ are assumed constant. The functional dependence on the speed variable is now denoted through the lower indices i and j ranging from 1 to N. As a result we obtain the transport equation

$$\mu\frac{\partial\psi_i}{\partial x}(x,\omega) + \sigma_i\psi_i(x,\omega) =$$
$$= (1/4\pi)\sum_{j=1}^{N} c_{ij}\int_\Omega \mathbf{p}_{ij}(\omega.\hat{\omega})\psi_j(x,\hat{\omega})d\hat{\omega} + q_i(x,\omega), \tag{4.1}$$

where $\mathbf{p}_{ij}(t)\geq 0$ and $\frac{1}{2}\int_{-1}^{1}\mathbf{p}_{ij}(t)=1$. The **one speed approximation** then arises if N=1. In the case of isotropic scattering we have $\mathbf{p}_{ij}(t)\equiv 1$ for all i,j and the transport equation reduces to the equation

$$\mu\frac{\partial\psi}{\partial x}i(x,\omega) + \sigma_i\psi_i(x,\omega) = (1/4\pi)\sum_{j=1}^{N} c_{ij}\int_\Omega \psi_j(x,\hat{\omega})d\hat{\omega} + q_i(x,\omega). \tag{4.2}$$

If we measure the distance in units of the largest mean free path among the N groups and order the N groups according to increasing mean free path, we obtain $\sigma_1>\sigma_2>\sigma_3>...>\sigma_N=1$. The number c_{ij} then denotes the average number of secondaries per collision for neutrons incident from group j and scattered to group i, and therefore $c_{ij}\geq 0$. If one defines the N×N-matrices $\mathbf{\Sigma}$, $\mathbf{C}$, $\mathbf{P}(\omega.\hat{\omega})$ and the N-vector-valued functions $\boldsymbol{\psi}(x,\omega)$ and $\mathbf{q}(x,\omega)$ by

$$\mathbf{\Sigma} = \mathrm{diag}(\sigma_1,...,\sigma_N), \tag{4.3a}$$

$$[\mathbf{C}]_{ij} = c_{ij}, \tag{4.3b}$$

$$[\mathbf{P}(\omega.\hat{\omega})]_{ij} = p_{ij}(\omega.\hat{\omega}), \tag{4.3c}$$

$$[\boldsymbol{\psi}(x,\omega)]_i = \psi_i(x,\omega), \tag{4.3d}$$

$$[\mathbf{q}(x,\omega)]_i = q_i(x,\omega), \tag{4.3e}$$

one obtains the multigroup neutron transport equation in vector form,

$$\mu\frac{\partial\boldsymbol{\psi}}{\partial x}(x,\omega) + \mathbf{\Sigma}\boldsymbol{\psi}(x,\omega) = (1/4\pi)\int_\Omega \mathbf{C}\otimes\mathbf{P}(\omega.\hat{\omega})\boldsymbol{\psi}(x,\hat{\omega})d\hat{\omega} + \mathbf{q}(x,\omega), \tag{4.4}$$

where $[\mathbf{A}\otimes\mathbf{B}]_{ij}=[\mathbf{A}]_{ij}[\mathbf{B}]_{ij}$ is the tensor product of the matrices $\mathbf{A}$ and $\mathbf{B}$. As boundary conditions we impose those specifying the angular densities in each group for neutrons incident at the boundaries of the medium, i.e.,

$$\boldsymbol{\psi}(0,\omega) = \boldsymbol{\varphi}(\omega), \quad \mu>0, \tag{4.5a}$$

$$\boldsymbol{\psi}(\tau,\omega) = \boldsymbol{\varphi}(\omega), \quad \mu<0, \tag{4.5b}$$

for $\tau<\infty$, and

$$\boldsymbol{\psi}(0,\omega) = \boldsymbol{\varphi}_+(\omega), \quad \mu>0, \tag{4.6a}$$

$$\lim_{x\to\infty} \|\boldsymbol{\psi}(x,\cdot)\| = 0, \tag{4.6b}$$

for $\tau=\infty$. For physical considerations, $[\varphi(\omega)]_i$, $[\varphi_+(\omega)]_i$, $[\mathbf{q}(x,\omega)]_i$ and $[\psi(x,\omega)]_i$ must be nonnegative, since all of these quantities represent angular densities for neutrons within group i.

Let us introduce the functional formulation. For $1\leq p\leq\infty$, we let H_p^N denote the Banach space of all measurable functions $\mathbf{h}:\Omega\to\mathbb{C}^N$ endowed with the weighted L_p-norms

$$\|\mathbf{h}\|_p = [\sum_{i=1}^{N} \sigma_i \int_\Omega |h_i(\omega)|^p d\omega]^{1/p}, \quad 1\leq p<\infty,$$

and

$$\|\mathbf{h}\|_\infty = \max_{1\leq i\leq N} \sigma_i \{\operatorname*{ess\,sup}_{\omega\in\Omega} |h_i(\omega)|\},$$

We define the operators T, B=I−A, Q_+, Q_- and J by

$$(T\mathbf{h})(\omega) = \Sigma^{-1}\mathbf{h}(\omega),$$

$$(B\mathbf{h})(\omega) = (1/4\pi)\Sigma^{-1}\int_\Omega \mathbf{C}\otimes\mathbf{P}(\omega.\hat{\omega})\mathbf{h}(\hat{\omega})d\hat{\omega},$$

$$(J\mathbf{h})(\omega) = \mathbf{h}(-\omega),$$

$$(Q_\pm\mathbf{h})(\omega) = \begin{cases} h(\omega), & \pm\mu>0, \\ 0, & \pm\mu<0, \end{cases}$$

and the vector-valued functions $\psi(x)$ and $\mathbf{q}(x)$ by

$$\psi(x)(\omega) = \psi(x,\omega),$$

$$\mathbf{q}(x)(\omega) = \mathbf{q}(x,\omega).$$

We then obtain the boundary value problems (4.4)–(4.5) for finite slabs and (4.6)–(4.6) for half spaces. The operator J is an invertible isometry satisfying the properties (1.29a)–(1.29d) of an inversion symmetry. On H_2^N the operators T, $Q_\pm$ and J are self adjoint, T has absolutely continuous spectrum [−1,1] and Q_+ and Q_- are the orthogonal projections onto maximal T-positive and T-negative T-invariant subspaces. On H_p^N, $1\le p<\infty$, T is a scalar-type spectral operator (cf. [105, 109]). As a consequence of Lemma 1.1, B is a compact operator on H_p^N. Moreover, denoting

$$[\mathbf{Y}^m_{\ell,k}]_i(\theta,\varphi) = Y^m_\ell(\theta,\varphi)\delta_{ik}, \quad \ell\ge 0, \quad \omega=(\theta,\varphi),$$

we have

$$[(B\mathbf{Y}^m_{\ell,k})(\omega)]_i = (2\ell+1)^{-1}\beta_\ell^{i,k}[\Sigma^{-1}C]_{ik}Y^m_\ell(\omega),$$

where

$$\beta_\ell^{i,j} = (\ell+\tfrac{1}{2})\int_{-1}^{1} p_{ij}(t)P_\ell(t)dt,$$

$$[\mathbf{B}_\ell]_{ij} = \beta_\ell^{i,j}.$$

In matrix form this reads

$$B\mathbf{Y}^m_{\ell,k} = (2\ell+1)^{-1}\sum_{i=1}^{N}\beta_\ell^{i,k}[\Sigma^{-1}C]_{ik}\mathbf{Y}^m_{\ell,i}.$$

Thus the eigenvalues of the compact operator B are the eigenvalues $\lambda_{\ell,1}$, $\lambda_{\ell,2},\dots,\lambda_{\ell,N}$ of the matrices $(2\ell+1)^{-1}\Gamma^\ell$, $\ell=0,1,2,\dots$, with

$$[\Gamma^\ell]_{ij} = \beta_\ell^{i,j}[\Sigma^{-1}C]_{ij} = [\mathbf{B}_\ell\otimes\Sigma^{-1}C]_{ij}.$$

The corresponding eigenvectors $\Xi^m_\ell(\omega)$ are given by

$$[\Xi^m_{\ell,s}(\omega)]_i = [\boldsymbol{\xi}_{\ell,s}]_i Y^m_\ell(\theta,\varphi),$$

where $1\le s\le N$, $\omega=(\theta,\varphi)$ and $\Gamma^\ell\boldsymbol{\xi}_{\ell,s}=\lambda_{\ell,s}\boldsymbol{\xi}_{\ell,s}$. In this way one indicates a complete set of eigenvectors of B in H_p^N, where "completeness" signifies that they span a dense subset of H_p^N for $1\le p<\infty$. It should be observed that $\lambda_{\ell,s}$ and $\boldsymbol{\xi}_{\ell,s}$ do not depend on m. In the case of isotropic scattering we have $\beta_0^{i,j}=1$, $\beta_\ell^{i,j}=0$, $\ell\ge 1$, so that the (nonzero) eigenvalues of B are the eigenvalues $\lambda_{0,1},\dots,\lambda_{0,N}$ of the matrix $\Sigma^{-1}C$ with corresponding eigenvectors $\Xi^0_{0,s}(\omega) = \boldsymbol{\xi}_{0,s}$, where

$$\Sigma^{-1}\mathbf{C}\xi_{0,s} = \lambda_{0,s}\xi_{0,s}, \quad s=1,2,...,N.$$

For later use we introduce C^N as the Banach space of continuous functions $\mathbf{h}:\Omega\to\mathbb{C}^N$ with the norm inherited from H^N_∞.

The operators B and A are self adjoint only in the case of symmetric multigroup models, where $\mathbf{p}_{ij}(t) = \mathbf{p}_{ji}(t)$ and $[\mathbf{C}]_{ij} = [\mathbf{C}]_{ji}$, $1\le i,j\le N$. Generally, scattering processes from groups having fast neutrons tend to dominate scattering processes from groups having slow neutrons, so that in most realistic situations one has a nonsymmetric multigroup model.

In the case of isotropic scattering one studies the half space problem $(0\le x<\infty)$

$$\mu\frac{\partial\psi_i}{\partial x}(x,\mu) + \sigma_i\psi_i(x,\mu) = \tfrac{1}{2}\sum_{j=1}^{N} C_{ij}\int_{-1}^{1}\psi_j(x,\hat{\mu})d\hat{\mu}, \quad -1\le\mu\le 1,$$

$$\psi_i(0,\mu) = \varphi_{+,i}(\mu), \quad 0\le\mu\le 1,$$

$$\int_{-1}^{1} |\psi_i(x,\mu)|^2 d\mu = O(1) \ (x\to\infty),$$

on the Hilbert space H of N-tuples $\mathbf{f}=\{f_i\}_{i=1}^N$ of square integrable functions $f_i:[-1,1]\to\mathbb{C}$, endowed with the inner product

$$(\mathbf{f},\mathbf{g}) = \sum_{i=1}^{N}\sigma_i\int_{-1}^{1} f_i(\mu)\overline{g_i(\mu)}d\mu.$$

Denoting the operators T, B and $Q_\pm$ by the expressions

$$(\mathbf{Tf})_i(\mu) = \sigma_i^{-1}\mu f_i(\mu),$$

$$(\mathbf{Bf})_i(\mu) = \tfrac{1}{2}\sigma_i^{-1}\sum_{j=1}^{N} C_{ij}\int_{-1}^{1} f_j(\hat{\mu})d\hat{\mu},$$

$$(Q_+\mathbf{f})_i(\mu) = \begin{cases} f_i(\mu), & \mu>0, \\ 0, & \mu<0, \end{cases}$$

we introduce also the resolution of the identity $\sigma(\tau)$ associated with T, given by

$$(\sigma(\tau)\mathbf{f})_i(\mu) = \begin{cases} f_i(\mu/\sigma_i), & \mu\in\tau\cap[-1,1], \\ 0, & \mu\notin\tau\cap[-1,1]. \end{cases}$$

We shall assume the existence of an albedo operator E such that $(E\varphi_+)_i(\mu) = \psi_i(0,\mu)$, $1\leq i\leq N$, where $\varphi_+ = \{\varphi_{+,i}\}_{i=1}^N$. This occurs, for instance, if $\Sigma^{-1}C$ has norm less than one, since in this case the operator B has norm less than one (cf. Theorem VII 3.4).

Let $\mathbf{e}_1,\ldots,\mathbf{e}_N$ be the orthogonal system of vectors in H, where

$$[\mathbf{e}_\ell]_i(\mu) = \delta_{i\ell}, \quad -1\leq\mu\leq1.$$

Then $\mathbb{B} = \text{span}\{\mathbf{e}_1,\ldots,\mathbf{e}_N\} \supset \text{Ran } B^*$. If j denotes the natural imbedding of $\mathbb{B}$ into H, then

$$\pi\mathbf{f} = \tfrac{1}{2}\sum_{\ell=1}^N \sigma_\ell^{-1}(\mathbf{f},\mathbf{e}_\ell)\mathbf{e}_\ell = \tfrac{1}{2}\sum_{\ell=1}^N \left(\int_{-1}^1 f_\ell(\hat{\mu})d\hat{\mu}\right)\mathbf{e}_\ell$$

is the orthogonal projection of H onto $\mathbb{B}$.

Let us compute the dispersion function. We find

$$[\pi\sigma(dt)Bj\mathbf{e}_k]_i(\mu) = \tfrac{1}{2}\sigma_i^{-1}C_{ik}dt, \quad dt\subset[-1,1],$$

whence

$$[\Lambda(z)\mathbf{e}_k]_i(\mu) = \delta_{ik} - \tfrac{1}{2}z\sigma_i^{-1}C_{ik}\int_{-1}^1 \frac{dt}{z-t}, \quad z\notin[-1,1].$$

Let us assume the existence of an albedo operator E and a corresponding factorization

$$\Lambda(z)^{-1} = \mathbf{H}_\ell(-z)\mathbf{H}_r(z)$$

for Re z=0. From Eq. (1.8) we now easily derive this albedo operator in the form

$$(E\varphi_+)_i(\mu) = \varphi_{+,i}(\mu)$$

for $0\leq\mu\leq1$, and

$$(E\varphi_+)_i(\mu) = \frac{1}{\sqrt{\pi}}\int_0^\infty \frac{\nu}{\nu-\mu}\Sigma^{-1}\mathbf{C}\mathbf{H}_\ell(-\mu)\mathbf{H}_r(\nu)\mathrm{col}[\varphi_{+,s}(\frac{\nu}{\sigma_s})]_{s=1}^N d\nu$$

for $-1\le\mu<0$, where $\mathrm{col}[\varphi_s]_{s=1}^N$ is the column vector with entries $\varphi_1,\dots,\varphi_N$. We also find the coupled set of H-equations

$$\mathbf{H}_\ell(z)^{-1} = \mathbf{I} - \tfrac{1}{2}z\int_0^1 (z+t)^{-1}\mathbf{H}_r(t)\Sigma^{-1}\mathbf{C}dt,$$

$$\mathbf{H}_r(z)^{-1} = \mathbf{I} - \tfrac{1}{2}z\int_0^1 (z+t)^{-1}\Sigma^{-1}\mathbf{C}\mathbf{H}_\ell(t)dt.$$

Formulas of the above type have been derived previously by Burniston et al. [58] for two group equations, by Bowden et al. [57] for the general N-group equation, and by Kelley [216] in a somewhat more abstract setting. A spectral analysis of the multigroup transport operator was carried out by Greenberg [156] for symmetric scattering kernels, and by Kaper and Lekkerkerker [210] more generally.

Using the theory of Chapter VI one may prove the finite slab and half space problems (4.4)-(4.5) and (4.4)-(4.6) to be equivalent to a convolution equation of the type

$$\psi(x) - \int_0^\tau \mathcal{H}(x-y)B\psi(y)dy = \varsigma(x), \quad 0<x<\tau, \tag{4.7}$$

where

$$[(\mathcal{H}(x)\mathbf{h})(\omega)]_i = \begin{cases}\sigma_i|\mu|^{-1}e^{-\sigma_i x/\mu}[\mathbf{h}(\omega)]_i, & x\mu>0,\\ 0, & x\mu<0,\end{cases}$$

is the propagator function and $\varsigma(x,\omega)$ is nonnegative whenever the incoming data vector $\varphi(\omega)$ (resp. $\varphi_+(\omega)$) and the internal source term $\mathbf{q}(x,\omega)$ are nonnegative. In fact, for finite slabs and uniformly Hölder continuous functions $\mathbf{q}:[0,\tau]\to H_p^N$, we have

$$[\varsigma(x,\omega)]_i = e^{-\sigma_i x/\mu}[\varphi(\omega)]_i + \int_0^x \sigma_i\mu^{-1}e^{-\sigma_i(x-y)/\mu}[\mathbf{q}(y,\mu)]_i dy, \quad \mu>0,$$

and

$$[\varsigma(x,\omega)]_i = e^{\sigma_i(\tau-x)/\mu}[\varphi(\omega)]_i - \int_x^\tau \sigma_i\mu^{-1}e^{-\sigma_i(x-y)/\mu}[\mathbf{q}(y,\mu)]_i dy, \quad \mu<0.$$

For half space and bounded uniformly Hölder continuous $\mathbf{q}:[0,\infty)\to H_p^N$ satisfying $\int_1^\infty x\|\mathbf{q}(x)\|_p dx<\infty$ we have a similar formula, where τ, $\varphi(\omega)$ for $\mu>0$ and $\varphi(\omega)$ for $\mu<0$ are replaced by ∞, $\varphi_+(\omega)$ and 0, respectively. The equivalence of Eq. (4.7) and the corresponding boundary value problem holds true if $\mathbf{p}_{ij}\in L_r[-1,1]$ for some $r>1$ and all $1\le i,j\le N$, since in this case $B[H_p^N]\subset|T|^\alpha[H_p^N]$ for $0<\alpha<(r-1)/pr$ (cf. Lemma 1.3). As in Section VI.2, we now introduce the Banach spaces $L_q(H_p^N)_a^b$ of strongly measurable functions $\psi:(a,b)\to H_p^N$ such that $\|\psi(\cdot)\|_p\in L_q(a,b)$, and the Banach space $C(H_p^N)_a^b$ of continuous functions from $[a,b]$ into H_p^N, endowed with the L_q-norm and the supremum norm, respectively. We define

$$(\mathcal{L}_\tau\psi)(x) = \int_0^\tau \mathcal{H}(x-y)B\psi(y)dy, \quad 0<x<\tau, \tag{4.8}$$

which is a bounded, and if τ is finite, a compact operator on $L_q(H_p^N)_0^\tau$ and on $C(H_p^N)_0^\tau$, provided $\mathbf{p}_{ij}\in L_r[-1,1]$ for some $r>1$ and all $1\le i,j\le N$. The convolution equation (4.7) may then be written in the vector form $(I-\mathcal{L}_\tau)\psi = \varsigma$.

THEOREM 4.1. Let $1\le p<\infty$, and let $\mathbf{p}_{ij}\in L_r[-1,1]$ for some $r>1$ and all $1\le i,j\le N$. Then either the spectral radius $r(\mathcal{L}_\tau)$ always vanishes, or it is a continuous, strictly monotonically increasing function from $(0,\infty)$ onto $(0,r(\mathcal{L}_\infty))$, which is a C^∞-function except at a discrete set of algebraic branch points $\tau_i\in(0,\infty)$. In particular, if the norm of B does not exceed (resp., is less than) unity, the convolution equation (4.7) is uniquely solvable on the spaces $L_q(H_p^N)_0^\tau$ and $C(H_p^N)_0^\tau$ for finite (resp., infinite) τ and the unique solution $\psi(x,\omega)$ is nonnegative whenever the right hand side $\varsigma(x)$ is nonnegative.

Proof: For finite τ we introduce the operator

$$(\hat{\mathcal{L}}_\tau\varphi)(x) = \tau\int_0^1 \mathcal{H}(\tau(x-y))B\varphi(y)dy, \quad 0<x<1,$$

on the spaces $L_q(H_p^N)_0^1$ and $C(H_p^N)_0^1$, which is similar to $\mathcal{L}_\tau$, depends analytically on τ on the open right half plane and is compact. The spectral radius $r(\hat{\mathcal{L}}_\tau)$ coincides with $r(\mathcal{L}_\tau)$ and is an eigenvalue of $\hat{\mathcal{L}}_\tau$, as a consequence of the positivity (in lattice sense) of $\hat{\mathcal{L}}_\tau$. Using analytic perturbation theory ([213], Theorem VI 1.8), we see that the eigenvalues of $\mathcal{L}_\tau$ (or $\hat{\mathcal{L}}_\tau$) are locally analytic functions of τ on the open right half plane, except for a discrete, finite or countably infinite, set of algebraic branch points. Hence, $r(\mathcal{L}_\tau)$ is a continuous function of τ on $(0,\infty)$, which is a

C^∞-function except for a discrete set of branch points.

If $r(\mathcal{L}_\tau)$ is constant and nonzero in a neighborhood of $\tau_0 \in (0,\infty)$, let $a,b \in (0,\infty)$ be the branch points of $r(\mathcal{L}_\tau)$ nearest to the left and right of τ_0, respectively. (If there are no branch points to the left or right, we take $a=0$ or $b=r(\mathcal{L}_\infty)$, respectively). Then $r(\mathcal{L}_\tau)$ is analytic on a neighborhood of (a,b) and therefore constant. Analytically continuing $r(\mathcal{L}_\tau)$ around a and b, we see that there are eigenvalues constant in τ on intervals $(a-\varepsilon,a]$ and $[b,b+\varepsilon)$ which cannot exceed $r(\mathcal{L}_\tau)$ on these intervals. Thus $r(\mathcal{L}_\tau)$ is constant in τ on intervals $a-\varepsilon<\tau\le a$ and $b\le\tau\le b+\varepsilon$. Since $r(\mathcal{L}_\tau)$ is monotonically nondecreasing in τ and $\lim_{\tau\to 0} r(\mathcal{L}_\tau)=0$, the continuity of $r(\mathcal{L}_\tau)$ as a function of τ implies the existence of $\delta>0$ such that $r(\mathcal{L}_\tau)<r(\mathcal{L}_{\tau_0})$ for $\tau\in(0,\delta)$. However, on going down from $\tau_0\in(0,\infty)$ until one reaches $(0,\delta)$, one passes through at most finitely many branch points. Thus $r(\mathcal{L}_\tau)=r(\mathcal{L}_{\tau_0})$ for some $\tau\in(0,\delta)$, which is a contradiction. Hence, $r(\mathcal{L}_\tau)$ is continuous and strictly monotonically increasing in τ.

Finally, as the spectral radius of $\mathcal{L}_\infty$ does not exceed the norm of B (cf. Theorem VII 3.4), we have $r(\mathcal{L}_\tau)<\|B\|$ for finite τ. Hence, if $\|B\|<1$ or if $\|B\|\le 1$ and $\tau<\infty$, Eq. (4.7) is uniquely solvable on all spaces $L_q(H_p^N)_0^\tau$ and $C(H_p^N)_0^\tau$. A simple Neumann series expansion implies the positivity of this unique solution for nonnegative $\varsigma(x,\omega)$. ∎

In the previous section the notion of u_0-positivity of linear operators leaving invariant a cone in a Banach space was sufficient to derive all criticality type results. For multigroup neutron transport an extension of these ideas is needed, since the relevant operators are not always u_0-positive. We shall exploit irreducibility, i.e., the non-existence of closed invariant nontrivial ideals.

THEOREM 4.2. Let $1\le p<\infty$, and let $\mathbf{p}_{ij}\in L_r[-1,1]$ for some $r>1$ and all $1\le i,j\le N$. Then $\mathcal{L}_\tau$ is irreducible on at least one (and hence all) of the Banach spaces $L_q(H_p^N)_0^\tau$ and $C(H_p^N)_0^\tau$, if and only if B is irreducible on H_p^N.

Proof: If B is reducible on H_p^N, there exists a closed ideal I in H_p^N satisfying $\{0\}\ne I\ne H_p^N$ and $B[I]\subset I$. However, as H_p^N is isomorphic to $L_p(\Omega\oplus\ldots\oplus\Omega)$, where N copies of Ω have been taken, there exist subsets D_1 and D_2 of positive measure of the direct sum $\Omega\oplus\ldots\oplus\Omega$ such that $D_1\cap D_2=\phi$, $D_1\cup D_2=\Omega\oplus\ldots\oplus\Omega$, $B[L_p(D_1)]\subset L_p(D_1)$. Now consider the closed ideal I_q of $L_q(H_p^N)_0^\tau$ consisting of all functions $\psi:(0,\tau)\to L_p(D_1)$ in this space. Then $B\psi(y)\in L_p(D_1)$ for $y\in(0,\tau)$ and therefore, using the diagonal nature of the propagator function, one obtains $\mathcal{H}(x-y)B\psi(y)\in L_p(D_1)$, whence $\mathcal{L}_\tau\psi\in I_q$. A similar argument applies to the space $C(H_p^N)_0^\tau$. Thus $\mathcal{L}_\tau$ is reducible on $L_q(H_p^N)_0^\tau$ and $C(H_p^N)_0^\tau$.

Conversely, let $\mathcal{L}_\tau$ be reducible on $L_q(H_p^N)_0^\tau$. If I_q is a closed ideal in $L_q(H_p^N)_0^\tau$ such that $\mathcal{L}_\tau[I_q]\subset I_q$ and $\{0\}\neq I_q\neq L_q(H_p^N)_0^\tau$, put $I_p=I_q\cap L_p(H_p^N)_0^\tau$ if $q\geq p$, and let I_p be the closure of I_q in $L_p(H_p^N)_0^\tau$ if $q\leq p$. Then I_p is a closed ideal in $L_p(H_p^N)_0^\tau$ such that $\mathcal{L}_\tau[I_p]\subset I_p$ and $\{0\}\neq I_p\neq L_p(H_p^N)_0^\tau$. However, $L_p(H_p^N)_0^\tau$ can be identified with $L_p((\Omega\oplus\ldots\oplus\Omega)\times(0,\tau))$, where N copies of Ω have been considered. Thus there exist subsets E_1 and E_2 of $(\Omega\oplus\ldots\oplus\Omega)\times(0,\tau)$ of positive measure such that $E_1\cap E_2=\phi$, $E_1\cup E_2=(\Omega\oplus\ldots\oplus\Omega)\times(0,\tau)$ and $\mathcal{L}_\tau[L_p(E_1)]\subset L_p(E_1)$.

Consider the closed ideal $I=\{\mathbf{h}\in I_p : \mathbf{h}$ does not depend on $x\}$ of H_p^N. For constant $\psi(x)\equiv\mathbf{h}$ we have

$$[Q_+-e^{-xT^{-1}}Q_+]B\mathbf{h} + [Q_--e^{(\tau-x)T^{-1}}Q_-]B\mathbf{h} = \int_0^\tau \mathcal{H}(x-y)B\mathbf{h}dzy \in I_p,$$

whence $B\mathbf{h}\in I$ as a consequence of the fact that

$$F(x) = [Q_+-e^{-xT^{-1}}Q_+] + [Q_--e^{(\tau-x)T^{-1}}Q_-]$$

is a positive multiplication operator. Hence, I is a closed ideal of H_p^N satisfying $B[I]\subset I$. The same conclusion, with the same ideal I as a result, is obtained by considering $C(H_p^N)_0^\tau$.

Next observe that an invariant closed ideal I_q cannot lead to a set $E_1\subset(\Omega\oplus\ldots\oplus\Omega)\times Z$ with Z and $(0,\tau)\setminus Z$ having positive measure, since on choosing nonnegative ψ from this ideal one gets

$$(\mathcal{L}_\tau\psi)(x) = \int_0^\tau \mathcal{H}(x-y)B\psi(y)dy = 0$$

for $x\notin Z$ and therefore $B\psi(y)\equiv 0$. But if Ker B contained nonzero nonnegative vectors, B would be reducible. Otherwise, $\psi(y)\equiv 0$, which contradicts $I_q\neq\{0\}$.

Now, if $I=H_p^N$, I_q will contain all constant functions. Then I_∞ will contain all functions in $L_\infty(H_p^N)_0^\tau$ and therefore $I_q=L_q(H_p^N)_0^\tau$, which is a contradiction. Thus $I\neq H_p^N$, and B is reducible. ∎

It is immediate from the above proof that the invariant closed ideals of $\mathcal{L}_\tau$

in $L_q(H_p^N)_0^\tau$ and $C(H_p^N)_0^\tau$ have the form $L_p(D_1)\oplus\ldots\oplus L_p(D_N)$, where $D_1,\ldots,D_N$ are measurable subsets of Ω, and thus it is clear that there is a one-to-one correspondence between the closed invariant ideals of B across different spaces H_p^N, where $1\le p<\infty$. In particular, if B is irreducible on one of the spaces H_p^N, with $1\le p<\infty$, then $\mathcal{L}_\tau$ is irreducible on all of the spaces $L_q(H_p^N)_0^\tau$ and $C(H_p^N)_0^\tau$.

Since the spaces $L_q(H_p^N)_0^\tau$ and $C(H_p^N)_0^\tau$ are Banach lattices and $\mathcal{L}_\tau$ is compact if τ is finite, we may apply Theorem I 4.5 and obtain the following result.

THEOREM 4.3. Let $1\le p<\infty$, and let $\mathbf{p}_{ij}\in L_r[-1,1]$ for some $r>1$ and $1\le i,j\le N$. Then for finite τ, and for $0<c_1<c_2<\ldots$ the finite or countably infinite set of numbers c for which $(I-c\mathcal{L}_\tau)\mathbf{h}=0$ has a nonzero positive solution, the convolution equation

$$\psi(x) - c\int_0^\tau \mathcal{H}(x-y)B\psi(y)dy = \varsigma(x), \quad 0<x<\tau,$$

where $c\ge0$ and $\varsigma\in L_q(H_p^N)_0^\tau$ is nonnegative, has a nonnegative solution $\mathbf{h}$, if and only if

$$\int_0^\tau (\varsigma(x),|\varphi(x)|)dx = 0$$

for every $\varphi\in L_{q'}(H_{p'}^N)_0^\tau$ such that $(i-c_i\mathcal{L}_\tau^*)^n\varphi=0$ for some $c_i\le c$ and $n\in\mathbb{N}$. Here $q'=q/(q-1)$, $p'=p/(p-1)$ and

$$(\mathcal{L}_\tau^*\varphi)(x) = \int_0^\tau B^*\mathcal{H}(y-x)\varphi(y)dy.$$

In particular, if $0\le c<c_1$, Eq. (4.10) has a unique solution, which is nonnegative.

Let us consider the case in which B is irreducible. Then $\mathcal{L}_\tau$ is irreducible on all Banach spaces of interest. The peripheral spectrum, which is the set $\{t\in\sigma(\mathcal{L}_\tau) : |t|=r(\mathcal{L}_\tau)\}$, consists of the (algebraically) simple eigenvalues $\varepsilon^k r(\mathcal{L}_\tau)$, $k=1,2,\ldots,m$, where ε is a primitive mth root of unity ([324], Theorem V 4.4). In that case $r(\mathcal{L}_\tau)$ can be continued to an analytic function on the open right half plane and therefore $r(\mathcal{L}_\tau)$ is a C^∞-function of τ. The corresponding eigenfunction is nonnegative and there are no other eigenvalues of $\mathcal{L}_\tau$ to which correspond nonnegative eigenfunctions ([324], Theorem V 4.2 and part (ii) of its corollary). As a result of Theorem 4.4 we have the following corollary.

COROLLARY 4.4. Let $1 \le p < \infty$, and let $p_{ij} \in L_r[-1,1]$ for some $r>1$ and all $1 \le i,j \le N$. Let B be irreducible. Then, for finite τ and nonnegative $\varsigma \in L_q(H_p^N)_0^\tau$, the convolution equation

$$\psi(x) - \int_0^\tau \mathcal{H}(x-y)B\psi(y)dy = \varsigma(x), \quad 0<x<\tau, \tag{4.9a}$$

has a unique solution which is nonnegative if $r(L_\tau)<1$, a one dimensional subspace of solutions if $r(L_\tau)=1$ and $\varsigma(x)\equiv 0$, and no nonnegative solutions if $r(L_\tau)=1$ and $\varsigma(x)\not\equiv 0$, or if $r(L_\tau)>1$.

In order to guarantee the absence of "competing" eigenvalues $\varepsilon^k r(\mathcal{L}_\tau)$ with $1 \le k \le m-1$, one could, for instance, require $\mathcal{L}_\tau$ to be u_0-positive. One may then prove that it is sufficient to demand that B be u_0-positive.

Finally, if $\tau=\infty$, then Corollary 4.4 can be generalized in the following fashion.

COROLLARY 4.5. Let $1 \le p < \infty$, and let $p_{ij} \in L_r[-1,1]$ for some $r>1$ and all $1 \le i,j \le N$. Let B be irreducible. Then for nonnegative $\varsigma \in L_q(H_p^N)_0^\infty$, the convolution equation

$$\psi(x) - \int_0^\infty \mathcal{H}(x-y)B\psi(y)dy = \varsigma(x), \quad 0<x<\infty, \tag{4.9b}$$

has a unique solution which is nonnegative if $\|B\|<1$, and does not have nonnegative solutions if $r(B)>1$.

Proof: According to Theorem 4.2, $r(\mathcal{L}_\infty) \le \|B\|$. Also, by Theorem 4.3, $\mathcal{L}_\infty$ is irreducible. Now choose finite τ with $r(\mathcal{L}_\tau)>1$. If Eq. (4.9b) has a nonnegative solution $\psi(x)$ while $r(B)>1$, write

$$\psi(x) - \int_0^\tau \mathcal{H}(x-y)B\psi(y)dy = \varsigma(x) + \int_\tau^\infty \mathcal{H}(x-y)B\psi(y)dy, \quad 0<x<\tau,$$

which contradicts the previous corollary and establishes the present result. ∎

Let us conclude with some general comments about these problems. First we should notice that we have not been able to prove in general that $r(\mathcal{L}_\infty)=r(B)$, which would considerably strengthen our results. This statement is known to hold true for many specific kinetic models. For an extensive class of multigroup equations on a slab with a collision operator satisfying a u_0-positivity condition, Borysiewicz and Mika [51, 52] argued that there is a unique critical eigenvalue parameter c which is independent of the L_p-setting. The unique solvability of the multigroup half space problem with isotropic scattering for $r(\Sigma^{-1}C)<1$ was established by Bowden et al. [57] in L_2 and Greenberg and Sancaktar [159] in L_p, $1<p<\infty$, using a factorization result published by Mullikin [273]. The latter results generalize previous two group results established by Siewert and Shieh [332] and Burniston et al. [58] Under the assumption of isotropic scattering and strict positivity of some iterate of the matrix $\Sigma^{-1}C$, Mullikin and Victory [275] have derived the above consequences of the irreducibility of B, such as Corollaries 4.4 and 4.5, and the strict continuous increase of the spectral radius of $\mathcal{L}_\tau$ as a function of τ. They also carried out a bifurcation analysis and obtained an asymptotic expansion for the critical eigenvalue.

5. The Boltzmann equation and BGK equation in rarefied gas dynamics

The distribution of molecules of a gas can be described by the Boltzmann equation. The equilibrium distribution, which is the Maxwell velocity distribution, is one of the solutions of this nonlinear integrodifferential equation. There are a variety of possible approaches to obtain a linearized equation, depending upon which aspects of the gas dynamics are to be maintained in the linearized model. Linearized Boltzmann equations will be the topic of the present and the next section.

Much of this section will be devoted to the linearized BGK equation, which is the linearized Boltzmann equation arising from the BGK model of the nonlinear Boltzmann equation. This model was first proposed by Bhatnagar, Gross and Krook [43], and Welander [388]. If we write the nonlinear Boltzmann equation as

$$\frac{\partial f}{\partial t} + \boldsymbol{\xi} \cdot \frac{\partial f}{\partial \mathbf{x}} = Q(f,f), \tag{5.1}$$

where t denotes time, $\boldsymbol{\xi}$ velocity and $\mathbf{x}$ position, and Q(f,f) is the nonlinear collision term, then the BGK model consists of replacing the quadratically nonlinear collision term Q(f,f) by a different, but still nonlinear, collision term J(f), which retains some of the qualitative and average properties of Q(f,f). In particular, the collision term J(f) has the special form

$$J(f) = \gamma[\Phi(\boldsymbol{\xi}) - f(\boldsymbol{\xi})], \tag{5.2}$$

where the collision frequency γ may depend on t and $\mathbf{x}$, and $\Phi(\boldsymbol{\xi})$ is the equilibrium (Maxwellian) distribution with the same density, velocity and temperature as the gas having the distribution function $f(\boldsymbol{\xi})$. Thus J(f) expresses the tendency of the gas to approach equilibrium. The nonlinearity of the collision term is exhibited by the fact that the density, velocity and temperature parameters in $\Phi(\boldsymbol{\xi})$ are themselves functions of f. (For a discussion of some of the mathematical intricacies of the BGK equation, cf. [157].)

The BGK collision term has the property

$$\int \gamma_\alpha J(f) d\boldsymbol{\xi} = 0, \quad \alpha=0,1,2,3,4, \tag{5.3}$$

where the functions γ_α are the (elementary) collision invariants $\gamma_0=1$, $(\gamma_1,\gamma_2,\gamma_3) = \boldsymbol{\xi}$, $\gamma_4=|\boldsymbol{\xi}|^2 = (\gamma_1)^2+(\gamma_2)^2+(\gamma_3)^2$. As a matter of fact, in the unmodified Boltzmann equation one has

$$\int \gamma_\alpha Q(f,f) d\boldsymbol{\xi} = 0, \quad \alpha=0,1,2,3,4,$$

which represents conservation of mass, the three components of momentum, and energy, respectively.

To derive the linearized BGK equation, one makes the Ansatz $f=f_0(1+\varepsilon h)$ for f_0 a Maxwellian distribution, and obtains the integrodifferential equation

$$\frac{\partial h}{\partial t} + \boldsymbol{\xi}\cdot\frac{\partial h}{\partial \mathbf{x}} = \nu\Big[\sum_{\alpha=0}^{4} \gamma_\alpha(\gamma_\alpha,h)-h\Big], \tag{5.4}$$

where the collision invariants are normalized by $(\gamma_\alpha,\gamma_\beta)=\delta_{\alpha\beta}$, $\alpha,\beta=0,1,2,3,4$, and

$$(g,h) = \int f_0(\boldsymbol{\xi})g(\boldsymbol{\xi})\overline{h(\boldsymbol{\xi})}d\boldsymbol{\xi}. \tag{5.5}$$

It can then easily be proved that

$$Lh = \nu\Big[\sum_{\alpha=0}^{4} (\gamma_\alpha,h)-h\Big] \tag{5.6}$$

is symmetric with respect to the inner product (5.5) and satisfies $(h,Lh)\leq 0$ and $\mathrm{Ker}\, L = \{\sum_{\alpha=0}^{4} c_\alpha \gamma_\alpha : c_\alpha \in \mathbb{C}\}$.

The theory of (linearizations of) the Boltzmann equation in rarefied gas dynamics has been treated in the two monographs of Cercignani [83, 84]. In particular, these monographs discuss the BGK model, the linearization procedure and the application of the Case–van Kampen eigenfunction method to the linearized BGK equation. The physical background in the kinetic theory of gases is discussed in the books of Kogan [222], Chapman and Cowling [92] and Ferziger and Kaper [121].

In the present section we discuss the existence and uniqueness theory for a variety of time independent BGK equations. Let us define $(2RT_0)^{1/2}$ as the unit of speed and $\nu^{-1}(2RT_0)^{1/2}$ as the unit of length, where T_0 is the temperature of the gas at equilibrium and R is the constant appearing in Boyle's law for ideal gases, and let us adopt one dimensional geometry with invariance of the macroscopic properties of the gas on rotation about the x–axis and symmetry with respect to the y–z plane, i.e., $(\xi_1,h)=0$. Eq. (5.4) may then be written

$$\xi_1\frac{\partial\psi}{\partial x} + \psi(x,\xi_1) =$$

$$= \pi^{-1/2}\int_{-\infty}^{\infty}\{1 + 2(\xi_1^2-1/2)(\xi_1'^2-1/2) + 2\xi_1\xi_1'\}\psi(x,\xi_1')e^{-\xi_1'^2}\, d\xi_1'. \tag{5.7}$$

A variant of this equation was studied in Section III.3. Note that the normalization $2RT_0=1$ leads to the Maxwell flux $\exp\{-\xi^2\}$, differing from the convention of Section III.4, but generally employed in rarefied gas dynamics.

Writing

$$h(x,\xi_1,\xi_2,\xi_3) = \psi_0(x,\xi_1) + (\xi_2^2+\xi_3^2-1)\psi_1(x,\xi_1) + 2\xi_2\psi_2(x,\xi_1) +$$

$$+ 2\xi_3\psi_3(x,\xi_1) + \psi_4(x,\xi_1,\xi_2,\xi_3),$$

one obtains (cf. [73, 83, 84]) the coupled system of equations

$$\xi_1 \frac{\partial\psi_0}{\partial x} + \psi_0(x,\xi_1) = \pi^{-1/2}\int_{-\infty}^{\infty}\psi_0(x,\xi_1')e^{-\xi_1'^2}d\xi_1' + 2/3\,\pi^{-1/2}(\xi_1^2-1/2)$$

$\times$

$$\times \{\int_{-\infty}^{\infty}(\xi_1'^2-1/2)e^{-\xi_1'^2}\psi_0(x,\xi_1')d\xi_1' + \int_{-\infty}^{\infty}e^{-\xi_1'^2}\psi_1(x,\xi_1')d\xi_1'\} \tag{5.8a}$$

and

$$\xi_1 \frac{\partial \psi_1}{\partial x} + \psi_1(x,\xi_1) = \tfrac{2}{3}\pi^{-1/2}\int_{-\infty}^{\infty} (\xi_1^{\,2}-\tfrac{1}{2})e^{-\xi_1^2}\,\psi_0(x,\xi_1)d\xi_1 + $$

$$+ \tfrac{2}{3}\pi^{-1/2}\int_{-\infty}^{\infty} e^{-\xi_1^2}\,\psi_1(x,\xi_1)d\xi_1, \tag{5.8b}$$

the uncoupled equations

$$\xi_1 \frac{\partial \psi_i}{\partial x} + \psi_i(x,\xi_1) = \pi^{-1/2}\int_{-\infty}^{\infty} e^{-\xi_1^2}\,\psi_i(x,\xi_1)d\xi_1, \tag{5.8c}$$

for i=2,3, and the trivial equation

$$\xi_1 \frac{\partial \psi_4}{\partial x} + \psi_4(x,\xi_1) = 0, \tag{5.8d}$$

whose solutions are the functions

$$\psi_4(x,\xi_1,\xi_2,\xi_3) = A(\boldsymbol{\xi})e^{-x/\xi_1} \tag{5.9}$$

with $A(\boldsymbol{\xi})$ orthogonal to the collision invariants γ_α, α=0,1,2,3, with respect to the inner product (5.5).

Equations (5.8a) and (5.8b) describe heat transfer effects and may be written

$$\xi \frac{\partial \psi}{\partial x} + \psi(x,\xi) = \pi^{-1/2}\int_{-\infty}^{\infty} \begin{bmatrix} 1+\tfrac{2}{3}(\xi^2-\tfrac{1}{2})(\xi^2-\tfrac{1}{2}) & \tfrac{2}{3}(\xi^2-\tfrac{1}{2}) \\ \tfrac{2}{3}(\xi^2-\tfrac{1}{2}) & 1 \end{bmatrix} \times$$

$$\times\ \psi(x,\xi)\exp\{-\xi^2\}d\xi, \tag{5.10}$$

where ψ is the vector $\{\psi_0,\psi_1\}$ and ξ_1 has been relabeled ξ. Eqs. (5.8c) model shear flow in the y- and z-direction, respectively. We will write them as

$$\xi \frac{\partial \psi}{\partial x} + \psi(x,\xi) = \pi^{-1/2}\int_{-\infty}^{\infty} e^{-\xi^2}\,\psi(x,\xi)d\xi. \tag{5.11}$$

In connection with the above equations a number of physically interesting boundary value problems can be considered. The Kramers problem or slip-flow problem consists of determining the velocity distribution of a gas filling the half space x>0 bounded by a physical wall and whose z-component of the mass velocity has a gradient along the x-axis tending to a constant as $x \to \infty$. Repeating the linearization of the nonlinear BGK equation about a Maxwellian with a translational velocity proportional to x in the z-direction, and also assuming diffuse reflection by the wall, one is led to the shear flow equation (5.11) for $x \in (0,\infty)$, with boundary conditions

$$\psi(0,\xi) = 0, \quad \xi \in (0,\infty), \tag{5.12a}$$

$$\lim_{x \to \infty} \{\psi(x,\xi)/x\} = -\hat{k}\xi, \tag{5.12b}$$

where $\hat{k}$ is a positive constant (cf. [73, 83, 84]).

Plane Poiseuille flow, i.e., the flow of a gas between two parallel plates at $x=0$ and $x=\tau$ induced by either a density gradient or a temperature gradient parallel to the plates in the z-direction, as well as plane Couette flow, i.e., the flow of a gas between parallel plates induced by moving them with opposite velocities parallel to the z-axis, can both be modeled by one of the equations (5.8c) to describe the shear flow effects or Eqs. (5.8a)-(5.8b) to describe the heat transfer effects, where $x \in (0,\tau)$ and, in the case of Poiseuille flow, an inhomogeneous term is added. The boundary value problem for shear flow then consists of the integrodifferential equation

$$\xi \frac{\partial \psi}{\partial x} + \psi(x,\xi) = \pi^{-1/2} \int_{-\infty}^{\infty} e^{-\xi^2} \psi(x,\xi) d\xi + q(x,\xi), \tag{5.13}$$

where $x \in (0,\tau)$, with boundary conditions

$$\psi(0,\xi) = \varphi(\xi), \quad \xi > 0, \tag{5.14a}$$

$$\psi(\tau,\xi) = \varphi(\xi), \quad \xi < 0. \tag{5.14b}$$

In the case of a Poiseuille flow one has $q(x,\xi)=\frac{1}{2}k$ for fixed $k>0$ and $\varphi(\xi)\equiv 0$; for Couette flow one has $q(x,\xi)\equiv 0$ and $\varphi(\xi)=\frac{1}{2}\mathrm{sgn}(\xi)U$, with U the speed of the moving plates.

In both the Kramers and the Couette/Poiseuille problems, it is possible to treat reflective boundary conditions. Introducing the scattering function $\Sigma(\xi' \to \xi)$, $\xi' < 0 < \xi$, satisfying the conditions (similar to the conditions III (4.2) - III (4.4), with attention to the difference in normalization of RT_0)

$$\Sigma(\xi\to\xi) \geq 0, \quad \xi<0<\xi, \tag{5.15}$$

$$\int_0^\infty \Sigma(\xi\to\xi)d\xi = 1, \quad \xi<0, \tag{5.16}$$

$$|\xi|e^{-\xi^2}\Sigma(\xi\to\xi) = |\xi|e^{-\xi^2}\Sigma(-\xi\to-\xi), \quad \xi<0<\xi, \tag{5.17}$$

one may replace (5.12a) for the Kramers problem by

$$\psi(0,\xi) =$$

$$= \alpha\psi(0,-\xi) + \beta\int_{-\infty}^0 |\xi/\xi|\exp\{\xi^2-\xi^2\}\Sigma(\xi\to\xi)\psi(0,\xi)d\xi, \tag{5.18a}$$

where $\xi\in(0,\infty)$ and α and β are accommodation coefficients satisfying $\alpha,\beta\geq 0$ and $\alpha+\beta\leq 1$. For $\alpha=0$ and $\beta=1$ one has diffuse reflection, and for $\alpha=1$ and $\beta=0$ (completely) specular reflection. In a similar way one may replace the boundary conditions (5.14b) for the Couette and Poiseuille problems by

$$\psi(\tfrac{1}{2}\tau\mp\tfrac{1}{2}\tau,\pm\xi) = \varphi(\pm\xi) + \alpha_\pm\psi(\tfrac{1}{2}\tau\mp\tfrac{1}{2}\tau,\mp\xi) +$$

$$+ \beta_\pm\int_{-\infty}^0 |\xi/\xi|\exp\{\xi^2-\xi^2\}\Sigma_\pm(\pm\xi\to\xi)\psi(\tfrac{1}{2}\tau\mp\tfrac{1}{2}\tau,\pm\xi)d\xi, \tag{5.18b}$$

where $\xi\in(0,\infty)$ and $\Sigma_\pm(\pm\xi\to\xi)$ are possibly distinct scattering functions satisfying the requirements (5.15)-(5.17).

Hitherto we have discussed linearized BGK equations for single species gases with one degree of freedom. For random mixtures of different species of gases the nonlinear BGK equation is somewhat more complicated. A linearization for binary gas mixtures was obtained by Cavalier and Greenberg [71]. One arrives at the coupled system of two integrodifferential equations

$$\xi\frac{\partial \mathbf{h}}{\partial x}(x,\xi) + \mathbf{\Sigma}\,\mathbf{h}(x,\xi) = \pi^{-1/2}\mathbf{D}\int_{-\infty}^\infty \mathbf{h}(x,\xi)\exp\{-\xi^2\}d\xi, \tag{5.19}$$

where

$$\mathbf{\Sigma} = \begin{bmatrix}\beta & O\\ 0 & \beta^*\end{bmatrix}, \tag{5.20}$$

$$\mathbf{D} = \begin{bmatrix} \beta-\alpha & (m/m^*)^{1/2}\alpha \\ (m^*/m)^{1/2}\alpha^* & \beta^*-\alpha^* \end{bmatrix}, \tag{5.21}$$

and a second copy of the same coupled system, as well as the coupled system of four integrodifferential equations

$$\xi \frac{\partial \mathbf{h}}{\partial x}(x,\xi) + \Sigma\mathbf{h}(x,\xi) =$$

$$= \pi^{-1/2} \int_{-\infty}^{\infty} [\mathbf{J}+\mathbf{Q}(\xi)\mathbf{L}(\xi)]\mathbf{h}(x,\xi)\exp\{-\xi^2\}d\xi, \tag{5.22}$$

where

$$\mathbf{J} = \begin{bmatrix} \beta & 0 & 0 & 0 \\ 0 & 0 & 0 & 0 \\ 0 & 0 & 0 & 0 \\ 0 & 0 & 0 & \beta^* \end{bmatrix}, \tag{5.23}$$

$$\mathbf{L}(\xi) = \begin{bmatrix} \xi & 0 & 0 & -(m/m^*)^{1/2}\xi \\ \xi^2-\frac{1}{2} & 1 & 0 & 0 \\ 0 & 0 & 1 & \xi^2-\frac{1}{2} \\ 0 & 0 & 0 & \xi \end{bmatrix} \tag{5.24}$$

and $\mathbf{Q}(\xi)$ is a matrix function with entries which are quadratic polynomials in ξ. Here m and m^* are the masses of the molecules and α, α^*, β, β^* are nonnegative parameters relating to their interaction. The various boundary value problems formulated above for a gas of single species can now be formulated for the binary system.

Cercignani [73] first obtained the solution of the Kramers problem (5.11)–(5.12) using the Case–van Kampen method of singular eigenfunction expansion. He extended the analysis to incompletely diffusing walls for the case $\beta=0$ [77], to plane Couette flow [76], and to plane Poiseuille flow [75], as well as to the analogous heat transfer problem [74]. Kriese, Chang, Siewert and Thomas solved the reflective boundary value problem (5.11)–(5.18) with $\beta=0$ [350], the Kramers problem [232] and two variations of the Kramers problem [333] for heat transfer using singular eigenfunction expansion. An application of the theory of generalized analytic functions to kinetic models with velocity dependent collision frequencies has been given by Cercignani [78, 79]. In the wake of developments toward more rigorous treatment of such problems, Kaper [207] solved the slip-flow problem for shear flow using the spectral method; the required eigenfunctions had been obtained also by Bowden and Garbanati [55] by resolvent integration. Using the abstract theory of Beals [32] a class of boundary value problems encompassing Couette and Poiseuille flow were solved by Kaper [208]. The solutions were expressed in analogs of the X and Y functions of radiative transfer by van der Mee [361]. Using resolvent integration Bowden and Cameron [54] analyzed the Kramers problem for heat transfer. On neglecting reflection, the existence and uniqueness results follow trivially from abstract kinetic equations theory (cf. [32, 160], for instance). When accounting for incompletely diffusing walls, these results follow from the theories given by Maslova [260] and van der Mee and Protopopescu [369], and from new results of Section V.4.

It was conjectured by Cercignani [85] that various BGK component equations, linearized about a drifting Maxwellian, would cease to have solutions if the drift velocity would reach or exceed a critical value corresponding to the speed of sound. This was subsequently proved by Arthur and Cercignani [15] using resolvent integration for an equation related to (5.7). Employing reduction to a Riemann–Hilbert problem and solving the latter explicitly, Siewert and Thomas obtained the same result as well as analytical solutions for this and similar equations with drift [334, 335]. Greenberg and van der Mee [162] established an abstract framework which allows one to derive non-existence results in terms of easily computable parameters for a large class of models including those above.

Cavalier and Greenberg [71] applied resolvent integration to the coupled equation (5.19) and obtained the solution in terms of Wiener–Hopf factors of a dispersion matrix. An equivalent expression for the solution at $x=0$ in terms of an explicit evaluation of the albedo operator will be derived shortly.

We shall now apply the abstract theory of Chapters III, V, and VII to study several of the models outlined above. Consider first the shear flow equation (5.11) or (5.13). Let us introduce the Banach space H_p of complex measurable functions on $(-\infty,\infty)$ bounded with respect to the weighted L_p-norm

$$\|h\|_p = \left[\pi^{-\frac{1}{2}}\int_{-\infty}^{\infty} |h(\xi)|^p \exp\{-\xi^2\} d\xi\right]^{1/p}, \quad 1\le p<\infty,$$

the operator

$$(Th)(\xi) = \xi h(\xi), \tag{5.25a}$$

where

$$D(T) = \{h \in H_p : \int_{-\infty}^{\infty} |\xi h(\xi)|^p \exp\{-\xi^2\} d\xi < \infty\},$$

the operators

$$(Q_\pm h)(\xi) = \begin{cases} h(\xi), & \pm\xi > 0, \\ 0, & \pm\xi < 0, \end{cases} \tag{5.25b}$$

and the vector $e(\xi) \equiv 1$. Let us also define the operators

$$(Bh)(\xi) = \pi^{-1/2} \int_{-\infty}^{\infty} h(\xi) \exp\{-\xi^2\} d\xi, \tag{5.25c}$$

$$(Jh)(\xi) = h(-\xi),$$

$$(R_\pm h)(\xi) = \alpha_\pm h(\xi) + \beta_\pm \int_{-\infty}^{\infty} |\hat{\xi}/\xi| \exp\{\xi^2 - \hat{\xi}^2)\} \Sigma(-\hat{\xi} \rightarrow \xi) h(\hat{\xi}) d\hat{\xi},$$

and $R = R_+Q_+ + R_-Q_-$, where Σ equals $\Sigma_\pm(\mp\hat{\xi} \rightarrow \pm\xi)$ in the finite medium situation and $\Sigma(\pm\hat{\xi} \rightarrow \mp\xi)$ for half spaces. We obtain the vector-valued differential equation

$$(T\psi)'(x) = -A\psi(x) + q(x), \qquad 0 < x < \tau \leq \infty. \tag{5.26}$$

For the Kramers problem we impose the boundary conditions

$$Q_+\psi(0) = RJQ_-\psi(0), \tag{5.27a}$$

$$\lim_{x \rightarrow \infty} \{\psi(x)/x\} = -\hat{k}e, \tag{5.27b}$$

$$q(x) \equiv 0. \tag{5.27c}$$

On the other hand, for Couette and Poiseuille flow, we impose the boundary conditions

$$Q_+\psi(0) = RJQ_-\psi(0) + Q_+\varphi, \tag{5.28a}$$

$$Q_-\psi(\tau) = RJQ_+\psi(\tau) + Q_-\varphi. \tag{5.28b}$$

One then easily obtains Ker A = span{e} = {constants}, and $Z_0(T^{-1}A) = \text{Ker}(T^{-1}A)^2$

= span $\{1,\xi\}$, where $(Te,e)=0$ in H_2. Let us introduce the Banach space $H_{p,T}$ of complex measurable functions on $(-\infty,\infty)$ bounded with respect to the norm

$$\|h\|_{p,T} = \left[\pi^{-\frac{1}{2}}\int_{-\infty}^{\infty} |\xi| \exp\{-\xi^2\} |h(\xi)|^p d\xi\right]^{1/p}, \tag{5.29}$$

where $1\leq p<\infty$, and the sesquilinear form

$$(h,k)_T = \pi^{-\frac{1}{2}}\int_{-\infty}^{\infty} |\xi| \exp\{-\xi^2\} h(\xi)\overline{k(\xi)} d\xi. \tag{5.30}$$

Then $H_{p,T}$ is the completion of $D(T)\subset H_p$ with respect to (5.30) and the operators T, $Q_\pm$, B, A and J can be extended from their restrictions on D(T) to $H_{p,T}$. For $h\in Q_+[H_{p,T}]$ and $k\in Q_+[H_{q,T}]$, $p^{-1}+q^{-1} = 1$, we estimate

$$|(Rh,k)_T| \leq$$

$$\leq \alpha_+\|h\|_{p,T} + \beta_+\pi^{-\frac{1}{2}}\int_0^{\infty}\int_0^{\infty} \xi \exp\{-\xi^2\}\Sigma(-\xi\to\xi) |h(\xi)\overline{k(\xi)}| d\xi d\xi \leq$$

$$\leq \alpha_+\|h\|_{p,T}\|k\|_{q,T} + \beta_+\pi^{-\frac{1}{2}}\left[\int_0^{\infty}\int_0^{\infty} \xi \exp\{-\xi^2\}\Sigma(-\xi\to\xi) |h(\xi)|^p d\xi d\xi\right]^{\frac{1}{p}} \times$$

$$\times \left[\int_0^{\infty}\int_0^{\infty} \xi \exp\{-\xi^2\}\Sigma(-\xi\to\xi) |k(\xi)|^q d\xi d\xi\right]^{\frac{1}{q}} \leq$$

$$\leq \alpha_+\|h\|_{p,T}\|k\|_{q,T} + \beta_+\pi^{-\frac{1}{2}}\left[\int_0^{\infty} \xi \exp\{-\xi^2\} |h(\xi)|^p d\xi\right]^{\frac{1}{p}} \times$$

$$\times \left[\int_0^{\infty} \xi \exp\{-\xi^2\} |k(\xi)|^q d\xi\right]^{\frac{1}{q}} = (\alpha_+ + \beta_+)\|h\|_{p,T}\|k\|_{q,T}.$$

Here we have used (5.15)-(5.17) as well as the Hölder inequality. Thus R is a contraction, and if $\alpha_+ + \beta_+ < 1$ even a strict contraction, on $H_{p,T}$. As a consequence of condition (5.16) we find $Re=(\alpha_+ + \beta_+)e$ and therefore

$$\|R\|_{H_{p,T}} = r(R)_{H_{p,T}} = \alpha_+ + \beta_+. \tag{5.31}$$

For the finite medium problems, one must allow for different accommodation coefficients at each wall, namely α_+ and β_+ for the wall at x=0 and α_- and β_- for the wall at $x=\tau$. One obtains, in this case, the estimate

$$\|R\|_{H_{p,T}} = r(R)_{H_{p,T}} \leq \max\{\alpha_+ + \beta_+, \alpha_- + \beta_-\}. \tag{5.32}$$

It is easily shown that for $1 \leq p < \infty$ and $0 < \alpha < (1/p)$,

$$B[H_p] \subset |T|^{\alpha}[H_p] \cap |T|^{-1-\alpha}[H_p]. \tag{5.33}$$

We may therefore obtain existence and uniqueness results on H_p if R is bounded on H_p. On $H_{2,T}$ such a boundedness assumption is not necessary since R is bounded on $H_{2,T}$.

THEOREM 5.1: If $R=0$, then the slip-flow problem (5.26)-(5.27) is uniquely solvable for all $\hat{k}$. The solution has its values in each of the Banach spaces H_p and $H_{p,T}$, where $1 \leq p < \infty$. If $\alpha_+ + \beta_+ < 1$, then the slip-flow problem (5.26)-(5.27) is uniquely solvable for all $\hat{k}$. If $\alpha_+ + \beta_+ = 1$, then the slip-flow problem (5.26)-(5.27) does not have solutions for $\hat{k} \neq 0$, and has a one dimensional space of solutions, namely the constant functions, if $\hat{k}=0$. In all cases the solution space is $H_{2,T}$, and also H_p if R is bounded on H_p and H_2.

Proof: As shown in Chapter III, the solutions have the form

$$\psi(x) = e^{-xT^{-1}A}\varphi_p + (I - xT^{-1}A)\varphi_0,$$

where $\varphi_0 \in M_{-,R} = [\mathrm{Ker}(Q_+ - RJQ_-) \oplus \mathrm{Ran}\, PP_+] \cap Z_0(T^{-1}A)$ satisfies $\varphi_0 = \varphi_- - \varphi_p$ for unique vectors $\varphi_- \in \mathrm{Ker}(Q_+ - RJQ_-)$ and $\varphi_p \in \mathrm{Ran}\, PP_+$. If $\alpha_+ + \beta_+ = r(R) < 1$, then $M_{-,R}$ is a maximal strictly negative subspace of $Z_0(T^{-1}A) = \mathrm{span}\{1, \xi\}$ with respect to $[h,k] = (Th,k)$ and thus one dimensional. On the other hand, if $\alpha_+ + \beta_+ = 1$, we easily see that $e \in \mathrm{Ker}(Q_+ - RJQ_-)$, while $(Q_+ - RJQ_-)Te = (I+R)TQ_+ e \neq 0$, whence $\mathrm{Ker}(Q_+ - RJQ_-) = \mathrm{span}\{e\} = \{\text{constants}\}$ and $M_{-,R} = \mathrm{span}\{e\}$. However, since one would then have $\psi(x) \equiv \varphi_0$, the problem is not solvable for $\hat{k} \neq 0$ and nonuniquely solvable for $k=0$. ■

THEOREM 5.2. Let $1 \leq p < \infty$, and let R be bounded on H_p and H_2. Suppose, for $0 \leq x, y \leq \tau$,

$$\int_{-\infty}^{\infty} e^{-\xi^2} |q(x,\xi)-q(y,\xi)|^p d\xi \leq M|x-y|^{\gamma p}$$

for some $\gamma\epsilon(0,1)$ and $M\epsilon(0,\infty)$. Then the boundary value problem (5.26)–(5.28) is uniquely solvable on H_p whenever $\max\{\alpha_++\beta_+,\alpha_-+\beta_-\}<1$ is satisfied. If $\alpha_++\beta_+=\alpha_-+\beta_-=1$, then the solution whenever existing is nonunique, and the constant functions satisfy the boundary value problem for $\varphi(\xi)\equiv0$ and $q(x,\xi)\equiv0$. If $p=2$ and $q(x)\equiv0$, the boundary value problem (5.26)–(5.28) is uniquely solvable on H_2 whenever $\max\{\alpha_++\beta_+,\alpha_-+\beta_-\}<1$, and nonuniquely solvable (when solvable) if $\alpha_++\beta_+=\alpha_-+\beta_-=1$. The constant functions satisfy the boundary value problem for $\varphi(\xi)\equiv0$.

The methods of Sections VIII.1 and VIII.2 may be used to construct the albedo operators. Consider, for example, the half space problem (5.26) on H_2 with boundary conditions

$$\psi(0,\xi) = \varphi_+(\xi), \quad 0\leq\xi<\infty,$$

$$\int_{-\infty}^{\infty} |\psi(x,\xi)|^2 e^{-\xi^2} d\xi = O(1) \ (x\to\infty).$$

We denote by $\sigma(\tau)$ the resolution of the identity associated with the unbounded self adjoint operator T,

$$(\sigma(\tau)f)(\xi) = \begin{cases} f(\xi), & \xi\in\tau\cap(-\infty,\infty), \\ 0, & \xi\notin\tau\cap(-\infty,\infty). \end{cases}$$

For $\mathbb{B}$ we take Ran B = span{e}, identify $\mathbb{B}$ with $\mathbb{C}$ and define $\tilde{\pi}:H\to\mathbb{B}$ and $j:\mathbb{B}\to H$ by

$$\tilde{\pi}f = \pi^{-1/2}\int_{-\infty}^{0} f(\hat{v})e^{-\hat{v}^2} d\hat{v},$$

$$j\xi = \xi e,$$

where $\xi\epsilon\mathbb{C}$. The dispersion function is a scalar function on $\mathbb{B}$ of the form

$$\Lambda(z) = 1 + \pi^{-1/2}\int_{-\infty}^{0} (t-z)^{-1}e^{-t^2}dt,$$

which is analytic outside the real line with continuations from either side through the real axis. It has a double zero at ∞ imbedded in the continuous spectrum of T. On the imaginary line it is continuous with $\Lambda(\pm i0)=1$. Using Muskhelishvili's theory of singular integral equations (see [276]) applied to the function $\Lambda(z)(1-z^{-2})$ without infinite zeros, we easily prove the existence of a unique function H(z) such that $\Lambda(z)^{-1} = H(-z)H(z)$ for Re $z=0$, which is analytic in the open right half plane, continuous in the closed right half plane and satisfies $zH(z) = O(1)$ ($z\to\infty$, Re$z\geq 0$). As in the example of isotropic neutron transport, a direct substitution of the data in (VIII 1.8) yields

$$(E\varphi_+)(\xi) = \pi^{-1/2}\int_0^\infty \xi(\xi-u)^{-1}H(-\xi)H(u)\varphi_+(u)e^{-u^2}du, \quad \xi<0.$$

This expression is readily found in the work of Cercignani [73]. Substitution in the H–equation (VIII 1.10) or (VIII 1.11) yields the integral equation

$$H(\xi)^{-1} = 1 - \xi\pi^{-1/2}\int_0^\infty (\xi+u)^{-1}H(u)e^{-u^2}du.$$

Next, let us consider the BGK equation for heat transfer (5.10), which one may write in the matrix form

$$\xi\,\frac{\partial\psi}{\partial x} + \psi(x,\xi) = \pi^{-1/2}\mathbf{Q}(\xi)\int_{-\infty}^{\infty}\tilde{\mathbf{Q}}(\xi)\psi(x,\xi)\exp\{-\xi^2\}d\xi, \tag{5.34a}$$

where

$$\mathbf{Q}(\xi) = \begin{bmatrix} (2/3)^{1/2}(\xi^2-1/2) & 1 \\ (2/3)^{1/2} & 0 \end{bmatrix} \tag{5.34b}$$

and tilde above a matrix denotes transposition. Introducing the spaces $H_p^{(2)}$ and $H_{p,T}^{(2)}$ as the ℓ_p–direct sums of two copies of H_p and $H_{p,T}$, respectively, and the operators and vectors

$$(T\mathbf{h})(\xi) = \xi\,\mathbf{h}(\xi),$$

$$(Q_\pm \mathbf{h})(\xi) = \begin{cases} \mathbf{h}(\xi), & \pm\xi>0, \\ 0, & \pm\xi<0, \end{cases}$$

$$(\mathbf{Bh})(\xi) = \frac{1}{\sqrt{\pi}}Q(\xi)\int_{-\infty}^{\infty}\tilde{Q}(\xi)\mathbf{h}(\xi)\exp\{-\xi^2\}d\xi, \tag{5.35}$$

$$(\mathbf{Jh})(\xi) = \mathbf{h}(-\xi),$$

$$\mathbf{e}_1(\xi) = \begin{bmatrix} 1 \\ 0 \end{bmatrix},$$

$$\mathbf{e}_2(\xi) = \begin{bmatrix} (2/3)^{1/2}(\xi^2-1/2) \\ (2/3)^{1/2} \end{bmatrix},$$

we consider the half space problem

$$(T\psi)'(x) = -A\psi(x) + \mathbf{q}(x), \quad 0<x<\infty, \tag{5.36}$$

$$Q_+\psi(0) = RJQ_-\psi(0) + Q_+\varphi_+, \tag{5.37a}$$

$$\lim_{x\to\infty}\{\psi(x)/x\} = -\hat{k}_1\mathbf{e}_1 - \hat{k}_2\mathbf{e}_2. \tag{5.37b}$$

and the finite slab problem

$$(T\psi)'(x) = -A\psi(x) + \mathbf{q}(x), \quad 0<x<\tau, \tag{5.38}$$

$$Q_+\psi(0) = RJQ_-\psi(0) + Q_+\varphi, \tag{5.39a}$$

$$Q_-\psi(\tau) = RJQ_+\psi(\tau) + Q_-\varphi. \tag{5.39b}$$

where we restrict ourselves to specular reflection with accommodation coefficients $\alpha_\pm$ at the finite boundaries, i.e., $R=\alpha_+Q_+ + \alpha_-Q_-$.

THEOREM 5.3. If $\alpha_+\epsilon[0,1)$, the slip-flow problem (5.36)-(5.37) with $\mathbf{q}\equiv 0$ is uniquely solvable for all $\hat{k}_1$ and $\hat{k}_2$. If $\alpha_+=1$ (specular reflection), the problem does not have solutions unless $\hat{k}_1 = \hat{k}_2 = 0$, in which case it has a two dimensional space of solutions, namely the linear span of collision invariants $\mathbf{e}_1$ and $\mathbf{e}_2$. The solutions have their values in all of the spaces $H_p^{(2)}$ and $H_{p,T}^{(2)}$ for $1\le p<\infty$.

Proof: By virtue of (5.34) and (5.35) and the orthonormality in $H_2^{(2)}$ of $\{\mathbf{e}_1,\mathbf{e}_2\}$, it is immediate that $\mathrm{Ker}\ A=\{\mathbf{e}_1,\mathbf{e}_2\}$, $Z_0(T^{-1}A)=\mathrm{span}\{\mathbf{e}_1,\mathbf{e}_2,T\mathbf{e}_1,T\mathbf{e}_2\}$ with $(T\mathbf{e}_1,\mathbf{e}_1)=(T\mathbf{e}_2,\mathbf{e}_2)=(T\mathbf{e}_1,\mathbf{e}_2)=0$. Then one may repeat the proof of Theorem 5.1, observing that the maximal strictly negative subspaces of $Z_0(T^{-1}A)$ with respect to $[\mathbf{h},\mathbf{k}]=(T\mathbf{h},\mathbf{k})$ are two dimensional. We conclude in noting that, for $\alpha_+=1$, $\{\mathbf{e}_1,\mathbf{e}_2\}\subset\mathrm{Ker}(Q_+-RJQ_-)$ while $T\mathbf{e}_1$ and $T\mathbf{e}_2$ do not belong to $\mathrm{Ker}(Q_+-RJQ_-)$. ∎

THEOREM 5.4. Let $1\le p<\infty$. Suppose that, for $0\le x,y\le\tau$,

$$\int_{-\infty}^{\infty}\exp\{-(\xi)^2\}\|\mathbf{q}(x,\xi)-\mathbf{q}(y,\xi)\|_p^p\,d\xi\ \le\ M|x-y|^{\gamma p}$$

for some $\gamma\in(0,1)$ and $M\in(0,\infty)$. For such a source term the boundary value problem (5.38)–(5.39) is uniquely solvable on $H_p^{(2)}$, if $\max\{\alpha_+,\alpha_-\}<1$. If $\alpha_+=\alpha_-=1$, then the solution whenever existing is nonunique and two solutions of the same problem differ by a function of the type $\psi(x,\xi)=c_1\mathbf{e}_1(\xi)+c_2\mathbf{e}(\xi)$.

Proof: The only relevant issue beyond what is immediate from Section V.4 is the solution of Eqs. (5.38)–(5.39) for $q(x,\xi)\equiv0$, $\varphi(\xi)\equiv0$ and $\alpha_+=\alpha_-=1$. For such a solution $\psi(x)$ we have $\frac{1}{2}(I-J)\psi(0)\in\mathrm{Ran}\ Q_-$ and $\frac{1}{2}(I-J)\psi(\tau)\in\mathrm{Ran}\ Q_+$, which imply $J\psi(0)=\psi(0)$ and $J\psi(\tau)=\psi(\tau)$. For $S_\tau=\exp[-\tau T^{-1}A]$ we get

$$S_{2\tau}\psi(0)\ =\ S_\tau\psi(\tau)\ =\ S_\tau J\psi(\tau)\ =\ JS_{-\tau}\psi(\tau)\ =\ J\psi(0)\ =\ \psi(0),$$

whence $\psi(0)\in\mathrm{Ker}\ A$ and likewise $\psi(\tau)=S_\tau\psi(0)\in\mathrm{Ker}\ A$. Hence, $\psi(x)\equiv\mathbf{h}$ with $\mathbf{h}=J\mathbf{h}\in\mathrm{Ker}\ A$, and the proof is complete. ∎

A similar result can be proved for $H_{2,T}^{(2)}$ if one assumes, for $0\le x,y\le\tau$,

$$\int_{-\infty}^{\infty}|\xi|\exp\{-(\xi)^2\}\|\mathbf{q}(x,\xi)-\mathbf{q}(y,\xi)\|_2^2\,d\xi\ \le\ M|x-y|^{2\gamma}$$

for some $\gamma\in(0,1)$ and $M\in(0,\infty)$.

Finally, let us discuss the binary gas problem (5.19) on the Banach spaces $H_p^{(2)}$ and $H_{p,T}^{(2)}$. We define the operators

$$(T\mathbf{h})(\xi)\ =\ \xi\Sigma^{-1}\mathbf{h}(\xi),$$

$$(Q_{\pm}\mathbf{h})(\xi) = \begin{cases} \mathbf{h}(\xi), & \pm\xi>0, \\ 0, & \pm\xi<0, \end{cases}$$

$$(B\mathbf{h})(\xi) = \pi^{-½}\Sigma^{-1}\mathbf{D}\int_{-\infty}^{\infty} \mathbf{h}(\xi)\exp\{-\xi^2\}d\xi.$$

It is straightforward to compute Ker A = span{**e**}, where

$$\mathbf{e} = \begin{bmatrix} 1 \\ (m^*/m)^{½} \end{bmatrix}.$$

It is also obvious that $\text{diag}((m^*)^{½}, m^{½})N\text{diag}((m^*)^{-½}, m^{-½})$ is self adjoint on $H_2^{(2)}$ if N is any of the operators T, $Q_{\pm}$, B and A. On computing $I-\Sigma^{-1}\mathbf{D}$ one obtains a matrix with zero determinant and trace 2, whence on symmetrization A becomes positive self adjoint on $H_2^{(2)}$ with span$\{\hat{e}\}$ as its kernel, where

$$(T\hat{e},\hat{e}) = \pi^{-½}\int_{-\infty}^{\infty} \xi\,\|\hat{e}(\xi)\|_2^2 \exp\{-\xi^2\}d\xi = 0.$$

Hence, defining the inversion symmetry

$$(J\mathbf{h})(\xi) = \mathbf{h}(-\xi),$$

and the surface reflection operator

$$(R\mathbf{h})(\xi) = \alpha_+(Q_+\mathbf{h})(\xi) + \alpha_-(Q_-\mathbf{h})(\xi),$$

the theory of Chapters III and V applies.

THEOREM 5.5. If $\alpha_+\in[0,1)$, the slip-flow problem for binary gas mixtures

$$T\psi'(x) = -A\psi(x), \quad 0<x<\infty, \tag{5.40}$$

$$Q_+\psi(0) = RJQ_-\psi(0), \tag{5.41a}$$

$$\lim_{x\to\infty} \{\psi(x)/x\} = -\hat{k}\mathbf{e} \tag{5.41b}$$

is uniquely solvable for all $\hat{k}$. If $\alpha_+=1$ (specular reflection), the problem does not have solutions for $\hat{k}\neq0$ and has a one dimensional manifold of solutions, namely the multiples of $\mathbf{e}$, if $\hat{k}=0$. The solutions have their values in all of the spaces $H_p^{(2)}$ and $H_{p,T}^{(2)}$ for $1\le p<\infty$.

THEOREM 5.6. Suppose that, for $0\le x,y\le\tau$,

$$\int_{-\infty}^{\infty}\exp\{-(\xi)^2\}\|\mathbf{q}(x,\xi)-\mathbf{q}(y,\xi)\|_p^p d\xi \le M|x-y|^{\gamma p}$$

for some $\gamma\in(0,1)$, $M\in(0,\infty)$, and $1\le p<\infty$. Then the boundary value problem for binary gas mixtures

$$T\psi'(x) = -A\psi(x) + \mathbf{q}(x), \quad 0<x<\tau, \tag{5.42}$$

$$Q_+\psi(0) = RJQ_-\psi(0) + Q_+\varphi, \tag{5.43a}$$

$$Q_-\psi(\tau) = RJQ_+\psi(\tau) + Q_-\varphi, \tag{5.43b}$$

is uniquely solvable on $H_p^{(2)}$ if $\max\{\alpha_+, \alpha_-\}<1$. If $\alpha_+=\alpha_-=1$, then the solution whenever existing is nonunique and two solutions differ by a function of the type $\psi(x,\xi)=c\mathbf{e}(\xi)$.

The proofs of these theorems follow those of Theorems 5.3 and 5.4. Let us turn to the construction of the albedo operator for the half space problem corresponding to Eq. (5.40) on $H_2^{(2)}$ along with boundary conditions

$$\psi(0,\mu) = \varphi_+(\mu), \quad 0\le\mu<\infty,$$

$$\sum_{i=1}^{2}\int_{-\infty}^{\infty}|\psi_i(x,\mu)|^2 e^{-\mu^2}d\mu = O(1) \ (x\to\infty).$$

Rather than utilize the similarity transformation $\mathrm{diag}((m^*)^{1/2}, m^{1/2})$, it is convenient to study this problem on the Hilbert space H of pairs $\mathbf{f} = \mathrm{col}(f_1,f_2)$ of measurable functions $f_i:(-\infty,\infty)\to\mathbb{C}$, endowed with the inner product

$$(\mathbf{f},\mathbf{g}) = \pi^{-1/2}\int_{-\infty}^{0}\{\beta f_1(\hat{\mu})\overline{g_1(\hat{\mu})}+\beta^* f_2(\hat{\mu})\overline{g_2(\hat{\mu})}\}e^{-\hat{\mu}^2}d\hat{\mu}.$$

The resolution of the identity of T has the form

$$(\sigma(\tau)\mathbf{f})_1(\xi) = \begin{cases} f_1(\xi/\beta), & \xi\in\tau\cap\mathbb{R}, \\ 0 \quad , & \xi\notin\tau\cap\mathbb{R}, \end{cases}$$

$$(\sigma(\tau)\mathbf{f})_2(\xi) = \begin{cases} f_2(\xi/\beta^*), & \xi\in\tau\cap\mathbb{R}, \\ 0 \quad , & \xi\notin\tau\cap\mathbb{R}. \end{cases}$$

Let $\mathbf{e}_1,\mathbf{e}_2$ be the orthonormal system of vectors in H, for which

$$[\mathbf{e}_\ell]_i(\xi) = \delta_{i\ell}, \quad -\infty<\xi<\infty,$$

let $j:\mathbb{B}\to H$ be the natural imbedding, and define $\tilde{\pi}:H\to\mathbb{B}$ by

$$\tilde{\pi}\mathbf{f} = \beta^{-1}(\mathbf{f},\mathbf{e}_1)\mathbf{e}_1 + (\beta^*)^{-1}(\mathbf{f},\mathbf{e}_2)\mathbf{e}_2.$$

Then $\mathbb{B} = \text{span}\{\mathbf{e}_1,\mathbf{e}_2\} \supset \text{Ran } B^*$. The dispersion function, in matrix form with respect to the basis $\{\mathbf{e}_1,\mathbf{e}_2\}$ of $\mathbb{B}$, is now easily computed. It is given by

$$[\Lambda(z)]_{ik} = \delta_{ik} - z\pi^{-1/2}\int_{-\infty}^{\infty}(z-t)^{-1}e^{-t^2}[\Sigma^{-1}\mathbf{D}]_{ik}dt, \quad i,k=1,2.$$

Corresponding to the albedo operator there is a factorization of the dispersion matrix $\Lambda(z)^{-1} = \mathbf{H}_\ell(-z)\mathbf{H}_r(z)$ for Re $z=0$, where $\mathbf{H}_\ell$ and $\mathbf{H}_r$ are analytic and continuous up to the boundary of the closed right half plane and assume invertible values there; the most singular behavior allowed is that $z\mathbf{H}_\ell(z)$ and $z\mathbf{H}_r(z)$ are bounded as $z\to\infty$ along the right half plane, while $\mathbf{H}_\ell(z)$ and $\mathbf{H}_r(z)$ are not. The albedo operator is easily found using Eq. (VIII 1.8), and reads

$$(E\varphi_+)(\xi) = \pi^{-1/2}\int_0^{\infty}\frac{\nu}{\nu-\xi}\,\mathbf{H}_\ell(-\xi)\mathbf{H}_r(\nu)\Sigma^{-1}\mathbf{D}\varphi_+(\nu)e^{-\nu^2}d\nu$$

for $-\infty<\xi<0$. A coupled set of H-equations appears of the form

$$\mathbf{H}_\ell(z)^{-1} = \mathbf{I} - z\pi^{-1/2}\int_0^{\infty}(z+t)^{-1}\mathbf{H}_r(t)\Sigma^{-1}\mathbf{D}e^{-t^2}dt,$$

$$\mathbf{H}_r(z)^{-1} = \mathbf{I} - z\pi^{-1/2}\int_0^{\infty}(z+t)^{-1}\Sigma^{-1}\mathbf{D}\mathbf{H}_\ell(t)e^{-t^2}dt.$$

The above expressions are equivalent to the half space solution formulas obtained by

Cavalier and Greenberg [71].

We have presented several models derived from linearization of the nonlinear BGK equation about a Maxwellian distribution. The BGK model has some undesirable characteristics (e.g., transport coefficients are not correctly represented) and much effort has been spent in studying more general models. One way to generalize the linearized BGK equation was proposed by Gross and Jackson [174], and consists of a systematic procedure for improving the model by adding a finite sum of rank one operators to the linear collision operator, corresponding to projection along additional eigenfunctions. For example, the eigenfunctions for the complete collision operator of a gas of Maxwell molecules have been calculated explicitly by Wang Chang and Uhlenbeck [387]. By expanding the collision operator into a series of these eigenfunctions, Eq. (5.4) may be replaced by the equation

$$\frac{\partial h}{\partial t} + \boldsymbol{\xi}\cdot\frac{\partial h}{\partial \mathbf{x}} + \nu_N h(\mathbf{x},\boldsymbol{\xi},t) = \sum_{j=0}^{N} (\nu_N+\lambda_j)\psi_j(\psi_j,h), \tag{5.44}$$

where ν_N is a collision frequency, the inner product (5.5) was used, and the functions $(\psi_j)_{j=0}^{N}$ form an orthonormal set with respect to (5.5). The linearized BGK equation arises as a special case if one chooses N=4, the collision invariants for the ψ_j, $\lambda_j=0$ for $0\leq j\leq N$ and $\nu_N=\nu$. Analogous finite rank approximative linearized equations with speed dependent collision frequencies $\nu=\nu(|\boldsymbol{\xi}|)$ have also been studied. This procedure can be followed for linearized Boltzmann equations of other types, provided the collision operator has a complete set of eigenfunctions. Such models are covered by the abstract theory of the previous chapters.

By expanding the solution of the nonlinear Boltzmann equation as $f(\mathbf{x},\boldsymbol{\xi},t) = f_0(\boldsymbol{\xi})(1+\varepsilon h(\mathbf{x},\boldsymbol{\xi},t))$, where f_0 is the equilibrium solution (the Maxwell velocity distribution), and retaining terms of order ε, one arrives at the linearized Boltzmann equation

$$\frac{\partial h}{\partial t} + \boldsymbol{\xi}\cdot\frac{\partial h}{\partial \mathbf{x}} = -(Lh)(\mathbf{x},\boldsymbol{\xi},t), \tag{5.45}$$

where for homogeneous media L is a symmetric nonnegative operator on the Hilbert space of square integrable functions of velocity $\boldsymbol{\xi}$ weighted by the Maxwellian factor $f_0(\boldsymbol{\xi})$, i.e., the Hilbert space of functions $h:\mathbb{R}^3\to\mathbb{C}$ endowed with the inner product (5.5). The linear operator L is given by

$$(Lh)(x,\boldsymbol{\xi},t) = \frac{1}{m}\int f_0(\hat{\boldsymbol{\xi}})(h_*' + h' - h_* - h)B(\theta,V)\,d\hat{\boldsymbol{\xi}}\,d\varepsilon\,d\theta. \tag{5.46}$$

Here $\boldsymbol{\xi}$ and $\hat{\boldsymbol{\xi}}$ are the outgoing velocities in a two-molecule interaction, h, h_*, h' and h'_* are the functions $h(\cdot)$ evaluated at arguments $\boldsymbol{\xi}$, $\hat{\boldsymbol{\xi}}$, $\boldsymbol{\xi}'$ and $\hat{\boldsymbol{\xi}}'$, where $\boldsymbol{\xi}'=\boldsymbol{\xi}-\mathbf{n}(V\cos\theta)$ and $\hat{\boldsymbol{\xi}}'=\hat{\boldsymbol{\xi}}+\mathbf{n}(V\cos\theta)$, $\mathbf{n}$ is a unit vector in the direction of the change in the incoming velocity $\boldsymbol{\xi}'$, θ and ε are the polar and azimuthal angles specifying the direction of the relative velocity $\mathbf{V}=\boldsymbol{\xi}-\hat{\boldsymbol{\xi}}$ measured from $\mathbf{n}$, and the kernel $B(\theta,V)$ contains the specific dynamics of the intermolecular interaction.

For some interactions the kernel B is of particularly simple form. For inverse power law intermolecular potentials of the form $U(r)\sim\alpha r^{-s}$ with $\alpha>0$ (repulsive force) and $s>0$, we have $B(\theta,V) = V^{\gamma}\beta(\theta)$, where $\gamma=(s-4)/s$, $V=|\mathbf{V}|$, $\beta(\theta)\sim\theta$ as $\theta\to0$ and $\beta(\theta)\sim(\frac{\pi}{2}-\theta)^{-(s+2)/s}$ as $\theta\to\pi/2$. For a gas of hard spheres of diameter $\sigma>0$, the kernel is given by $B(\theta,V) = \sigma^2 V\sin\theta\cos\theta$.

Because $B(\theta,V)$ has a nonintegrable singularity at $\theta=\frac{1}{2}\pi$, which reflects the contribution of grazing collisions, it is customary to cut off the effect of such collisions and thus to modify $B(\theta,V)$ in such a way that the singularity at $\theta=\frac{1}{2}\pi$ either disappears or becomes integrable. Two cut-off procedures are commonly used. Angular cut-off was introduced by Grad [155] and consists of putting $B(\theta,V)=0$ for $\theta_0<\theta\leq\frac{1}{2}\pi$. Radial potential cut-off was introduced by Cercignani [80] and consists of the assumption that the two-body interaction has finite range σ: the potential vanishes identically if the two molecules are out of range. Both of these cut-offs have the effect of making the collision frequency $\nu(\boldsymbol{\xi})$ finite, where

$$\nu(\boldsymbol{\xi}) = \frac{2\pi}{m}\int f_0(\hat{\boldsymbol{\xi}})B(\theta,V)d\hat{\boldsymbol{\xi}}\,d\theta.$$

A particularly interesting case arises when the rarefied gas interacts with an equilibrium distribution (cf. [84]). Then the collision operator may be written as $L = \nu-K = \nu-K_2+K_1$, with

$$K_1h = \frac{2\pi}{m}\int_{\mathbb{R}^3}\int_0^{\pi} f_0(\hat{\boldsymbol{\xi}})h(\boldsymbol{\xi})B(\theta,V)d\theta\,d\hat{\boldsymbol{\xi}},$$

$$K_2h = \frac{2}{m}\int_{\mathbb{R}^3}\int_0^{2\pi}\int_0^{\pi} f_0(\hat{\boldsymbol{\xi}})h(\boldsymbol{\xi})[B(\theta,V)+B(\tfrac{1}{2}\pi-\theta,V)]d\theta\,d\varepsilon\,d\hat{\boldsymbol{\xi}},$$

and Eq. (5.45) becomes

$$\frac{\partial h}{\partial t} + \boldsymbol{\xi}\cdot\frac{\partial h}{\partial \mathbf{x}} = -\nu(\boldsymbol{\xi})h(\mathbf{x},\boldsymbol{\xi},t) + (Kh)(\mathbf{x},\boldsymbol{\xi},t). \tag{5.47}$$

Under assumptions of time independence and plane parallel symmetry, the linearized Boltzmann equation (5.47) becomes

$$\xi_x \frac{\partial h}{\partial x}(x,\boldsymbol{\xi}) + h(x,\boldsymbol{\xi}) = -\nu(\boldsymbol{\xi})h(x,\boldsymbol{\xi}) + (Kh)(x,\boldsymbol{\xi}), \tag{5.48}$$

where $x\in(0,\tau)$, $\xi=|\boldsymbol{\xi}|$ and $\xi_x\in(-\infty,\infty)$.

Let us consider the case when the collision frequency depends only on the speed $\xi=|\boldsymbol{\xi}|$ and is bounded away from zero. This is a property shared by Maxwell molecules (s=4) and hard intermolecular potentials (s>4) with cut-off. Using physical units yielding the equilibrium Maxwellian distribution $\pi^{-3/2}e^{-\xi^2}$, we introduce the Hilbert spaces H and H_T of measurable functions $h:\mathbb{R}^3\to\mathbb{C}$ bounded relative to the respective norms

$$\|h\| = \left[\pi^{-3/2}\int_{\mathbb{R}^3}\nu(\xi)|h(\boldsymbol{\xi})|^2 e^{-\xi^2} d^3\boldsymbol{\xi}\right]^{1/2}$$

and

$$\|h\|_T = \left[\pi^{-3/2}\int_{\mathbb{R}^3}|h(\boldsymbol{\xi})|^2|\xi_x|e^{-\xi^2} d^3\boldsymbol{\xi}\right]^{1/2}.$$

We define the operators T, $Q_\pm$, A and J by

$$(Th)(\boldsymbol{\xi}) = \frac{1}{\nu(\xi)}\xi_x h(\boldsymbol{\xi}),$$

$$(Ah)(\boldsymbol{\xi}) = \nu(\xi)^{-1}(Lh)(\boldsymbol{\xi}) = h(\boldsymbol{\xi}) - \frac{1}{\nu(\xi)}(Kh)(\boldsymbol{\xi}),$$

$$(Jh)(\boldsymbol{\xi}) = h(-\xi_x,\xi_y,\xi_z)$$

$$(Q_\pm h)(\boldsymbol{\xi}) = \begin{cases} h(\boldsymbol{\xi}), & \pm\xi_x>0, \\ 0, & \pm\xi_x<0. \end{cases}$$

We shall assume that $\nu(\xi)^{-1}K$ is bounded on H, and that the five-fold degenerate eigenvalue of A at zero is isolated. These properties are always satisfied for hard potentials (s≥4) with cutoff [66, 155]. Since A is bounded on H, positive and Fredholm, the existence and uniqueness theory can be developed in H_T (cf. Section III.3). On observing that J anticommutes with T and commutes with A, we define a surface reflection operator R on H_T satisfying $\|Rh\|_T\le\|h\|_T$.

THEOREM 5.7. Under the assumptions indicated above, for all functions $\varphi \epsilon H_T$ there exists a unique solution in H_T of the linearized Boltzmann equation (5.48) on a finite slab $x \epsilon (0,\tau)$, which satisfies the boundary conditions

$$h(0,\xi) = (RJQ_-h(0))(\xi) + \varphi(\xi), \quad \xi_x > 0,$$

$$h(\tau,\xi) = (RJQ_+h(\tau))(\xi) + \varphi(\xi), \quad \xi_x < 0,$$

where both $\|RQ_+\|_T < 1$ and $\|RQ_-\|_T < 1$.

For specular reflection $h(x,\xi) = a + b\xi^2 + c\xi_y + d\xi_z$ is a solution of the corresponding homogeneous problem, and therefore for this case non-uniqueness is clear.

As a consequence of Proposition III 4.1 and Corollary III 4.3 we have

THEOREM 5.8. For all functions $\varphi_+ \epsilon Q_+[H_T]$ there exists at least one solution in H_T of the linearized Boltzmann equation (5.48) on the half space $x \epsilon (0,\infty)$ which satisfies the boundary conditions

$$h(0,\xi) = (RJQ_-h(0))(\xi) + \varphi_+(\xi), \quad \xi_x > 0,$$

$$\limsup_{x\to\infty} \|h(x)\|_T < \infty.$$

If $\|RQ_+\|_T < 1$, there is a one dimensional manifold of solutions to the corresponding homogeneous problem.

Guiraud [177, 178], Maslova [260, 261] and Bardos et al. [24] have shown wellposedness of the boundary value problem for the linearized Boltzmann equation in H, under conditions which restrict the intermolecular potential to "hard potentials." In general, regularity conditions such as III (2.12) will fail. As indicated in Section VII.4, regularity restrictions have been removed for (modified) collision operators which are trace class perturbations of the identity operator, and it is hoped that those methods may be extended to compact perturbations of the identity.

6. A Boltzmann equation for phonon and electron transport

The transport of phonons in crystalline solids and electrons in metals and semiconductors is often described by the so-called relaxation time approximation. On considering the electron scattering case, suppose an electron is accelerated by a constant electric field directed along a metal or semiconductor strip. Sooner or later, by running into an impurity or for some other reason, the electron will lose its drift velocity. The above approximation then consists of the assumptions that (i) there is a characteristic time interval between successive collisions, the relaxation time τ, (ii) the electron comes to a complete stop on collision and "forgets" its previous motion, and (iii) the probability of a collision in an infinitesimal period dt equals $(dt)/\tau$, whence $P(t)=e^{-t/\tau}$ is the probability of collisionless travel over a period t and $\int_0^\infty tP(t)dt=\tau$ is the expected free travel time of an electron. This approximation can be expected to break down near boundaries, where surface scattering becomes important. For the description of electron diffusion in metals and semiconductors and phonon transport in crystalline solids we refer the reader the textbooks on solid state physics, such as the monographs [217, 404, 405].

In spite of the expected inaccuracy of the relaxation time approximation near boundaries, this method has been used frequently, even when treating boundary effects (see, for instance, [288, 321] and references therein). A linearized Boltzmann equation intended to remedy the unreliability of the relaxation time approximation near boundaries was derived by Nonnenmacher and Zweifel [289]. In their derivation they assumed the existence of a quasi-equilibrium distribution function, valid far from the boundaries, which has the form

$$\hat{f}_0(z,k) = (\exp[(\varepsilon(k)-\lambda)/T(z)]\pm 1)^{-1}. \tag{6.1}$$

Here $\varepsilon(k)$ is the electron or phonon energy as a function of the wave number $k=|\mathbf{k}|$, λ is the chemical potential and T(z) is a local temperature ($\varepsilon(k)=\hbar vk$, $\lambda=0$ for phonons; $\varepsilon(k)=\hbar^2k^2/2m=mv^2/2$, $\lambda\neq 0$ for electrons, with $2\pi\hbar=h$ Planck's constant, v the phonon/electron velocity, m the electron mass). For phonons which obey Bose–Einstein statistics we take the minus sign, for electrons which are subject to Fermi–Dirac statistics we take the plus sign in Eq. (6.1). By substitution of $f = f_0\{1+h\}$, with $f_0(z,k,\omega) = \hat{f}_0(|\mathbf{k}-\mathbf{k}_1|)$, into the linearized Boltzmann equation, they arrived at the equation

$$\mu\frac{\partial h}{\partial x}(x,\omega) + \sigma(z,k,\omega)h(x,\omega) = \int_\Omega \sigma(z,k,\hat{\omega}\to\omega)h(x,\hat{\omega})d\hat{\omega}, \tag{6.2}$$

where $\mu=\cos\theta$ is the direction cosine of propagation and the angular direction ω is parametrized using the polar angle θ and the azimuthal angle φ. $\mathbf{k}_1$ accounts for the flow of particles induced by the electric field (for electrons) or the temperature gradient (for phonons), which lies in the positive z-direction. We also have

$$\sigma(z,k,\omega) = \frac{1}{v}\{\nu_0+(\hat{f}_1/\hat{f}_0)k_1[\int K(\hat{\omega}\cdot\omega,k)\beta'\,d\hat{\omega}-\nu_0\beta]\},$$

$$\sigma(z,k,\hat{\omega}\to\omega) = \frac{1}{v}K(\hat{\omega}\cdot\omega)\{1+(\hat{f}_1/\hat{f}_0)k_1[\beta'-\beta]\}.$$

Here $\beta=\sin\theta\cos\varphi$ corresponds to ω, similarly β' corresponds to $\hat{\omega}$, $K(\hat{\omega}\cdot\omega,k)$ is the elastic scattering frequency which yields the collision frequency ν_0, and

$$\hat{f}_1 = \gamma_\pm\{\varepsilon(k)\hat{f}_0^2/mT(z)\}\exp\{(\varepsilon(k)-\lambda)/T(z)\},$$

with $\gamma_\pm=1$ for electrons, $\gamma_\pm=2$ for phonons.

On inspecting (6.2), one observes that the longitudinal (parallel to the electric field or the temperature gradient) position variable z and the wave number k appear in the equations as parameters, different values of which remain uncoupled. Thus in the remainder of this section we shall not explicitly display any z or k dependence.

Let $H_{p,\sigma}$ be the Banach space of measurable functions $h:\Omega\to\mathbb{C}$ which are bounded with respect to the norm

$$\|h\| = \left[\int_{-1}^{1}\int_{0}^{2\pi} \sigma(\hat{\mu},\hat{\varphi})\,|h(\hat{\mu},\hat{\varphi})|^p d\hat{\varphi}d\hat{\mu}\right]^{1/p},$$

where $1\le p<\infty$. We define the operators T, $Q_\pm$, B, J and R by

$$(Th)(\mu,\varphi) = \frac{\mu}{\sigma(\mu,\varphi)}h(\mu,\varphi),$$

$$(Jh)(\mu,\varphi) = h(-\mu,\varphi+\pi),$$

$$(Q_\pm h)(\mu,\varphi) = \begin{cases} h(\mu,\varphi), & \pm\mu>0, \\ 0, & \pm\mu<0, \end{cases}$$

$$(Bh)(\mu,\varphi) = \frac{1}{\sigma(\mu,\varphi)} \int_{-1}^{1} \int_{0}^{2\pi} \sigma(\hat{\omega}\to\omega) h(\hat{\mu},\hat{\varphi}) d\hat{\varphi} d\hat{\mu},$$

$$(Rh)(\mu,\varphi) = ([\alpha_0 Q_+ + \alpha_d Q_-]h)(\mu,\varphi).$$

The integro-differential equation (6.2) then reads

$$Th'(x) = -Ah(x), \qquad 0<x<d, \tag{6.3}$$

where $d\epsilon(0,\infty]$ is the thickness of the metal or semiconductor strip (perpendicular to the electric field or the temperature gradient) and (strong) differentiation is applied with respect to x. We consider the (partial) specular reflection boundary condition, given by

$$Q_+h(0) = RJQ_-h(0) + Q_+\varphi. \tag{6.4}$$

For semi-infinite metal or semiconductor strips ($d=\infty$), which may be thought of as the large thickness limit of a finite strip, one may impose the condition

$$\limsup_{x\to\infty} \|h(x)\| < \infty. \tag{6.5}$$

If d is finite, one may have (another) accommodation coefficient α_d at the surface z=d, and impose the boundary condition

$$Q_-h(d) = RJQ_+h(d) + Q_-\varphi. \tag{6.6}$$

LEMMA 6.1. Let $K(\omega\cdot\hat{\omega})\not\equiv 0$ be nonnegative, and assume $\int_{-1}^{1} K(t)dt<\infty$ and

$$|(\hat{f}_1/\hat{f}_0)k_1[\frac{1}{\nu_0}\int K(\hat{\omega}\cdot\omega)\beta' d\hat{\omega}-\beta]| \le 1-\delta$$

for some $\delta=\delta(x,k)\epsilon(0,1)$. Then B has spectral radius one and Ker A consists of the constant functions.

Proof: We clearly have

$$0 \le \frac{1}{v}(1-\delta)K(\hat{\omega}\cdot\omega) \le \sigma(\hat{\omega}\to\omega) \le \frac{1}{v}(1+\delta)K(\hat{\omega}\cdot\omega), \tag{6.7}$$

whence [cf. (7.4)]

$$0 < \frac{2\pi}{v}(1-\delta)\int_{-1}^{1} K(t)dt = \sigma(\omega) \leq \frac{2\pi}{v}(1+\delta)\int_{-1}^{1} K(t)dt \tag{6.8}$$

is strictly positive, and therefore T is bounded on $H_{p,\sigma}$ if $1 \le p < \infty$. As a consequence of the bound $\int_{-1}^{1} K(t)dt < \infty$ and (6.7)–(6.8), the operator B can be proved u_0-positive on $H_{p,\sigma}$ and compact on $H_{p,\sigma}$ (cf. Proposition 3.1 and Lemma 1.1). Hence, the spectral radius of B is positive and is an algebraically simple eigenvalue with corresponding positive eigenfunction, while there is no other eigenvalue with the same modulus corresponding to a positive eigenfunction (cf. Theorem I 4.3). One easily computes that the constant functions belong to Ker A, where A=I−B, and therefore Ker A consists of the constant functions only. ∎

For symmetric scattering kernels satisfying the reciprocity condition $\sigma(\hat{\omega}\to\omega)=\sigma(-\omega\to-\hat{\omega})$, we conclude that A is positive self adjoint with one dimensional null space Ker A=span $\{\hat{e}\}$, where $(T\hat{e},\hat{e})=0$. The existence and uniqueness theory of the boundary value problems (6.3)–(6.4)–(6.5) and (6.3)–(6.4)–(6.6) then is a straightforward application of the theory of Sections III.2 and V.2.

THEOREM 6.2. Let $\sigma(\hat{\omega}\to\omega)=\sigma(\omega\to\hat{\omega})=\sigma(-\omega\to-\hat{\omega})$. Then there exists a unique solution of the boundary value problems (6.3)–(6.4)–(6.5) and (6.3)–(6.4)–(6.6) on $H_{p,T}$, $1 \le p < \infty$, if $0 \le \alpha_0 < 1$ and $0 \le \alpha_d < 1$ for the finite strip, $0 \le \alpha_0 < 1$ for the semi infinite strip. For $\alpha_0=\alpha_d=1$ (finite strip) or $\alpha_0=1$ (semi infinite strip) the solution when it exists is nonunique.

Proof: It should be observed that, for $h \in H_{2,\sigma}$,

$$(|T|h,h)_{2,\sigma} = \|h\|^2_{2,T}.$$

We may therefore apply the H_T-theory of Chapters III and V, since the norms

$$\|RQ_\pm\|_{2,\sigma} = \alpha_{\frac{1}{2}d\pm\frac{1}{2}d} \quad \text{for a finite strip,}$$

$$\|R\|_{2,\sigma} = \alpha_0 \quad \text{for a semi-infinite strip,}$$

satisfy the requirements necessary for unique solvability. The above nonuniqueness statement is straightforward. ∎

To compute the albedo operator for the half space problem with R=0, it is convenient to rewrite the (incoming flux) boundary value problem as

$$(\cos\theta)\frac{\partial h}{\partial x}(x,\theta,\varphi) + \Sigma(\theta,\varphi)h(x,\theta,\varphi) =$$

$$= \frac{1}{4\pi}\int_{-1}^{1}\int_{0}^{2\pi}\Sigma_s(\hat{\theta},\hat{\varphi})h(x,\hat{\theta},\hat{\varphi})d\hat{\varphi}d(\cos\hat{\theta}), \quad 0\le\theta\le\pi, \quad 0\le\varphi\le2\pi,$$

$$h(0,\theta,\varphi) = f_+(\theta,\varphi), \quad \cos\theta\ge0,$$

$$\int_{-1}^{1}\int_{0}^{2\pi}\Sigma(\hat{\theta},\hat{\varphi})\Sigma_s(\hat{\theta},\hat{\varphi})\,|\,h(x,\hat{\theta},\hat{\varphi})\,|^2 d\hat{\varphi}d(\cos\hat{\theta}) = O(1) \ (x\to\infty),$$

where $\Sigma=\sigma(\omega)$ and $\Sigma_s=4\pi\sigma(\hat{\omega}\to\omega)$ are measurable functions satisfying $\Sigma \ge \Sigma_s \ge \varepsilon > 0$. We also impose the regularity condition

$$\exists\alpha>0: \int_{-1}^{1}\int_{0}^{2\pi}(\cos\hat{\theta})^{-2\alpha}\Sigma(\hat{\theta},\hat{\varphi})^{2\alpha-1}\Sigma_s(\hat{\theta},\hat{\varphi})d\hat{\varphi}d(\cos\hat{\theta}) < \infty \tag{6.9}$$

and the reciprocity conditions

$$\Sigma(\theta,\varphi) = \Sigma(\theta+\pi,\pi-\varphi),$$

$$\Sigma_s(\theta,\varphi) = \Sigma_s(\theta+\pi,\pi-\varphi).$$

We first introduce

$$\psi(x,\theta,\varphi) = \Sigma(\theta,\varphi)^{1/2}\,\Sigma_s(\theta,\varphi)^{1/2}\,h(x,\theta,\varphi),$$

$$\Upsilon_+(\theta,\varphi) = \Sigma(\theta,\varphi)^{1/2}\,\Sigma_s(\theta,\varphi)^{1/2}\,f_+(\theta,\varphi),$$

$$\Gamma(\theta,\varphi) = \Sigma_s(\theta,\varphi)/\Sigma(\theta,\varphi),$$

and rewrite the problem as follows:

$$\frac{\cos\theta}{\Sigma(\theta,\varphi)}\frac{\partial\psi}{\partial x}(x,\theta,\varphi) + \psi(x,\theta,\varphi) =$$

$$= \frac{1}{4\pi}\Gamma(\theta,\varphi)^{1/2}\int_{-1}^{1}\int_{0}^{2\pi}\Gamma(\hat{\theta},\hat{\varphi})^{1/2}\,\psi(x,\hat{\theta},\hat{\varphi})d\hat{\varphi}d(\cos\hat{\theta}),$$

$$\psi(0,\theta,\varphi) = \Upsilon_{+}(\theta,\varphi), \quad \cos\theta\geq 0,$$

$$\int_{-1}^{1}\int_{0}^{2\pi}|\psi(x,\hat{\theta},\hat{\varphi})|^2 d\hat{\varphi}d(\cos\hat{\theta}) = O(1) \;\; (x\to\infty).$$

On the Hilbert space $H = L_2[-1,1]$ we relabel the operators T, B and $Q_{\pm}$ and define the vector ρ by

$$(Tf)(\theta,\varphi) = \frac{\cos\theta}{\Sigma(\theta,\varphi)}\, f(\theta,\varphi),$$

$$(Bf)(\cos\theta) = \frac{\rho(\theta,\varphi)}{4\pi}\int_{-1}^{1}\int_{0}^{2\pi}\rho(\hat{\theta},\hat{\varphi})f(\hat{\theta},\hat{\varphi})d\hat{\varphi}d(\cos\hat{\theta}),$$

$$(Q_{+}f)(\theta,\varphi) = \begin{cases} f(\theta,\varphi), & \cos\theta>0, \\ 0, & \cos\theta<0, \end{cases}$$

$$\rho(\theta,\varphi) = \Gamma(\theta,\varphi)^{1/2}.$$

We denote the resolution of the identity associated with T by

$$(\sigma(\tau)f)(\theta,\varphi) = \begin{cases} f(\theta,\varphi), & \cos\theta/\Sigma(\theta,\varphi) \in \tau, \\ 0, & \cos\theta/\Sigma(\theta,\varphi) \notin \tau. \end{cases}$$

Let us observe that B has the one dimensional range of scalar multiples of ρ. We put $\mathbb{B} = \mathrm{span}\{\rho\}$ identified in a natural way with $\mathbb{C}$ and define $\tilde{\pi}:H\to\mathbb{B}$ and $j:\mathbb{B}\to H$ by

$$\tilde{\pi}f = \rho(\theta,\varphi)\int_{-1}^{1}\int_{0}^{2\pi}\rho(\hat{\theta},\hat{\varphi})f(\hat{\theta},\hat{\varphi})d\hat{\varphi}d(\cos\hat{\theta}) \times$$
$$\times \left[\int_{-1}^{1}\int_{0}^{2\pi}|\rho(\hat{\theta},\hat{\varphi})|^2 d\hat{\varphi}d(\cos\hat{\theta})\right]^{-1},$$

$$j(\xi) = \xi\rho,$$

where $\xi\in\mathbb{C}$. For the dispersion function we find the scalar expression

$$\Lambda(z) = 1 - \frac{z}{4\pi}\int_{-\infty}^{\infty}\int_{-1}^{1}\int_{0}^{2\pi}(z-t)^{-1}\delta(t-\cos\hat{\theta}/\Sigma(\hat{\theta},\hat{\varphi}))\Gamma(\hat{\theta},\hat{\varphi})d\hat{\varphi}d(\cos\hat{\theta})dt =$$

$$= 1 + \frac{z}{4\pi}\int_{-1}^{1}\int_{0}^{2\pi}[\Sigma_s(\hat{\theta},\hat{\varphi})/(\cos\hat{\theta} - z\Sigma(\hat{\theta},\hat{\varphi}))]d\hat{\varphi}d(\cos\hat{\theta}).$$

This function is analytic on the Riemann sphere cut along a bounded subset of the real line, is even, and has a zero at infinity if and only if $\Sigma(\theta,\varphi) = \Sigma_s(\theta,\varphi)$ almost everywhere; if this condition is fulfilled, the zero at infinity has multiplicity two. Hence, using the regularity condition (6.9) and the evenness of $\Lambda(z)$, one easily proves the existence of a unique function $H(z)$, analytic on the open right half plane, continuous on its closure, and having at most a simple pole at infinity, such that $\Lambda(z)^{-1} = H(-z)H(z)$ for $\mathrm{Re}\, z=0$. In terms of the original boundary value problem, we now obtain for the albedo operator the formula

$$(\hat{E}f_+)(\theta,\varphi) = \frac{1}{4\pi}\int_0^1\int_0^{2\pi}\Gamma(\hat{\theta},\hat{\varphi})^{1/2}[\Sigma(\theta,\varphi)\cos\hat{\theta} - \Sigma(\hat{\theta},\hat{\varphi})\cos\theta]^{-1} \times$$

$$\times\ \Sigma(\theta,\varphi)H(-\cos\theta/\Sigma(\theta,\varphi))H(\cos\hat{\theta}/\Sigma(\hat{\theta},\hat{\varphi}))f_+(\hat{\theta},\hat{\varphi})(\cos\hat{\theta})d\varphi d(\cos\hat{\theta})$$

for $\cos\theta<0$. From VIII (1.10) or VIII (1.11) we obtain the H-equation

$$H(z)^{-1} = 1 - \frac{z}{4\pi}\int_{-1}^{1}\int_{0}^{2\pi}(\cos\hat{\theta} + z\Sigma(\hat{\theta},\hat{\varphi}))^{-1}H(\cos\hat{\theta}/\Sigma(\hat{\theta},\hat{\varphi}))\Sigma_s(\hat{\theta},\hat{\varphi})d\hat{\varphi}d(\cos\hat{\theta}).$$

Under more restrictive hypotheses and for azimuthally-independent Σ and Σ_s, this expression was derived before by Williams [395] using the classical Wiener-Hopf method.

Employing the Banach space theory of Chapters VI and VII and assuming that $\int_{-1}^{1}K(t)^r dt<\infty$ for some $r>1$, one may prove Theorem 6.2 for the spaces $H_{p,\sigma}$ where $1\le p<\infty$. In nonsymmetric cases the theory is more complicated, since no selfadjointness assumptions can be applied. However, in this case one can adapt the reasoning of the proofs of the multigroup results Theorem 4.1 and Corollary 4.5 to prove that Eqs. (6.3)-(6.4)-(6.5) are uniquely solvable on $H_{p,\sigma}$ $(1\le p<\infty)$, if $\int_{-1}^{1}K(t)^r dt<\infty$ for some $r>1$ and $\alpha_0=\alpha_d=0$.

Chapter X

INDEFINITE STURM–LIOUVILLE PROBLEMS

1. Kinetic equations of Sturm–Liouville type

In this chapter we shall discuss in some detail partial differential equations associated with self adjoint Sturm–Liouville boundary value problems with indefinite weights. An example of such an equation is the Fokker–Planck equation for Brownian motion,

$$v\frac{\partial\psi}{\partial x}(x,v) = \frac{\partial^2\psi}{\partial v^2}(x,v) - v\frac{\partial\psi}{\partial v}(x,v), \tag{1.1}$$

where v is a velocity variable and x is a position coordinate. Boundary conditions are provided both with respect to the position coordinate and the velocity variable. This equation was first used by Fokker [122] and Planck [309] to describe the Brownian motion by a particle of relatively large mass immersed in a fluid. Like the (time independent) Boltzmann equation, this Fokker–Planck equation describes the stationary state of the system at the macroscopic (or kinetic) scale. For such systems one no longer assumes that the interactions only involve two particles at the same time, as is done consistently for the Boltzmann equation, but rather that a large number of particles interact simultaneously. Transport processes in such media may be described by separating off the long range many body interaction via a so-called self consistent approximation, and treating the residual Coulomb interaction via a stochastic model. If the interactions are weak, one can in fact expand the Boltzmann collision term in orders of the interaction potential and obtain the Fokker–Planck collision term. Since the resulting equation appears as the limit case of weak intermolecular potentials and large mass, the Fokker–Planck collision operator essentially describes grazing collisions in which the variation of the momentum of the observed particles is small, whence the dissipation mechanism is exhibited as a diffusion in momentum space.

More generally, we wish to study equations of the form

$$w(\mu)\frac{\partial\psi}{\partial x}(x,\mu) = \frac{\partial}{\partial\mu}\left(p(\mu)\frac{\partial\psi}{\partial\mu}(x,\mu)\right) - q(\mu)\psi(x,\mu), \tag{1.2}$$

where μ ranges over an open subset I of the real line, and the real valued function $w(\mu)$ changes sign on I. The equation will be endowed with forward-backward spatial boundary conditions in half space geometry,

$$\psi(0,\mu) = \varphi_+(\mu) \text{ for those } \mu \text{ where } w(\mu)>0, \tag{1.3a}$$

$$\|\psi(x,\cdot)\| = O(1) \text{ or } o(1) \ (x\to\infty). \tag{1.3b}$$

in an appropriate Hilbert space setting. In addition, all solutions $\psi(x,\mu)$ are required to satisfy certain self adjoint boundary conditions. Formally, separation of variables of the type $\psi(x,\mu) = e^{-\lambda x}y(\mu)$ then leads to the Sturm-Liouville boundary value problem

$$-[(py')' - qy] = \lambda wy \tag{1.4}$$

with the same self adjoint boundary conditions. For this reason we refer to these as indefinite Sturm-Liouville problems. We shall present an analysis along lines followed by Beals [34] in order to settle existence and uniqueness issues, and then we shall study the representation of solutions and discuss several applications.

Let us denote by H the Hilbert space $L_2(I,d\mu)$ of square integrable Lebesgue measurable complex-valued functions, with norm and inner product denoted by $\|\cdot\|$ and $(\cdot,\cdot)$. We suppose that I is an open subset of $\mathbb{R}$ and that the indefinite weight $w:I\to\mathbb{R}$ has the following properties:

(i) $I_\pm \equiv \{\mu\in I : \pm w(\mu)>0\}$ are nonempty finite unions of open intervals, and $I_0 \equiv \{\mu\in I : w(\mu)=0\}$ has finite cardinality,

(ii) w is continuous on $I_+\cup I_-$,

(iii) In a neighborhood of each sign change $\mu_0\in I$ of the weight w there is a C^1-function m satisfying $w(\mu) = \operatorname{sign}(\mu-\mu_0)|\mu-\mu_0|^\alpha m(\mu)$ with $\alpha>-\frac{1}{2}$ and $m(\mu_0)\neq 0$.

In addition we assume that the functions $p:I\to\mathbb{R}$ and $q:I\to\mathbb{R}$ satisfy

(iv) p is locally absolutely continuous and strictly positive on I,

(v) q is continuous on I.

Finally, let us denote by D_1 the linear subspace of functions $h \in H$ that are absolutely continuous on I and whose (almost everywhere defined measurable) derivatives h' satisfy $p|h'|^2 \in L_1(I,d\mu)$ and $q|h|^2 \in L_1(I,d\mu)$. For $h,g \in D_1$ we define the sesquilinear form

$$(h,g)_A = \int_I p(\mu)h'(\mu)\overline{g}'(\mu)d\mu + \int_I q(\mu)h(\mu)\overline{g}(\mu)d\mu. \tag{1.5}$$

Then we assume the existence of a linear subspace $D \subset D_1$ containing the compactly supported C^1-functions on I and a finite dimensional subspace $N_0 \subset D$ with the following properties:

(vi) $(wh,wh) \leq c(h,h)_A$ for some constant c and all $h \in D$ such that $h \perp N_0$ in H,
(vii) $(h,h)_A \geq 0$ for all $h \in D$,
(viii) $(h,h)_A = 0$ for all $h \in N_0$,
(ix) $\|h\|^2 \leq c(h,h)_A$ for some constant c and all $h \in D$ such that $h \perp N_0$ in H.

These last assumptions will be utilized to impose the necessary self adjoint boundary conditions on (1.2).

We will now show that the boundary value problem (1.2)–(1.3) may be written as an abstract kinetic equation

$$(T\psi)'(x) = -A\psi(x), \quad 0<x<\infty,$$

$$Q_+\psi(0) = \varphi_+,$$

$$\|\psi(x)\| = O(1) \text{ or } o(1) \ (x\to\infty),$$

and thus is subject to the results of the previous chapters. On H we define the (possibly unbounded) self adjoint operator T with $D(T)=\{h \in H : wh \in H\}$ by $(Th)(\mu) = w(\mu)h(\mu)$ for $\mu \in I$. The orthogonal projections Q_+ and Q_- of H onto maximal T-positive and T-negative T-invariant subspaces have the form

$$(Q_\pm h)(\mu) = \begin{cases} h(\mu), & \mu \in I_\pm, \\ 0, & \mu \in I_\mp. \end{cases}$$

The definition of A is somewhat more involved. Let us define on D the inner product

$$(h,g)_1 = (h,g)_A + (h,g). \tag{1.6}$$

We denote by H_A the completion of D with respect to this inner product. By virtue of the finite dimension of N_0 and condition (ix), H_A is continuously and densely imbedded in H. Let H_A' be the dual of H_A with respect to H. If we extend the inner product (1.5) to H_A, we may realize an operator A_0 from H_A into H_A' by $(A_0h,g) = (h,g)_A$ for $h,g \in H_A$. We now define $D(A)=\{h \in H_A : A_0h \in H\}$ with $A=A_0$ on D(A). Using the Friedrichs' representation theorem for symmetric bilinear forms [213] and (i)–(ix), we have the following lemma.

LEMMA 1.1. The operator $A_0:H_A \to H_A'$ is a contraction, and the operator A is positive self adjoint on H.

LEMMA 1.2. The operator $T^{-1}A$ with domain $\{h \in D(A) : Ah \in T[H_A]\}$ is closed with respect to the H_A-topology.

Proof: Consider a sequence $\{h_n\}_{n=1}^{\infty}$ in D(A) satisfying $Ah_n=Tk_n$ for some $k_n \in H_A$, as well as $\|h_n-h\|_A \to 0$ and $\|k_n-k\|_A \to 0$. By assumption (vi), T acts as a bounded operator from H_A into H, and so we have $\|Ah_n-Tk\| = \|Tk_n-Tk\| \to 0$. Because of Lemma 1.1 we have $\|A_0h_n-A_0h\|_{H_A'} \to 0$, and therefore $A_0h=Tk \in T[H_A] \subset H$. We thus conclude that $h \in D(A)$ and $Ah \in T[H_A]$. ■

The proof of Lemma 1.2 used the fact that $T:H_A \to H$ is bounded. This boundedness further implies the condition

$$Z_0(T^{-1}A) = \bigcup_{n=1}^{\infty} \mathrm{Ker}(T^{-1}A)^n \subset D(T)$$

for the zero root linear manifold, since $D(A) \subset H_A \subset D(T)$. Such a condition (for T unbounded) was assumed in Chapters III and IV. Under the additional assumption that T acts as a compact operator from H_A into H_A', the operator $T^{-1}A$ on H_A has compact resolvent and its spectrum is discrete. This is satisfied, in particular, if the operator A has compact resolvent, since in this case the natural imbedding of H into H_A' is compact.

The next result is an adaptation, due to Beals [34], of a lemma proved by Baouendi and Grisvard [22] and extended to the Fokker–Planck model (1.1) by Beals and Protopopescu [35].

LEMMA 1.3. There exist bounded linear operators X and Y on H_A satisfying

$$XQ_+ = Q_+, \tag{1.7a}$$

$$|T|X = Y^*T \tag{1.7b}$$

on H_A, with Y^* the adjoint of Y in H. A similar result holds with Q_- in place of Q_+.

Proof: Suppose first that $I_+=(0,\infty)$ and $I_-=(-\infty,0)$. Let $\varphi:\mathbb{R}\to\mathbb{C}$ be a C^1-function of compact support satisfying $\varphi(0)=1$, and put

$$(Xh)(\mu) = \begin{cases} h(\mu), & \mu\in I_+, \\ \varphi(\mu)\{\alpha_1 t_1 h(-t_1\mu)+\alpha_2 t_2 h(-t_2\mu)\}, & \mu\in I_-, \end{cases}$$

where t_1 and t_2 are two distinct positive numbers and α_1 and α_2 are to be selected. If X is not to introduce a jump at $\mu=0$, we must require

$$\alpha_1 t_1 + \alpha_2 t_2 = 1. \tag{1.8}$$

In order that the equalities (1.7) hold true, we must have

$$(Y^*h)(\lambda) = \begin{cases} h(\mu), & \mu\in I_+, \\ \varphi(\mu)\{\alpha_1 g_1(\mu)h(-t_1\mu)+\alpha_2 g_2(\mu)h(-t_2\mu)\}, & \mu\in I_-, \end{cases}$$

where

$$g_j(\mu) = -w(\mu)/w(-t_j\mu).$$

By assumption there is an interval $(-\varepsilon,\varepsilon)$ on which $w(\mu)=\text{sign}(\mu)|\mu|^{\alpha}m(\mu)$ for some C^1-function $m(\mu)$ with $m(0)\neq 0$ and $\alpha>-½$. This implies

$$g_j(\mu) = (t_j)^{-\alpha} m(\mu)/m(-t_j\mu), \quad -\varepsilon<\mu<\varepsilon.$$

If we now choose the numbers t_1 and t_2 such that $t_1\varepsilon$ and $t_2\varepsilon$ do not belong to the convex hull of the (compact) support of $\varphi(\mu)$, the functions g_1 and g_2 will be continuously differentiable on an open interval containing the support of $\varphi(\mu)$. As a result, φg_1 and φg_2 are continuously differentiable up to $\mu=0$. We then have

$$(Yh)(\mu) = \begin{cases} h(\mu)+\alpha_1(\varphi g_1 h)(-\mu/t_1)+\alpha_2(\varphi g_2 h)(-\mu/t_2), & \mu\in I_+, \\ 0, & \mu\in I_-. \end{cases}$$

In order that Y map H_A into H_A, we must require

$$\alpha_1 g_1(0^-) + \alpha_2 g_2(0^-) = -1. \tag{1.9}$$

Because of the regularity assumption on $w(\mu)$ near $\mu=0$, it is possible to solve Eqs. (1.8) and (1.9) uniquely for t_1 and t_2, since the determinant of the system is

$$t_1 g_2(0^-) - t_2 g_1(0^-) = t_1 t_2 (t_2^{-1-\alpha} - t_1^{-1-\alpha}) \neq 0.$$

The general case is easily proved using a C^1-partition of unity. Let $(U_i)_{i\in I}$ be an open cover of I consisting of open intervals U_i such that $w(\mu)$ changes sign on U_i at most once. Let $(\varphi_i)_{i\in I}$ be a C^1-partition of unity subordinated to the cover $(U_i)_{i\in I}$ (cf. [385] for its existence). This means that each function φ_i is a nonnegative C^1-function on I with its support contained in U_i such that for every $\mu\in I$ we have $\varphi_i(\mu)\neq 0$ for only finitely many i, while $\sum_{i\in I}\varphi_i(\mu) = 1$, $\mu\in I$. For every $i\in I$ we construct the operators X_i and Y_i on H_A of the above type, as if the weight function changes sign in U_i. (The latter occurs at most once.) We then define

$$(Xh)(\mu) = \sum_{i\in I}\varphi_i(\mu)(X_i h)(\mu),$$

$$(Yh)(\mu) = \sum_{i\in I}\varphi_i(\mu)(Y_i h)(\mu),$$

which satisfies the lemma. ∎

As discussed in detail in Chapters III and IV, there exists a strictly positive self adjoint operator A_β which has the same domain as A and coincides with A on a subspace of finite co-dimension (in D(A), H_A and H). Moreover, the completion of D(A) with respect to the positive definite inner product

$$(h,k)_{A_\beta} = (A_\beta h,k) \tag{1.10}$$

coincides with H_A and the operator $S_\beta = A_\beta^{-1}T$ is compact and self adjoint on H_A (relative to (1.10)). We then define P_+ and P_- to be the $(\cdot,\cdot)_{A_\beta}$-orthogonal projections of H_A onto maximal S_β-positive and negative S_β-invariant subspaces. Let H_T denote the Hilbert space of measurable functions $h:I\to\mathbb{C}$ which are bounded in the norm

$$\|h\|_T = [\int_I |h(\mu)|^2 \cdot |w(\mu)|\,d\mu]^{1/2}, \tag{1.11}$$

and let $(\cdot,\cdot)_T$ denote the corresponding inner product on this space. Define H_S to be the completion of H_A with respect to the inner products

$$(h,k)_S = (A_\beta^{-1}T(P_+-P_-)h,k)_{A_\beta}, \tag{1.12}$$

which are equivalent for different β. The projections Q_+ and Q_- extend continuously from D(T) to orthogonal projections on H_T, and likewise the projections P_+ and P_- extend to H_S.

The next result is due to Beals [34].

THEOREM 1.4. The inner products $(\cdot,\cdot)_T$ and $(\cdot,\cdot)_S$ are equivalent on H_A and therefore $H_T \simeq H_S$.

Proof: Given $g=A_\beta^{-1}TP_+h$ where $h\in H_A$, we define

$$\psi(x) = \exp\{-xT^{-1}A_\beta\}g = \exp\{-xT^{-1}A_\beta\}A_\beta^{-1}TP_+h, \quad 0\le x<\infty.$$

Then $\psi(x)\to 0$ in H_A-norm as $x\to\infty$, ψ is H_A-continuously differentiable on $(0,\infty)$ and H_A-continuous on $[0,\infty)$, while $T\psi'(x) = -A_\beta\psi(x)$, $0<x<\infty$. Introducing the bounded operators X and Y on H_A via Lemma 1.3, we obtain the estimate

$$\|Q_+g\|_T^2 \le \|Q_+g\|_T^2 + \|XQ_-g\|_T^2 = (|T|Xg,Xg) = -\int_0^\infty \frac{d}{dx}(|T|X\psi(x),X\psi(x))dx =$$

$$= -2\int_0^\infty (|T|X\psi', X\psi)dx = -2\int_0^\infty (T\psi',YX\psi)dx = 2\int_0^\infty (A_\beta\psi,YX\psi)dx \le$$

$$\le c\int_0^\infty \|\psi\|_{A_\beta}^2 dx = c\int_0^\infty (e^{-2xT^{-1}A_\beta} g,g)_{A_\beta} dx = \tfrac{1}{2}c\|g\|_S^2.$$

Likewise, using Lemma 1.3 with Q_+ replaced by Q_-, we obtain $\|Q_-g\|_T^2 \le \tfrac{1}{2}c\|g\|_S^2$ for $g = A_\beta^{-1}TP_+h \in H_A$. Similar estimates can be obtained if one replaces P_+ by P_-, yielding a constant d instead of c. Hence, on using all four estimates we find

$$\|g\|_T^2 = \|P_+g+P_-g\|_T^2 \le 2\|P_+g\|_T^2+2\|P_-g\|_T^2 \le m(\|P_+g\|_S^2+\|P_-g\|_S^2) = m\|g\|_S^2,$$

where m is some constant. Since $A_\beta^{-1}T[H_A]$ is dense in H_A, we conclude $H_S\subset H_T$. The converse inclusion follows from the proof of Theorem II 4.5, since $H_A\subset D(T)$. ∎

The existence and uniqueness theory for abstract kinetic equations may now be applied. Since an eigenfunction h of A at the eigenvalue λ satisfies the homogeneous second order differential equation (1.4) with $p(\mu)>0$ and $w(\mu)\equiv 1$ for $\mu\in I$, every eigenvalue of A has multiplicity at most two. In most cases of interest the multiplicities of these eigenvalues are, in fact, one. This is certainly true in the case of **separated** boundary conditions [94], since this imposes a linear constraint on the eigenfunctions. Throughout the remainder of this chapter we will assume that Ker A is either trivial or one dimensional, i.e., dim $N_0\le 1$.

We have as a consequence of Theorems III 2.2 and III 2.3 the following result.

THEOREM 1.5. If Ker A = {0} or if Ker A = span$\{\varphi_0\}$ with $(T\varphi_0,\varphi_0) \ge 0$, there exists a unique solution of the differential equation

$$T\psi'(x) = -A\psi(x), \qquad 0<x<\infty, \tag{1.13}$$

with boundary conditions

$$Q_+\psi(0) = \varphi_+, \tag{1.14a}$$

$$\limsup_{x\to\infty} \|\psi(x)\|_T < \infty. \tag{1.14b}$$

This solution satisfies

$$\exists \psi_\infty \in \text{Ker } A,\ \exists M,r>0:\ \|\psi(x)-\psi_\infty\|_T \leq Me^{-rx}. \tag{1.15}$$

On the other hand, if Ker A $= \{0\}$ or if Ker A $= \text{span}\{\varphi_0\}$ with $(T\varphi_0,\varphi_0) < 0$, there exists a unique solution of the differential equation (1.13) with boundary conditions

$$Q_+\psi(0) = \varphi_+, \tag{1.16a}$$

$$\lim_{x\to\infty} \|\psi(x)\|_T = 0. \tag{1.16b}$$

This solution satisfies

$$\exists M,r>0:\ \|\psi(x)\|_T \leq Me^{-rx}. \tag{1.17}$$

In the next several sections we shall present methods to compute the albedo operator for an indefinite Sturm-Liouville problem. Some work in this direction was already done by Pagani [291, 292, 293, 294, 295], Beals and Protopopescu [35, 36], Kaper et al. [209], and Marshall and Watson [257], using techniques from the theory of partial differential equations and employing special functions to arrive at (the equivalent of) the albedo operator for various simple models. However, no general procedure was given. Recently, two general methods were employed by Klaus et al. [220] (announced in [219]). The first method relies on eigenfunction expansions and the solution of the resulting integral equation. The second method consists of the derivation of an equivalent integral equation, which can be solved by Wiener-Hopf factorization. In Section 2 we shall present the first method. The second method will be pursued in Sections 3 and 4.

2. Half-range solutions by eigenfunction expansion

In this section we shall construct solutions of the boundary value problem (1.6)-(1.7) by an eigenfunction expansion method. We continue the notation and assumptions of the previous section.

Following Chapter III, H_A may be decomposed as $H_A = Z_0(T^{-1}A)\oplus Z_1$, and this decomposition reduces the operator $T^{-1}A$. We will write P for the projection of H_A onto Z_1 along $Z_0(T^{-1}A)$, which extends to H_S. On H_S let us define the projection $\tilde{P}_+$ as follows: $\tilde{P}_+f = P_+f$ if Ker A={0} or if Ker A=span{φ_0} with $(\varphi_0,\varphi_0)<0$; $\tilde{P}_+f = P_+f + (I-P)f$ if Ker A=span{φ_0} with $(T\varphi_0,\varphi_0)>0$; and $\tilde{P}_+f = P_+f + \{(f,T\psi_0)/(\varphi_0,T\psi_0)\}\varphi_0$ if Ker A=span{φ_0} with $(\varphi_0,\varphi_0) = 0$. In the last case we choose ψ_0 in such a way that $A\psi_0=T\varphi_0$ and $(T\psi_0,\psi_0)=0$; as a result we then have $(\varphi_0,T\psi_0)=(A\psi_0,\psi_0)>0$, $\tilde{P}_+\varphi_0=\varphi_0$ and $\tilde{P}_+\psi_0=0$. In all three cases the unique solution ψ of the boundary value problem described in Theorem 1.5 can then be represented as

$$\psi(x) = \exp\{-xT^{-1}A\}E_+\varphi_+, \quad 0\le x<\infty,$$

where E_+ is the projection of H_T onto the range of $\tilde{P}_+$ along $Q_-[H_T]$. Following the conventions of the previous chapters we may write $E_+=EQ_+$, where E is the appropriate albedo operator.

LEMMA 2.1. The vector $g=Q_-E_+\varphi_+$ is the unique solution in $Q_-[H_T]$ of the equation

$$(I+K_-)g = K_+\varphi_+, \tag{2.1}$$

where $K_\pm = \pm Q_-\tilde{P}_+Q_\pm: Q_\pm[H_T] \to Q_-[H_T]$.

Proof: One easily calculates that $g=Q_-E_+\varphi_+$ satisfies Eq. (2.1):

$$K_+\varphi_+-K_-g = Q_-\tilde{P}_+(\varphi_++g) = Q_-\tilde{P}_+(Q_+E_+\varphi_++Q_-E_+\varphi_+) = Q_-\tilde{P}_+E_+\varphi_+ = g.$$

Conversely, if g is a solution of Eq. (2.1) in $Q_-[H_T]$, then

$$Q_-(g-\tilde{P}_+(\varphi_++g)) = g-K_+\varphi_++K_-g = 0,$$

and thus there exists $\psi_+\in Q_+[H_T]$ satisfying

$$g-\tilde{P}_+(\varphi_++g) = \psi_+-\varphi_+.$$

This in turn implies $\psi_+\in Q_+[H_T] \cap (I-\tilde{P}_+)[H_S] = \{0\}$, whence $\varphi_++g\in\tilde{P}_+[H_S]$. As a result we have $g=Q_-E_+\varphi_+$. ∎

We may compute E_+ in principle by solving Eq. (2.1) and putting

$$E_+\varphi_+ = \varphi_+ + (I+K_-)^{-1}K_+\varphi_+. \tag{2.2}$$

Let us assume $T:H_A\to H_A'$ is compact. Since $A_\beta^{-1}T$ is then compact, self adjoint and injective on H_A, it has a complete H_A-orthonormal set of eigenfunctions $\{\varphi_n^\beta\}_{n=-\infty}^{\infty}$ (where n=0 is skipped in the numbering) satisfying $A_\beta\varphi_n^\beta = \lambda_n^\beta T\varphi_n^\beta$ with $(\varphi_n^\beta,\varphi_m^\beta)_{A_\beta} = \delta_{nm}$ and

$$\dots \le \lambda_{-3}^\beta \le \lambda_{-2}^\beta \le \lambda_{-1}^\beta<0<\lambda_1^\beta \le \lambda_2^\beta \le \lambda_3^\beta \le \dots .$$

The eigenfunctions form a complete orthogonal set in H_S, with normalization

$$(\varphi_n^\beta,\varphi_m^\beta)_{S_\beta} = |\lambda_n^\beta|^{-1}\delta_{nm}. \tag{2.3}$$

We also have the orthogonality relation

$$(T\varphi_n^\beta,\varphi_m^\beta) = \int_I w(\mu)\varphi_n^\beta(\mu)\overline{\varphi_m^\beta}(\mu)d\mu = (\lambda_n^\beta)^{-1}\delta_{nm}. \tag{2.4}$$

Under the conditions of the oscillation theorem for eigenvalues of Sturm-Liouville differential operators [94], the eigenvalues are simple. However, the multiplicity of the nonzero eigenvalues of A will not play a role in the construction. Note that if a non-strictly positive Sturm-Liouville operator A is modified to obtain A_β, only the multiplicities of finitely many eigenvalues and eigenfunctions will be affected.

In order to specify Eq. (2.1) further, we use the identity (cf. Section II.4)

$$(f,g)_S = (f,(2V-I)g)_T = ((Q_+-Q_-)f,(P_{1,+}-P_{1,-})g)_T,$$

where $\{f,g\}\subset P[H_S]\subset H_S\cong H_T$. As a consequence we obtain, for Ker A={0},

$$K_-h = -\sum_{n>0}\lambda_n((Q_+-Q_-)Q_-h,\ (P_+-P_-)\varphi_n)_TQ_-\varphi_n = \sum_{n>0}\lambda_n(h,Q_-\varphi_n)_TQ_-\varphi_n,$$

$$K_+h = +\sum_{n>0}\lambda_n((Q_+-Q_-)Q_+h,\ (P_+-P_-)\varphi_n)_TQ_-\varphi_n = \sum_{n>0}\lambda_n(h,Q_+\varphi_n)_TQ_-\varphi_n.$$

If Ker $A \neq \{0\}$, we also consider $\lambda_0 = 0$ and Ker $A = \mathrm{span}\{\varphi_0\}$. We then obtain the same formulas if $(T\varphi_0, \varphi_0) < 0$ and formulas corrected by one term if $(T\varphi_0, \varphi_0) \geq 0$.

COROLLARY 2.2. Let $g = Q_- E_+ \varphi_+$ and assume $T: H_A \to H_A'$ is compact. Then:

(i) If Ker $A = \{0\}$ or if Ker $A = \mathrm{span}\{\varphi_0\}$ with $(T\varphi_0, \varphi_0) < 0$, then

$$g + \sum_{n>0} \lambda_n (g, Q_- \varphi_n)_T Q_- \varphi_n = \sum_{n>0} \lambda_n (\varphi_+, Q_+ \varphi_n)_T Q_- \varphi_n.$$

(ii) If Ker $A = \mathrm{span}\{\varphi_0\}$ with $(T\varphi_0, \varphi_0) > 0$, then

$$g + \sum_{n>0} \lambda_n (g, Q_- \varphi_n)_T Q_- \varphi_n + \frac{(g, TQ_- \varphi_0)}{(\varphi_0, T\varphi_0)} Q_- \varphi_0 =$$

$$= \sum_{n>0} \lambda_n (\varphi_+, Q_+ \varphi_n)_T Q_- \varphi_n + \frac{(\varphi_+, TQ_+ \varphi_0)}{(\varphi_0, T\varphi_0)} Q_- \varphi_0.$$

(iii) If Ker $A = \mathrm{span}\{\varphi_0\}$ with $(T\varphi_0, \varphi_0) = 0$, then

$$g + \sum_{n>0} \lambda_n (g, Q_- \varphi_n)_T Q_- \varphi_n + \frac{(g, TQ_- \psi_0)}{(\varphi_0, T\psi_0)} Q_- \varphi_0 =$$

$$= \sum_{n>0} \lambda_n (\varphi_+, Q_+ \varphi_n)_T Q_- \varphi_n + \frac{(\varphi_+, TQ_+ \psi_0)}{(\varphi_0, T\psi_0)} Q_- \varphi_0.$$

3. Reduction to a modified Sturm–Liouville problem

In the present and the next section we shall present a second method to compute the albedo operator for Sturm–Liouville diffusion problems. We restrict ourselves to Sturm–Liouville operators on intervals $I = (a,b)$ with domains D which lead to separated boundary conditions of Neumann–Dirichlet type. That is to say, if $p(x)$ extends to a continuous strictly positive function and $q(x)$ to a real continuous function at each finite endpoint a,b (regular endpoints), then the Sturm–Liouville operator A is a self adjoint operator on H with boundary conditions

$$\cos\alpha g(a) - p(a)\sin\alpha g'(a) = 0, \tag{3.1}$$

$$\cos\beta g(b) - p(b)\sin\beta g'(b) = 0, \tag{3.2}$$

for certain $\alpha,\beta \in [0,\pi)$. If one of the endpoints is singular, then the corresponding boundary condition is replaced either by a limit condition as μ approaches the endpoint (limit circle case) or by no endpoint boundary condition (limit point case). If both endpoints are singular, one may consider two subproblems on (a,c) and (c,b), where the boundary condition at an arbitrary intermediate point $c \in (a,b)$ has the form

$$g(c)\cos\gamma - p(c)g'(c)\sin\gamma = 0 \tag{3.3}$$

for any $\gamma \in [0,\pi)$. One may then distinguish between the limit point case and limit circle case at each of the singular endpoints of the respective subintervals (a,c) and (c,b). For a thorough treatment of these boundary conditions we refer to [18, 94, 195].

The resolvent for any of these Sturm–Liouville operators is given by

$$[(A-\lambda)^{-1}f](\mu) = \frac{1}{W(\lambda)}\{\psi(\mu,\lambda)\int_a^{\mu}\varphi(\nu,\lambda)f(\nu)d\nu + \varphi(\mu,\lambda)\int_{\mu}^{b}\psi(\nu,\lambda)f(\nu)d\nu\}.$$

Here

$$W(\lambda) \equiv p(\mu)\{\varphi'(\mu,\lambda)\psi(\mu,\lambda) - \psi'(\mu,\lambda)\varphi(\mu,\lambda)\} \tag{3.4}$$

and the functions φ and ψ are nontrivial solutions of the eigenvalue equation

$$(Ah)(\mu) = \lambda h(\mu), \qquad \mu \in I, \tag{3.5}$$

locally square integrable and satisfying the boundary condition at a and b, respectively. These functions are both analytic on the open upper and the open lower half plane, and hence the resolvent operator can be analytically continued to the open upper and lower half planes and an open subset of the real line. The (simple and real) eigenvalues of $T^{-1}A$ then appear as simple zeros of $W(\lambda)$.

Next, let us explain the ramifications of decomposing the Sturm–Liouville operator into a problem on (a,c) and a problem on (c,b). Let $\chi(\mu)$ be a non-trivial solution of Eq. (3.5) on (a,b) that satisfies the boundary condition (3.3). The resolvents of the Sturm–Liouville operators on (a,c) and (c,b) have the respective forms

$$[(\hat{A}_\ell-\lambda)^{-1}f](\mu) = \frac{1}{W_\ell(\lambda)}\{\chi(\mu,\lambda)\int_a^\mu \varphi(\nu,\lambda)f(\nu)d\nu+\varphi(\mu,\lambda)\int_\mu^c \chi(\nu,\lambda)f(\nu)d\nu\},$$

where

$$W_\ell(\lambda) = p(\mu)\{\varphi'(\mu,\lambda)\chi(\mu,\lambda) - \chi'(\mu,\lambda)\varphi(\mu,\lambda)\},$$

and

$$[(\hat{A}_r-\lambda)^{-1}f](\mu) = \frac{1}{W_r(\lambda)}\{\psi(\mu,\lambda)\int_c^\mu \chi(\nu,\lambda)f(\nu)d\nu + \chi(\mu,\lambda)\int_\mu^b \psi(\nu,\lambda)f(\nu)d\nu\},$$

where

$$W_r(\lambda) = p(\mu)\{\chi'(\mu,\lambda)\psi(\mu,\lambda) - \psi'(\mu,\lambda)\chi(\mu,\lambda)\}.$$

Writing

$$\chi(\mu,\lambda) = c_\ell(\lambda)\varphi(\mu,\lambda) + c_r(\lambda)\psi(\mu,\lambda) \tag{3.6}$$

and using the formulas

$$c_\ell(\lambda) = W_r(\lambda)/W(\lambda), \tag{3.7a}$$

$$c_r(\lambda) = W_\ell(\lambda)/W(\lambda), \tag{3.7b}$$

obtained by substituting (3.6) into the formulas for the Wronskian expressions W_ℓ and W_r, we finally get

$$[(A-\lambda)^{-1}f-(\hat{A}_\ell-\lambda)^{-1}f](\mu) = -\frac{W_r(\lambda)}{W_\ell(\lambda)W(\lambda)}\varphi(\mu,\lambda)\int_a^c \varphi(\nu,\lambda)f(\nu)d\nu,$$

where $\mu\in(a,c)$ and $f\in L_2(a,c)$, and

$$[(A-\lambda)^{-1}f-(\hat{A}_r-\lambda)^{-1}f](\mu) = -\frac{W_\ell(\lambda)}{W_r(\lambda)W(\lambda)}\psi(\mu,\lambda)\int_c^b \psi(\nu,\lambda)f(\nu)d\nu,$$

where $\mu\in(c,b)$ and $f\in L_2(c,b)$. Thus if λ belongs to the resolvent set of all three Sturm–Liouville operators and $\hat{A}$ denotes the direct sum of the Sturm–Liouville operators on $L_2(a,c)$ and $L_2(c,b)$ (viewed as an operator on $L_2(a,b)$), we have, for

$f \in L_2(a,b)$,

$$[(A-\lambda)^{-1}f-(\hat{A}-\lambda)^{-1}f](\mu) =$$

$$= -\frac{W_\ell(\lambda)W_r(\lambda)}{W(\lambda)} k(\mu,\lambda)\int_a^b k(\nu,\lambda)f(\nu)d\nu, \tag{3.8}$$

where

$$k(\mu,\lambda) = \begin{cases} W_\ell(\lambda)^{-1}\varphi(\mu,\lambda), & \mu\in(a,c), \\ W_r(\lambda)^{-1}\psi(\mu,\lambda), & \mu\in(c,b). \end{cases} \tag{3.9}$$

Hence, the difference of the resolvents is an operator of rank one.

If A is positive self adjoint with spectrum $\sigma(A)\subset\{0\}\cup[\varepsilon,\infty)$ for some $\varepsilon>0$, it is possible to choose the constant γ in condition (3.3) in such a way that the resulting operator $\hat{A}$ is strictly positive self adjoint. Let us normalize $\chi(x,\lambda)$ by requiring $\chi(\mu,\lambda)=p(c)\sin\gamma$ and $\chi'(\mu,\lambda)=\cos\gamma$. We then easily derive the identities

$$W_\ell(\lambda) = +p(c)\{p(c)\sin\gamma\varphi'(c,\lambda)-\cos\gamma\varphi(c,\lambda)\},$$

$$W_r(\lambda) = -p(c)\{p(c)\sin\gamma\psi'(c,\lambda)-\cos\gamma\psi(c,\lambda)\},$$

from (3.6) and (3.7) and the derivative of (3.6) with respect to μ (for λ real and $W(\lambda)\neq0$), and observe that $W_\ell(\lambda)=W_r(\lambda)=0$ implies $W(\lambda)=0$. The latter then implies the existence of unique and distinct $\gamma_\ell=\gamma_\ell(\lambda)$ and $\gamma_r=\gamma_r(\lambda)$ in $[0,\pi)$ satisfying $W_\ell(\lambda)=0$ for $\gamma=\gamma_\ell(\lambda)$ and $W_r(\lambda)=0$ for $\gamma=\gamma_r(\lambda)$. This in turn gives the existence of an interval of values of γ where $W_\ell(\lambda)W_r(\lambda)W(\lambda)^{-1}>0$ and an interval of values of γ where $W_\ell(\lambda)W_r(\lambda)W(\lambda)^{-1}<0$. On inspecting (3.8) it is evident that we may choose γ in such a way that

$$(\hat{A}-\lambda)^{-1} \geq (A-\lambda)^{-1} \geq 0$$

for an interval of values of γ. If A is positive with isolated (simple) zero eigenvalues, then $W(0)=0$, $W_\ell(0)\neq0$ and $W_r(0)\neq0$, except for the unique $\gamma=\gamma_0$ satisfying the equality $\varphi'(c,0)p(c)\sin\gamma_0 = \varphi(c,0)\cos\gamma_0$. Under the conditions of the usual oscillation theorems we have $\varphi(\mu,0)\neq0$ for $\mu\in I$ and therefore $\gamma_0\in(0,\pi)$. If we exclude γ_0 from this interval, we find an interval of γ (the same or a smaller one) where $\sigma(\hat{A})\subset(0,\infty)$. Hence, under these conditions the constant γ in condition (3.3) can be chosen so as to make $\hat{A}$ strictly positive.

THEOREM 3.1. Let us consider a weight function $w(\mu)$ on (a,b) with finitely many sign changes at $c_1,\ldots,c_N$ $(a<c_1<\ldots<c_N<b)$. Let A be positive with zero as an isolated eigenvalue, or strictly positive, and if $\operatorname{Ker} A\neq\{0\}$, assume that the zero eigenfunction does not have zeros in (a,b). At the sign changes $c_1,\ldots,c_N$ let us add boundary conditions of the type

$$g(c_j)\cos\gamma_j - p(c_j)g'(c_j)\sin\gamma_j = 0, \quad j=1,2,\ldots,N,$$

and obtain strictly positive Sturm–Liouville operators $\hat{A}_0,\hat{A}_1,\ldots,\hat{A}_N$ on the respective subintervals $(a,c_1),(c_1,c_2),\ldots,(c_N,b)$. Then the direct sum $\hat{A} = \hat{A}_0\oplus\hat{A}_1\oplus\ldots\oplus\hat{A}_N$ is strictly positive self adjoint on H. If the Sturm–Liouville operator A is strictly positive, then $C = A^{-1}T-\hat{A}^{-1}T$ is a bounded operator on H_T of rank N.

Proof: The proof follows easily by induction on N if the weight function is bounded. First we construct $\hat{A}_0$ by using Eq. (3.8) for $c=c_1$. On (c_1,b) we further split up the Sturm–Liouville operator obtained at $c=c_2$ and apply (3.8) again. After N steps we have constructed $\hat{A}_0,\ldots,\hat{A}_N$ and $(A-\lambda)^{-1}-(\hat{A}-\lambda)^{-1}$ is an operator of rank N for $\lambda\notin\sigma(A)\cup\sigma(\hat{A}) = \sigma(A)\cup\sigma(\hat{A}_0)\cup\ldots\cup\sigma(\hat{A}_N)$. If the weight function $w(\mu)$ is bounded, then $C(\lambda) = \{(A-\lambda)^{-1}-(\hat{A}-\lambda)^{-1}\}T$ $(\lambda\notin\sigma(A)\cup\sigma(\hat{A}))$ is bounded on H_T. If the weight function $w(\mu)$ is unbounded, it must be unbounded near $a,c_1,\ldots,c_N$ or b. Let $\varphi(\mu)$ be a positive C^∞-function on I with compact support within $I\setminus\{c_I,\ldots,c_N\}$. For each of the functions $k_j(\mu,\lambda)$, $\lambda\notin\sigma(A)\cup\sigma(\hat{A})$, constructed in (3.9) the function $\varphi(\mu)k_j(\mu,\lambda)$ is continuous and has compact support on which $w(\mu)$ is bounded; thus,

$$\int_I |w(\mu)|^2\cdot|\varphi(\mu)k_j(\mu,\lambda)|^2 d\mu < \infty.$$

Since $(1-\varphi(\mu))k_j(\mu,\lambda)$ is continuous on (a,b) and satisfies the boundary conditions (or local square integrability), $w(\mu)=O(|\mu-c_j|^{\alpha_j})$ $(\mu\to c_j)$ for some $\alpha_j>-\frac{1}{2}$, and $D(A)\subset D(T)$, one may show that the function $(1-\varphi(\mu))k_j(\mu,\lambda)$ is square integrable on I with weight $|w(\mu)|$. Hence, $C(\lambda) = \{(A-\lambda)^{-1}-(\hat{A}-\lambda)^{-1}\}T$ is a bounded operator on H_T, and has rank N if $\lambda\notin\sigma(A)\cup\sigma(\hat{A})$. ∎

As a result of the above theorem, the operator $T_1 = \hat{A}^{-1}T$ is bounded on H and compact if A has a compact resolvent. Also, T_1 commutes with the projections $Q_\pm$ defined by

$$(Q_\pm h)(\mu) = \begin{cases} h(\mu), & \pm w(\mu) > 0 \\ 0, & \pm w(\mu) < 0. \end{cases}$$

If we denote by $H_{\hat{A}} = D(\hat{A}^{1/2})$ the completion of $D(\hat{A})$ with respect to $(\cdot,\cdot)_{\hat{A}} = (\hat{A}\cdot,\cdot)$, then T_1 is self adjoint on $H_{\hat{A}}$, Q_+ and Q_- are $H_{\hat{A}}$-orthogonal projections onto maximal T_1-positive and negative T_1-invariant subspaces and $|T_1| = \hat{A}^{-1}|T|$ is the $H_{\hat{A}}$-positive absolute value of T_1. Thus,

$$(h,k)_{T_1} = (|T_1|h,k)_{\hat{A}} = (\hat{A}^{-1}|T|h,k)_{\hat{A}} = (h,k)_T. \tag{3.10}$$

Also, defining $A_1 = \hat{A}^{-1}A$ as an $H_{\hat{A}}$-positive operator, we have

$$T_1^{-1}A_1 = T^{-1}A$$

and, if Ker A={0},

$$(h,k)_{S_1} = (|A_1^{-1}T_1|h,k)_A = (A_1^{-1}T_1(P_+ - P_-)h,k)_A = (h,k)_S. \tag{3.11}$$

Hence, the Sturm-Liouville diffusion equation on $H_T \simeq H_S$ has precisely the same solutions as the modified problem

$$T_1\psi'(x) = -A_1\psi(x), \quad 0<x<\infty, \tag{3.12}$$

$$Q_+\psi(0) = \varphi_+, \tag{3.13}$$

$$\|\psi(x)\|_T = O(1) \text{ or } o(1) \ (x\to\infty), \tag{3.14}$$

and the uniqueness properties of the latter problem can be described by Theorem 1.5. We have thus obtained the problem (3.12)-(3.14) of the same type and with the same solutions as the original one, but now the resolvents of $T^{-1}A$ and $T_1^{-1}\hat{A}$ have a rank

N difference. As we shall see, it is exactly this rank condition which guarantees a further reduction to a matrix integral equation and factorization.

4. The integral form of Sturm–Liouville diffusion problems and factorization

In this section we shall reformulate the modified Sturm–Liouville problem (3.12)–(3.14) as an integral equation of convolution type, which for strictly positive Sturm–Liouville operators A will be solved by factorization. Throughout we assume A strictly positive, but at the end of this section we will indicate how this assumption may be relaxed in order to keep at least a part of the results. We shall need the following technical assumption:

$$\exists 0<\alpha<1: \quad \text{Ran } C \subset |T_1|^{\alpha}[H_T] \cap |A^{-1}T|^{\alpha}[H_T], \tag{4.1}$$

where $|T_1| = \hat{A}^{-1}|T|$ and $|A^{-1}T|$ are the absolute values of the operators T_1 and $A^{-1}T$ with respect to the inner products $(\cdot,\cdot)_{\hat{A}}$ and $(\cdot,\cdot)_A$, respectively. In the appendix of [220] one may find sufficient conditions for (4.1) to be true. We shall exploit the estimates

$$\| \, |T_1|^{\alpha} \mathcal{H}_1(x) \|_{H_T} = O(|x|^{\alpha-1}) \ (x\to 0), \tag{4.2a}$$

$$\| \, |A^{-1}T|^{\alpha-1} \exp\{-xT^{-1}A\} P_+ \|_{H_T} = O(|x|^{\alpha-1}) \ (x\to 0), \tag{4.2b}$$

where $\mathcal{H}_1(x)$ is the propagator function, defined by

$$\mathcal{H}_1(x) = \begin{cases} +T_1^{-1}\exp\{-xT_1^{-1}\}Q_+, & 0<x<\infty, \\ -T_1^{-1}\exp\{-xT_1^{-1}\}Q_-, & -\infty<x<0. \end{cases}$$

LEMMA 4.1. Suppose that $\varphi_+ \in Q_+[H_T]$. Then the vector function $\varphi(x) = T^{-1}A\exp\{-xT^{-1}A\}E_+\varphi_+$ is the unique solution of the integral equation

$$\varphi(x) + \frac{d}{dx}\int_0^{\infty} \mathcal{H}_1(x-y)C\varphi(y)dy = \mathcal{H}_1(x)\varphi_+, \quad 0<x<\infty, \tag{4.3}$$

satisfying $\int_{\varepsilon}^{\infty} e^{rx}\|\varphi(x)\|_T dx<\infty$ for some $r>0$.

Proof: Using (4.1) and (4.2) one easily proves that for every $\varepsilon\leq x<\infty$

$$\int_{\varepsilon}^{\infty}\|\mathcal{H}(x-y)C\varphi(y)\|_T dy \leq M(\varepsilon) < \infty, \tag{4.4}$$

so that $\int_0^{\infty}\mathcal{H}_1(x-y)C\varphi(y)dy$ is an absolutely convergent Bochner integral for $0<x<\infty$. We compute

$$T_1\int_0^{\infty}\mathcal{H}_1(x-y)C\varphi(y)dy = \int_0^x \frac{d}{dy}\{e^{-(x-y)T^{-1}}Q_+e^{-yT^{-1}A}E_+\varphi_+\}dy +$$
$$+\int_x^{\infty}-\frac{d}{dy}\{e^{-(x-y)T^{-1}}Q_-e^{-yT^{-1}A}E_+\varphi_+\}dy = \psi(x) - e^{-xT^{-1}}\varphi_+,$$

for $0<x<\infty$, where we have used the identity $Q_+E_+=Q_+$. By differentiation and using $\psi'(x)=-\varphi(x)$, we get (4.3).

Conversely, since $\int_1^{\infty}e^{rx}\|\varphi(x)\|_T dx<\infty$ for some $r>0$, we may put $\psi(x) = \int_x^{\infty}\varphi(y)dy$, integrate (4.3) and obtain

$$\psi(x) - \int_0^{\infty}\mathcal{H}_1(x-y)C\varphi(y)dy = e^{-xT_1^{-1}}\varphi_+, \quad 0<x<\infty, \tag{4.5}$$

where the integral term is strongly differentiable for $x\in(0,\infty)$. Let us write

$$T_1^2\psi(x) = \int_0^x T_1e^{-(x-y)T_1^{-1}}Q_+C\varphi(y)dy - \int_x^{\infty}T_1e^{-(x-y)T_1^{-1}}Q_-C\varphi(y)dy +$$
$$+ T_1^2e^{-xT_1^{-1}}\varphi_+.$$

Because of the estimate (4.4) and dominated convergence (for Bochner integrals; cf. [401]) we may differentiate this equation in the following manner:

$$T_1^2\psi'(x) + \int_0^x e^{-(x-y)T_1^{-1}}Q_+C\varphi(y)dy - T_1Q_+C\varphi(x) +$$
$$+\int_x^{\infty}-e^{-(x-y)T_1^{-1}}Q_-C\varphi(y)dy - T_1Q_-C\varphi(x) = -T_1e^{-xT_1^{-1}}\varphi_+,$$

whence

$$T_1\psi'(x) - C\varphi(x) + \int_0^\infty \mathcal{H}_1(x-y)C\varphi(y)dy = -e^{-xT_1^{-1}}\varphi_+, \quad 0<x<\infty. \tag{4.6}$$

On adding (4.5) and (4.6) we obtain

$$T_1\psi'(x) - C\varphi(x) + \psi(x) = 0, \quad 0<x<\infty.$$

By virtue of $\psi'(x)=-\varphi(x)$ and $T_1+C=A^{-1}T$, this in turn implies (3.12). Using (4.5) we conclude

$$Q_+\psi(x) = e^{-xT_1^{-1}}\varphi_+ + \int_0^x \mathcal{H}_1(x-y)C\varphi(y)dy, \tag{4.7}$$

which gives (3.13). ■

Let $C^\dagger$ denote the H_T-adjoint of C, i.e., $C^\dagger = (Q_+-Q_-)C(Q_+-Q_-)$. Let j denote the natural imbedding of Ran $C^\dagger$ into H_T, and π the H_T-orthogonal projection of H_T onto Ran $C^\dagger$ (as an operator $\pi:H_T\rightarrow\text{Ran } C^\dagger$). Then j and π are adjoints and $Cj\pi = C$. These operators can be used to obtain a reduction of order similar to that in Section VII.4. Putting $\varsigma(x)=\pi\varphi(x)$, we obtain

$$\varphi(x) = \mathcal{H}_1(x)\varphi_+ - \frac{d}{dx}\int_0^\infty \mathcal{H}_1(x-y)Cj\varsigma(y)dy, \quad 0<x<\infty, \tag{4.8}$$

where

$$\varsigma(x) + \frac{d}{dx}\int_0^\infty \pi\mathcal{H}_1(x-y)Cj\varsigma(y)dy = \pi\mathcal{H}_1(x)\varphi_+, \quad 0<x<\infty. \tag{4.9}$$

We then have $\int_\varepsilon^\infty e^{rx}\|\varsigma(x)\|dx<\infty$ for some $r>0$. Conversely, if $\varsigma(x)$ is a solution of Eq. (4.9) satisfying $\int_\varepsilon^\infty e^{rx}\|\varsigma(x)\|dx<\infty$ for some $r>0$ and $\varphi(x)$ is given by (4.8), we may premultiply (4.8) by π and subtract the resulting equation from (4.9), whence $\varsigma(x)=\pi\varphi(x)$. Substituting the latter in (4.8) and employing (4.7) we get (4.3). Hence, Eq. (4.3) is equivalent to the system of equations (4.8) and (4.9), but the latter are formulated on the finite dimensional space Ran $C^\dagger$. Next, let us transform (4.9) into a Riemann-Hilbert problem, which we shall then solve by factorization. Using (4.4) it is not difficult to establish that the vector $\int_0^\infty \pi\mathcal{H}_1(x-y)C_j\varsigma(y)dy$ is (strongly) differentiable for $x\epsilon(-\infty,0)$. Let us put

$$\varsigma(x) = -\frac{d}{dx}\int_0^\infty \pi\mathcal{H}_1(x-y)Cj\varsigma(y)dy, \quad -\infty<x<0.$$

Since $\varsigma(x)=\pi\varphi(x)=\pi T^{-1}A\exp\{-xT^{-1}A\}E_+\varphi_+$ and (4.1) is satisfied, we have $\int_0^\infty e^{rx}\|\varsigma(x)\|dx<\infty$ and therefore

$$\int_{-\infty}^\infty e^{-s|x|}\left\|\int_0^\infty \pi\mathcal{H}_1(x-y)Cj\varsigma(y)dy\right\|dx < \infty$$

for some $s>0$. We may define the Laplace transforms

$$\hat{\varsigma}_\pm(\lambda) = \pm\int_0^{\pm\infty} e^{\lambda x}\varsigma(x)dx, \tag{4.10a}$$

$$\hat{k}(\lambda) = \int_{-\infty}^\infty e^{\lambda x}\pi\mathcal{H}_1(x)Cjdx, \tag{4.10b}$$

$$\hat{\omega}(\lambda) = \int_0^\infty e^{\lambda x}\pi\mathcal{H}_1(x)\varphi_+dx = \pi(I-\lambda T_1)^{-1}T_1\varphi_+, \tag{4.10c}$$

where $\mathrm{Re}\lambda=0$. Using the formula

$$\int_{-\infty}^\infty e^{\lambda x}f'(x)dx = [e^{\lambda x}f(x)]_{x=-\infty}^{0^-} + [e^{\lambda x}f(x)]_{x=0^+}^{\infty} - \lambda\int_{-\infty}^\infty e^{\lambda x}f(x)dx,$$

we obtain the Riemann-Hilbert problem

$$[I-\lambda\hat{k}(\lambda)]\hat{\varsigma}_+(\lambda) + \hat{\varsigma}_-(\lambda) = \hat{\omega}(\lambda), \quad \mathrm{Re}\lambda=0. \tag{4.11}$$

This problem must be solved using (4.10) and a factorization of the **dispersion function**

$$\Lambda(\lambda) = I-\lambda\hat{k}(\lambda) = I-\lambda\pi(I-\lambda T_1)^{-1}Cj, \quad \mathrm{Re}\lambda=0.$$

LEMMA 4.2. For Ker $A=\{0\}$ the dispersion function

$$\Lambda(1/\xi) = I-\pi(\xi-T_1)^{-1}Cj \tag{4.12}$$

allows a factorization of the form

$$\Lambda(1/\xi)^{-1} = H_\ell(-1/\xi)H_r(1/\xi), \tag{4.13}$$

where the factors satisfy the following conditions:

(i) $H_\ell(z)$, $H_\ell(z)^{-1}$, $H_r(z)$ and $H_r(z)^{-1}$ are continuous on the closed right half plane (except at infinity) and analytic on the open right half plane.

(ii) $H_\ell(0^+)$ and $H_r(0^+)$ are the identity operator, where the limits at $z=0$ are taken from the closed right half plane.

(iii) $\|H_\ell(z)\|$, $\|H_\ell(z)^{-1}\|$, $\|H_r(z)\|$ and $\|H_r(z)^{-1}\|$ are all of order $o(z)$ as $z\to\infty$, $\mathrm{Re}\,z\geq 0$.

If the dispersion function has two factorizations of the type (4.13) with the factors satisfying the properties (i)–(iii), then the factorizations are related by the formulas

$$H_\ell^{(2)}(z) = H_\ell^{(1)}(z)(I+zD),$$

$$H_r^{(2)}(z) = (I+zD)H_r^{(1)}(z),$$

where $D^2=0$ and for $i=1,2$ the expressions $\|H_\ell^{(i)}(z)D\|$, $\|H_\ell^{(i)}(z)^{-1}D\|$, $\|H_r^{(i)}(z)D\|$ and $\|H_r^{(i)}(z)^{-1}D\|$ vanish as $z\to\infty$ from the closed right half plane.

Proof: The operator function (4.12) can be factorized using Theorem VII 3.1, where $\mathbb{A}=T_1$, $\mathbb{B}=Cj$, $\mathbb{C}=-\pi$, $\mathbb{D}=I$ and $\mathbb{A}^\times=\mathbb{A}-\mathbb{B}\mathbb{D}^{-1}\mathbb{C}=T_1+Cj\pi=A^{-1}T$. Then $E_+=EQ_+$ is a bounded projection on H_T whose range $P_+[H_S]$ is invariant under $\mathbb{A}^\times=A^{-1}T$ and whose kernel $Q_-[H_T]$ is invariant under $\mathbb{A}=T_1$. As a result we find the factorization (4.13), where

$$H_\ell(-1/\xi) = I + \pi(\xi-A^{-1}T)^{-1}E_+Cj, \tag{4.14}$$

$$H_\ell(-1/\xi)^{-1} = I - \pi E_+(\xi-T_1)^{-1}Cj, \tag{4.15}$$

$$H_r(1/\xi) = I + \pi(I-E_+)(\xi-A^{-1}T)^{-1}Cj, \tag{4.16}$$

$$H_r(1/\xi)^{-1} = I - \pi(\xi-T_1)^{-1}(I-E_+)Cj. \tag{4.17}$$

It is easily seen that the factors (4.14)–(4.17) have property (i). Properties (ii) and (iii) follow from the Spectral Theorem. Indeed, if S is a strictly positive self adjoint operator, then for every vector h we have

$$\lim_{\xi\to 0,\,\mathrm{Re}\,\xi\le 0} \|\xi(\xi-S)^{-1}h\| = 0.$$

Using similar reasoning one may show that, for all $\omega\in(0,\frac{1}{2}\pi)$, $\|\Lambda(z)\| = o(z)$ and $\|\Lambda(z)^{-1}\| = o(z)$ as $z\to\infty$ with $|\frac{1}{2}\pi-\arg z|\le\omega$.

Finally, if the dispersion function has two factorizations of the type (4.13) (denoted by superscripts (1) and (2)), put

$$F(z) = H_\ell^{(2)}(-z)^{-1}H_\ell^{(1)}(-z) = H_r^{(2)}(z)H_r^{(1)}(z)^{-1}, \quad \mathrm{Re}\,z=0.$$

It then is evident that $F(z)$ and $F(z)^{-1}$ are entire functions satisfying $\|F(z)\|=o(z^2)$ and $\|F(z)^{-1}\|=o(z^2)$ as $z\to\infty$. Using Liouville's theorem and property (ii) one obtains $F(z)=I+zD$ and $F(z)^{-1}=I-zD$, where $D^2=0$. ∎

In the case when the weight $w(\mu)$ has more than one sign change, the factorization (4.13) may be nonunique. However, this does not affect the form of the (unique) albedo operator.

Using a factorization of the type (4.13) we reduce the algebraic equation (4.11) to the Riemann-Hilbert problem

$$H_\ell(-\lambda)^{-1}\hat{\varsigma}_+(\lambda) + H_r(\lambda)\hat{\varsigma}_-(\lambda) = H_r(\lambda)\hat{\omega}(\lambda), \quad \mathrm{Re}\,\lambda=0. \tag{4.18}$$

LEMMA 4.3. The Riemann-Hilbert problem (4.18) has precisely one solution $\hat{\varsigma}$ of the following type:

(i) $\hat{\varsigma}_\pm(\lambda)$ is analytic in the open left/right half plane and continuous on the closed left/right half plane,

(ii) $\lim_{\lambda\to\infty} \lambda\hat{\varsigma}_+(\lambda)$ exists as λ approaches infinity from the closed left half plane,

(iii) $\lim_{\lambda\to\infty} \lambda^c\hat{\varsigma}_-(\lambda)$ exists for all $0<c<1$, as λ approaches infinity from the closed right half plane.

This solution of the Riemann-Hilbert problem (4.18) leads to a unique solution of the boundary value problem (3.12)-(3.14).

Proof: Put $h_\pm = \lim_{\lambda\to\infty} \lambda\hat{\varsigma}_\pm(\lambda)$ as λ approaches infinity from the closed left/right half plane. Then the conditions on H_ℓ and H_r imply

$$\lim_{\lambda\to\infty,\,\mathrm{Re}\lambda\le 0} H_\ell(-\lambda)^{-1}\hat{\varsigma}_+(\lambda) = 0,$$

$$\lim_{\lambda\to\infty,\,\mathrm{Re}\lambda\ge 0} H_r(\lambda)\hat{\varsigma}_-(\lambda) = 0.$$

From (4.10c) we also have

$$\lim_{\lambda\to\infty,\,\mathrm{Re}\lambda=0} \lambda\hat{\omega}(\lambda) = -\varphi_+,$$

$$\lim_{\lambda\to\infty,\,\mathrm{Re}\lambda=0} H_r(\lambda)\hat{\omega}(\lambda) = 0.$$

Given a Hölder continuous function $\hat{h}(\lambda)$ on the extended imaginary line satisfying $\hat{h}(\pm i\infty)=0$, we can find unique functions $\hat{h}_\pm(\lambda)$ that are analytic on the open left/right half plane, are continuous on the closed left/right half plane and satisfy $\hat{h}_\pm(i\infty)=0$ (when approached from the appropriate half plane) such that $\hat{h}(\lambda) = \hat{h}_+(\lambda) + \hat{h}_-(\lambda)$ for $\mathrm{Re}\lambda=0$ (cf. [276]). We therefore obtain

$$\hat{\varsigma}_+(\lambda) = H_\ell(-\lambda)(H_r\hat{\omega})_+(\lambda),$$

$$\hat{\varsigma}_-(\lambda) = H_r(\lambda)^{-1}(H_r\hat{\omega})_-(\lambda).$$

As a consequence of Lemma 4.3, these formulas do not depend on the particular choice of H_ℓ and H_r.

Finally, the solution of problem (4.11) is given by

$$\hat{\varsigma}_-(\lambda) = \int_{-\infty}^{0} e^{\lambda x}\varsigma(x)dx = -\pi(I-\lambda T_1)^{-1}Q_-E_+\varphi_+, \quad \mathrm{Re}\lambda=0.$$

To obtain this we have employed the expression

$$\varsigma(x) = -\frac{d}{dx}\int_0^\infty \pi\mathcal{H}_1(x-y)C\varphi(y)dy = \frac{d}{dx}[\pi e^{-(x-y)T_1^{-1}}Q_-e^{-yT^{-1}A}E_+\varphi_+]_{y=0}^{\infty} =$$

$$= -\frac{d}{dx}\pi e^{-xT_1^{-1}}Q_-E_+\varphi_+ = -\pi\mathcal{H}_1(x)E_+\varphi_+, \quad -\infty<x<0.$$

From this and the estimate $\|H_r(z)\| = O(z^{1-\alpha})$ ($z\to\infty$, $\mathrm{Re}z\ge 0$), the properties (i), (ii) and (iii) are clear. ■

THEOREM 4.4. The albedo operator is given by

$$E_+\varphi_+ = \varphi_+ + \int_{-\infty}^{0} \sigma_1(d\mu)\int_0^\infty \frac{\mu\nu}{\mu-\nu} CjH_\ell(-\mu)H_r(\nu)\pi\sigma_1(d\nu)\varphi_+, \tag{4.19}$$

where $\varphi_+ \in Q_+[H_T]$ and $\sigma_1(\cdot)$ is the resolution of the identity of $T^{-1}A$.

Proof: Choose $\varphi_+ \in Q_+[H_T]$ and some factorization of the type (4.13). Then

$$E_+\varphi_+ = \psi(0) = \int_0^\infty \varphi(x)dx = \int_0^\infty \{\mathcal{H}_1(x)\varphi_+ - \frac{d}{dx}\int_0^\infty \mathcal{H}_1(x-y)Cj\varsigma(y)dy\}dx =$$

$$= \varphi_+ + \int_0^\infty \mathcal{H}_1(-y)Cj\varsigma(y)dy = \varphi_+ + \int_0^\infty dy \int_{-\infty}^0 \sigma_1(d\mu)(-\mu)e^{\mu y}Cj\varsigma(y) =$$

$$= \varphi_+ + \int_{-\infty}^0 \sigma_1(d\mu)(-\mu)Cj\hat{\varsigma}_+(\mu). \tag{4.20}$$

The right hand side of (4.18), with $\lambda=\mu$, can be written as

$$H_r(\mu)\hat{\omega}(\mu) = H_r(\mu)\int_0^\infty dx\; e^{\mu x}\int_0^\infty \nu e^{-\nu x}\pi\sigma_1(d\nu)\varphi_+ =$$

$$= H_r(\mu)\int_0^\infty \frac{\nu}{\nu-\mu}\pi\sigma_1(d\nu)\varphi_+ = \int_0^\infty \nu\left\{\frac{H_r(\nu)}{\nu-\mu} - \frac{H_r(\nu)-H_r(\mu)}{\nu-\mu}\right\}\sigma_1(d\nu)\varphi_+,$$

whence

$$H_\ell(-\mu)^{-1}\hat{\varsigma}_+(\mu) = \int_0^\infty \nu\frac{H_r(\nu)}{\nu-\mu}\pi\sigma_1(d\nu)\varphi_+. \tag{4.21}$$

The identity (4.19) now follows immediately from (4.20) and (4.21). ∎

When specialized to the situation that $A^{-1}T$ is compact on H_A and $w(\mu)$ has one sign change (i.e., $w(\mu)<0$ on (a,c) and $w(\mu)>0$ on (c,b)), Theorem 4.4 is particularly straightforward. Let us denote by $\{\lambda_n\}_n$ and $\{\varphi_n\}_n$ the eigenvalues and corresponding eigenfunctions of $T^{-1}A$, where $0\neq n\in\mathbb{Z}$ and $(A\varphi_n,\varphi_m)=\delta_{nm}$. Similarly, let us denote by $\{\varsigma_n\}_n$ and $\{\psi_n\}_n$ the eigenvalues and corresponding eigenfunctions of $T^{-1}\hat{A}$, where $0\neq n\in\mathbb{Z}$ and $(\hat{A}\psi_n,\psi_m)=\delta_{nm}$, and we continue the convention

$$\ldots<\lambda_{-2}<\lambda_{-1}<0<\lambda_1<\lambda_2<\ldots ,$$

$$\ldots<\varsigma_{-2}<\varsigma_{-1}<0<\varsigma_1<\varsigma_2<\ldots .$$

We then have

$$(\psi_n,\psi_m)_T = |\varsigma_n|^{-1}\delta_{nm},$$

and therefore $\{|\varsigma_n|^{1/2}\psi_n\}_n$ is an orthonormal basis in H_T. Similarly, $\{|\lambda_n|^{1/2}\varphi_n\}_n$ is an orthonormal basis in $P[H_S]$. If we now define $k(\mu)=k(\mu,0)$ by (3.9) and put $k_n=(k,|\varsigma_n|^{1/2}\psi_n)_T$, $\tilde{k}_n=(k,|\lambda_n|^{1/2}\varphi_n)_S$, $0\neq n\in\mathbb{Z}$, then the regularity assumptions $k\in|T_1|^\alpha[H_T]$ and $Pk\in|S_1|^\alpha[H_S]$ are satisfied if and only if, for some $0<\alpha<1$,

$$\sum_{0\neq n\in\mathbb{Z}} |\varsigma_n|^{2\alpha}|k_n|^2 < \infty, \tag{4.22a}$$

$$\sum_{0\neq n\in\mathbb{Z}} |\lambda_n|^{2\alpha}|\tilde{k}_n|^2 < \infty. \tag{4.22b}$$

In the appendix of [220] it has been shown that these conditions are satisfied for $w(\mu)=\mathrm{sgn}(\mu)|\mu|^\gamma$, which will cover applications to electron scattering and the Fokker–Planck equation, among others.

THEOREM 4.5. Let A be strictly positive, let $T:H_A\to H_A'$ be compact with $w(\mu)$ having one sign change at $c\in(a,b)$, and suppose (4.22) hold true. Then the albedo operator is given by the expression

$$(E_+\varphi_+)(\mu) = \varphi_+(\mu)$$

for $\mu\in(c,b)$, and

$$(E_+\varphi_+)(\mu) = -Z\sum_{n=1}^{\infty}\sum_{m=1}^{\infty}\frac{\varsigma_{-m}\varsigma_n}{\varsigma_{-m}-\varsigma_n}\, k_{-m}k_n g_m^\ell g_n^r \varphi_{+n}|\varsigma_{-m}|^{1/2}\psi_{-m}(\mu)$$

for $\mu\in(a,c)$, where $Z = W_\ell(0)W_r(0)/W(0)\|k\|_T^2$, $g_n^\ell = H_\ell(-\varsigma_{-n})$, $g_n^r = H_r(\varsigma_n)$ and $\varphi_{+n} = (\varphi_+,|\varsigma_n|^{1/2}\psi_n)_T$, and where $H_\ell(z)$ and $H_r(z)$ are the unique functions appearing in the factorization

$$H_r(z)^{-1}H_\ell(-z)^{-1} = 1 + z\sum_{n=1}^{\infty}\{\varsigma_n(\varsigma_n-z)^{-1}|k_n|^2-\varsigma_{-n}(\varsigma_{-n}-z)^{-1}|k_{-n}|^2\}.$$

These factors have the following properties:

(i) $H_\ell(z)$, $H_\ell(z)^{-1}$, $H_r(z)$ and $H_r(z)^{-1}$ are continuous on the closed right half plane (except at infinity) and analytic on the open right half plane.

(ii) $H_\ell(0^+) = H_r(0^+) = 1$.

(iii) $|H_\ell(z)|$, $|H_\ell(z)^{-1}|$, $|H_r(z)|$ and $|H_r(z)^{-1}|$ are all of order $o(z)$ as $z\to\infty$, $\mathrm{Re}z\geq 0$.

Proof: The uniqueness of the factorization (4.23) is a direct consequence of Lemma 4.3 if there is only one sign change. Let $k(\mu)$ be the function in (3.9) (with $\lambda=0$) and put $\mathcal{W}=W_\ell(0)W_r(0)/W(0)$. Then j is the imbedding of the one dimensional space spanned by $\mathrm{sgn}(\mu)k(\mu)$ into $L_2(a,b)$, while π is the projection $\pi f = \|k\|_T^{-2}\int_a^b \mu f(\mu)k(\mu)d\mu$ with $\|k\|_T^2 = \int_a^b |w(\mu)k(\mu)^2|d\mu$. We compute

$$\Lambda(z) = 1 + z\mathcal{W}\sum_{n=1}^{\infty}\{\varsigma_n(\varsigma_n-z)^{-1}|k_n|^2-\varsigma_{-n}(\varsigma_{-n}-z)^{-1}|k_{-n}|^2\}.$$

We now obtain

$$Q_-E_+\varphi_+ = -\frac{\mathcal{W}}{\|k\|_T^2}\sum_{n=1}^{\infty}\sum_{m=1}^{\infty}\frac{\varsigma_{-m}\varsigma_n}{\varsigma_{-m}-\varsigma_n}k_{-m}k_n g_m^\ell g_n^r\varphi_{+n}|\varsigma_{-m}|^{1/2}\psi_{-m},$$

which reduces to the above expression for E_+. ∎

Let us consider the special case when $\varphi_+(\mu)=|\varsigma_p|^{1/2}\psi_p(\mu)$ for some $n\in\mathbb{N}$. Then $\varphi_{+n}=\delta_{np}$, as a result of the orthonormality of the functions $\{|\varsigma_n|^{1/2}\psi_n\}_{0\neq n\in\mathbb{Z}}$ in H_T. We obtain, for $\mu\in(c,b)$,

$$(E_+\varphi_+)(\mu) = -Z\sum_{m=1}^{\infty}\varsigma_{-m}\varsigma_p(\varsigma_{-m}-\varsigma_p)^{-1}k_{-m}k_p g_m^\ell g_p^r|\varsigma_{-m}|^{1/2}\psi_{-m}(\mu).$$

Clearly, we must have

$$C_p^2 \equiv \sum_{m=1}^{\infty}\left|\frac{\varsigma_{-m}}{\varsigma_{-m}-\varsigma_p}k_{-m}g_m^\ell\right|^2 = -|\varsigma_p k_p g_p^r|^{-2}\int_a^c w(\mu)|(E_+\varphi_+)(\mu)|^2 d\mu < \infty,$$

and therefore $\varsigma_p k_p g_p^r C_p = O(1)$ $(\ell\to\infty)$.

Next, let us consider the special case when $I=(-d,d)$ is a (finite or infinite) interval symmetric about the origin and the weight and Sturm–Liouville operator satisfy the conditions $w(-\mu) = w(\mu)$ and $(Af)(\mu) = (A\ \mathrm{sgn}(\mu)f)(-\mu)$. The eigenvalues and eigenfunctions will then satisfy both $\varsigma_{-n}=-\varsigma_n$ and $\psi_{-n}(\mu)=\psi_n(-\mu)$, whence $k(-\mu)=-k(\mu)$ and $g_n^\ell=g_n^r=g_n$. In this case we obtain the simplified expressions

$$\Lambda(z) = 1 + 2z^2 W \sum_{n=1}^{\infty} \varsigma_n(\varsigma_n^2-z^2)^{-1}$$

and

$$(E_+\varphi_+)(\mu) = -Z \sum_{n=1}^{\infty}\sum_{m=1}^{\infty} \frac{\varsigma_m\varsigma_n}{\varsigma_m+\varsigma_n} k_m k_n g_m g_n \varphi_{+n}(\varsigma_m)^{1/2}\psi_m(-\mu)$$

for $\mu\in(-d,0)$. Here we have the summability condition (where $g_n>0$)

$$\sup_{n\in\mathbb{N}} \varsigma_n |k_n| g_n \Big[\sum_{m=1}^{\infty} (\varsigma_m(\varsigma_m+\varsigma_n)^{-1}k_m g_m)^2\Big]^{1/2} < \infty.$$

Finally, let us consider the case when A is positive with isolated (simple) zero eigenvalue. We denote by S_1 the bounded inverse of the restriction of $T^{-1}A$ to $P[H_S]$, and define C by $C = (S_1-\hat{A}^{-1}T)P$, which is a bounded operator on H_T of rank at most N. We must replace the regularity assumption (4.1) by the conditions

$$\exists 0<\alpha<1\text{: Ran } C \subset |T_1|^\alpha[H_T] \text{ and } \mathrm{Ran}PC \subset |S_1|^\alpha[H_T].$$

By construction, the function $\varphi(x) = T^{-1}A\exp\{-xT^{-1}A\}E_+\varphi_+$, $0<x<\infty$, has its values in $P[H_T]\oplus\mathrm{Ker}\ A$. We may then repeat the proof of Lemma 4.1 and the derivation of (4.8) and (4.9), with the adaptation that $\psi(x)-\psi(\infty) = \int_\infty^x\varphi(y)dy$ (and similarly for $\varsigma(x)$). We then obtain the Riemann–Hilbert problem (4.11) as a result. The existence proof for the factorization of the dispersion function, Lemma 4.2, will fail, however, in this case. It is not clear whether for this case the albedo operator can be written in the form (4.19).

5. The Fokker–Planck equation

As we have noted in Section 1, the Fokker–Planck equation first arose in the study of Brownian motion. Its applicability, however, goes far beyond this problem, and includes electron diffusion in an external field, chemical reaction models, Ornstein–Uhlenbeck processes, lasers and superionic conductors. The reader interested in the solution methods for, and the physical applications of, the Fokker–Planck equation is referred to the monographs of Risken [320] and van Kampen [372], and the articles of Chandrasekhar [88] and Chen Wang and Uhlenbeck [93]. Recently, there appears to be renewed interest in computing certain aspects of the solution of the stationary Fokker–Planck equation related to the boundary layer solutions, density profiles near the wall, validity of Fick's law, etc. (cf. [59, 60, 187, 262, 351])

Let us recall the Fokker–Planck equation

$$v\frac{\partial \Phi}{\partial x}(x,v) = \frac{\partial^2 \Phi}{\partial v^2} - v\frac{\partial \Phi}{\partial v}, \tag{5.1}$$

where $x\in(0,\tau)$ is a position coordinate and $v\in(-\infty,\infty)$ is the velocity variable. In order to adopt the framework of the previous sections, we consider the auxiliary diffusion problem

$$v\frac{\partial \psi}{\partial x}(x,v) = \frac{\partial^2 \psi}{\partial v^2} + (\tfrac{1}{2}-\tfrac{1}{4}v^2)\psi(x,v), \qquad 0<x<\tau, \quad v\in\mathbb{R}, \tag{5.2}$$

which reduces to (5.1) using the relationship

$$\psi(x,v) = e^{-\frac{1}{4}v^2}\Phi(x,v). \tag{5.3}$$

We introduce the Hilbert space $H=L_2(\mathbb{R},dv)$ and the multiplication operator

$$(Th)(v) = vh(v),$$

as well as the positive self adjoint operator

$$(Ah)(v) = -h''(v) - (\tfrac{1}{2}-\tfrac{1}{4}v^2)h(v) \tag{5.4}$$

with domain $D(A)=\{f\in H : pf'$ absolutely continuous, $(pf')'\in H\}$. (The differential operator A is limit-point at both singular endpoints $-\infty$ and $+\infty$; as a result one does not have to impose additional boundary conditions on A.) It is easily seen that A has discrete spectrum with eigenvalues $n=0,1,2,\ldots$ and eigenfunctions $\exp(-v^2/4)H_n(v/\sqrt{2})$, where

$$H_n(z) = (-1)^n e^{z^2} \frac{d^n}{dz^n} e^{-z^2}$$

is the Hermite polynomial of degree n (cf. [1]; 22.6.21 & 22.5.18). Thus A has a nonzero kernel spanned by the vector $\varphi_0(v)=\exp(-\frac{1}{4}v^2)$ and, in the usual inner product of H, $(T\varphi_0,\varphi_0)=0$.

Now one may see that this problem satisfies the assumptions of Theorem 1.5. As Ker A $=$ span$\{\varphi_0\}$ with $(T\varphi_0,\varphi_0)=0$, we have immediately the following theorem (cf. [35]).

THEOREM 5.1. For every $\tilde{\varphi}_+ \in L_2(\mathbb{R}_+, |v| \exp\{-\frac{1}{2}v^2\}dv)$ there exists a unique solution of Eq. (5.1) satisfying $\tilde{\psi}(x,\cdot) \in L_2(\mathbb{R}, |v| \exp\{-\frac{1}{2}v^2\}dv)$ with boundary conditions

$$\tilde{\psi}(0,v) = \tilde{\varphi}_+(v), \qquad 0<v<\infty, \tag{5.5a}$$

$$\limsup_{x\to\infty} \int_{-\infty}^{\infty} |v| e^{-\frac{1}{2}v^2} |\tilde{\psi}(x,v)|^2 dv < \infty. \tag{5.5b}$$

Using results from Section III.4, it is also possible to analyze the boundary value problem (5.1) with reflective boundary conditions

$$\tilde{\psi}(0,v) = \tilde{\varphi}_+(v) + \alpha\tilde{\psi}(0,-v) +$$

$$+ \beta\int_{-\infty}^{0} |\hat{v}/v| \exp\{-\tfrac{1}{2}(\hat{v}^2-v^2)\}\Sigma(\hat{v}\to v)\tilde{\psi}(0,\hat{v})d\hat{v}, \qquad 0<v<\infty, \tag{5.6a}$$

$$\limsup_{x\to\infty} \int_{-\infty}^{\infty} |v| \exp\{-\tfrac{1}{2}v^2\} |\tilde{\psi}(x,v)|^2 dv < \infty, \tag{5.6b}$$

where the scattering function satisfies

$$\Sigma(\hat{v}\to v) \geq 0, \qquad \hat{v}<0<v, \tag{5.7}$$

$$\int_0^{\infty} \Sigma(\hat{v}\to v)dv = 1, \qquad \hat{v}<0, \tag{5.8}$$

$$|\hat{v}|\exp\{-\tfrac{1}{2}\hat{v}^2\}\Sigma(\hat{v}\to v) = |v|\exp\{-\tfrac{1}{2}v^2\}\Sigma(-v\to-\hat{v}), \quad \hat{v}<0<v. \tag{5.9}$$

As a result of the reciprocity condition (5.9), the surface reflection operator

$$(Rh)(v) = \alpha h(v) + \beta\int_0^\infty (\hat{v}/v)\exp\{-\tfrac{1}{2}(\hat{v}^2-v^2)\}\Sigma(-\hat{v}\to v)h(\hat{v})d\hat{v}$$

is self adjoint on $Q_+[H_T]=L_2(\mathbb{R}_+,|v|\exp\{-\tfrac{1}{2}v^2\}dv)$, and for $\alpha,\beta\geq 0$ we have

$$0 \leq (Rh,h)_T \leq (\alpha+\beta)\|h\|_T^2,$$

where we have used (5.7) and (5.8). The details of this estimate can be found in Section VII.5. Hence, if $\alpha,\beta\geq 0$ and $\alpha+\beta<1$ (medium with absorbing wall), Eq. (5.1) with boundary conditions (5.6a) and (5.6b) is uniquely solvable. For $\alpha,\beta\geq 0$ and $\alpha+\beta=1$ there always exists a solution, which may be nonunique (as it is, for instance, for $\alpha=1$ and $\beta=0$). We thus recover results obtained by Beals and Protopopescu [35, 36]. (A redundant condition needed for the existence of solutions was removed in [369].)

Let us compute the albedo operator following the method of Section 2. We note that the unique vector ψ_0 satisfying $A\psi_0=T\varphi_0$ and $(T\psi_0,\psi_0)=0$ is given by $\psi_0(v) = v\exp\{-\tfrac{1}{4}v^2\}$. Using the normalization properties of Hermite polynomials, one sees that the functions

$$\varphi_{\pm n}(v) = [(2\pi n)^{1/2}\cdot 2^{n+1}\cdot(n!)]^{-1/2}\, e^{-\frac{1}{4}v^2} e^{\pm v n^{1/2}}\, H_n((v\mp 2n^{1/2})/\sqrt{2}),$$

for $n\in\mathbb{N}$, satisfy the normalization condition

$$(A\varphi_{\pm n},\varphi_{\pm n})_H = \lambda_{\pm n}(T\varphi_{\pm n},\varphi_{\pm n}) = \lambda_{\pm n}\int_{-\infty}^\infty v|\varphi_{\pm n}(v)|^2 dv = 1,$$

with $\lambda_0=0$ and $\lambda_{\pm n} = \pm n^{1/2}$. The albedo operator E related to Eq. (5.1) is then given by the equations

$$(E\varphi_+)(v) = \begin{cases} \varphi_+(v), & 0<v<\infty, \\ g(v), & -\infty<v<0, \end{cases}$$

and

$$(2\pi)^{1/2}g(v) + \int_{-\infty}^{0} \exp\{-\tfrac{1}{2}\hat{v}^2\}K_-(v,\hat{v})g(\hat{v})d\hat{v} = \int_0^{\infty} \exp\{-\tfrac{1}{2}\hat{v}^2\}K_+(v,\hat{v})\varphi_+(\hat{v})d\hat{v},$$

where

$$K_{\pm}(v,\hat{v}) = \hat{v} \pm \sum_{n=1}^{\infty} (n^{1/2}2^{n+1}n!)^{-1}\exp\{(v+\hat{v})n^{1/2}\}\, H_n((v-2n^{1/2})/\sqrt{2})\, H_n((v-2n^{1/2})/\sqrt{2}).$$

6. Electron scattering

As a second application we consider the equation

$$\mu\frac{\partial\psi}{\partial x}(x,\mu) = \frac{\partial}{\partial\mu}((1-\mu^2)\frac{\partial\psi}{\partial\mu}), \tag{6.1}$$

where $x\epsilon(0,\tau)$ denotes position and $\mu\epsilon[-1,1]$ the direction cosine of propagation. The equation was derived by Bothe [53] to describe electron scattering. A formal solution, using eigenfunction expansion, was given by Bethe et al. [42]. Using a variational method, without recourse to the functional analytic approach presented herein, Beals [31] proved the existence of a unique solution of Eq. (6.1) with appropriate boundary conditions for finite τ and justified its expansion in eigenfunctions. For $\tau=\infty$ this result is also due to Beals [34]. Different proofs were given by Kaper et al. [209] and Degond and Mas-Gallic [99]. The result can be formulated in the following fashion.

THEOREM 6.1. For every $\varphi_+\epsilon L_2([0,1],|\mu|d\mu)$ there exists a unique solution $\psi(x,\cdot)\epsilon L_2([-1,1],|\mu|d\mu)$ of Eq. (6.1), which satisfies the boundary conditions

$$\psi(0,\mu) = \varphi_+(\mu), \quad 0\le\mu\le 1, \tag{6.2a}$$

$$\psi(x,\mu) = O(1) \quad (\mu\to\pm 1), \tag{6.2b}$$

$$\limsup_{x\to\infty} \int_{-1}^{1} |\mu|\cdot|\psi(x,\mu)|^2 d\mu < \infty. \tag{6.2c}$$

Proof: The differential operator

$$(Ah)(\mu) = \frac{-d}{d\mu}((1-\mu^2)\frac{dh}{d\mu})$$

is limit-circle at both endpoints -1 and $+1$, and thus its domain is taken to consist of functions which remain bounded as $\mu\to\pm1$, along with the usual smoothness requirements (see [109], Ch. XIII; also [115]). It is then clear that A is positive selfadjoint, and that it has a discrete spectrum consisting of the numbers $n(n+1)$, $n=0,1,2,\ldots$, with the Legendre polynomials $P_n(\mu)$ as the corresponding eigenfunctions. Also, Ker A = span$\{\varphi_0\}$ with $\varphi_0(\mu)\equiv1$. If we define $(Th)(\mu) = \mu h(\mu)$ in order to apply Theorem 1.5, we have $(T\varphi_0,\varphi_0)=0$, and the unique vector ψ_0 satisfying $A\psi_0=T\varphi_0$ and $(T\psi_0,\psi_0)=0$ is given by $\psi_0(\mu)=\frac{1}{2}\mu$. The assumptions of Theorem 1.5 are clearly fulfilled and the result follows. ■

The albedo operator E may be written explicitly in terms of a set of eigenfunctions, using either the method of Section 2 or that of Section 4. For the method of Section 2 we have to solve the eigenvalue problem

$$(1-\mu^2)\varphi_n''(\mu) - 2\mu\varphi_n'(\mu) + \lambda_n\mu\varphi_n(\mu) = 0,$$

$$\varphi_n(\mu) = O(1) \ (\mu\to\pm1).$$

For the method of Section 4 we have to solve the eigenvalue problem

$$(1-\mu^2)\psi_n''(\mu) - 2\mu\psi_n'(\mu) + \varsigma_n\mu\psi_n(\mu) = 0,$$

$$\psi_n(0) = 0,$$

$$\psi_n(\mu) = O(1) \ (\mu\to+1),$$

(or the related problem on $(-1,0)$ with boundary condition as $\mu\to-1$ and ς_n replaced by $-\varsigma_n$). In both cases there is a discrete spectrum of simple eigenvalues. We have to satisfy the normalization conditions

$$\int_{-1}^{1} \mu\,|\varphi_{\pm n}(\mu)|^2 d\mu = 1/\lambda_{\pm n}, \quad n\in\mathbb{N},$$

with $\ldots<\lambda_{-2}<\lambda_{-1}<\lambda_0=0<\lambda_1<\lambda_2<\ldots$, and

$$\int_0^{\pm 1} \mu |\psi_{\pm n}(\mu)|^2 d\mu = 1/\varsigma_n, \quad n \in \mathbb{N},$$

with $\ldots<\varsigma_{-2}<\varsigma_{-1}<0<\varsigma_1<\varsigma_2<\ldots$. Since the differential equation has regular singular points at $\mu=\pm1$ with equal and vanishing exponents, the eigenfunctions $\varphi_n(\mu)$ and $\psi_n(\mu)$ can be analytically continued to entire functions that do not vanish at $\mu=\pm1$. The point at infinity is an irregular singular point.

It is not known if the eigenvalues and eigenfunctions can be obtained in terms of known special functions. The best results in this direction are due to Veling [374], who derived the following asymptotic formula:

$$\lambda_n = \{\Gamma(\tfrac14)^4/96\pi^2\}\{12\pi(n+\tfrac12)^2 - 5 + O(n^{-2})\}$$

This formula provides a reasonable (within 2 decimal places for n=3 and within 4 decimal places for n=33) approximation of the actual eigenvalues.

Chapter XI

TIME DEPENDENT KINETIC EQUATIONS: METHOD OF CHARACTERISTICS

1. Introduction

Time dependent linear kinetic equations arise in a number of diverse applications in biology, chemistry and physics, as well as in various other modeling problems. Due to a tradition deeply rooted in classical mathematical physics and reinforced by the successes of quantum mechanics, such time dependent problems were initially attacked using the eigenfunction method. Yet, this method met with a relative lack of success, due to the nonnormal nature of the operators occurring in these kinetic problems, and it was supplanted by the semigroup approach, which for decades became the dominant method of time dependent kinetic theory. Despite its virtues, the semigroup approach is somewhat indirect, and is not naturally suited for treating linear evolution problems with time dependent operators, phase spaces and boundary conditions. This chapter will be devoted to an approach to these problems based on the method of characteristics.

The deterministic nature of the phenomena described by this type of equation, like most time dependent evolution equations in classical physics, leads naturally to the concept of a well posed initial value problem. The mathematical principle that guides the proper formulation of such problems is that there should be one and only one solution for every initial state and that the solution should depend continuously on the initial state.

For any fixed value of the variable t, certain functions of the other variables, the so-called phase space variables, describe an instantaneous state of the physical system. These functions will be denoted by u, which is regarded as a vector in a function space $\mathcal{X}$. As time elapses, the vector $u=u(t)$ moves through $\mathcal{X}$ describing a trajectory which corresponds to the evolution of the system. The linearity of the problem, the completeness of its mathematical description and the physically relevant normalization of its solutions, all of which derive from basic physical requirements, select certain Banach spaces as the natural framework for the mathematical description of the evolution of the system.

Time dependent linear kinetic equations describe various physical, chemical or biological phenomena, such as the evolution of neutral or charged fluids under conditions of rarefaction, interaction and closeness to equilibrium, the evolution of planetary or stellar atmospheres under conditions of single scattering and polarizability, the diffusion of reactants in solutions, and the growth of cell populations. An extensive literature is available, presenting various realizations of these equations for different geometries and boundary conditions. These realizations include the neutron transport equation, equations of radiative transfer, Bhatnagar–Gross–Krook (BGK) related equations in rarefied gas dynamics, the linearized Boltzmann equation with various intermolecular potentials, the Fokker–Planck, Landau–Balescu and linearized Vlasov equations, and equations modeling growing cell populations, reaction diffusion processes, traffic flow, etc.

We consider here a general time dependent kinetic problem for a distribution function u depending on position x, velocity ξ, and time t in the form

$$\frac{Du}{Dt}(x,\xi,t) + (Au)(x,\xi,t) = f(x,\xi,t). \tag{1.1}$$

The independent variables (x,ξ,t) take values in a set Σ, which is called the **phase space** of the problem. Sometimes this is referred to as the Boltzmann or reduced phase space, as opposed to the Gibbs phase space used when considering the Liouville equation. A typical (but not the most general) situation occurs when Σ can be written as the Cartesian product of the space, velocity and time domains. The term Du/Dt gives the total time derivative along the trajectory, and is defined by

$$\frac{Du}{Dt} \equiv \frac{\partial u}{\partial t} + \xi\cdot\frac{\partial u}{\partial x} + a\cdot\frac{\partial u}{\partial \xi} \equiv \frac{\partial u}{\partial t} + Xu \equiv Yu, \tag{1.2}$$

where in the definition of the operators involved the separate terms in (1.2) may not make sense independently. The **acceleration** $a=a(x,\xi,t)$ describing the total force acting on particles in the system is composed of two contributions. The first contribution stems from an exterior force field satisfying the laws of Newtonian dynamics. The second one is of intermolecular origin and would in principle introduce nonlinearities. Usually, this contribution is either neglected or approximated via so–called self–consistent or stochastic schemes, and modeled accordingly. Thus, the force will be considered here as external and given, in such a way that the problem will remain linear. For technical reasons we also consider it to be local in space, velocity and time.

This locality condition excludes from our description nonMarkovian transport processes and many realistic plasma problems.

Even for $a \equiv 0$, the first order partial differential operator of hyperbolic type, $\xi \cdot \frac{\partial}{\partial x}$, is rather complicated, since, in general, the two variables x and ξ decouple only for one dimensional geometries (in position and velocity). Otherwise, the vector $\xi \neq 0$ defines a characteristic direction at each position x, and $\xi \cdot \frac{\partial u}{\partial x}$ is interpreted as the directional derivative of u with respect to x in the direction of ξ. An extra complication arises from the singular nature of the operator $\xi \cdot \frac{\partial}{\partial x}$; the characteristic direction is not defined for $\xi = 0$. The collision term Au describes the change of u due to scattering, absorption, fission and similar events. This term is not essential for the "transport" nature of the equation (1.1), which is induced by the operator Y. Traditionally, transport theory has been the study of processes characterized by mean free paths which are much longer than the distance over which a collision takes place (dilute systems), such as neutron transport, radiative transfer, and rarefied gas dynamics. It is in such systems that the transport nature of the equation dominates the collision and single scattering processes.

In addition to the evolution equation, supplementary conditions must be specified on the boundary of the phase space. This amounts to specifying initial and boundary conditions which account for the initial state of the system and the "incoming fluxes". Thus, in a possibly time dependent region Ω_t and velocity domain V_t one seeks a solution of the initial-boundary value problem

$$\frac{Du}{Dt}(x,\xi,t) + (Au)(x,\xi,t) = f(x,\xi,t), \tag{1.3a}$$

where $x \in \Omega_t$, $\xi \in V_t$ and $0<t<T$,

$$u(x,\xi,0) = u_0(x,\xi), \qquad (1.3b),$$

where $x \in \Omega_0$ and $\xi \in V_0$, and

$$u_-(x,\xi,t) = (Ku_+)(x,\xi,t) + g(x,\xi,t), \tag{1.3c}$$

where $(x,\xi) \in D_-$ and $0<t<T$. Here the initial distribution u_0, the internal source f, and the incident flux g are given, D_- (resp. D_+) is the (possibly time dependent) part of the phase space boundary corresponding to the incoming (resp. outgoing) "fluxes", $u_\pm$ denotes the restriction of u to $D_\pm$ and K is an operator describing boundary processes

such as absorption, partial and/or diffuse reflection, etc. In general, the operators A and K depend on time. We shall assume them to be local in time.

In this chapter we shall provide a quite general treatment of the basic questions of existence, uniqueness, dissipativity and positivity for the abstract time dependent kinetic problem (1.3), without special assumptions on the geometry, the boundary conditions, or the exact form of the operators. By direct analysis it will be shown that the problem (1.3) is well posed in the space of L_p-functions, $1 \leq p < \infty$, and in the space of bounded measures on phase space, under general and rather mild assumptions.

Historically, the L_2-setting was chosen for mathematical convenience, although the interpretation of u as a one particle distribution function or an intensity makes the L_1-setting physically more relevant. As observed by Vidav [375], the argument of physical relevance applies with perhaps even more force to the space of bounded measures. In particular, this space allows one to consider pointlike sources and distributions and unidirectional beams of particles and radiation. Yet, Vidav found the space of measures to be inconvenient for technical reasons, and worked in the mathematically more convenient L_p-setting. Later, Suhadolc and Vidav [346] did include measures in their analysis, using the semigroup approach, but they encountered some technical complications not appearing when using the L_p-setting. For example, the domain of the generator is not dense in the space of bounded measures. It will turn out that the trajectory approach is well suited for spaces of bounded measures (see Section 6).

Most of the existence and uniqueness results for time dependent linear kinetic equations were obtained by semigroup or spectral methods, well suited when the acceleration a, the operators A and K and the domains Ω and V are independent of time. Time dependent operators and domains were considered only occasionally, limited to particular situations and dealt with by perturbation methods within the semigroup framework (cf. Belleni-Morante and Farano [40], Palczewski [297] and Wenzel [389]). On the other hand, the trajectory method on vector fields was traditionally applied to Liouville evolutions in the Gibbs phase space of the system (cf. Schnute and Shinbrot [326] and Marchioro et al. [256]), but almost neglected for evolutions in the Boltzmann phase space. We note, however, Reed [316, 317], who applied Kato's "evolution equation method" (cf. [215]) to a neutron transport problem. Some years later, Bardos [23] undertook a rather systematic study of Boltzmann-like kinetic equations with acceleration term, using methods from the theory of first order hyperbolic equations. Recently, Babovsky [20], Asano [17] and Ukai [353] used the trajectory method for different concrete situations connected with the nonlinear Boltzmann equation. A unified formulation in a general abstract setting has been given by Beals and Protopopescu [37], which we will follow closely.

2. The functional formulation

Let Σ be an open subset of $\mathbb{R}^n \times \mathbb{R}^n \times (0,T)$ with boundary $\partial\Sigma$. Let Y be a real vector field on Σ of the form

$$Y = \tau \cdot \frac{\partial}{\partial \varsigma} = \frac{\partial}{\partial t} + \xi \cdot \frac{\partial}{\partial x} + a(x,\xi,t) \cdot \frac{\partial}{\partial \xi}, \tag{2.1}$$

where $\varsigma = (t,x,\xi)$ is a vector in $\mathbb{R}^{2n+1}$ and a is assumed to be Lipschitz continuous on the closure $\bar{\Sigma}$ of Σ (in $\mathbb{R}^{2n+1}$). We shall write $X = \xi \cdot \frac{\partial}{\partial x} + a(x,\xi,t) \cdot \frac{\partial}{\partial \xi}$. Let μ be the Borel measure on Σ given by

$$d\mu(x,\xi,t) = \begin{cases} dx\ d\rho(\xi)\ dt, & a \equiv 0, \\ dx\ d\xi\ dt, & a \not\equiv 0, \end{cases} \tag{2.2}$$

where $\rho(\xi)$ is a positive Borel measure on $\mathbb{R}^n$ such that all bounded Lebesgue measurable sets have finite ρ-measure. The difference between the two definitions in (2.2) stems from the fact that for $a \equiv 0$ the vector field acts only on the variables t and x, while ξ remains an independent variable. This situation creates a certain freedom in choosing the measure $d\mu$. Throughout we shall impose the following two conditions on Y:

(i) Y is divergence free: $\sum_{j=1}^{n} \frac{\partial}{\partial \xi_j} a_j(x,\xi,t) = 0$;

(ii) Each integral curve of the vector field $Y \equiv X + \frac{\partial}{\partial t}$ remains in a bounded region of Σ.

We do not require Σ to be convex. Simple, and typical, examples of Σ have the form

$$\Sigma = \Omega \times V \times (0,T), \tag{2.3}$$

where Ω is open in $\mathbb{R}^n$ and V is a finite union of balls or open spherical shells in $\mathbb{R}^n$. The divergence free condition (i) on the vector field Y is satisfied for most of the commonly considered force fields (for example, the Lorentz force $a(x,\xi,t) = E(x,t) + \xi \times B(x,t)$). Such an assumption is related to Liouville's theorem in statistical mechanics, which expresses the conservation of the measure μ under the dynamical flow in the absence of collisions. Certain results for non-divergence free fields can be found in [23].

The second condition on the vector field, which leads to the unique solvability of the system of differential equations

$$\frac{dx}{ds} = \xi, \qquad \frac{d\xi}{ds} = a(x,\xi,t), \qquad \frac{dt}{ds} = 1, \tag{2.4}$$

with initial condition $(x(s^*), \xi(s^*), t(s^*)) \in \Sigma$ and $(x(s), \xi(s), t(s))$ extended over the maximal s-interval for which the curve lies in Σ (an interval of length at most T), implies that no trajectory reaches infinity in finite time. Obviously, this condition is trivially satisfied if Σ is bounded. It is satisfied also if the vector-valued acceleration $a(x,\xi,t)$ satisfies the bound $|a(x,\xi,t)| \le C(1+|x|+|\xi|)$. Indeed, let $f(x,\xi,t)=1+|x|^2+|\xi|^2$. On an integral curve, this bound and the Schwarz inequality imply

$$\left|\frac{df}{ds}\right| = \left|2x\cdot\xi + 2\xi\cdot a\right| \le 2|f|^{1/2}\{|\xi|^2 + C^2(1+|x|+|\xi|)^2\}^{1/2} \le C_1 f.$$

Gronwall's inequality then yields $|f(s)| \le |f(0)|\exp(C_1 s)$, whence

$$f(x,\xi,t) \le \{1+|x(0)|^2+|\xi(0)|^2\}\exp(C_1 t).$$

Therefore f remains bounded for bounded time on any integral curve.

Under the above assumptions, each integral curve of Y defined on a maximally extended interval (s_0, s_1) will have a limit at s_0 and a limit at s_1, each lying in $\partial\Sigma$. Referring to these as the left and right endpoints, respectively, we define the Borel sets $D^\pm \subset \partial\Sigma$, where D^- (resp. D^+) is the set of all left (resp. right) endpoints of integral curves of Y. (In the next section we shall prove that the sets $D^\pm$ are indeed Borel sets). In general, these sets are not disjoint and do not exhaust $\partial\Sigma$. However, if $\partial\Sigma$ is piecewise C^1, both $D^+ \cap D^-$ and $\partial\Sigma\setminus(D^+ \cup D^-)$ are negligible, in the sense that the union of all associated integral curves in Σ has μ-measure zero (see Section 3).

We will show in Section 3 that there are unique positive Borel measures $\nu^{\pm}$ on $D^{\pm}$ such that the Green's identity

$$\int_{\Sigma} Yv d\mu = \int_{D^+} v d\nu^+ - \int_{D^-} v d\nu^-$$

is valid for every v in a space Φ of test functions. To see what this means in a representative example, suppose Σ has the form (2.3) with $V=\mathbb{R}^n$ and that Y has the form (2.1). Suppose Ω has a piecewise C^1 boundary $\partial\Omega$. At a point x of a C^1 section, let n(x) denote the unit outer normal and set

$$C_{\pm} = \{(x,\xi) \in \partial\Omega \times V : \pm\xi \cdot n(x) > 0\},$$

$$D_t = \{(x,\xi,t) : (x,\xi) \in \Omega \times V\}.$$

Then, writing $A \approx B$ if $\nu^{\pm}(A \cup B \setminus A \cap B) = 0$, one has

$$D^- \approx [C_- \times (0,T)] \cup D_0, \tag{2.5a}$$

$$D^+ \approx [C_+ \times (0,T)] \cup D_T. \tag{2.5b}$$

Moreover, if $d\sigma$ denotes the surface measure on $\partial\Omega$, then

$$d\nu^{\pm} = \begin{cases} |\xi \cdot n(x)| d\sigma(x) d\rho(\xi) dt & \text{on } C_{\pm} \times (0,T) \\ dx d\rho(\xi) & \text{on } D_0 \cup D_T, \end{cases}$$

where $d\rho(\xi)$ is to be replaced by $d\xi$ if $a \neq 0$.

In the general case, when Σ does not have the form (2.3), we introduce decompositions of $D^{\pm}$ into spatial and temporal pieces as in (2.5):

$$D_{\pm} = \{(x,\xi,t) \in D^{\pm} : 0<t<T\}, \tag{2.6a}$$

$$D_0 = \{(x,\xi,0) \in D^-\}, \tag{2.6b}$$

$$D_T = \{(x,\xi,T)\in D^+\}, \tag{2.6c}$$

whence $D^- = D_- \cup D_0$ and $D^+ = D_+ \cup D_T$. If u belongs to the (real) function space $L_p(\Sigma,d\mu)$, $1\le p<\infty$, it may be considered in the usual way as a distribution of the form

$$<u,v> \equiv \int_\Sigma uv\,d\mu, \quad v\in C_0^\infty(\Sigma),$$

where $C_0^\infty(\Sigma)$ is the space of all infinitely differentiable functions on Σ that vanish on the boundary. We shall replace this space of test functions by larger ones Φ_0 and Φ (to be specified in the next section) satisfying $C_0^\infty(\Sigma) \subset \Phi_0 \subset \Phi$. The distributional derivative is defined in the usual way by the formula

$$<Yu,v> = -<u,Yv>, \quad v\in\Phi_0.$$

If u and Yu belong to $L_p(\Sigma, d\mu)$, we shall define a **trace** for u as a pair of functions $u^\pm$ in $L_{p,loc}(D^\pm, d\nu^\pm)$ such that the extended Green's identity is valid:

$$<Yu,v> + <u,Yv> = \int_{D^+} u^+ v\,d\nu^+ - \int_{D^-} u^- v\,d\nu^-$$

for all $v\in\Phi$. The precise definition of the space $L_{p,loc}(D^\pm,d\nu^\pm)$ will be given in the next section. Corresponding to the decomposition (2.6) of $D^\pm$, we decompose the measures and the traces as follows:

$$d\nu^- = (d\nu_-,d\nu_0), \quad d\nu^+ = (d\nu_+,d\nu_T),$$

$$\nu_\pm = \nu^\pm|_{D_\pm}, \quad \nu_0 = \nu^-|_{D_0}, \quad \nu_T = \nu^+|_{D_T},$$

$$u^- = (u_-,u_0), \quad u^+ = (u_+,u_T).$$

Then we have

$$d\nu_0 = dx\,d\rho \quad \text{on } D_0,$$

$$d\nu_T = dxd\rho \quad \text{on } D_T.$$

Let us now consider a nonnegative Lebesgue measurable function h on Σ, which is Lebesgue integrable on all bounded Lebesgue measurable subsets of Σ, and two bounded linear operators $J:L_p(\Sigma,d\mu)\rightarrow L_p(\Sigma,d\mu)$ and $K:L_p(D_+,d\nu_+)\rightarrow L_p(D_-,d\nu_-)$, $1\leq p<\infty$, with the following properties:

(i) For every bounded continuous function r of t alone we have $J(ru)=rJu$ and $K(ru)=rKu$.

(ii) $\overline{Ju}=J\overline{u}$ and $\overline{Ku}=K\overline{u}$, where the bar denotes complex conjugation.

The boundedness condition on J excludes from our analysis such transport processes as electron scattering in metals or Fokker-Planck type diffusion for which the collision operator A is a second order differential operator of Sturm-Liouville type. Assumption (i) above excludes processes which are nonMarkovian in time (nonlocality in space is allowed), while assumption (ii) will allow us to consider these problems in both real and complex Banach spaces.

We shall call $-X$ the **free streaming operator**, $-(X+h) \equiv S$ the **streaming operator** and $B \equiv S + J$ the (full) **transport operator**. The operator $-A \equiv -h + J$ will be called the **collision operator**.

With the previous notation, definitions and assumptions we shall study the following abstract time dependent linear kinetic problem:

$$Yu + hu - Ju = f \quad \text{on } \Sigma, \tag{2.7a}$$

$$u_0 = g_0 \quad \text{on } D_0, \tag{2.7b}$$

$$u_- = Ku_+ + g_- \quad \text{on } D_-. \tag{2.7c}$$

The initial condition (2.7b) and the boundary condition (2.7c) can be written with $\mathcal{K}\equiv(0,K):L_p(D^+,d\nu^+)\rightarrow L_p(D^-,d\nu^-)$ as the single equation

$$u^- = \mathcal{K}u^+ + g^- \quad \text{on } D^-.$$

We shall seek the solutions of this initial-boundary value problem in a linear space E_p. For all $1\leq p<\infty$ we shall denote by E_p the space of functions $u\in L_p(\Sigma,d\mu)$ such that $(Y+h)u$ belongs to $L_p(\Sigma,d\mu)$ and the traces $u^{\pm}$ of u belong to $L_p(D^{\pm},d\nu^{\pm})$. In the next section we shall introduce all our function spaces more rigorously and prove the existence of a trace for every solution of the problem (2.7). Sections 4 and 5 are devoted to the wellposedness and positivity properties of this problem, while in Section 6 this problem is analyzed on the space of bounded measures on Σ.

3. Vector fields, function spaces, and traces

In this section we shall consider a general setting which includes the vector field of interest. We shall assume Σ is a C^{∞} manifold with piecewise C^1 boundary, embedded in $\mathbb{R}^d$. (It will be evident that the regularity assumption on the boundary $\partial\Sigma$ may be relaxed so long as endpoints of integral curves of Y which are both left and right endpoints may be considered as distinct boundary points.) Σ is assumed to be equipped with a positive Borel measure μ for which bounded Lebesgue measurable sets have finite μ-measure. Let Y be a Lipschitz continuous vector field defined on Σ which extends Lipschitz continuously to $\bar{\Sigma}$ and does not vanish at any point of the closure. The vector field Y is assumed to be real and divergence free with respect to μ, in the sense that

$$\int_{\Sigma} Yv\,d\mu = 0, \qquad v \in C_c^1(\Sigma). \tag{3.1}$$

Here $C_c^1(\Sigma)$ is the space of continuously differentiable functions defined on Σ having compact support in Σ. Given a point $y_0 \in \Sigma$, there is a unique maximal integral curve for Y passing through y_0, specified by

$$\frac{d}{ds}y(s) = Yy(s), \quad y(0) = y_0.$$

Since Y is Lipschitz continuous on Σ, the integral curves of Y do not intersect. By the **length** of this curve, we mean the length of the maximal s-interval over which the curve remains in Σ. Note that $\frac{dt}{ds}=1$ implies that the length of an integral curve is the travel time rather than the arc length. We shall further assume:

Every maximal integral curve for Y in Σ has a length bounded by a fixed finite constant T and has left and right limits.

Since Y is nonvanishing on $\bar{\Sigma}$, the limits are points of $\partial\Sigma$. Moreover, since Y is Lipschitz continuous up to $\partial\Sigma$, each boundary point is a left (resp. right) limit of at most one integral curve.

Let D^- (resp. D^+) be the subset of $\partial\Sigma$ consisting of all left (resp. right) endpoints of maximal integral curves for Y in Σ. As already noted, since $\partial\Sigma$ is piecewise continuously differentiable, the union of all integral curves which intersect $D^+\cap D^-$ has μ-measure zero. This is a consequence of Sard's theorem [101, 280]. Indeed, arguing as in [23], we remark that the trajectories that meet a point of $\partial\Sigma$ where the unit normal either does not exist or is discontinuous form a set of Lebesgue measure (and hence μ-measure) zero. Consider a fully C^1-portion $\mathcal{C}$ of the remainder of $\partial\Sigma$, and define, for $x\in\mathcal{C}$ and $s\in\mathbb{R}$, $\varphi(x,s)$ as the point obtained from x by moving on the trajectory passing through x to the right over a "distance" s. On choosing a local coordinate system on $x\in U\subset\mathcal{C}$, we easily derive that the Jacobian of $\varphi(x,s)$ vanishes if and only if $x\in D^-\cap D^+$. According to Sard's theorem, the set of such $\varphi(x,s)$ must have Lebesgue measure (and hence μ-measure) zero, which settles the issue. Thus it will follow from Proposition 3.2 below that $D^+\cap D^-$ has $\nu^{\pm}$-measure zero.

The above argument also shows that $D^-\cap D^+$ is a Borel set if $\partial\Sigma$ is piecewise C^1. This property of $D^{\pm}$ in fact is true in general. We may represent Σ in the following way. Given a point $x\in D^-$ there is a unique integral curve with x as its left endpoint. If the length of this curve is denoted by $\ell(x)$, where $0<\ell(x)\le T$, the curve may be parametrized by points of the interval $(0,\ell(x))$. Thus we obtain the identification

$$\Sigma \approx \{(x,s) : x\in D^-, \quad 0<s<\ell(x)\}. \tag{3.2}$$

Let us first notice that D^- and D^+ are Borel sets and $\ell(x)$ is a Borel function of position. Indeed, consider a Borel set $M\subset\Sigma$ that contains exactly one point from every trajectory. Such a set can be constructed as follows. Take a countable dense subset N of Σ, and, for every $x\in N$, an open neighborhood U_x and a Borel set $M_x\subset U_x$ containing at most one point from every trajectory. The latter is possible, because the vector field does not vanish within Σ. Put

$$T_x = \{(y,s) : y\in D^-, \quad 0<s<\ell(y), \quad (y,\hat{s})\in M_x \text{ for some } \hat{s}\in(0,\ell(y))\}.$$

Then Σ can be written as the union of neighborhoods U_{x_n}, where $n \in \mathbb{N}$, while

$$M = M_{x_1} \cup \bigcup_{j=2}^{\infty} \{M_{x_j} \setminus \bigcup_{i=1}^{j-1} T_{x_i}\}.$$

Since the distance from $x \in M$ to the boundary when moving along the trajectory to the right/left is upper semicontinuous, it is clear that $D^{\pm}$ are Borel sets and $\ell(x)$ is a Borel function, which proves the assertion. Hence, the identification (3.2) represents a measure preserving transformation between Borel algebras and implies the correspondence $Y \approx \frac{\partial}{\partial s}$. Also we will abuse notation and make the identifications

$$D^- \approx \{(x,0) : x \in D^-\},$$

$$D^+ \approx \{(x,\ell(x)) : x \in D^-\}.$$

The above identifications will be used consistently throughout this section. As a result the bounded (measurable) subsets of Σ are exactly the (measurable) subsets of sets of the form

$$\{(x,s) : x \in E, \ \ 0<s<\ell(x)\},$$

where E is a bounded (measurable) subset of D^-

We shall now define the test function space Φ as the space of all Borel functions v on Σ with the following properties:

(i) v is continuously differentiable along each integral curve (but not necessarily even continuous in other directions).

(ii) v and Yv are bounded.

(iii) The support of v is bounded and there is a positive lower bound to the lengths of the integral curves which meet the support of v.

It is clear from the boundedness of Yv and the continuous differentiability of v along trajectories that every $v \in \Phi$ can be extended to be continuous at the endpoints of each integral curve. We shall then define the test function space Φ_0 as the subspace of Φ for which the functions have limit zero at the endpoints of each integral curve.

LEMMA 3.1. There are unique positive Borel measures $\nu^{\pm}$ on $D^{\pm}$ such that

$$\int_{\Sigma} Yu d\mu = \int_{D^+} u d\nu^+ - \int_{D^-} u d\nu^-, \quad u \in \Phi. \tag{3.3}$$

Proof: We shall prove this lemma first for a class of functions u that are linear along trajectories. We will next establish the lemma for functions u that are finite linear combinations of products of functions in $C_c^1(\Sigma)$ and bounded Borel functions that are constant on each trajectory. The information thus obtained will be used to prove Lemma 3.2. With the help of this lemma we will then complete the proof of the present result.

Let us suppose that E_0 is a bounded subset of D^- such that $\ell(x) \geq \delta > 0$ for $x \in E_0$. Suppose that w is a bounded Borel function defined on the set

$$E = \{(x,0) : x \in E_0\} \cup \{(x,\ell(x)) : x \in E_0\}.$$

We extend w by zero on the rest of $D^- \cup D^+$, and for $x \in D^-$ we set

$$\hat{w}(x,s) = [1 - \ell(x)^{-1}s]w(x,0) + \ell(x)^{-1}s\, w(x,\ell(x)). \tag{3.4}$$

Then $\hat{w}$ belongs to Φ, because of the condition $\ell(x) \geq \delta > 0$ on E_0, and

$$\sup |Y\hat{w}| \leq 2\delta^{-1} \sup |\hat{w}|.$$

It follows that the map $w \to \int_{\Sigma} Y\hat{w} d\mu$ is a bounded linear functional on the set of such functions. Varying E_0, we see that there are unique Borel measures $\nu^{\pm}$ on $D^{\pm}$ such that, for every w,

$$\int_{\Sigma} Y\hat{w} d\mu = \int_{D^+} w d\nu^+ - \int_{D^-} w d\nu^-. \tag{3.5}$$

Indeed, if E_0 is a subset of D^- as above and $E_\ell = \{(x,\ell(x)) : x \in E_0\}$, we define

$$\nu^-(E_0) = \int_{\Sigma} Y\hat{w} d\mu, \quad w = -\chi_{E_0},$$

$$\nu^{+}(E_{\ell}) = \int_{\Sigma} Y\hat{w}d\mu, \quad w=\chi_{E_{\ell}},$$

where χ_A denotes the characteristic function of the set A. Then (3.5) is certainly true for positive step functions w of the type allowed. Since every nonnegative w of the indicated type is a monotonically increasing limit of such positive step functions, and $\nu^{\pm}$ are positive σ-additive Borel measures, Eq. (3.5) is clear for all such w. Moreover, if $w(x,\ell(x))-w(x,0)$ is nonnegative, then $Y\hat{w}$ is nonnegative, and therefore ν^{+} and ν^{-} are positive measures. At this point we observe that $Y\hat{w} = \ell(x)^{-1}\{w(x,\ell(x))-w(x,0)\}$.

It suffices to prove that $\int_{\Sigma}Yvd\mu = 0$ for all $v\in\Phi_0$, since every $v\in\Phi$ can be written as the sum of a function in Φ_0 and a function $w\in\Phi$ that is linear along trajectories. We shall give the proof for a special class Ψ of functions $v\in\Phi$ and exploit it to establish Lemma 3.2, which is a decomposition of the measure μ as the product of the measure $\nu^{\pm}$ and the arc length measure along trajectories. With this decomposition in hand, we will have immediately

$$\int_{\Sigma} Yud\mu = \int_{D^-} \{u(x,\ell(x))-u(x,0)\}d\nu^{-}(x), \quad u\in\Phi_0,$$

which will complete the proof.

Let us observe that (3.1) holds true for all $v\in C_c^1(\Sigma)$, by assumption. This implies the validity of (3.1) for all $v\in\Phi$ of the form

$$v(x,s) = \tau(x,s)w(x), \quad x\in D^{-}, \quad 0<s<\ell(x), \tag{3.6}$$

where $\tau\in C_c^1(\Sigma)$ and w is a bounded Borel function on D^-. It should be noted that the compactness of the support of τ implies that $\ell(x)$ is bounded below for all integral curves meeting the support of τ, so that one indeed has $v\in\Phi_0$. Then (3.3) is true on the class Ψ_0 of finite linear combinations of functions of the type (3.6) and therefore on the class Ψ of sums of a function in Ψ_0 and a function in Φ that is linear along trajectories. We may apply the validity of (3.3) on Ψ to prove Lemma 3.2 and hence to finish the proof of the present lemma. ∎

LEMMA 3.2. In the realization (3.2) we have

$$d\mu = d\nu^-ds, \qquad d\mu = d\nu^+ds.$$

Proof: Consider a general $w\in\Psi$, and set

$$v_1(x,s) = -\int_s^{\ell(x)} w(x,t)dt.$$

Then $Yv_1=w$, and $v_1=0$ on D^+. Thus Lemma 3.1, proven above for all $w\in\Psi$, together with Fubini's theorem, implies

$$\int_\Sigma wd\mu = \int_{D^+} v_1(x,\ell(x))d\nu^+(x) - \int_{D^-} v_1(x,0)d\nu^-(x) = \int_\Sigma w(x,s)d\nu^-(x)ds.$$

On the other hand, putting

$$v_2(x,s) = \int_0^s w(x,t)dt,$$

we get $Yv_2 = w$ and $v = 0$ on D^-, and therefore

$$\int_\Sigma wd\mu = \int_{D^+} v_2(x,\ell(x))d\nu^+(x) - \int_{D^-} v_2(x,0)d\nu^-(x) = \int_\Sigma w(x,s)d\nu^+(x)ds. \tag{3.7}$$

Observe that Ψ is dense in $L_p(\Sigma,d\nu^\pm ds)$ for $p\in[1,\infty)$ and that both μ and the above product measures are Borel measures. Then every $w\in L_p(\Sigma,d\nu^\pm ds)$ has the same integral over Σ with respect to the product measure as well as with respect to μ, which completes the proof. ∎

We are now in a position to derive two important propositions on which the proofs of the existence theorems of the next section are based. For $1\le p<\infty$ we denote by $L_{p,loc}(\Sigma,d\mu)$ the linear vector space of all μ-measurable functions u on Σ with the property that $|u|^p$ is μ-integrable on every bounded μ-measurable subset of Σ on which $\ell(x,s)\equiv\ell(x)$ is bounded away from zero.

PROPOSITION 3.3. Suppose u and $(Y+h)u$ belong to $L_p(\Sigma,d\mu)$, $1\le p<\infty$. Then:

(i) u has a unique trace $u^\pm$.

(ii) If u^- belongs to $L_p(D^-,d\nu^-)$, then u^+ belongs to $L_p(D^+,d\nu^+)$, in which case $h|u|^p$ and $|u|^{p-1}Yu$ are integrable and

$$\int_{D^+} |u^+|^p d\nu^+ + p\int_\Sigma h|u|^p d\mu = \int_{D^-} |u^-|^p d\nu^- + p\int_\Sigma \mathrm{sgn}(u)|u|^{p-1}(Y+h)u d\mu.$$

Proof: Suppose first that u and Yu belong to $L_{p,loc}(\Sigma,d\mu)$. Writing $u=u_0+u_1$, where $u_0(x,s) = [1-\ell(x)^{-1}s]u(x,s)$, we deduce that

$$v_0 \equiv Yu_0 = [1-\ell(x)^{-1}s]Yu - \ell^{-1}(x)u$$

belongs to $L_{p,loc}(\Sigma,d\mu)$. It follows from Lemma 3.2 that, for almost every $x\in D^-$, $v_0(s,\cdot)\in L_p((0,\ell(x)),ds)$. Indeed, if $v_0\in L_{p,loc}(\Sigma,d\mu)$ and χ_S denotes the characteristic function of a set S, then $v_0\chi_S\in L_p(\Sigma,d\mu)$ for every S of the form $S = \{(x,s) : x\in E_0,\ 0<s<\ell(x)\}$, where the set $E_0\subset D^-$ is bounded and $\ell(x)$ is bounded away from zero on E_0. Hence, for every such x,

$$\int_\Sigma |v_0\chi_S|^p d\mu = \int_{E_0}\int_0^{\ell(x)} |v_0(x,s)|^p ds d\nu^- < \infty,$$

whence $v_0(x,.)\in L_p(0,\ell(x))$ almost everywhere on D^-, which proves our assertion. For such x set

$$u_0^*(x,s) = -\int_s^{\ell(x)} v_0(x,t)dt. \tag{3.8}$$

Assume $\varphi\in\Phi$ and set

$$w_0(x,s) = \int_0^s \varphi(x,t)dt.$$

Then $w_0\in\Phi$ and $w_0=0$ on D^-, so $[1-\ell(x)^{-1}]w_0 \in \Phi_0$ and we have

$$\langle u_0^*,\varphi\rangle = \int_{D^-}\int_0^{\ell(x)} u_0^*\varphi ds d\nu^- = \int_{D^-} \{[u_0^* w_0]_0^{\ell(x)} - \int_0^{\ell(x)} v_0 w_0 ds\} d\nu^- =$$

$$= -\langle(1-\ell^{-1}s)Yu,w_0\rangle + \langle\ell^{-1}u,w_0\rangle = \int_{D^-} \{-[(1-\ell^{-1}s)uw_0]_0^{\ell(x)} +$$

$$+ \int_0^{\ell(x)} [-\ell^{-1}uw_0+(1-\ell^{-1}s)u\varphi]ds\}d\nu^- + \langle\ell^{-1}u,w_0\rangle = \langle u_0,\varphi\rangle.$$

Thus we may identify u_0 with $\overset{*}{u}_0$. For any $z\in\Phi$ an integration by parts gives

$$\langle u_0,Yz\rangle + \langle Yu_0,z\rangle = \langle\overset{*}{u}_0,Yz\rangle + \langle v_0,z\rangle = \int_{D^-} [\overset{*}{u}_0 z]_{s=0}^{\ell(x)} d\nu^- =$$

$$= -\int_{D^-} u_0(x,0)z(x,0)d\nu^-(x).$$

This shows that u_0 has traces $u_0^- = \overset{*}{u}_0(\cdot,0)$ and $u_0^+\equiv 0$. The Hölder inequality applied to (3.8) shows that $u^- = u_0^-$ is in $L_{p,loc}(D^-,d\nu^-)$.

Next, put $u_1=\ell(x)^{-1}su$, $v_1\equiv Yu_1=\ell^{-1}sYu+\ell^{-1}u$ and

$$\overset{*}{u}_1(x,s) = \int_0^s v_1(x,t)dt. \tag{3.9}$$

Taking $\varphi\in\Phi$ and putting $w_1(x,s)=-\int_s^{\ell(x)}\varphi(x,t)dt$, we get $w_1\in\Phi$, and $w_1=0$ on D^+; thus $(sw_1/\ell(x))\in\Phi_0$. As before we obtain

$$\langle\overset{*}{u}_1,\varphi\rangle = -\langle v_1,w_1\rangle = -\langle\ell^{-1}sYu,w_1\rangle - \langle\ell^{-1}u,w_1\rangle = \langle u_1,\varphi\rangle,$$

whence we may identify u_1 and $\overset{*}{u}_1$. Thus for any $z\in\Phi$ we obtain

$$\langle u_1,Yz\rangle + \langle Yu_1,z\rangle = \int_{D^+} u_1(x,\ell(x))d\nu^+.$$

Hence, u_1 has as its trace $u_1^-=0$, $u_1^+=\overset{*}{u}_1(\cdot,\ell(x))$, while $u^+=u_1^+\in L_{p,loc}(D^+,d\nu^+)$ as a consequence of the Hölder inequality applied to (3.9).

To prove the uniqueness of the trace we start with a function w on $D^-\cup D^+$ as in the proof of Lemma 3.1. Then we must have

$$\int_{D^+} wu^+ d\nu^+ - \int_{D^-} wu^- d\nu^- = <u,Y\hat{w}> + <Yu,\hat{w}>. \tag{3.10}$$

The right hand side of (3.10) is uniquely determined by w and u, and therefore $u^\pm$ are unique.

Now suppose that u and (Y+h)u belong to $L_p(\Sigma,d\mu)$, and set

$$\hat{u}(x,s) = u(x,s) \exp\{\int_0^s h(x,t)dt\} \equiv u(x,s)H(x,s).$$

Then $Y\hat{u} = H(Y+h)u$, so that $\hat{u}$ and $Y\hat{u}$ belong to $L_{p,loc}(\Sigma,d\mu)$. From the first part of the proof it follows that $\hat{u}$ has a trace. To complete the proof, we note first that a straightforward calculation of $-<|u|^p,Yv>$ for $v\in\Phi_0$ implies that $|u|^p$ has the distributional derivative

$$Y(|u|^p) = p(\text{sgn } u)|u|^{p-1}Yu,$$

which, together with $|u|^p$, belongs to $L_{1,loc}(\Sigma,d\mu)$. For the former function we apply the Hölder inequality and the property $Yu\in L_{p,loc}(\Sigma,d\mu)$; for the latter we have to repeat the reasoning of the beginning of this proof with u replaced by $|u|^p$, where u_0^* and u_1^* are replaced by $|u_0^*|^p$ and $|u_1^*|^p$, respectively. If we apply the Green's identity to $w\in\Phi$ defined as the characteristic function of the set $\Sigma_0 = \{(x,s) : x\in C^-\subset D^-,\ 0<s<\ell(x)\}$, then the identity in the statement of the proposition holds with $D^\pm$ replaced by a Borel set $C^\pm$ on which ℓ^{-1} is bounded, and with Σ replaced by Σ_0. Furthermore, since $u\in L_p(\Sigma,d\mu)$ and $(Y+h)u\in L_p(\Sigma,d\mu)$, one has $(\text{sgn } u)|u|^{p-1}(Y+h)u\in L_1(\Sigma,d\mu)$, and therefore its integral over Σ converges. Hence, if $u^-\in L_p(D^-,d\nu^-)$, then the above equation implies that $u^+\in L_p(D^+,d\nu^+)$ and $h|u|^p\in L_1(\Sigma,d\mu)$. Thus we may take Σ as the union of such sets Σ_0 and pass to the limit to obtain the identity. ∎

The boundary spaces $L_p(D^\pm,d\nu^\pm)$, which are natural for the physical problem involved, do not allow for stronger results than part (i) of Proposition 3.3. These can be obtained, however, if instead of $L_p(D^\pm,d\nu^\pm)$ one considers $L_p(D^\pm,\ell d\nu^\pm)$, where the weight function $\ell(x)$ is the length of the characteristic curve. Results in this direction have been obtained by Cessenat [87] for $a\equiv0$ and Ukai [353] for a large class of force fields. Indeed, let $u,Yu\in L_p(\Sigma,d\mu)$ for some $1\le p<\infty$, and put $q=p/(p-1)$. Following Ukai [353] we then have

$$u^-(x) = u(x,s) - \int_0^s (Yu)(x,\hat{s})d\hat{s}, \quad x \in D^-.$$

Using the Hölder inequality as well as the inequality $c^{1/p}+d^{1/p} \leq 2^{1/q}(c+d)$ for $c=|u(x,s)|^p$ and $d=s^{p-1}\int_0^s |(Yu)(x,\hat{s})|^p d\hat{s}$ we obtain

$$|u^-(x)|^p \leq \{|u(x,s)| + s^{1/q}[\int_0^{\ell(x)} |(Yu)(x,\hat{s})|^p d\hat{s}]^{1/p}\}^p \leq$$

$$\leq 2^{p-1}\{|u(x,s)|^p + s^{p-1}\int_0^{\ell(x)} |(Yu)(x,\hat{s})|^p d\hat{s}\}.$$

Integrating over the interval $(0,\ell(x))$ and using that $\ell(x)\leq T$, one obtains

$$\ell(x)|u^-(x)|^p \leq 2^{p-1}\{\int_0^{\ell(x)} |u(x,s)|^p ds + \frac{T^p}{p}\int_0^{\ell(x)} |(Yu)(x,\hat{s})|^p d\hat{s}\},$$

whence

$$\int_{D^-} \ell(x)|u^-(x)|^p d\nu^-(x) \leq 2^{p-1}\{\|u\|^p + \frac{T^p}{p}\|Yu\|^p\}.$$

This in turn implies

$$\|u^-\|_\ell \leq 2^{1/q}\{\|u\| + Tp^{-1/p}\|Yu\|\} \leq C(\|u\| + \|Yu\|),$$

where $C=2(1+T)$ and $\|.\|_\ell$ denotes the norm in $L_p(D^\pm,\ell d\nu^\pm)$. A similar estimate holds true if D^-, ν^- and u^- are replaced by D^+, ν^+ and u^+. If the function h is essentially bounded on Σ, we obtain the important estimate

$$\|u^\pm\|_\ell \leq 2(1 + T + \|h\|_\infty)(\|u\| + \|(Y+h)u\|), \tag{3.11}$$

which proves that $u^\pm \in L_p(D^\pm,\ell d\nu^\pm)$ whenever $u\in L_p(\Sigma,d\mu)$, $(Y+h)u\in L_p(\Sigma,d\mu)$, and $h\in L_\infty(\Sigma,\mu)$.

The next result is an existence, uniqueness and positivity theorem for the solution of the initial-boundary value problem (2.7) in the absence of scattering (J=0) and boundary reflection (K=0). We recall from Section 2 the definition of the solution space E_p.

PROPOSITION 3.4. Given $f \in L_p(\Sigma, d\mu)$ and $g \in L_p(D^-, d\nu^-)$, $1 \le p < \infty$, there is a unique function $u \in E_p$ such that $(Y+h)u = f$ in Σ and $u^- = g$ on D^-. Moreover, if f and g are nonnegative, then u is nonnegative.

Proof: Given $f \in L_p(\Sigma, d\mu)$ and $g \in L_p(D^-, d\nu^-)$, set

$$u(x,s) = \exp\{-\int_0^s h(x,\sigma)d\sigma\}g(x) + \int_0^s \exp\{-\int_t^s h(x,\sigma)d\sigma\}f(x,t)dt \equiv$$

$$\equiv g(x) + u_0(x,s). \tag{3.12}$$

Clearly, u is nonnegative if g and f are, and an integration by parts gives

$$<u,(Y-h)v> + <f,v> = \int_{D^+} u(x,\ell(x))v(x,\ell(x))d\nu^+(x) - \int_{D^-} g(x)v(x,0)d\nu^-(x)$$

for all $v \in \Phi$. Hence, $(Y+h)u = f$ in the distributional sense. To see that $u \in L_p(\Sigma, d\mu)$ we note that $\|u\| \le T^{1/p}\|g\| + \|u_0\|$ and

$$\|u_0\|^p \le \int_{D^-}\int_0^{\ell(x)} \{s^{p-1}\int_0^s |f(x,t)|^p dt\}ds d\nu^-(x) \le p^{-1}T^p\|f\|^p.$$

The uniqueness of u given by (3.12) follows from the fact that, for almost every $x \in D^-$, $u(x,\cdot)$ must satisfy an ordinary differential equation on $(0,\ell(x))$ with given initial condition, so (3.12) is necessary. ∎

4. Existence, uniqueness, dissipativity and positivity in L_p.

In this section we shall prove the main existence, uniqueness and positivity results on the abstract time dependent kinetic problem (2.7). Throughout we shall assume partial absorption at the boundary of the physical system, i.e., we assume that $\|K\| < 1$. In the next section we will deal with the case $\|K\| = 1$. We begin with two auxiliary results. The first result will enable us to introduce a parameter λ, which will be chosen conveniently later on; the second will be a useful estimate.

PROPOSITION 4.1. The problem (2.7) has a unique solution if and only if for $\lambda \in \mathbb{R}$ the modified problem

$$(Y + h - J + \lambda)u_\lambda = f_\lambda \quad \text{on } \Sigma, \tag{4.1a}$$

$$u_{\lambda,0} = g_{\lambda,0} \quad \text{on } D_0, \tag{4.1b}$$

$$u_{\lambda,-} = Ku_{\lambda,+} + g_{\lambda,-} \quad \text{on } D_-, \tag{4.1c}$$

has a unique solution. Moreover, nonnegative solutions of (2.7) correspond to nonnegative solutions of (4.1).

Proof: Set $u_\lambda(x,\xi,t) = e^{-\lambda t}u(x,\xi,t)$, $f_\lambda(x,\xi,t) = e^{-\lambda t}f(x,\xi,t)$ and $g_\lambda(x,\xi,t) = e^{-\lambda t}g(x,\xi,t)$. Because of the assumption on J and K of locality in time, u solves (2.7) with data (f,g) if and only if u_λ solves (4.1) with data (f_λ,g_λ). ∎

We remark that the locality of A and K in time was used here in an essential way. If A and K are not local in time, one cannot shift the problem in such a transparent way.

PROPOSITION 4.2. Suppose u_λ belongs to E_p and satisfies (4.1a), where $\lambda \geq 0$ and $f_\lambda \in L_p(\Sigma,d\mu)$. Then

$$\|u_\lambda^+\|^p + \lambda\|u_\lambda\|^p \leq \|u_\lambda^-\|^p + \lambda^{1-p}\|Ju_\lambda + f_\lambda\|^p. \tag{4.2}$$

Proof: If p=1, (4.2) follows from (3.7), since $Yu_\lambda = -(h-J+\lambda)u_\lambda + f_\lambda$ and $h \geq 0$. If $p>1$, (3.7) gives

$$\|u_\lambda^+\|^p + \lambda p\|u_\lambda\|^p \leq \|u_\lambda^-\|^p + p\int_\Sigma |u_\lambda|^{p-1}|Ju_\lambda + f_\lambda|\,d\mu. \tag{4.3}$$

Take $q=p(p-1)^{-1}$ and use the inequality $ab \leq p^{-1}a^p + q^{-1}b^q$ for $a,b \geq 0$. Then with $a = \lambda^{-1/q}|Ju_\lambda + f_\lambda|$ and $b = \lambda^{1/q}|u_\lambda|^{p-1}$, an integration gives

$$p\int_\Sigma |u_\lambda|^{p-1}|Ju_\lambda + f_\lambda|\,d\mu \leq \lambda^{1-p}\int_\Sigma |Ju_\lambda + f_\lambda|^p d\mu + (p-1)\lambda\int_\Sigma |u_\lambda|^p d\mu.$$

The inequality (4.2) follows from this estimate and (4.3). ■

We shall now present the main result of this section, which answers in the affirmative the wellposedness of the time dependent problem, if the boundary operator K satisfies $\|K\|<1$. The special case of $\|K\|=1$, relating to conservative boundary processes, will be discussed separately in Section 5; the proof for this case requires an additional hypothesis as well as the weakening of the notion of solution.

THEOREM 4.3. Suppose that J and K are bounded linear operators $J:L_p(\Sigma,d\mu)\to L_p(\Sigma,d\mu)$ and $K:L_p(D_+,d\nu_+)\to L_p(D_-,d\nu_-)$, $1\le p<\infty$. Let us suppose further that K has operator norm $\|K\|<1$. Then for any vectors $f\in L_p(\Sigma,d\mu)$ and $g=(g_-,g_0)\in L_p(D^-,d\nu^-)$, the abstract time dependent linear kinetic problem

$$Yu + hu - Ju = f \quad \text{on } \Sigma, \tag{4.4a}$$

$$u_0 = g_0 \quad \text{on } D_0, \tag{4.4b}$$

$$u_- = Ku_+ + g_- \quad \text{on } D_-, \tag{4.4c}$$

has a unique solution u in E_p. Moreover, there is a constant $C=C(p,A,K)$ such that the L_p-norms of u and u_T satisfy

$$\|u\| + \|u_T\| \le C(\|f\| + \|g\|). \tag{4.5}$$

Proof: The strategy for proving the present theorem and the next theorem (related to positivity of solutions) is to give a direct construction of the solution for $J=0$ and $K=0$ and then to use a Green's identity to obtain estimates which permit passage to the general case by a perturbation analysis. Thus, let us consider the modified problem (4.1) with $J=0$, $K=0$, and $\lambda>0$:

$$(Y + h + \lambda)u_\lambda = f_\lambda, \quad u_\lambda^- = g_\lambda. \tag{4.6}$$

Propositions 3.4 and 4.1 imply the existence of a unique solution of (4.6), which we denote by

$$u_\lambda = S_\lambda(f_\lambda,g_\lambda). \tag{4.7}$$

The inequality (4.2) implies

$$\|u_\lambda^+\|^p + \lambda\|u_\lambda\|^p \le \|g_\lambda\|^p + \lambda^{1-p}\|f_\lambda\|^p.$$

Therefore, one has the inequalities

$$\|S_\lambda(f_\lambda,0)\| \le \lambda^{-1}\|f_\lambda\|, \tag{4.8}$$

$$\|S_\lambda(f_\lambda,0)^+\| \le \lambda^{-1+1/p}\|f_\lambda\|, \tag{4.9}$$

$$\|S_\lambda(0,g_\lambda)\| \le \lambda^{-1/p}\|g_\lambda\|, \tag{4.10}$$

$$\|S_\lambda(0,g_\lambda)^+\| \le \|g_\lambda\|. \tag{4.11}$$

Next, consider the case $K=0$, $\lambda > \|J\|$:

$$(Y + h - J + \lambda)u_\lambda = f_\lambda, \quad u_\lambda^- = g_\lambda. \tag{4.12}$$

We look for a solution having the form $u_\lambda = S_\lambda(f_\lambda^*, g_\lambda)$ with $f_\lambda^* \epsilon L_p(\Sigma, d\mu)$, where f_λ^* is to be determined. The necessary and sufficient condition for u_λ to solve (4.12) is

$$f_\lambda^* - JS_\lambda(f_\lambda^*, g_\lambda) = f_\lambda. \tag{4.13}$$

As $S_\lambda(f_\lambda^*, g_\lambda) = S_\lambda(f_\lambda^*, 0) + S_\lambda(0, g_\lambda)$, (4.13) becomes

$$(I + L_\lambda)f_\lambda^* = f_\lambda + JS_\lambda(0, g_\lambda),$$

with

$$L_\lambda f_\lambda^* = -JS_\lambda(f_\lambda^*, 0). \tag{4.14}$$

Now, (4.8) implies that the operator norm satisfies $\|L_\lambda\| \le \lambda^{-1}\|J\| < 1$. Thus (4.13) has the unique solution

$$f_\lambda^* = \sum_{m=0}^{\infty} (-L_\lambda)^m [f_\lambda + JS_\lambda(0,g_\lambda)]. \tag{4.15}$$

We denote the solution of (4.12) by $u_\lambda = T_\lambda(f_\lambda, g_\lambda)$. In particular, for $f_\lambda = 0$ the identity (4.15) and the inequalities (4.8)–(4.11), along with the norm estimate above, give the estimates

$$\begin{aligned} \|T_\lambda(0,g_\lambda)^+\| &= \|S_\lambda(f_\lambda^*, g_\lambda)^+\| = \|S_\lambda(0,g_\lambda)^+ + S_\lambda(f_\lambda^*,0)^+\| \leq \\ &\leq \|g_\lambda\| + \lambda^{-1+1/p}\|f_\lambda^*\| \leq [1 + (\lambda - \|J\|)^{-1}\|J\|]\|g_\lambda\|. \end{aligned} \tag{4.16}$$

The final case to consider is the general case

$$(Y + h - J + \lambda)u_\lambda = f_\lambda, \tag{4.17a}$$

$$u_\lambda^- = \mathcal{K}u_\lambda^+ + g_\lambda, \tag{4.17b}$$

where $\mathcal{K}=(0,K)$. Let us look for a solution of (4.17) having the form $u_\lambda = T_\lambda(f_\lambda, g_\lambda^*)$, where $g_\lambda^* \in L_p(D^-, d\nu^-)$ is to be determined. The necessary and sufficient condition is

$$g_\lambda^* = g_\lambda + \mathcal{K}(T_\lambda(f_\lambda, g_\lambda^*)^+), \tag{4.18}$$

so we require

$$(I - M_\lambda)g_\lambda^* = g_\lambda + \mathcal{K}(T_\lambda(f_\lambda,0)^+),$$

where $M_\lambda g_\lambda^* = \mathcal{K}(T_\lambda(0,g_\lambda^*)^+)$. The inequality (4.16) implies $\|M_\lambda\| < 1$, provided $\lambda > \|J\|$ and $[1 + (\lambda - \|J\|)^{-1}\|J\|]\|K\| < 1$. The second condition can be satisfied because $\|K\| < 1$. Thus, for such λ, (4.17) has the unique solution $u_\lambda = T_\lambda(f_\lambda, g_\lambda^*)$, if the function g_λ^* in (4.18) is given by

$$g_\lambda^* = \sum_{m=0}^{\infty} M_\lambda^m [g_\lambda + \mathcal{K}(T_\lambda(f_\lambda,0)^+)]. \tag{4.19}$$

Finally, we convert (4.19) to a solution of the problem with $\lambda=0$ by the method of Proposition 4.1. This completes the proof. ∎

In most cases, kinetic equations are evolution equations for (one-particle) distribution functions or radiative intensities on a phase space. As such, the solution of a kinetic equation is supposed, on physical grounds, to be nonnegative if the data are nonnegative. An important exception is related to those kinetic equations, for example in rarefied gas dynamics, which are not written for the distributions themselves, but for differences between the actual distribution and an equilibrium distribution.

We call the problem (4.4) **positive**, if the solution u is nonnegative whenever the data (f,g) are nonnegative. An operator L between spaces of real functions (or measures) is called **positive** (in the lattice sense) if $u\geq 0$ implies $Lu\geq 0$ (cf. Section I.4). Throughout the present chapter and the next two chapters the concept of a positive operator will always be used in the lattice sense, and not in the sense of positive selfadjointness (as in Chapters II to V).

THEOREM 4.4. In addition to the assumptions of Theorem 4.3, suppose $K\geq 0$ and suppose, for some real λ_0, $\lambda_0 I+J\geq 0$. Then problem (4.4) is positive.

Proof: We assume that $K\geq 0$, $(\lambda_0 I+J)\geq 0$, $f\geq 0$ and $g\geq 0$. In view of Proposition 4.1 we may replace J by $J+\lambda_0 I$ and assume $J\geq 0$. We now follow the proof of Theorem 4.3. Using Propositions 4.1 and 4.2, we obtain the positivity of the solution: $S_\lambda(f_\lambda,g_\lambda)\geq 0$ if $f_\lambda\geq 0$ and $g_\lambda\geq 0$.

We proceed similarly for K=0 and $\lambda>\|J\|$. The solution $T_\lambda(f_\lambda,g_\lambda)$ of (4.12) is given by $u_\lambda=T_\lambda(f^*_\lambda,g_\lambda)$ with f^*_λ given by (4.15), while the operator $(-L_\lambda)$ of (4.14) is positive. Using the positivity of J and the solution operator, we obtain $JS_\lambda(0,g_\lambda)\geq 0$ if $g_\lambda\geq 0$. Therefore, if $f_\lambda\geq 0$ and $g_\lambda\geq 0$, the series (4.15) implies $f^*_\lambda\geq 0$, which immediately implies $u_\lambda=S_\lambda(f^*_\lambda,g_\lambda)\geq 0$. Summarizing, we have $T_\lambda(f_\lambda,g_\lambda)\geq 0$ if $f_\lambda\geq 0$ and $g_\lambda\geq 0$.

Finally, we consider the general problem (4.17). It is evidently sufficient to prove $g^*_\lambda\geq 0$. But (4.19) and the positivity of K imply $g^*_\lambda\geq 0$. ∎

Actually, in most kinetic problems the inequality $\lambda_0 I+J\geq 0$ is satisfied for $\lambda_0=0$, implying that the gain term is nonnegative for nonnegative distributions.

In addition to positivity, many kinetic problems have a dissipativity property. In physical terms, a system is dissipative if relevant quantities are non increasing or actually decreasing in time because of the general conservation laws and loss mechanisms: absorptive boundaries, dissipation by friction, etc. We call the problem (4.4) **dissipative** if $\|u_t\|\leq\|u_0\|$ for $0\leq t\leq T$ whenever $f\equiv 0$ and $g_-=0$. We remark that this inequality makes sense since, in addition to its "traces" u_0 and u_T, a solution of (4.4) has a "trace" u_t on each slice $\{(x,\xi) : (x,\xi,t)\epsilon\Sigma\}$.

THEOREM 4.5. In addition to the assumptions of Theorem 4.3, suppose that for every $u \epsilon L_p(\Sigma, d\mu)$,

$$\int_\Sigma \operatorname{sgn}(u)\ |u|^{p-1}(hu-Ju)d\mu \geq 0. \tag{4.20}$$

Then problem (4.4) is dissipative.

Proof: Suppose $u \epsilon E_p$ satisfies

$$(Y + h - J)u = 0, \quad u_- = Ku_+. \tag{4.21}$$

The identity (3.7) gives

$$\int_{D_T} |u_T|^p d\nu_T = \int_{D_0} |u_0|^p d\nu_0 - p\int_\Sigma \operatorname{sgn}(u)|u|^{p-1}(hu-Ju)d\mu +$$

$$+ \int_{D_-} |Ku_+|^p d\nu_- - \int_{D_+} |u_+|^p d\nu_+.$$

Since $\|K\|<1$, the inequality (4.20) implies $\|u_T\| \leq \|u_0\|$. The result for the intermediate values u_t follows by replacing the time interval (0,T) by the smaller time interval (0,t), and using the uniqueness of the solution on the smaller interval. ∎

For $p=1$ and under the previous positivity assumption, it is sufficient to assume (4.20) only for nonnegative u to obtain dissipativity.

THEOREM 4.6. Suppose $p=1$ and suppose that, in addition to the assumptions of Theorem 4.4, (4.20) holds for every nonnegative $u \epsilon E_1$. Then problem (4.4) is both positive and dissipative.

Proof: In the positive case, the preceding argument gives a suitable bound on $\|u_T\|$ for u satisfying (4.21) and $u_0 \geq 0$, provided (4.20) holds for nonnegative u. For p=1 this suffices to obtain the appropriate bound on any solution. In fact, let us write $u_0 = u_0^* - u_0^{**}$ with $u_0^* \geq 0$, $u_0^{**} \geq 0$ and $u_0^* u_0^{**} \equiv 0$. Then the solution u can be written as the difference $u = u^* - u^{**}$, where u^* and u^{**} are nonnegative and

$$\|u_T\| \leq \|u_T^*\| + \|u_T^{**}\| \leq \|u_0^*\| + \|u_0^{**}\| = \|u_0\|,$$

which completes the proof. ∎

5. The conservative case

The case $\|K\|=1$ describes conservative boundary conditions. A typical example is purely specular reflection, where K has the form

$$Ku(x,\xi,t) = u(x,\varphi(x,\xi,t),t)$$

with $\varphi(x,\xi,t) = \xi - 2(\xi \cdot n(x,t))n(x,t)$ and n(x,t) the unit outer normal to the possibly time dependent surface of Ω_t.

Examination of the case $\|K\|=1$ shows that it is necessary, in general, to drop the requirement $u \in E_p$ in order to guarantee the existence of a solution. From a physical point of view this can be understood as follows. If an integral curve of Y reaches the boundary at a point $(x,\xi,t) \in D_+$, let it be "continued" with starting point $(x,\varphi(x,\xi,t),t) \in D_-$, where $\varphi(x,\xi,t)$ is as above. Except for plane parallel domains and $a \equiv 0$, there is no finite upper bound to the number of times such continued curves reach the boundary in the interval $0 \leq t \leq T$. Thus it is possible that the "solution" of the kinetic problem will not belong to E_p, i.e. the trace $u_\pm$ will not have finite norm. On the other hand, if there is a finite upper bound to the number of collisions with the boundary, then the existence is guaranteed for boundary conditions with $\|K\|=1$ and $K \geq 0$, as we shall prove shortly. For a discussion of various examples with $\|K\|=1$ in which the existence of a solution in E_p is not valid for general L_p-data, see the work of Voigt [380], who treats the situation $L_p = L_1$, $A \equiv 0$, $a \equiv 0$, and K, Ω and V independent of time. A similar situation, but with time dependent nonlocal boundary conditions, has been studied by Babovsky [20], who treated reflecting walls through probability measures on the boundary. Interesting examples of conservative boundary conditions (designed actually for the full Liouville equation rather than for a kinetic equation) are discussed by Schnute and Shinbrot [326].

Below we shall prove the existence, uniqueness and nonnegativity of the solution of problem (2.7), when the boundary conditions are conservative. In the proof of Theorem 5.2 below we shall first take the incoming current at the boundary to be zero, after which we will obtain the general case by subtracting a particular solution and using a stability argument. We will therefore consider the abstract kinetic problem

$$(Y + h - J)u = f \quad \text{on } \Sigma, \tag{5.1a}$$

$$u_0 = g_0 \quad \text{on } D_0, \tag{5.1b}$$

$$u_- = Ku_+ \quad \text{on } D_-. \tag{5.1c}$$

We first prove the uniqueness of the solutions of problem (2.7).

THEOREM 5.1. Suppose J and K satisfy the conditions of Theorem 4.3 and $\|K\|=1$. Then, given $f \in L_p(\Sigma, d\mu)$ and $g^- = (g_0, g_-) \in L_p(D_0, d\nu_0)$, the problem (2.7a)–(2.7c) has at most one solution $u \in E_p$.

Proof: By Proposition 4.1 we may replace J by $J - \lambda I$ for arbitrary $\lambda > 0$. It is sufficient to show that the problem has only the trivial solution if $f=0$ and $g_0=0$. For such a solution the identity (3.7) gives

$$\|u_{\lambda,T}\|^p + \lambda p \|u_\lambda\|^p =$$

$$= \|Ku_{\lambda,+}\|^p + p \int_\Sigma \operatorname{sgn}(u)\, |u_\lambda|^{p-1} Ju_\lambda \, d\mu \le p\|J\| \; \|u_\lambda\|^p. \tag{5.2}$$

If $\lambda > \|J\|$, this inequality implies $u_\lambda = 0$. ■

To obtain satisfactory existence results we strengthen the assumptions on the operators J and K and weaken the requirement on a solution. Specifically, we will impose the conditions $K \ge 0$ and $\lambda_0 I + J \ge 0$ for some $\lambda_0 \in \mathbb{R}$. The positivity condition on K implies that K can be extended to a larger class of functions than the (global)

L_p-functions. Indeed, suppose $\{v_m\}_{m=1}^{\infty}$ is a nondecreasing sequence of nonnegative functions in $L_p(D_+,d\nu_+)$ with pointwise limit v. Then $\{Kv_m\}_{m=1}^{\infty}$ is also nondecreasing and has a strong limit depending only on v, so we may extend K by setting $Kv = \lim_{m\to\infty} Kv_m$. A natural procedure for trying to solve (5.1) is to replace K by $K_m=\alpha_m K$, m=1,2,..., where $\{\alpha_m\}_{m=1}^{\infty}$ is an increasing sequence of nonnegative scalars with limit 1. Then $\|K_m\|=\alpha_m<1$, and thus problem (5.1) with K replaced by K_m has a unique solution u_m.

THEOREM 5.2. Suppose the operators J and K satisfy the assumptions of Theorem 4.3 along with the conditions $K\geq 0$ and $\lambda_0 I+J\geq 0$ for some $\lambda_0\in\mathbb{R}$, and suppose $f\in L_p(\Sigma,d\mu)$ and $g^-=(g_0,g_-)\in L_p(D_0,d\nu_0)$ are nonnegative. Let $K_m=\alpha_m K$, where $\{\alpha_m\}_{m=1}^{\infty}$ is increasing with limit 1, $0<\alpha_m<1$, and let u_m be the solution of (2.7) with K replaced by K_m. Then the sequence $\{u_m\}_{m=1}^{\infty}$ is nondecreasing and converges in the norm to a function $u\in L_p(\Sigma,d\mu)$. This function satisfies (2.7a) and has a trace satisfying (2.7b) and also (2.7c) with K extended as above. Moreover, u is independent of the choice of the sequence $\{\alpha_m\}_{m=1}^{\infty}$.

Proof: We shall first consider problem (5.1), where $g_-=0$. Let $T_m(f,g)$ denote the solution of the approximate problem

$$(Y + h - J)u = f \quad \text{on } \Sigma, \tag{5.3a}$$

$$u_0 = g_0 \quad \text{on } D_0, \tag{5.3b}$$

$$u_- = K_m u_+ + g_- \quad \text{on } D_-. \tag{5.3c}$$

Fix m′, and suppose m≥m′. Let us look at the solution of (5.3) in the form

$$u_m = T_{m'}(f,g_m^*).$$

Since $K_m-K_{m'}$ is positive, we may (after introducing a suitable parameter $\lambda>0$) repeat the argument of the proof of Theorem 4.4 and obtain $g_m^*\geq(0,g_0)$. As $T_{m'}$ is positive, this gives $u_m\geq u_{m'}$ for $m\geq m'$. In order to show that the pointwise limit of the sequence $\{u_m\}_{m=1}^{\infty}$ belongs to L_p and is the limit in the norm, it is sufficient to obtain a uniform L_p-bound for this sequence, since it is nondecreasing and positive. Once again, we may replace J by $J-\lambda I$ for $\lambda>\|J\|$ and use (3.7) with J replaced by

$J-\lambda I$ to obtain

$$\|(u_{m,\lambda})_T\|^p + \|(u_{m,\lambda})_+\|^p = \|g_{\lambda,0}\|^p + \|\alpha_m K(u_{m,\lambda})_+\|^p + \tag{5.4}$$

$$+ p\int_\Sigma \{|u_{m,\lambda}|^{p-1}f_\lambda - h|u_{m,\lambda}|^p - \lambda|u_{m,\lambda}|^p + |u_{m,\lambda}|^{p-1}Ju_{m,\lambda}\}d\mu \le$$

$$\le \|g_{\lambda,0}\|^p + \|\alpha_m K(u_{m,\lambda})_+\|^p + p\|u_{m,\lambda}\|^{p-1}\|f_\lambda\| + p(-\lambda+\|J\|)\|u_{m,\lambda}\|^p.$$

Since $\lambda > \|J\|$ and $\alpha_m\|K\| < 1$, we have

$$\|u_{m,\lambda}\|^p \le \frac{1}{p}(\lambda-\|J\|)^{-1}\{\|g_{\lambda,0}\|+p\|u_{m,\lambda}\|^{p-1}\|f_\lambda\|\},$$

which, on using $ab \le p^{-1}a^p + q^{-1}b^q$ with $q=p/(p-1)$, $a=\|f_\lambda\|$ and $b=p\|u_{m,\lambda}\|^{p-1}$, reduces to an inequality from which we easily derive

$$\|u_{m,\lambda}\| \le C_{\lambda,p}[\|g_{\lambda,0}\|^p + \|f_\lambda\|^p]^{1/p}. \tag{5.5}$$

Performing a similar transformation to that in Proposition 4.1, we obtain (5.5) for $\lambda=0$, though with a different constant. By repeating the chain of inequalities leading to (5.5) we do not get a uniform finite upper bound for $\{(u_m)_\pm\}_{m=1}^\infty$, since $\{1-\alpha_m\|K\|\}_{m=1}^\infty$ is not bounded away from zero; so we cannot conclude that this sequence converges in E_p.

In the derivation of the rightmost member of (5.4) we have deleted the term $-p\int_\Sigma h|u_{m,\lambda}|^p d\mu$. However, if we retain this term in all of our subsequent estimates, we obtain a uniform finite upper bound for $\int_\Sigma h|u_{m,\lambda}|^p d\mu$, and therefore $u_{m,\lambda}$ converges monotonically in the norm of $L_p(\Sigma,hd\mu)$ to a limit, which must be u_λ. Then the boundedness of J and the equation $(Y+h+\lambda-J)u_{m,\lambda}=f_\lambda$ imply that $Yu_{m,\lambda}$ converges in the norm of $L_p(\Sigma,d\mu)$. The latter implies that Yu_λ exists as a distributional derivative equal to $f_\lambda-(h+\lambda-J)u_\lambda$. Moreover, there is a uniform finite upper bound for $\{(u_{m,\lambda})_T\}_{m=1}^\infty$ as a result of (5.4); so $(u_{m,\lambda})_T$ converges to $u_{\lambda,T}$ in the norm of $L_p(D_T,d\nu_T)$. It is also immediate that the suprema satisfy

$$u_{\lambda,-} = \sup\,(u_{m,\lambda})_- = \sup\,K_m(u_{m,\lambda})_+ = \sup\,\alpha_m K(u_{m,\lambda})_+ = \sup\,K(u_{m,\lambda})_+.$$

Further, if $\{\alpha_m^*\}_{m=1}^\infty$ is a different sequence of scalars and $\{u_m^*\}_{m=1}^\infty$ is the sequence of corresponding solutions of (5.3), the above argument shows that for a sequence $\{\alpha_m^{**}\}_{m=1}^{\infty}$ satisfying $\alpha_m^{**} \geq \max(\alpha_m, \alpha_m^{*})$ and for the corresponding solutions $\{u_m^{**}\}_{m=1}^{\infty}$, we have $u_m \leq u_m^{**}$ and $u_m^{*} \leq u_m^{**}$, whence uniqueness follows.

Finally, let us consider the original problem (2.7) for arbitrary g_-, where we put

$$\hat{g}(x,s) = \{1-\ell(x)^{-1}s\}g_-(x).$$

Setting $\hat{u}=u-\hat{g}$, we obtain problem (5.1) with u replaced by $\hat{u}$, which has a unique solution in $L_p(\Sigma,d\mu)$ with traces in $L_{p,loc}(D^{\pm},d\nu^{\pm})$. Hence, the original problem (2.7) has a unique solution in $L_p(\Sigma,d\mu)$ with traces in $L_{p,loc}(D^{\pm},d\nu^{\pm})$. In order to prove the nonnegativity of the latter and the monotone approximation statement of the theorem for this general situation, we observe that the solution of the approximative problems for arbitrary g_- are nonnegative and converge to the unique solution of the present problem in the strong operator topology, the latter because the approximative solutions of problem (2.7) differ from those of problem (5.1) by a fixed function $\hat{g}$. ■

COROLLARY 5.3. Suppose that the lengths $\ell(x)$ of the integral curves of the vector field Y satisfy $\ell(x)\geq\delta$, for some constant $\delta>0$, and let h be essentially bounded. Let the operators J and K satisfy the assumptions of Theorem 4.3 along with the conditions $K\geq 0$, $\|K\|=1$ and $\lambda_0 I+J\geq 0$ for some $\lambda_0\in\mathbb{R}$. Then for every $f\in L_p(\Sigma,d\mu)$ and any $g=(g_-,g_0)\in L_p(D^-,d\nu^-)$, the problem (2.7a)–(2.7c) has a unique solution in E_p, which is nonnegative whenever f and g are nonnegative.

Proof: In view of Theorem 5.2, it remains to prove that every solution u of problem (2.7) in the weak sense of Theorem 5.2 has its trace $u^{\pm}$ in $L_p(D^{\pm},d\nu^{\pm})$. However, the estimate (3.11) implies that $u^{\pm}\in L_p(D^{\pm},\ell d\nu^{\pm})$. Since $\delta \leq \ell(x) \leq T$ for all $x\in D^-$, the result is immediate. ■

6. Existence and uniqueness results on spaces of measures

In the previous two sections we have obtained all of the existence and uniqueness results in the spaces $L_p(\Sigma,d\mu)$, where $1\leq p<\infty$, assuming that the source term $f\in L_p(\Sigma,d\mu)$ and the incident flux $g_-\in L_p(D_-,d\nu_-)$. However, in many derivations in neutron transport, radiative transfer, and other fields of interest to kinetic theory, one is accustomed to use Dirac's δ-function to represent point sources and unidirectional beams. For all of these cases the bulk of the existence and uniqueness theory has been developed in an L_p or C space setting and is therefore not able to account directly for these "concentrated" data. For this reason it has been argued that spaces of measures, rather than L_p or C spaces, are the physically relevant spaces in which to formulate kinetic equations. Suhadolc and Vidav [346] were the first to study a class of time dependent equations on the space of bounded measures on the position-velocity region, though under conditions that exclude many physically interesting phenomena. Because spaces of measures bring about mathematical problems distinct from those in the L_p-settings, these spaces never became part of the mainstream of kinetic theory. In this section we shall give a full account of the existence and uniqueness theory for time dependent kinetic equations in spaces of measures, however restricted to positive models.

Let $M(\Sigma)$ be the linear vector space of bounded signed Borel measures on Σ. For a positive measure $\nu\in M(\Sigma)$ we define its norm $\|\nu\|$ by $\|\nu\| = \nu(\Sigma)$. For a signed measure the definition of the norm is more involved. Given a signed measure $\nu\in M(\Sigma)$, we write $\nu=\nu_+-\nu_-$, where $\nu_+,\nu_-\in M(\Sigma)$ are the unique positive measures having the property that $\nu_\pm\le\nu^*_\pm$ for any pair of positive measures $\nu^*_\pm\in M(\Sigma)$ such that $\nu=\nu^*_+-\nu^*_-$. We then define $|\nu|=\nu_++\nu_-$ and put

$$\|\nu\| = \|\,|\nu|\,\| = |\nu|(\Sigma) = \nu_+(\Sigma) + \nu_-(\Sigma).$$

With these definitions $M(\Sigma)$ becomes a Banach lattice, which has the following property in common with $L_p(\Sigma,d\mu)$ for $1\le p<\infty$:

> If $0\le\mu_1\le\mu_2\le\ldots\le\nu$ is a monotonically increasing sequence of positive measures $\{\mu_m\}_{m=1}^{\infty}$ in $M(\Sigma)$ bounded above by some $\nu\in M(\Sigma)$, then there exists a measure $\mu\in M(\Sigma)$ such that
>
> $$\lim_{m\to\infty} \|\mu-\mu_m\| = \lim_{m\to\infty} \{\mu(\Sigma)-\mu_m(\Sigma)\} = 0.$$

Another useful property of $M(\Sigma)$ is that it contains $L_1(\Sigma,d\mu)$ as a closed Banach sublattice. Indeed, on choosing the fixed positive Borel measure μ on Σ as in Section 2, we associate with every function $\varphi\in L_1(\Sigma,d\mu)$ a measure $\nu_\varphi\in M(\Sigma)$ by

$$\nu_\varphi(E) = \int_E \varphi d\mu, \quad E \subset \Sigma \text{ Borel.}$$

Let Φ and Φ_0 be the test function spaces defined in Section 2. We define the bilinear form

$$<u,v> = \int_\Sigma v du, \quad u \in M(\Sigma), \quad v \in \Phi. \tag{6.1}$$

If $u=\nu_\varphi$ for some $\varphi \in L_1(\Sigma,d\mu)$, we obtain

$$<u,v> = \int_\Sigma v d\nu_\varphi = \int_\Sigma \varphi v d\mu = <\varphi,v>,$$

and therefore the form (6.1) extends the definition of the previous sections. We now define the distributional derivative Yu for measures $u \in M(\Sigma)$ by

$$<Yu,v> = -<u,Yv>, \quad v \in \Phi_0, \tag{6.2}$$

in which manner we extend the previous definition of Yu. A pair of Borel measures $u^\pm$ on $D^\pm$ is called a **trace** for $u \in M(\Sigma)$, if $Yu \in M(\Sigma)$ and the extended Green's identity

$$<Yu,v> + <u,Yv> = \int_{D^+} v du^+ - \int_{D^-} v du^-, \quad v \in \Phi,$$

is valid. It should be observed that one does not necessarily have $u^\pm \in M(D^\pm)$.

PROPOSITION 6.1. Suppose that u and (Y+h)u belong to $M(\Sigma)$. Then u has a unique trace $u^\pm$, and $u \geq 0$ implies $u^\pm \geq 0$. If $u \geq 0$ and $u^- \in M(D^-)$, then $u^+ \in M(D^+)$ and $Yu \in M(\Sigma)$, while

$$u^+(D^+) - u^-(D^-) = [Yu](\Sigma). \tag{6.3}$$

Proof: For a given Borel subset of D^- on which $\ell(x)^{-1}$ is bounded, let w be a bounded Borel function on the set

$$E = \{(x,0) : x \in E_0\} \cup \{(x,\ell(x)) : x \in E_0\},$$

extend w by zero on the rest of the "disjoint" union $D^- \cup D^+$, and define $\hat{w}(x,s)$ as in (3.4). Suppose u and Yu are locally finite measures in the sense that $|u|(E)$ and $[|Yu|](E)$ are finite for every Borel subset E of Σ on which $\ell(x)^{-1}$ is bounded. We compute

$$<Yu,\hat{w}> + <u,Y\hat{w}> = \int_\Sigma \hat{w}d(Yu) + \int_\Sigma (Y\hat{w})du =$$

$$= \int_\Sigma w(x,\ell(x))\{\frac{du}{\ell(x)} + \frac{s}{\ell(x)}d(Yu)\} - \int_\Sigma w(x,0)\{\frac{du}{\ell(x)} - [1-\ell(x)^{-1}s]d(Yu)\}.$$

Since $\ell(x)^{-1}$ is bounded on the support of w, the latter is easily written as

$$<Yu,\hat{w}> + <u,Y\hat{w}> = \int_{D^+} wdu^+ - \int_{D^-} wdu^-, \tag{6.4}$$

where w is varied over the set of all such functions. Now suppose $z \in \Phi$, and let w be the restriction of z to $D^- \cup D^+$ and $\hat{w}$ its extension of the form (3.4). Then $v=z-\hat{w} \in \Phi_0$, and therefore $\int_\Sigma Yvd\mu=0$. Hence (cf. (6.2)),

$$<Yu,z> + <u,Yz> = \{<Yu,v>+<u,Yv>\} + \{<Yu,\hat{w}>+<u,Y\hat{w}>\} =$$

$$= \int_{D^+} wdu^+ - \int_{D^-} wdu^- = \int_{D^+} zdu^+ - \int_{D^-} zdu^-,$$

and consequently $u^\pm$ is a trace for u.

If u is a positive measure and if $w \geq 0$ on D^+ and $w \leq 0$ on D^-, then the extension $\hat{w}$ as in (3.4) may be replaced by a sequence of extensions $w_m \in \Phi$ with the properties $Yw_m \geq 0$, $w_m \to 0$ pointwise on Σ, and $\sup_{m,\Sigma} |w_m| \leq C$, where C is a fixed constant. Then (6.4) and $w_m \in \Phi$ imply

$$<Yu,w_m> + <u,Yw_m> = \int_{D^+} wdu^+ - \int_{D^-} wdu^-,$$

which has a nonnegative limit as $m\to\infty$. On varying w one sees that $u^{\pm}\geq 0$ whenever $u\geq 0$.

Finally, suppose $u\geq 0$ and that u and $(Y+h)u$ belong to $M(\Sigma)$. Suppose also that u^- is a bounded measure. In (6.4) we take w to be the characteristic function of a set $E_{0,m}$ of the above type, where $\{E_{0,m}\}_{m=1}^{\infty}$ is an increasing sequence of such sets with union D^-. On writing $E_{\ell,m}=\{(x,\ell(x)) : x\in E_{0,m}\}$ and $E_m=\{(x,s) : x\in E_{0,m}, s\in(0,\ell(x))\}$ we get from (6.4)

$$[(Y+h)u](E_m) + u^-(E_{0,m}) = u^+(E_{\ell,m}) + (hu)(E_m).$$

Since the left hand side has a limit as $m\to\infty$ and both terms on the right hand side are nonnegative and increasing in m, the measures u^+ and hu, and hence also Yu, are finite and (6.3) holds true. ∎

As in Section 2, we may decompose the trace $u^{\pm}$ as $u^- = (u_-,u_0)$ and $u^+ = (u_+,u_T)$, in order to display the temporal part. Let E_M be the set of nonnegative bounded Borel measures $u\in M(\Sigma)$ such that $(Y+h)u$ belongs to $M(\Sigma)$ and the trace $u^{\pm}$ consists of bounded measures. We have the following analog of Proposition 3.4.

PROPOSITION 6.2. Given $f\in M(\Sigma)$ and $g\in M(D^-)$ with $f\geq 0$ and $g\geq 0$, there is a unique measure u in E_M such that

$$(Y+h)u = f \quad \text{on } \Sigma,$$

$$u^- = g \quad \text{on } D^-.$$

Proof: Given $w\in\Phi$, set

$$v(x,s) = -\int_s^{\ell(x)} w(x,t)dt. \tag{6.5}$$

Then $Yv=w$ and $v=0$ on D^+, while

$$\sup_{\Sigma} |v(y)| \leq T \sup_{\Sigma} |w(y)|,$$

where T is the length of the fixed time interval. Using that $f\in M(\Sigma)$ and

$g\in M(D^-)$, we can prove the existence of a unique $u\in M(\Sigma)$ such that

$$<u,w> = -\int_{D^-} v\,dg - <f,v>, \qquad w\in\Phi. \tag{6.6}$$

Since $v\leq 0$ for $w\geq 0$, one has $u\geq 0$ if $f\geq 0$ and $g\geq 0$. For every $v\in\Phi$ vanishing on D^+ we find a function $w\in\Phi$ satisfying (6.5) as well as the identity

$$<Yu,v> = -<u,Yv> = -<u,w> = <f,v> + \int_{D^-} v\,dg,$$

whence $Yu=f$ and $u^-=g$. ∎

We are now prepared to prove the existence and uniqueness results for the abstract time dependent kinetic problem in this measure space setting. We consider the bounded linear operators $J:M(\Sigma)\to M(\Sigma)$ and $K:M(D_+)\to M(D_-)$, which are again assumed local in time (i.e. $J(ru)=rJ(u)$ and $K(rg)=rK(g)$ for every bounded continuous function $r(t)$ of time alone) and real. In addition, we assume $K\geq 0$ and $(\lambda_0 I+J)\geq 0$ for some $\lambda_0>0$.

THEOREM 6.3. Under the assumptions above, for $\|K\|<1$ the problem (4.4) has a unique solution $u\in E_M$ for each pair of nonnegative measures $f\in M(\Sigma)$ and $g\in M(D^-)$. Moreover, there is a constant $C=C(h,J,K)$ such that (4.5) holds. If, in addition,

$$\int_\Sigma (h\,du - dJu) \geq 0 \tag{6.7}$$

for every nonnegative $u\in M(\Sigma)$, then the problem (4.4) is dissipative.

Proof: Once again one may consider the modified problem (4.1), where $\lambda>\|J\|$. Let us denote the solution $u_\lambda\in E_M$ of the problem (4.1) with $J=0$ and $K=0$ by $u_\lambda=S_\lambda(f_\lambda,g_\lambda)$. On using (6.3), $u_\lambda^\pm\in M(D^\pm)$ and $u_\lambda^\pm\geq 0$ we find

$$\|u_\lambda^+\| + \lambda\|u_\lambda\| + \|hu_\lambda\| = \|f_\lambda\| + \|g_\lambda\|, \tag{6.8}$$

where $\|S_\lambda(f_\lambda,0)\| \le \lambda^{-1}\|f_\lambda\|$.

Next we look for the solution of (4.1) for the given J but for K=0, and seek it in the form $u_\lambda = S_\lambda(f_\lambda^*, g_\lambda)$. As in the proof of Theorem 4.3 we reduce the latter to a vector equation, perform an estimate and derive the unique solution f^*, whence problem (4.1) with K=0 has a unique solution $u_\lambda = T_\lambda(f_\lambda, g_\lambda^*) \in E_M$. Finally, on seeking the solution of problem (4.1) in the form $u_\lambda = T_\lambda(f_\lambda, g_\lambda^*)$ for general J and K with $\|K\|<1$, one again obtains a vector equation for g_λ, which for λ sufficiently large can be obtained uniquely using the contraction mapping principle. Here we use again $\|K\|<1$ in an essential way. The dissipativity of the solution under the assumption (6.7) can be shown as in the proof of Theorem 4.6. ∎

THEOREM 6.4. Under the assumptions of the the previous theorem, suppose that $\|K\|=1$. In addition, suppose $f \in M(\Sigma)$ and $g^- = (g_0, g_-) \in M(D^-)$ satisfy $f \ge 0$ and $g \ge 0$. Let $K_m = \alpha_m K$, and let $u_m \in E_M$ be the unique solution of the approximate problem (5.3). Then the sequence $\{u_m\}_{m=1}^\infty$ is nondecreasing and converges in the norm of $M(\Sigma)$ to a measure $u \in M(\Sigma)$. The measure u is independent of the choice of $\{\alpha_m\}_{m=1}^\infty$, satisfies (2.7a) and has a trace which satisfies (2.7b). Finally, in the sense of weak*-convergence of measures, $u_- = \lim_{m\to\infty} K(u_m)_+$ on D_-.

Proof: We first consider the initial-boundary value problem for $g_-=0$. As in the proof of Theorem 5.2 one may show that $\{u_{m,\lambda}\}_{m=1}^\infty$ is an increasing sequence of measures in $M(\Sigma)$, where one once again considers the modified problem (4.1) for $\lambda > \|J\|$, but with K replaced by K_m. Using (6.3) one obtains

$$\|u_{m,\lambda}^+\| + \lambda\|u_{m,\lambda}\| + \|hu_{m,\lambda}\| = \|f_\lambda\| + \|g_\lambda\| + \|Ju_{m,\lambda}\|. \tag{6.9}$$

On decomposing the measures $u_{m,\lambda}^+$ and g_λ in their spatial and temporal parts and using that

$$\|u_{m,\lambda}^+\| = \|(u_{m,\lambda})_+\| + \|(u_{m,\lambda})_T\|,$$

$$\|g_\lambda\| = \|g_{\lambda,0}\| + \|g_{\lambda,-}\|,$$

due to the positivity of these measures, one may rewrite (6.9) as

$$\|(u_{m,\lambda})_+\| + \|(u_{m,\lambda})_T\| + (\lambda - \|J\|)\|u_{m,\lambda}\| \le$$

$$\leq \|f_\lambda\| + \|g_{\lambda,0}\| + \|K_m(u_{m,\lambda})_+ + g_{\lambda,-}\|.$$

However, since $K_m(u_{m,\lambda})_+ \geq 0$ and $g_{\lambda,-} \geq 0$, we may use the additivity of the norm of $M(\Sigma)$ and the norm estimate $\|K_m\| = \alpha_m < 1$ to prove

$$\|(u_{m,\lambda})_+\| + \|(u_{m,\lambda})_T\| + (\lambda - \|J\|)\|u_{m,\lambda}\| \leq$$

$$\leq \|f_\lambda\| + \|g_{\lambda,0}\| + \|K_m(u_{m,\lambda})_+\| + \|g_{\lambda,-}\| \leq$$

$$\leq \|f_\lambda\| + \|g_{\lambda,0}\| + \|(u_{m,\lambda})_+\| + \|g_{\lambda,-}\|,$$

whence

$$\|(u_{m,\lambda})_T\| + (\lambda - \|J\|)\|u_{m,\lambda}\| \leq \|f_\lambda\| + \|g_{\lambda,0}\| + \|g_{\lambda,-}\|.$$

Hence, the measures $u_{m,\lambda}$ and $(u_{m,\lambda})_T$ have the uniform (in m) bounds

$$\|u_{m,\lambda}\| \leq (\lambda - \|J\|)^{-1}\{\|f_\lambda\| + \|g_{\lambda,0}\| + \|g_{\lambda,-}\|\},$$

$$\|(u_{m,\lambda})_T\| \leq \|f_\lambda\| + \|g_{\lambda,0}\| + \|g_{\lambda,-}\|.$$

As a consequence of these bounds and the monotonicity in m, there exist measures $u_\lambda \in M(\Sigma)$ and $u_{\lambda,T}$ such that

$$\lim_{m\to\infty} \|u_\lambda - u_{m,\lambda}\|_{M(\Sigma)} = 0,$$

$$\lim_{m\to\infty} \|u_{\lambda,T} - (u_{m,\lambda})_T\|_{M(D_T)} = 0.$$

Since $(Y+h)u_{m,\lambda} = f_\lambda - (\lambda - J)u_{m,\lambda}$, we find the uniform bound

$$\|(Y+h)u_{m,\lambda}\|_{M(\Sigma)} \leq \|f_\lambda\| + \frac{\lambda + \|J\|}{\lambda - \|J\|}\{\|f_\lambda\| + \|g_{\lambda,0}\| + \|g_{\lambda,-}\|\},$$

and thus $(Y+h)u_\lambda \in M(\Sigma)$. It is then clear that u_λ satisfies (4.1a) and (4.1b). Moreover, as in the proof of Theorem 5.2, one may show that u_λ is independent of he approximating sequence $\{\alpha_m\}_{m=1}^{\infty}$.

The above argument does not yield a uniform bound on $\|(u_{m,\lambda})_-\|$, and herefore the trace $u_{\lambda,\pm}$ need not belong to $M(D_\pm)$. In order to give proper sense o the boundary condition (4.1c), we have to extend the definition of K. Let us onsider a nondecreasing sequence of positive measures $\{v_m\}_{m=1}^{\infty}$ in $M(D_+)$ with veak*-limit v. Then $\{Kv_m\}_{m=1}^{\infty}$ is also nondecreasing and has a weak*-limit depending only on v; so we may extend K by setting

$$Kv = \text{weak*-}\lim_{m\to\infty} Kv_m,$$

which implies (4.1c).

Finally, one transforms the results easily to the unmodified problem (4.4), and then extends the result to problem (2.7) with arbitrary $g_- \in M(D_-)$ as in the last paragraph of the proof of Theorem 5.2. ■

It is not difficult to derive the analog of (3.11) in the $M(\Sigma)$ setting, and thus to derive the analog of Corollary 5.3. Then the essential boundedness of h and the estimate $\ell(x) \geq \delta$ for all $x \in D^-$ allow one to extend Theorem 6.3 to the case where $\|K\|=1$.

Chapter XII

TIME DEPENDENT KINETIC EQUATIONS: SEMIGROUP APPROACH

1. Introduction and historical remarks

In the previous chapter we have solved time dependent kinetic problems using integration along trajectories and perturbation techniques. In many applications, however, the phase space, the transport operator and the boundary conditions do not depend on time, in which case a semigroup approach is natural. In this chapter we shall discuss the semigroup approach in detail. Throughout we maintain the notation and terminology introduced in Chapter XI with minor modifications. In particular, the open set $\Lambda \subset \mathbb{R}^n \times \mathbb{R}^n$ specifying the position-velocity domain will now be referred to as the **phase space** of the system. (We recall that in the previous chapter $\Sigma \subset \mathbb{R}^n \times \mathbb{R}^n \times (0,T)$ was taken as the phase space, in order to treat time in a somewhat symmetric fashion with spatial and velocity variables). Similarly, μ will be a Borel measure on Λ such that subsets of Λ with finite Lebesgue measure have finite μ-measure, $\nu_{\pm}$ will be appropriate Borel measures on the parts $D_{\pm}$ of $\partial\Lambda$ corresponding to the outgoing (resp. incoming) "fluxes", J and K are bounded linear operators defined on $L_p(\Lambda,d\mu)$ and from $L_p(D_+,d\nu_+)$ into $L_p(D_-,d\nu_-)$, respectively, and $h(x,\xi)$ is a nonnegative Lebesgue measurable function on Λ that is integrable on each subset of finite Lebesgue measure. Define

$$B = -\xi \cdot \frac{\partial}{\partial x} - a(x,\xi) \cdot \frac{\partial}{\partial \xi} - h(x,\xi) + J,$$

for $(x,\xi) \in \Lambda t$. Then we call the vector field $-X$, where

$$X = \xi \cdot \frac{\partial}{\partial x} + a(x,\xi) \cdot \frac{\partial}{\partial \xi},$$

the free streaming operator, and, as before, we call $S=-(X+h)$ the streaming operator, $-A=-h+J$ the collision operator, and $B=S+J$ the (full) transport operator.

Let us suppose for the moment that the problem is homogeneous, i.e., there are no incident fluxes at the boundary and no internal sources. We know from the previous chapter that the abstract Cauchy problem

$$\frac{\partial u}{\partial t}(x,\xi,t) = (Bu)(x,\xi,t) \quad \text{on } \Lambda x(0,T), \tag{1.1a}$$

$$u(x,\xi,0) = u_0(x,\xi) \quad \text{on } \Lambda, \tag{1.1b}$$

$$u_-(x,\xi,t) = (Ku_+)(x,\xi,t) \quad \text{on } D_- \times (0,T) \tag{1.1c}$$

is well posed in a suitable L_p-setting. However, in order to prove that the solution $u(t)$, for $u_0 \in D(B)$, is continuous for $0 \le t \le T$ and continuously differentiable for $0 < t < T$, one has to show that the closed operator B defined on a domain of functions satisfying (1.1c) generates a C_0-semigroup of bounded operators $U(t)$. For characterizations of generators of C_0-semigroups we refer to Section I.3. In principle, one may show directly that the operator B on a suitable domain satisfies the assumptions of the Hille-Yosida theorem and therefore generates a C_0-semigroup. In the present chapter, however, we shall exploit the analysis of unique solvability of the problem developed in Chapter XI in order to derive the C_0-semigroup property.

During the past 30 years, the application of the Hille-Yosida and Hille-Phillips theorems and of related perturbation results has been the main strategy in proving existence and uniqueness for time dependent kinetic problems of the form (1.1). An exhaustive bibliography of such equations would contain hundreds of references ranging from engineering and physical approaches to rigorous mathematics. Virtually all the existence and uniqueness results have been proved for special situations: specific collision operators deduced from (classical or quantum) mechanics or modeled ad hoc, regular and stationary geometries, purely absorbing (vacuum) or specular reflection boundary conditions, and specific function spaces (usually L_1 or L_2). The richness of the literature stems from the large variety of concrete situations encountered in practical kinetic problems. We outline a few high points in the historical development.

The first successful implementation of the semigroup strategy for a kinetic equation was carried out in the 1950's by Lehner and Wing [245, 246] (see also [244]), who analyzed a neutron transport problem in L_2 in one dimensional geometry with monoenergetic particles and purely absorbing boundary conditions ($K=0$), obtaining results well beyond existence and uniqueness. Jörgens [204] extended the analysis to three dimensional geometry with $K=0$. Cercignani [81, 82] studied a velocity dependent model with more general boundary conditions, though still in L_2. Vidav [375] assumed

K=0, but generalized the setting to any L_p, $1 \leq p < \infty$. A more refined analysis of the spectral properties of the transport operator in L_1 was carried out by Larsen and Zweifel [242] for K=0, three dimensional geometry, and rather general collision operators, but for a compact velocity domain V. This article contains a good review of the literature up to 1974. Under the same assumptions, namely K=0 and V compact, and with some regularity conditions on the collision operator, Shikhov and Shkurpelov [327] obtained existence and uniqueness results in any L_p as well as numerous spectral properties of the transport operator and of the semigroup it generates. An account of existence, uniqueness and asymptotics in L_1 with K=0 and a neutron transport setting is given in [190] and Chapters 11–13 of [211].

The linearized Boltzmann equation in the spatial domain $\mathbb{R}^n$ was studied in detail by Arsenyev [14]. Particular considerations of this problem have resulted from studies of the linearized problem as a prerequisite for solving nonlinear kinetic problems [20, 287, 354] in the manner first proposed by Grad [154]. The extensive paper of Shizuta [328] contains infinite medium results in any L_p, an excellent review of perturbation methods, and up-to-date references.

An external force ($a \neq 0$) in the kinetic equation was taken into account, among others, by Scharf [325], Molinet [267] and Drange [107]. Drange proved existence and uniqueness for infinite geometry and any L_p. This analysis was extended by Bartholomäus and Wilhelm [28] to take boundaries into account. Many results for models with a force term have actually been obtained in connection with the nonlinear Vlasov or nonlinear Boltzmann equation (see [352] and the related work cited therein). Recent results of this type connected to the Boltzmann equation are due to Asano [17] and Tabata [349] in a Hilbert space setting.

A careful and complete analysis of the collisionless kinetic problem, which is generated by $\xi \cdot \frac{\partial}{\partial x}$ alone, under general boundary conditions in L_1 is due to Voigt [380], who discussed thoroughly the Green's identity which is crucial for the analysis. Earlier generalizations of the existence and uniqueness results to three dimensional geometry usually overlooked difficulties related to this issue and to the case of conservative boundaries. We mention, among the exceptions, Guiraud [179] and Schnute and Shinbrot [326], who carefully assumed the existence of traces and the existence of global solutions for any initial configuration. By limiting the number of collisions with the boundary via regularity assumptions, Busoni and Frosali [64] obtained a similar result for $\xi \cdot \frac{\partial}{\partial x}$ in L_1 with diffuse, mixed, and purely specular reflection. Based on his earlier results, Voigt [381] subsequently extended his analysis to the full (Boltzmann) transport operator in any L_p, but only for K=0 (and $a \equiv 0$). Similar results appeared independently in a paper by Greiner [171].

In Section 2 we shall generalize the existence and uniqueness theorems to $K\neq0$ and $a\neq0$. These results were obtained by Beals and Protopopescu [37] in a more general setting, as detailed in the previous chapter. The crux of the matter is to show that the streaming operator S generates a C_0–semigroup, and for this we shall rely on the results obtained in Chapter XI. In Section 3 we shall extend this theory to stationary kinetic problems in one and three dimensional geometry and derive the wellposedness of the time dependent counterparts of stationary problems studied in Chapters II, III, V and X. In Section 4 we shall develop some general properties of positive semigroups in Banach lattices, and in the next section apply this to obtain information on the long time behavior of the solutions to time dependent kinetic problems.

2. Existence, uniqueness, dissipativity and positivity in L_p

In this section we will specialize the results of Chapter XI to the case of a time independent phase space, transport operator and boundary reflection law. These results will be reformulated in a semigroup context. We consider a C^∞–manifold Λ imbedded as an open subset of $\mathbb{R}^n\times\mathbb{R}^n$. Let X be a vector field on Λ of the form

$$X = \xi\cdot\frac{\partial}{\partial x} + a(x,\xi)\cdot\frac{\partial}{\partial\xi},$$

where a is assumed to be Lipschitz continuous on the closure $\bar{\Lambda}$ of Λ, and μ is the Borel measure given by $d\mu(x,\xi) = dx\ d\rho(\xi)$; here ρ is the Lebesgue measure on Λ if $a\neq0$, but an arbitrary positive Borel measure on Λ such that all bounded Lebesgue measurable subsets of $\partial\Lambda$ have finite ρ–measure if $a\equiv0$. In addition, we shall assume as before that the vector field X satisfies the divergence free condition

$$\sum_{j=1}^{n}\frac{\partial}{\partial\xi_j}\,a_j(x,\xi,t) = 0$$

as well as the requirement that every maximal integral curve of X whose extension to the left or right is finite has a corresponding left or right endpoint belonging to

$$\partial\Lambda' = \{(x,\xi)\in\partial\Lambda : |\xi|^2+|a(x,\xi)|^2\neq0\}.$$

We do not make any convexity assumptions on the phase space.

The above endpoint condition on X guarantees that the solutions $(x(s), \xi(s))$ of the system of differential equations

$$\frac{dx}{ds} = \xi(s), \qquad \frac{d\xi}{ds} = a(x,\xi),$$

with initial conditions $(x(s^*),\xi(s^*))\in\Lambda$ extended over the maximal s-interval (s_0,s_1) for which the curve lies in Λ approaches neither infinity nor the closure of $\Lambda_0 = \{(x,\xi)\in\Lambda : \xi=0 \text{ and } a(x,\xi)=0\}$ in a finite s-interval. This condition is automatically fulfilled, for example, if $a(x,\xi)\cdot\xi\equiv 0$, since in this case $\frac{d}{ds}\{|\xi|^2\}=0$ and therefore $|\xi|$ is constant on each integral curve. As a result, $|\frac{d}{ds}\{|x|^2\}| \leq 2C|x|$ for some constant C on each integral curve, and the statement follows from Gronwall's inequality. It is important to note, here and throughout, that the parameter s measures the "time" along a trajectory, and not the arc length (cf. Section XI.3).

In Chapter XI we have defined the Borel sets $D_\pm$ as subsets of the sets $D^\pm$ of left and right endpoints of maximal integral curves of $Y=\frac{\partial}{\partial t}+X$. Since in the present chapter the phase space does not display any time dependence and definitions will be altered accordingly, we will, to avoid confusion, denote the Borel sets $D_\pm$ of Chapter XI by $D_{(\pm)}$ and, similarly, we shall write $\nu_{(\pm)}$ for the boundary measures on $D_{(\pm)}$; we shall only need this notation briefly. Let us define the Borel sets D_- and D_+ as the sets of all left and right endpoints of maximal integral curves of X whose interval of definition is finite to the left and right, respectively. We then have the identities

$$D_{(\pm)} = D_\pm \times (0,T), \tag{2.1a}$$

$$d\nu_{(\pm)} = d\nu_\pm \, dt. \tag{2.1b}$$

It is immediate from the results of Section XI.3 that there are unique Borel measures $d\nu_\pm$ on $D_\pm$ such that the Green's identity

$$\int_\Lambda Xv\,d\mu = \int_{D_+} v\,d\nu_+ - \int_{D_-} v\,d\nu_- \tag{2.2}$$

holds for all $v \in \Phi$. The test function space Φ in this setting is taken to be the space of Borel functions v on Λ with the following properties:

(i) v is continuously differentiable along each integral curve of X.

(ii) v and Xv are bounded.

(iii) The support of v is bounded and there is a positive lower bound to the lengths of the integral curves which meet the support of v.

Here the length of an integral curve is again the length of the corresponding s-interval.

By $L_{p,loc}(\Lambda,d\mu)$ we denote the linear vector space of μ-measurable functions on Λ such that $|u|^p$ is μ-integrable on every bounded μ-measurable subset of Λ on which the length of the integral curves passing through it is bounded away from zero. The following existence result for the "trace" is immediate from (2.1) in combination with Proposition XI 3.1.

PROPOSITION 2.1. Suppose u and (X+h)u belong to $L_p(\Lambda,d\mu)$, $1 \le p < \infty$. Then u has a unique trace $u_\pm$ in $L_{p,loc}(D_\pm,d\nu_\pm)$ such that the Green's identity

$$\langle Xu,v\rangle + \langle u,Xv\rangle = \int_{D_+} u_+ v\, d\nu_+ - \int_{D_-} u_- v\, d\nu_- \tag{2.3}$$

is valid for all $v \in \Phi$. Further, if $u_- \in L_p(D_-,d\nu_-)$, then $u_+ \in L_p(D_+,d\nu_+)$, in which case $h|u|^p$ and $|u|^{p-1}Xu$ are μ-integrable and

$$\int_{D_+} |u_+|^p d\nu_+ + p\int_\Lambda h|u|^p d\mu =$$

$$= \int_{D_-} |u_-|^p d\nu_- + p\int_\Lambda (\mathrm{sgn}\, u)|u|^{p-1}(Xu+hu)d\mu. \tag{2.4}$$

Let us denote by F_p, $1 \le p < \infty$, the space of functions $u \in L_p(\Lambda,d\mu)$ such that $(X+h)u \in L_p(\Lambda,d\mu)$ and the trace $u_\pm \in L_p(D_\pm,d\nu_\pm)$. We consider the bounded linear operators $J: L_p(\Lambda,d\mu) \longrightarrow L_p(\Lambda,d\mu)$ and $K: L_p(D_+,d\nu_+) \longrightarrow L_p(D_-,d\nu_-)$ and define the operator $B_K \equiv -(X+h-J)_K$ to be the restriction of the transport operator B to the domain $F_{p,K} \equiv \{u \in F_p : u_- = Ku_+\}$. As in Chapter XI, the "derivative" $X \equiv \frac{d}{ds}$ is considered as a distributional derivative along integral curves.

We state now general existence and uniqueness results for the abstract time dependent kinetic problem (1.1). We pay special attention to the positivity of the solution and consider separately the dissipative boundary condition $\|K\|<1$ and the conservative boundary condition $\|K\|=1$.

THEOREM 2.2. Suppose $\|K\|<1$ and $1 \le p < \infty$. Then:

(i) The transport operator B_K generates a C_0-semigroup $\{U(t)\}_{t \ge 0}$ on $L_p(\Lambda,d\mu)$, with $\|U(t)\| \le \exp\{t\|J\|\}$ for $t \ge 0$.

(ii) If $\int_\Lambda (\text{sgn}u)|u|^{p-1}(hu-Ju)d\mu \ge 0$ for all $u \in F_p$, then B_K generates a contraction semigroup on $L_p(\Lambda,d\mu)$.

(iii) If $K \ge 0$ and $(\lambda_0 I+J) \ge 0$ for some $\lambda_0 \in \mathbb{R}$, then B_K generates a positive semigroup on $L_p(\Lambda,d\mu)$.

(iv) If $\int_\Lambda (hu-Ju)d\mu \ge 0$ for all nonnegative $u \in F_1$, then B_K generates a positive contraction semigroup on $L_1(\Lambda,d\mu)$.

Proof: We extend X, J, h and K to act on complex functions. By the Hille-Yosida theorem it is sufficient to show that for any $\lambda > \|J\|$ and any $f_\lambda \in L_p(\Lambda,d\mu)$ the modified problem

$$(\lambda I - B_K)u_\lambda = f_\lambda \tag{2.5}$$

has a unique solution $u_\lambda \in F_{p,K}$ and that

$$\|u_\lambda\| \le (\lambda - \|J\|)^{-1}\|f_\lambda\|. \tag{2.6}$$

The argument used to prove Theorem XI 4.3 shows that (2.5) has a unique solution $u_\lambda \in F_{p,K}$, provided

$$[1+(\lambda-\|J\|)^{-1}\|J\|]\|K\| < 1 < \lambda\|J\|^{-1}. \tag{2.7}$$

Suppose λ satisfies (2.7) and u_λ is the corresponding solution of (2.5). Since $\|K\|<1$, the identity (2.4) applied to $h+\lambda$ instead of h gives

$$\lambda \|u_\lambda\|^p \le \int_\Lambda (h+\lambda)|u_\lambda|^p d\mu + \frac{1}{p}\int_{D_+} |u_{\lambda,+}|^p d\nu_+ - \frac{1}{p}\int_{D_-} |u_{\lambda,-}|^p d\nu_- =$$

$$= \int_\Lambda (\mathrm{sgn}\, u_\lambda)|u_\lambda|^{p-1}(f_\lambda + Ju_\lambda)d\mu \le \|u_\lambda\|^{p-1}\|f_\lambda + Ju_\lambda\|,$$

which implies (2.6). To extend the result to the range $\lambda_1 > \|J\|$, we choose λ satisfying (2.7) as well as $\lambda \ge \lambda_1$, and write $\lambda_1 = \lambda - \lambda_0$. One sees easily that if N is an invertible operator and if $|\lambda_0| < \|N^{-1}\|^{-1}$, then $\lambda_0 I - N$ has a bounded inverse with norm

$$\|(\lambda_0 I-N)^{-1}\| \le \|N^{-1}\|(1-|\lambda_0|\ \|N^{-1}\|)^{-1}. \tag{2.8}$$

Since $|\lambda_1| < |\lambda_0| < \|B_K^{-1}\|^{-1}$, we may apply (2.8) to $\lambda_1 I - B_K$ and get

$$\|(\lambda_1 I-B_K)^{-1}\| \le \|(\lambda I-B_K)^{-1}\| \cdot (1-|\lambda_0|\,\|(\lambda I-B_K)^{-1}\|)^{-1} \le$$

$$\le (\lambda-\|J\|)^{-1}(1-|\lambda_0|(\lambda-\|J\|)^{-1})^{-1} = (\lambda_1-\|J\|)^{-1},$$

which is the desired estimate for part (i). Hence, B_K generates the strongly continuous evolution semigroup of the kinetic problem (1.1).

Suppose u_0 belongs to $D(B_K)$. Then $u(\cdot,t) = U(t)u_0(\cdot)$ is a C^1-function on $(0,\infty)$ in the strong topology of $L_p(\Lambda,d\mu)$ with values in $D(B_K)$. Thus, u is a solution of (XI 2.7) on $\Sigma \equiv \Lambda\times(0,T)$ for any $T>0$. Here we have extended B and K to functions on Σ and $D_-\mathrm{x}(0,T)$ by setting $[Bf](\cdot,t) = B[f(\cdot,t)]$ and $[Kg](\cdot,t) = K[g(\cdot,t)]$. Now the remaining parts of the theorem follow at once from Theorems XI 4.4, XI 4.5 and XI 4.6. Actually, contractivity and positivity are obtained for $U(t)u_0$ with $u_0 \in D(B_K)$, but this domain is dense in $L_p(\Lambda,d\mu)$. ∎

We mention that for positive contraction semigroups in L_1, Voigt [383] provided a proof which extends results of the type above to a special class of unbounded collision operators J. This allows for including more general scattering kernels, as, for instance, the so-called free gas kernel [269].

THEOREM 2.3. Let us suppose that $\|K\|=1$ and $K\ge 0$, and that $(\lambda_0 I+J)\ge 0$ on

$L_p(\Lambda,d\mu)$ $(1\leq p<\infty)$ for some $\lambda_0\in\mathbb{R}$. Set $K_m=\alpha_m K$, where $\{\alpha_m\}_{m=1}^{\infty}$ is an increasing sequence of positive numbers with limit 1, and let $\{U_m(t)\}_{t\geq 0}$ be the semigroup generated by B_{K_m}. Then, for each $t\geq 0$, $U_m(t)$ converges in the strong topology to $U(t)$ and $\{U(t)\}_{t\geq 0}$ is a C_0-semigroup in $L_p(\Lambda,d\mu)$ satisfying $\|U(t)\| \leq \exp\{t\|J\|\}$.

Proof: For each $t\geq 0$, the estimate of part (i) of the previous theorem provides a uniform bound for the operators $\{U_m(t)\}_{t\geq 0}$. Therefore, it is sufficient to prove the convergence of $U_m(t)u_0$ as $m\to\infty$ for each u_0 in a dense subset of $L_p(\Lambda,d\mu)$. We may treat the two cases $u_0\geq 0$ and $u_0\leq 0$ separately, and thus reduce the problem to proving the convergence for $u_0\in C_c^1(\Lambda)$, $u_0\geq 0$, where $C_c^1(\Lambda)$ is the set of C^1-functions on Λ of compact support. The functions $u_m(\cdot,t)=U_m(t)u_0$ are solutions of (XI 5.1) with $f=0$ and $g_0=u_0$. As shown in the proof of Theorem XI 5.2, these functions converge in the norm of $L_p(\Lambda,d\mu)$ and the convergence is uniform with respect to t on bounded intervals. ∎

The Hille–Yosida argument of the proof of Theorem 2.2 leads to the estimate

$$\|(\lambda I-B_K)^{-1}\| \leq (\lambda-\|J\|)^{-1}, \quad \lambda>\|J\|,$$

where Ran $(\lambda I-B_K)^{-1}\subset F_{p,K}$. In this proof we have used the property $\|K\|<1$. On the other hand, if $K\geq 0$ and $\|K\|=1$, and if $(\lambda_0 I+J)\geq 0$ for some $\lambda_0\in\mathbb{R}$, then the generator $\hat{B}_K$ of the C_0-semigroup $\{U(t)\}_{t\geq 0}$ also satisfies

$$\|(\lambda I-\hat{B}_K)^{-1}\| \leq (\lambda-\|J\|)^{-1}, \quad \lambda>\|J\|,$$

because of a monotone approximation argument. The same approximation argument implies that $D(B_K) = D(\hat{B}_K)\cap F_{p,K}$. In general, we will not find Ran $(\lambda I-\hat{B}_K)^{-1}\subset F_{p,K}$. However, if $D(\hat{B}_K)\subset F_{p,K}$, the operator B_K will still generate the semigroup $\{U(t)\}_{t\geq 0}$, even if $\|K\|=1$ and $K\geq 0$. Conversely, if B_K generates this semigroup, then $B_K=\hat{B}_K$ and Ran $(\lambda I-\hat{B}_K)^{-1}\subset F_{p,K}$. This situation occurs, in particular, if the lengths of the integral curves of the vector field $Y=\frac{\partial}{\partial t}+X$ on $\Sigma=\Lambda\times(0,T)$ are bounded away from zero on Σ (see Corollary XI 5.3).

One cannot expect similar results in the space of bounded measures on phase space, where even for the simplest examples, such as $\Lambda=\mathbb{R}\times\mathbb{R}$ and $a\equiv 0$, the associated semigroup is not strongly continuous. However, as Suhadolc and Vidav [346] pointed out, under certain restrictions on J one can find a proper subspace of the original space of measures on which the transport operator is densely defined and generates a C_0-semigroup. Yet, the conditions on J are restrictive and are not met in most physically interesting situations.

Finally, we note that once the solution of the homogeneous system (1.1) is found, inhomogeneous versions of the kinetic problem can be solved rather easily. Indeed, suppose an inhomogeneous term $f(x,\xi,t)$ is added to the right hand side of (1.1a) and (1.1c) is replaced by

$$u_-(x,\xi,t) = (Ku_+)(x,\xi,t) + g_-(x,\xi) \qquad \text{on } D_-\times(0,T).$$

Then we may extend $g_+=0$ on D_+ and g_- on D_- to $\hat{u}\in L_p(\Lambda,d\mu)$ by suitable interpolation along integral curves (where $\hat{u}=0$ on integral curves without endpoints and on closed integral curves) and consider problem (1.1) for $v=u-\hat{u}$, where an inhomogeneous term $\hat{f}(x,\xi,t)$ appears on the right hand side of (1.1a). We thus obtain an inhomogeneous abstract Cauchy problem with B_K arising as the semigroup generator, which can be solved using the variation of constants formula. From the positivity results of Sections XI.4 and XI.5 it is immediate that, under the hypotheses of either parts (iii) and (iv) of Theorem 2.2 or under those of Theorem 2.3, nonnegative g_- and f lead to a nonnegative solution.

3. Connection between stationary and time dependent kinetic equations

In Chapters II to X we have developed a comprehensive theory of stationary kinetic equations, both abstract approaches and specific applications. In all cases we have considered plane parallel spatial domains of the form $(0,\tau)$, where $0<\tau\leq\infty$, in the absence of external forces. For this case the operator X of the previous section becomes $X=\xi_1\frac{\partial}{\partial x_1}$, where one has an ordinary (and not an inner) product of the velocity variable and the spatial gradient. Then one may analyze the relationship between the spectral properties of the operator T and those of the operator $T^{-1}A$, in conjunction with the contractivity properties of K, where $(Tu)(\xi)=\xi u(\xi)$ and A is the collision operator. It is not evident how one may generalize such a spectral approach beyond one dimensional spatial domains. On the other hand, the spectral approach is amenable to the derivation of closed form solutions for many concrete problems.

In this section we shall apply the time dependent theory of Section 2 to derive existence and uniqueness results for stationary kinetic equations on general (multidimensional) spatial domains for bounded collision operators. Afterwards, we will apply the stationary existence and uniqueness theory developed in the earlier chapters of this monograph to derive the wellposedness of the corresponding time dependent problem for one dimensional spatial domains which are spatially homogeneous.

We consider first the stationary kinetic boundary value problem

$$\xi \cdot \frac{\partial u}{\partial x}(x,\xi) + a(x,\xi) \cdot \frac{\partial u}{\partial \xi}(x,\xi) + h(x,\xi)u(x,\xi) - (Ju)(x,\xi) =$$

$$= f(x,\xi) \qquad \text{on } \Lambda, \tag{3.1a}$$

$$u_-(x,\xi) = (Ku_+)(x,\xi) + g_-(x,\xi) \qquad \text{on } D_-, \tag{3.1b}$$

where domains and functions are assumed to have the properties indicated in Section 2. This may be formulated as the abstract boundary value problem

$$(X + h - J)u = f \qquad \text{on } \Lambda, \tag{3.2a}$$

$$u_- = Ku_+ + g_- \qquad \text{on } D_-. \tag{3.2b}$$

In accordance with Theorem 2.2 (for $\|K\|<1$) and Theorem 2.3 (for $\|K\|=1$ and $K\geq 0$), we consider separately dissipative and conservative boundary conditions.

PROPOSITION 3.1. Let $1\leq p<\infty$, and let $\|K\|<1$. Then there exists a unique solution $u\in F_p$ of problem (3.2) for every $f\in L_p(\Lambda,d\mu)$ and $g_-\in L_p(D_-,d\nu_-)$ if and only if $0\notin\sigma(B_K)$. In this case the solution is given by

$$u = -B_K{}^{-1}(f+B\hat{u}) + \hat{u}, \tag{3.3a}$$

where

$$\hat{u}(x,\xi) = \{1-\ell(x,\xi)^{-1}s\}F(x,\xi,s)\; g_-(x,\xi) \tag{3.4}$$

with $\ell(x,\xi)$ the length of the integral curve starting at $(x,\xi)\in D_-$. Here

$F(x,\xi,s)$ is an arbitrary nonnegative bounded Borel function on Λ which is continuously differentiable with respect to s, whose p-th power is integrable along trajectories, and which satisfies $F(x,\xi,0) = 1$ on D_-.

Proof: The function $\hat{u}$ belongs to $L_p(\Lambda,d\mu)$ if g_- is in $L_p(D_-,d\nu_-)$, as one sees from the identity $d\mu = d\nu_- ds$, which follows from (2.1b) and Proposition XI 3.2. Moreover, $(\hat{u})_- = g_-$ and $(\hat{u})_+ = 0$. Thus $u \in F_p$ is a solution of problem (3.2) if and only if $v = u - \hat{u}$ is a solution of the vector equation

$$B_K v = -f - B\hat{u},$$

which yields the proposition. ∎

Suitable choices for $F(x,\xi,s)$ are $F(x,\xi,s) = 1$ if $\ell(x,\xi)$ is bounded above on Λ, and $F(x,\xi,s) = e^{-s}$ in general. These choices for $F(x,\xi,s)$ are adequate to prove the existence of a unique stationary solution, but, as we will see shortly, one must make a different choice to prove positivity for positive data.

An analogous proof for conservative boundary conditions gives the following result. Note that in this case one may not conclude that the solution u is contained in F_p.

PROPOSITION 3.2. Let $1 \le p < \infty$, and let $\|K\| = 1$ and $K \ge 0$. Then there exists a unique solution $u \in L_p(\Lambda,d\mu)$ of problem (3.2) for every $f \in L_p(\Lambda,d\mu)$ and $g_- \in L_p(D_-,d\nu_-)$ if and only if $0 \notin \sigma(\hat{B}_K)$. In this case the solution is given by

$$u = -\hat{B}_K^{-1}(f + B\hat{u}) + \hat{u}, \tag{3.3b}$$

where $\hat{u}$ satisfies (3.4).

THEOREM 3.3. Let $1 \le p < \infty$, and let either $\|K\| < 1$, or $\|K\| = 1$ and $K \ge 0$. If for some $\varepsilon > 0$ the condition

$$\int_\Lambda (\operatorname{sgn} u)\,|u|^{p-1}(hu - Ju)\,d\mu \ \ge\ \varepsilon\,\|u\|_p$$

is satisfied, then problem (3.2) is uniquely solvable for any $f \in L_p(\Lambda,d\mu)$ and

$g_-\epsilon L_p(D_-,d\nu_-)$. Moreover, if $K\geq 0$ and $(\lambda_0 I+J)\geq 0$ for some $\lambda_0\epsilon\mathbb{R}$, then the solution is nonnegative if f and g_- are nonnegative. Finally, for p=1 the condition

$$\int_\Lambda (hu-Ju)d\mu \geq \epsilon \|u\|_p$$

for all nonnegative $u\epsilon L_p(\Lambda,d\mu)$ implies that problem (3.2) has a unique nonnegative solution if f and g_- are nonnegative.

Proof: The hypotheses of the theorem imply, as a result of Theorems 2.2 and 2.3, that the evolution semigroup of the time dependent problem, with $h(x,\xi)$ replaced by $h(x,\xi)+\epsilon$, is a contraction semigroup. This in turn implies that the evolution semigroup of the corresponding time dependent problem itself has order equal or less than $-\epsilon$, and thus its generator B_K (in the case $\|K\|<1$) or $\hat{B}_K$ (in the case $K\geq 0$ and $\|K\|=1$) has its spectrum in the half plane Re $\lambda\leq-\epsilon$. From Propositions 3.1 and 3.2 it then follows that problem (3.2) is uniquely solvable. The positivity of $-B_K{}^{-1}$ under the appropriate positivity and contractivity assumptions follows directly from the corresponding hypotheses in Theorems 2.2 and 2.3.

It remains to prove that under the above positivity and contractivity assumptions the function $\hat{u}$ in (3.3b) is nonnegative if g_- and f are nonnegative. In order to prove this statement, it suffices to show that

$$z = \hat{u} - B_K{}^{-1}B\hat{u}$$

is nonnegative for nonnegative g_-. With this in mind we will make a suitable choice for the function $F(x,\xi,s)$ in (3.4). For nonnegative λ_0 such that $(\lambda_0 I+J)\geq 0$ we define

$$F(x,\xi,s) = \exp\left\{-\int_0^s [h(x,\xi,s)+\lambda_0+1]ds'\right\}.$$

Then $\hat{u}$ belongs to $L_p(\Lambda,d\mu)$, is nonnegative if $g_-\geq 0$ and satisfies the estimate

$$G(x,\xi,s) = -F'(x,\xi,s)/F(x,\xi,s) \geq h(x,\xi,s) + \lambda_0 - \{\ell(x,0)-s\}^{-1},$$

where F' is the partial s-derivative. This in turn implies that

$$B\hat{u} = -\frac{\partial\hat{u}}{\partial s} - (h+\lambda_0)\hat{u} + (\lambda_0+J)\hat{u} =$$

$$= \{\ell^{-1} - (h+\lambda_0)(1-\ell^{-1}s) + (1-\ell^{-1}s)G(x,\xi,s)\}F(x,\xi,s) + (\lambda_0 I+J)\hat{u} \geq 0.$$

Hence, due to the positivity of $-B_K^{-1}$, the function z is nonnegative, and so is the solution u. ■

If the lengths of the integral curves of the vector field $Y=\frac{\partial}{\partial t}+X$ on $\Sigma=\Lambda\times(0,T)$ have a uniform positive lower bound, then one may apply Corollary XI 5.3 to prove that B_K (rather than $\hat{B}_K$) generates the strongly continuous semigroup $\{U(t)\}_{t\geq 0}$, even if $K\geq 0$ and $\|K\|=1$. For this case we obtain solutions $u\in F_p$ even if $\|K\|=1$ and $K\geq 0$.

Results analogous to Theorem 3.2 can be obtained for problems with nonnegative data in the space of measures. This was done by Nelson and Victory [285] for neutron transport.

Next, let us derive the unique solvability of one dimensional time dependent problems from the unique solvability of their stationary counterparts. Suppose that T is an injective self adjoint operator and A is a positive self adjoint operator with closed range and finite dimensional kernel. Suppose $D(T)\cap D(A)$ is dense in H. Let H_A be the completion of D(A) with respect to the inner product

$$(u,v)_A = (Au,v).$$

We define H_T as the completion of D(T) with respect to the inner product

$$(u,v)_T = (|T|u,v). \tag{3.5}$$

Writing Z_0 for the zero root manifold of $T^{-1}A$ and $Z_1=\{u\in H_A : (Tu,v)=0 \text{ for all } v\in Z_0\}$, we denote by H_S the direct sum of Z_0 and the completion of Z_1 with respect to the inner product

$$(u,v)_S = (|A^{-1}T|u,v)_A. \tag{3.6}$$

We shall assume that $H_A\subset D(T)$ and that the inner products (3.5) and (3.6) are equivalent on D(T), i.e., that $H_T\cong H_S$. We may therefore identify H_T and H_S.

The time dependent kinetic equation now has the form

$$\frac{\partial u}{\partial t} + T\,\frac{\partial u}{\partial x} = -Au + f, \qquad 0<x<\tau, \quad t\geq 0, \tag{3.7a}$$

with initial condition

$$u(x,0) = g_0(x), \qquad 0<x<\tau, \tag{3.7b}$$

and boundary conditions

$$Q_+u(0,t) = Q_+\hat{g}(0,t) + RJQ_-u(0,t), \tag{3.7c}$$

$$Q_-u(\tau,t) = Q_-\hat{g}(\tau,t) + RJQ_+u(\tau,t), \tag{3.7d}$$

if $0<\tau<\infty$, and the boundary condition (3.7c) along with

$$\|u(x,t)\|_T = O(1) \quad (x\to\infty) \tag{3.7e}$$

if $\tau=\infty$. Here $Q_\pm$ are orthogonal projections onto T-positive and T-negative T-invariant subspaces, J is an inversion symmetry (i.e., an invertible isometry satisfying $TJ=-JT$ and $AJ=JA$) and R describes the boundary reflection processes.

It is instructive to consider for a moment the typical situation, where the Hilbert space $H=L_2(V,d\rho)$ for some one dimensional velocity domain $V\subset\mathbb{R}$ with Borel measure ρ and T is the multiplication operator $(Tu)(\xi)=\xi u(\xi)$. Then the projections $Q_\pm$ and the inversion symmetry J are given by

$$(Q_\pm u)(\xi) = \begin{cases} u(\xi), & \pm\xi>0, \\ 0, & \pm\xi<0, \end{cases}$$

$$(Ju)(\xi) = u(-\xi).$$

The Green's identity (2.3) can be specified further as follows:

$$\int_0^\tau \int_V \{\xi\frac{\partial u}{\partial x}v + \xi u\frac{\partial v}{\partial x}\}d\rho(\xi)dx = \int_V \xi\{u(\xi,\tau)v(\xi,\tau) - u(\xi,0)v(\xi,0)\}d\rho(\xi),$$

whence

$$\nu_{\pm}(E\times W) = \int_W |\xi|\, d\rho(\xi)$$

for $E=\{0\}$ or $E=\{\tau\}$ and for any ρ-measurable $W\subset V$. As a result, we may identify $L_2(D_\pm, d\nu_\pm)$ with $L_2(V, |\xi|\, d\rho(\xi))$, which is the Hilbert space H_T. On defining K $= RJQ_+ + RJQ_-$, we may show that K has the property $\|K\| = \|R\|_{H_T}$. This gives an indication that the stationary existence and uniqueness theory in H_T (rather than in H) should play a role in time dependent problems.

We now return to the abstract kinetic equation. The existence theory for the stationary equation detailed in Chapters II, III, and V leads to the following result.

THEOREM 3.4. Let $1\le p<\infty$, and let $\|R\|_{H_T} \le 1$. Under the above conditions on T and A with $\tau\le\infty$, the initial-boundary value problem (3.7) has a unique bounded solution $u:[0,\infty)\to L_p(H_T)_0^\tau$ for every $\hat{g}\in H_T$, every $g_0\in L_p(H_T)_0^\tau$, and every bounded continuous function $f:[0,\infty)\to L_p(H_T)_0^\tau$ that is strongly continuously differentiable on $(0,\infty)$.

Proof: On the Banach space $L_p(H_T)_0^\tau$ we define the operator

$$(\hat{B}u)(x) = -T\frac{\partial u}{\partial x}(x) - Au(x), \quad 0<x<\tau,$$

and the linear manifolds

$$F_{p,\tau} = \{u\in L_p(H_T)_0^\tau : u(0)\in H_T,\ u(\tau)\in H_T\}, \quad 0<\tau<\infty,$$

$$F_{p,\infty} = \{u\in L_p(H_T)_0^\infty : u(0)\in H_T\}.$$

We define the operator $\hat{B}_R$ as the operator B restricted to the domains

$$D(\hat{B}_R) = \{u\in F_{p,\tau} : Q_+u(0)=RJQ_-u(0),\ Q_-u(\tau)=RJQ_+u(\tau)\}, \quad 0<\tau<\infty,$$

$$D(\hat{B}_R) = \{u\in F_{p,\tau} : Q_+u(0)=RJQ_-u(0)\}, \quad \tau=\infty.$$

According to Section II.3, the stationary boundary value problem

$$(\lambda I-\hat{B})u = T\frac{\partial u}{\partial x} + (\lambda I+A)u = f, \tag{3.8a}$$

$$Q_+u(0) = RJQ_-u(0), \tag{3.8b}$$

$$Q_-u(\tau) = RJQ_+u(\tau), \tag{3.8c}$$

where for $\tau=\infty$ condition (3.8c) is replaced by

$$\|u(x)\|_T = O(1) \quad (x\to\infty), \tag{3.8d}$$

has a unique solution in H_T whenever $\|R\|_{H_T} \leq 1$. Moreover, for $u=u_\lambda$ we have the estimate

$$\|u_\lambda\|_{L_p(H_T)_0^\tau} \leq \frac{M}{\lambda}\|f\|_{L_p(H_T)_0^\tau},$$

for a dense set of inhomogeneous terms f. As a consequence of the Hille–Yosida theorem, we see that $\hat{B}_R$ generates a C_0–semigroup on $L_p(H_T)_0^\tau$. ■

4. Spectral properties of positive semigroups

Under rather general conditions we have shown that problem (1.1) has a unique positive solution $u(t)=U_K(t)u_0$ described by the semigroup $\{U_K(t)\}_{t\geq 0}$. In principle, the semigroups $\{U_K(t)\}_{t\geq 0}$ and $\{U_{0,K}(t)\}_{t\geq 0}$ generated by $\hat{B}_K$ and $S_K=\hat{B}_K-J$, respectively, can be computed from each other using the Hille–Dyson–Phillips expansion

$$U_K(t) = \sum_{n=0}^{\infty} U_K^{(n)}(t), \qquad t\geq 0, \tag{4.1a}$$

where

$$U_K^{(0)}(t) = U_{0,K}(t) \tag{4.1b}$$

and

$$U_K^{(n)}(t) = \int_0^t U_K^{(n-1)}(t-s)JU_{0,K}(s)ds, \quad n \in \mathbb{N}. \tag{4.1c}$$

Sometimes, explicit knowledge of $U_{0,K}(t)$ and the truncation of the series (4.1a) after finitely many nonzero terms lead to an explicit expression for the solution semigroup $\{U_K(t)\}_{t\geq 0}$. In most applications, however, no such truncation occurs and the series expansion (4.1a) is so complicated that one cannot even extract from it the asymptotic behavior of the solution as $t \to \infty$. Yet, one might expect that the asymptotic behavior of the solution, with its obvious implications for the approach of the physical system to equilibrium as time elapses, is related to the rightmost part of the spectrum of the generator $\hat{B}_K$, and indeed this will turn out to be the case.

As we have seen, the semigroups which arise in kinetic theory are defined generally on Banach lattices. For this reason we shall examine positive semigroups $\{U(t)\}_{t\geq 0}$ on Banach lattices E with order relation $\geq$ (cf. Section I.4). Throughout the remainder of the chapter we shall repeatedly employ the classification of points of the spectrum $\sigma(B)$ of a closed linear operator B as point or eigenvalue spectrum $\sigma_p(B)$, continuous spectrum $\sigma_c(B)$, and residual spectrum $\sigma_r(B)$, and also as approximate point spectrum $\sigma_{ap}(B)$. For definitions we refer to Section I.3. In this section we will introduce and discuss some additional notions of this type.

Consider a closed linear operator B. By the **spectral bound** of B is meant the (extended) real number

$$s(B) = \sup \{\text{Re } \lambda : \lambda \in \sigma(B)\}.$$

The **peripheral spectrum** of B is the set

$$\sigma_{per}(B) = \{\lambda \in \sigma(B) : \text{Re } \lambda = s(B)\}.$$

We recall that a closed operator B is Fredholm if B has closed range, Ker B has finite dimension and Ran B has finite co-dimension. The **essential spectrum** of B is the set

$$\sigma_{ess}(B) = \{\lambda \in \mathbb{C} : (\lambda I - B) \text{ is not Fredholm}\}.$$

By the **essential spectral bound** is meant the (extended) real number

$$s_{ess}(B) = \sup \{\text{Re } \lambda : \lambda \in \sigma_{ess}(B)\}.$$

The **asymptotic spectrum** of B is the set

$$\sigma_{as}(B) = \{\lambda \in \sigma(B) : \text{Re } \lambda > s_{ess}(B)\}.$$

The operator B is said to be **additively cyclic** if $\alpha + i\beta \in \sigma_{per}(B)$ implies that $\alpha + ik\beta \in \sigma_{per}(B)$ for all $k \in \mathbb{Z}$.

Although these spectral concepts are defined for arbitrary closed operators, they are clearly gauged toward generators of semigroups, and, more specifically, toward generators of positive semigroups in Banach lattices. Indeed, it is common to refer to s(B) as the spectral bound of the semigroup $\{U(t)\}_{t\geq 0}$, where B is its generator, and analogously for the other notions. Unfortunately, this can lead to some confusion. In addition, the notions of peripheral spectrum and asymptotic spectrum, as well as additive cyclicity, have dual definitions in semigroup theory, and the usage must be discerned from the context. Let us assume $\{U(t)\}_{t\geq 0}$ is a C_0 semigroup. The **peripheral spectrum** of U(t) is defined to be the set

$$\sigma_{per}(U(t)) = \{\lambda \in \sigma(U(t)) : |\lambda| = r(U(t))\}.$$

By the **Fredholm radius** or **essential spectral radius** of U(t) is meant the real number

$$r_{ess}(U(t)) = \sup \{|\lambda| : (\lambda I - U(t)) \text{ is not Fredholm}\}.$$

The **asymptotic spectrum** of U(t) is the set

$$\sigma_{as}(U(t)) = \{\lambda \in \sigma(U(t)) : |\lambda| > r_{ess}(U(t))\}.$$

The bounded operator U(t) is said to be **additively cyclic** if $|\lambda| e^{i\tau} \in \sigma_{per}(U(t))$ implies that $|\lambda| e^{ik\tau} \in \sigma_{per}(U(t))$ for all $k \in \mathbb{Z}$.

The relationship between the spectrum of the generator of a semigroup and the spectrum of the semigroup operators themselves is given by the Spectral Mapping Theorem for C_0 semigroups. For the proof we refer to any standard text on semigroup theory (e.g., [194, 213, 301]).

THEOREM 4.1. Let $\{U(t)\}_{t\geq 0}$ be a C_0-semigroup on a Banach space X with infinitesimal generator B.

(i) If $\lambda \in \sigma_p(B)$ then $e^{\lambda t} \in \sigma_p(U(t))$, and if $e^{\lambda t} \in \sigma_p(U(t))$ then there exists $k \in \mathbb{N}$ such that $\lambda_k = \lambda + \frac{2\pi i k}{t} \in \sigma_p(B)$, i.e.,

$$e^{t\sigma_p(B)} \subset \sigma_p(U(t)) \subset e^{t\sigma_p(B)} \cup \{0\}.$$

The multiplicity of λ_k is at most equal to the multiplicity of $e^{\lambda_k t}$.

(ii) If $\lambda \in \sigma_r(B)$ and if none of the numbers $\lambda_n = \lambda + \frac{2\pi i n}{t}$, $n \in \mathbb{N}$, is in $\sigma_p(B)$, then $e^{\lambda t} \in \sigma_r(U(t))$; if $e^{\lambda t} \in \sigma_r(U(t))$, then none of the numbers $\lambda_n = \lambda + \frac{2\pi i n}{t}$, $n \in \mathbb{N}$, is in $\sigma_p(B)$, and there exists $k \in \mathbb{N}$ such that $\lambda_k \in \sigma_r(B)$. This implies

$$\sigma_r(U(t)) \subset e^{t\sigma_r(B)} \cup \{0\}.$$

(iii) If $\lambda \in \sigma_c(B)$ and if none of the numbers $\lambda_n = \lambda + \frac{2\pi i n}{t}$, $n \in \mathbb{N}$, is in $\sigma_p(B) \cup \sigma_r(B)$, then $e^{\lambda t} \in \sigma_c(U(t))$.

In terms of the spectral bound s(B) and the type $\omega_0(U)$ of a semigroup $\{U(t)\}_{t\geq 0}$, the Spectral Mapping Theorem expresses that, in general, $-\infty \leq s(B) \leq \omega_0(U) < +\infty$. Several explicit examples have been constructed for which $s(B) < \omega_0(U)$. The first such example was constructed by Hille and Phillips ([194], Section 23) with the help of fractional integration theory. Less complicated examples followed (see [5, 191, 173, 279, 398, 403]), mostly as by-products of a still unsuccessful attempt to find necessary and sufficient conditions in order to have $s(B) = \omega_0(U)$ for any Banach space or for positive semigroups on any Banach lattice. Greiner et al. [173] were the first to construct a positive semigroup $\{U(t)\}_{t\geq 0}$ such that $s(A) < \omega_0$. A simple example of such a semigroup was given by Altomare and Nagel [5]. Following their account, we consider the Banach lattice E of all real continuous functions on $[0,\infty)$ that vanish at infinity and are integrable with respect to the weighted measure $e^x dx$, endowed with the norm

$$\|u\| = \sup_{x\geq 0} |u(x)| + \int_0^\infty |u(x)| e^x dx.$$

For the semigroup we take $(U(t)u)(x) = u(x+t)$, whose generator is the operator $(Bu)(x) = \frac{du}{dx}$ on the domain of those $u \in E$ that are continuously differentiable on $[0,\infty)$ with derivative in E. Then it is easily verified that $\omega_0(U) = 0$ and $s(A) = -1$. In fact, every λ with Re $\lambda < -1$ is an eigenvalue of B with corresponding eigenvector $e^{\lambda x}$.

In some settings it is possible to demonstrate the equality of the type of a semigroup and the spectral bound of its generator. An important result in this direction has been provided by Gearhart [131], who derived sufficient conditions in a (separable) Hilbert space, in terms of properties of the resolvent of the generator (see also [337, 192] for related results). It is important to note that the equality $s(B)=\omega_0(U)$ is valid for all positive semigroups in L_1-, L_2-, L_∞- and C-spaces. (See [102] for all of these results except for L_2-space; for the latter see [172].) The matter is still open for positive semigroups in the remaining L_p-spaces. A partial result in this direction has been provided by Voigt [384] by using interpolation theory for positive operators.

The next result is difficult to reference concisely, since it is due to quite a number of authors and most of it was developed in stages. Additive cyclicity of positive semigroups on Banach lattices (part (iii) of Theorem 4.3 below) was first proved by Derndinger [102]. At that time there already existed such a result in Banach lattices of continuous functions (see [103]). The multiplicity structure of an isolated eigenvalue at the spectral radius of a positive operator (part (iv)) was studied by Lotz [251] (also [324], Theorem V 4.9). The group structure of the peripheral spectrum of the generator of a positive semigroup (parts (v) and (vi)) was studied in detail by Greiner [169] and Greiner et al. [173] Special situations were already found by Lotz [251].

For the proof of part (iv) of Theorem 4.3, we will need a lemma due to Derndinger [102].

LEMMA 4.2. Let X be a (real or complex) Banach space, and let $\{U(t)\}_{t\geq 0}$ be a C_0-semigroup with generator B. Consider the Banach quotient space

$$\hat{X} = \{\{x_n\}_{n=1}^{\infty} \in \ell_\infty(X) : \lim_{t\downarrow 0} \|U(t)x_n - x_n\| = 0 \text{ uniformly in } n \in \mathbb{N}\}/c_0(X),$$

where $c_0(X)$ is the subspace of sequences in X norm convergent to zero. Let $[\{x_n\}_{n=1}^{\infty}]$ denote the equivalence class of the sequence $\{x_n\}_{n=1}^{\infty}$. Then the operators $\hat{U}(t)$ on $\hat{X}$ defined by $\hat{U}(t)[\{x_n\}_{n=1}^{\infty}] = [\{U(t)x_n\}_{n=1}^{\infty}]$ form a C_0-semigroup on $\hat{X}$, whose generator $\hat{B}$ is given by

$$\hat{B}[\{x_n\}_{n=1}^{\infty}] = [\{Bx_n\}_{n=1}^{\infty}],$$

where $D(\hat{B})$ consists of those $[\{x_n\}_{n=1}^{\infty}]$ such that $x_n \in D(B)$, $\{Bx_n\}_{n=1}^{\infty}$ is a bounded sequence and $\lim_{t\downarrow 0} \|U(t)Bx_n - Bx_n\| = 0$ uniformly in $n \in \mathbb{N}$. Moreover, the operator $\hat{B}$ has the following properties:

$$\sigma_p(\hat{B}) = \sigma_{ap}(\hat{B}) = \sigma_p(B) \cup \sigma_c(B), \tag{4.2a}$$

$$\sigma_c(\hat{B}) = \phi, \tag{4.2b}$$

$$\sigma_r(\hat{B}) = \sigma_r(B). \tag{4.2c}$$

Proof: It is easy to check that $\hat{B}$ and $\hat{U}(t)$ are well-defined. It is also easy to check that $U(t)$ and $\hat{U}(t)$ have the same norm, whence $\omega_0(U) = \omega_0(\hat{U})$. Using the Laplace transform it follows that $(\lambda I - \hat{B})^{-1}[\{x_n\}_{n=1}^{\infty}] = [\{(\lambda I - B)^{-1}x_n\}_{n=1}^{\infty}]$, which implies the expression for $D(\hat{B})$. Next, if $\lambda \in \sigma_p(\hat{B})$, then there exists a bounded sequence $\{x_n\}_{n=1}^{\infty}$ in $D(B)$ such that

$$\lim_{n\to\infty} \|Bx_n - \lambda x_n\| = 0, \tag{4.3}$$

whence $\lambda \in \sigma_{ap}(B)$. Conversely, if $\lambda \in \sigma_{ap}(B)$, then there exists a bounded sequence $\{x_n\}_{n=1}^{\infty}$ in X satisfying (4.3). Clearly,

$$\|U(t)x_n - x_n\| \le \left\|\int_0^t e^{\lambda(t-s)} U(s)(B-\lambda I)x_n \, ds\right\| + |e^{\lambda t} - 1| \, \|x_n\|$$

converges to zero as $t \downarrow 0$ uniformly in $n \in \mathbb{N}$, whence the equivalence class of $\{x_n\}_{n=1}^{\infty}$ is an eigenvector of $\hat{B}$ corresponding to the eigenvalue λ and $\lambda \in \sigma_p(\hat{B})$.

If $\lambda \in \sigma_r(\hat{B})$, then $\lambda \notin \sigma_p(\hat{B})$ and there exist $\varepsilon > 0$ and a bounded sequence $\{z_n\}_{n=1}^{\infty}$ such that

$$\|Bx_n - \lambda x_n - z_n\| > \varepsilon \tag{4.4}$$

for any bounded sequence $\{x_n\}_{n=1}^{\infty}$. Therefore, for all $x \in X$ there exists $n \in \mathbb{N}$ such that, for $z = z_n$,

$$\|Bx - \lambda x - z\| \ge \varepsilon, \tag{4.5}$$

whence either $\lambda \in \sigma_r(B)$ or $\lambda \in \sigma_p(B)$. The latter is excluded because of (4.2a).

Conversely, if $\lambda \in \sigma_r(B)$, then $\lambda \notin \sigma_p(B)$ and there exists $z \in X$ such that (4.5) holds true for every $x \in X$. Then, for $z_n \equiv z$, one has (4.4) for every bounded sequence $\{x_n\}_{n=1}^{\infty}$ in X such that $\|U(t)x_n - x_n\|$ vanishes as $t \downarrow 0$ uniformly in $n \in \mathbb{N}$, whence either $\lambda \in \sigma_r(\hat{B})$ or $\lambda \in \sigma_p(\hat{B})$. The latter is excluded because of (4.2a). ■

THEOREM 4.3. Let E be a complex Banach lattice with dual lattice E^*, and let $\{U(t)\}_{t \geq 0}$ be a positive semigroup on E with generator B. Then the following statements hold true:

(i) $r(U(t)) \in \sigma(U(t))$.

(ii) $s(B) \in \sigma(B)$ if $s(B) > -\infty$.

(iii) If $z = r(U(t))$ is a pole of $(zI-U(t))^{-1}$, then $\sigma_{per}(U(t))$ is cyclic and $r(U(t))^{-1}\sigma_{per}(U(t))$ consists solely of roots of unity.

(iv) $\sigma_{per}(B)$, $\sigma_{per}(B) \cap \sigma_p(B)$, $\sigma_{per}(B) \cap \sigma_{ap}(B)$ and $\sigma_{per}(B) \cap \sigma_r(B)$ are additively cyclic.

(v) If $s(B) > s_{ess}(B)$ and $s(B)$ is an eigenvalue of B with finite algebraic multiplicity, then $(\sigma_{per}(B)-s(B))$ is a finite union of additive subgroups of $i\mathbb{R}$ and consists only of eigenvalues of B of finite algebraic multiplicity.

(vi) If $s(B)$ is an eigenvalue of B of algebraic multiplicity 1, then $\{\sigma_{per}(B)-s(B)\}$ is an additive subgroup of $i\mathbb{R}$ and consists only of eigenvalues of B of algebraic multiplicity 1.

Proof: Part (i) follows immediately from Theorem I 4.2. In order to prove part (ii), one first derives the Laplace transform

$$(\lambda I-B)^{-1}u = \int_0^{\infty} e^{-\lambda t}U(t)u\,dt, \quad u \in E, \quad \text{Re } \lambda > s(B). \tag{4.6}$$

Indeed, for nonnegative $u \in E$ and nonnegative $\hat{u} \in E^*$ and $s(B) < \lambda \leq \mu$ one has

$$<(\lambda I-B)^{-1}u,\hat{u}> = \sum_{n=0}^{\infty} \int_0^{\infty} (\mu-\lambda)^n t^n \frac{1}{n!} e^{-\mu t} <U(t)u,\hat{u}> dt.$$

Using monotone convergence one may interchange the order of summation and integration and obtain (4.6) in the weak sense. One proves the integral in (4.6) to converge in the strong operator topology with the help of a Cauchy sequence argument, which settles the assertion. Now, choosing a sequence $\{\lambda_n\}_{n=1}^{\infty}$ with limit s(B) such that $\{\mathrm{Re}\ \lambda_n\}_{n=1}^{\infty}$ is monotonically decreasing and $\|(\lambda_n I-B)^{-1}\|\to\infty$ as $n\to\infty$, the Banach–Steinhaus theorem implies the existence of a nonnegative vector $u\in E$ for which $\{\| |(\lambda_n I-B)^{-1}u| \|\}_{n=1}^{\infty}$ diverges. From this statement one gets the divergence of $\{\|(\mathrm{Re}\ \lambda_n-B)^{-1}u\|\}_{n=1}^{\infty}$, which in turn implies that $s(B)\in\sigma(B)$ when finite.

In order to prove part (iv), we apply Lemma 4.2. Then $\{\hat{U}(t)\}_{t\geq 0}$ is a strongly continuous positive semigroup on $\hat{E}$ with generator $\hat{B}$. Now, assuming $\sigma_p(B)$ nonempty and putting $C=B-s(B)I$, we take $0\neq u\in D(B)$ with $Cu=\lambda u$ and $\mathrm{Re}\ \lambda=0$. Then according to part (i) of Theorem 4.1, we have $U(t)u=e^{\lambda t}u$ and hence the equality $U(t)|u|=|U(t)u|=|e^{\lambda t}u|=|u|$ for $t>0$. We may identify $E_{|u|}$ (those $v\in E$ for which $|v|$ is dominated by some multiple of $|u|$) with C(X) for some compact Hausdorff space X ([324], Corollary to Theorem II 7.2), while the restrictions of the operators $e^{-s(B)t}U(t)$ to this ideal correspond to Banach algebra *–homomorphisms on C(X) (cf. [324], III 9.1, Ex. 25). As a result, there must be a sequence $\{v_n\}_{n=-\infty}^{\infty}$ such that $e^{-s(B)t}U(t)v_n=e^{n(-s(B)+\lambda)t}v_n$ for all $t>0$. Hence, $\sigma_{per}(B)\cap\sigma_p(B)$ is additively cyclic. Using (4.2a) we immediately get the additive cyclicity of $\sigma_{per}(B)\cap\sigma_{ap}(B)$. The additive cyclicity of the residual part of $\sigma_{per}(B)$ follows by using the adjoint. All these statements together yield the additive cyclicity of $\sigma_{per}(B)$.

From part (iv) it follows directly that the peripheral spectrum of U(t) consists only of isolated points whenever r(U(t)) is an isolated eigenvalue, while $r(U(t))^{-1}\sigma_{per}(U(t))$ consists only of roots of unity.

In order to prove (v), let $s(B) > s_{ess}(B)$. Then $(\lambda I-B)$ is a Fredholm operator for all $\lambda\in\sigma_{per}(B)$ and $(\lambda I-B)^{-1}$ is bounded and defined on E for such λ except, perhaps, for countably many. These countably many λ are eigenvalues of finite algebraic multiplicity (because of the continuity in λ of the Fredholm index), while $(\sigma_{per}(B)-s(B))$ is the union of additive subgroups of $i\mathbb{R}$. Finally, to derive (vi) from (v), it is sufficient to show that s(B) is the eigenvalue with the maximal pole order among the eigenvalues in the peripheral spectrum of B. Indeed, assuming that s(B) is a pole of $(\lambda I-B)^{-1}$ of order m, we have $\lim_{\lambda\to s(B)}|\lambda-s(B)|^n(\lambda I-s(B))^{-1}=0$ for all $n\geq m$. Using (4.6) and taking a nonnegative $u\in E$ we have, for another pole $s(B)+i\rho$ in $\sigma_{per}(B)$,

$$|\lambda-s(B)-i\rho|^n\left|\int_0^{\infty}e^{-\lambda t}<U(t)u,v>dt\right| \leq |\lambda-s(B)-i\rho|^n<(\lambda I-B)^{-1}u,v>$$

for any nonnegative $v \in E^*$. Since $s(B)+i\rho$ is a pole of $(\lambda I-B)^{-1}$, its order cannot exceed m. ■

In Section I.4 we have introduced the notions of real and complex Banach lattices and discussed some of the properties of positive operators on such spaces. A related concept is irreducibility, which we define here in a semigroup context. A semigroup $\{U(t)\}_{t\geq 0}$ on a Banach lattice is called **irreducible** if, for each nonzero u in the positive cone of the Banach lattice E and for each nonzero u^* in the positive cone of the dual Banach lattice E^*, there exists $t_0=t_0(u,u^*)\geq 0$ such that $\langle U(t_0)u, u^*\rangle > 0$. If t_0 does not depend on u^*, then $\{U(t)\}_{t\geq 0}$ is called **strongly positive**. If t_0 does not depend on either u or u^*, then $\{U(t)\}_{t\geq 0}$ is called **positivity improving**.

In time dependent kinetic theory, where B is the full transport operator (actually the operator B_K of Section 2) and $\{U(t)\}_{t\geq 0}$ is the evolution semigroup of the system, the spectral bound s(B) is expected to be an isolated eigenvalue which determines the long time behavior of the system. If s(B) is finite, algebraically simple with eigenvector φ_0, and the only point of $\sigma_{per}(B)$, we expect from physical intuition a long time behavior of the system of the type $u(x,\xi,t)= e^{s(B)t}\varphi_0(x,\xi)+o(e^{\lambda t})$ for some $\lambda<s(B)$ as $t\to\infty$. Further, we would expect the eigenfunction $\varphi_0(x,\xi)$ to be nonnegative. The situation turns out to be more complicated. First, we give some definitions. By a **leading eigenvalue** we mean $s(B)\in\sigma_p(B)$, where at least one of the corresponding eigenvectors is nonnegative. We call λ a **dominant eigenvalue** if $\lambda\in\sigma_p(B)\cap\sigma_{per}(B)$ and λ is algebraically simple. Finally, we call λ a **strictly dominant eigenvalue** if λ is a dominant eigenvalue and, at the same time, $\sup\{\text{Re}\,\mu : \mu\in\sigma(B), \mu\neq\lambda\} < \lambda$.

The next theorem describes some properties of dominant eigenvalues. It has important applications to kinetic equations, where the time evolution semigroup can often be proven positive and irreducible. Part (i) was first obtained by Vidav [375] for a specific transport problem. His proof was based on duality rather than irreducibility arguments. An early version of parts (ii) and (iii) was worked out by Angelescu and Protopopescu [11] for another transport problem. The subsequent generalization to abstract positive semigroups is due to Voigt [378] and Greiner et al. [173]

THEOREM 4.4. Let E be a complex Banach lattice with dual lattice E^*, and let $\{U(t)\}_{t\geq 0}$ be a positive semigroup on E with generator B. Then the following statements hold true:

(i) If $r_{ess}(U(t)) < r(U(t))$, then $s(B)$ is a dominant eigenvalue with finite algebraic multiplicity and $\sigma_{per}(U(t))=\{e^{s(B)t}\}$.

(ii) If $r_{ess}(U(t)) < r(U(t))$, then $\{U(t)\}_{t\geq 0}$ is irreducible if and only if the spectral bound $s(B)$ is an algebraically simple, strictly dominant eigenvalue of B and the associated spectral projection $P_{s(B)}$ is positivity improving.

(iii) If $r_{ess}(U(t)) < r(U(t))$, then $\{U(t)\}_{t\geq 0}$ is irreducible if and only if $\lambda(t)=e^{s(B)t}$ is an algebraically simple eigenvalue of $U(t)$ for all $t>0$ and the associated spectral projection $P_{\lambda(t)}$ is positivity improving.

Proof: In order to settle part (i), denote by P_0 the degenerate projection corresponding to $\sigma_{per}(U(1))$. The restriction $\{U_0(t)\}_{t\geq 0}$ of $\{U(t)\}_{t\geq 0}$ to Ran P_0 can be thought of as a semigroup of matrices, and therefore there are $\beta_1,...,\beta_k$ such that

$$\sigma_{per}(U(t)) = \sigma(\hat{U}_0(t)) = \{e^{t\beta_1}, ... ,e^{t\beta_k}\}.$$

Then $\beta_j=\varepsilon_j+in_j$ and $|e^{t\beta_j}|=e^{t\omega_0(U)}$ imply $\varepsilon_j=\omega_0(U)$. Part (iii) of Theorem 4.3 implies that $\exp(itn_j)$ is a root of unity. Therefore $\frac{t}{2\pi}n_j$ is rational. Since this has to be true for all $t>0$, we conclude $n_j=0$, and thus $\beta_1=...=\beta_k$. As a result of part (i) of Theorem 4.1 we get $\beta_1=...=\beta_k=s(B)$. Then, denoting the restriction of B to Ran P_0 by B_0, we can apply matrix algebra to obtain $s(B)=s(B_0)=\omega_0(U_0)$.

Let us prove (ii) and (iii). The existence of a positive eigenvalue $\mu_0(t)$ with a nonnegative eigenfunction $\varphi_0\in L_p$ follows from Vidav's generalization [375] of the Krein–Rutman theorem [223, 229]. Moreover, following Kato ([213], Section III 6.6), one may show that the adjoint semigroup $U^*(t)$ also has $\mu_0(t)$ as an eigenvalue with nonnegative eigenfunction $\psi_0\in E^*$. Since $U(t)$ (and therefore $U^*(t)$) is irreducible, one obtains

$$0 < \langle U(t_1)\varphi_0,v\rangle = e^{\mu_0(t_1)}\langle\varphi_0,v\rangle$$

for some $t_1>0$ and all nonnegative $0\neq v\in E^*$, and

$$0 < \langle u,U(t_2)^*\psi\rangle = e^{\mu_0(t_2)}\langle u,\psi_0\rangle$$

for some $t_2>0$ and all nonnegative $0\neq u\in E$. Hence, φ_0 and ψ_0 are strictly positive vectors.

Suppose now that φ is another eigenvector of $U(t)$ corresponding to $\mu_0(t)$

and choose it (taking, if necessary, a linear combination with φ_0) such that $<\varphi,\psi_0>=0$. Then φ cannot be nonnegative, and therefore $|U(t)\varphi| < U(t)|\varphi|$. Thus,

$$\mu_0(t)<|\varphi|,\psi_0> = <|U(t)\varphi|,\psi_0> < <U(t)|\varphi|,\psi_0> = <|\varphi|,U(t)^*\psi_0> =$$

$$= \mu_0(t)<|\varphi|,\psi_0>,$$

which is a contradiction, and $\mu_0(t)$ is geometrically simple.

Suppose $\mu_0(t)$ were not algebraically simple. Then, there exists $\varphi\neq 0$ such that $(U(t)-\mu_0(t))\varphi=\varphi_0$. But this is impossible in view of the estimate

$$0 < <\varphi_0,\psi_0> = <(U(t)-\mu_0(t)I)\varphi,\psi_0> = <\varphi,(U^*(t)-\mu_0(t)I)\psi_0> = 0.$$

Theorem 4.1 together with the geometric simplicity of $\mu_0(t)$ show that the equation $e^{\lambda t}=\mu_0(t)$ has only one solution λ_0. The eigenvalue λ_0 is algebraically simple, because $(B-\lambda_0 I)\varphi=\varphi_0$ implies $(U(t)-\mu_0(t)I)\varphi=t\exp(\lambda_0 t)\varphi_0$ which contradicts the algebraic simplicity of $\mu_0(t)$.

In order to prove that P_0 is positivity improving, it suffices to show that $<P_0u,v> > 0$ for all nonnegative $0\neq u\in E$ and $0\neq v\in E^*$. However, if this is false, there exists a nonnegative $0\neq u\in E$ such that $P_0u=0$. Using (4.6) we also have

$$P_0u = \int_0^\infty e^{-\lambda t}U(t)u\,dt, \qquad \mathrm{Re}\ \lambda>s(B), \tag{4.7}$$

whence $U(t)u=0$ for all $t>0$. The latter obviously contradicts the irreducibility of $\{U(t)\}_{t\geq 0}$, whence P_0 is positivity improving.

Conversely, assume that P_0 is positivity improving and has rank one. Let $0\neq u\in E$ and $0\neq v\in E^*$ be nonnegative. Then $<P_0u,v> > 0$ and

$$e^{-s(B)t}<U(t)u,v> = <P_0u,v> + <e^{-s(B)t}U(t)(I-P_0)u,v>,$$

which tends to $<P_0u,v>$ for large t. Therefore, $<U(t)u,v> >0$ for all t large enough, which implies that $\{U(t)\}_{t\geq 0}$ is irreducible. ■

One issue that was left open in the above theorem is the question of whether one has the decomposition

$$U(t) = U_{remainder}(t) \oplus U_{as}(t) \tag{4.8}$$

in the case when $s_{ess}(B)<s(B)$. Here $U_{as}(t)$ accounts for the contribution from the asymptotic spectrum of the generator, while $U_{remainder}(t)$ is the remaining contribution, having a type strictly less than $\omega_0(U)$. Such a result would be relevant to the study of the long time behavior of a kinetic system governed by a positive semigroup, since it would imply that the asymptotic spectrum of the generator determines the long time behavior of the kinetic system. Using the asymptotic spectrum of the generator, one could then write down an asymptotic series for the semigroup for $t\to\infty$, containing the eigenvectors and generalized eigenvectors of the generator corresponding to the eigenvalues in its asymptotic spectrum. At the same time one would know that the remainder contribution to the semigroup has a smaller exponential increase in time than the asymptotic series. Unfortunately, this is not the case for general positive semigroups. Indeed, if $\{V(t)\}_{t\geq 0}$ is a positive semigroup on a Banach lattice F with its generator C satisfying $s(C)<\omega_0(V)$, then on choosing λ with $s(C)<\lambda<\omega_0(V)$ and defining $E=F\oplus\mathbb{C}$ and $U(t)=V(t)\oplus\{e^{\lambda t}\}$ one obtains a positive semigroup $\{U(t)\}_{t\geq 0}$ with its generator B satisfying $s_{ess}(B)<s(B)<\omega_0(U)$, while $\|V(t)\|$ grows faster as $t\to\infty$ than $e^{\lambda t}$. The real issue, however, is whether $s_{ess}(B)<s(B)$ for a positive semigroup on an L_p-space with $1\leq p<\infty$ implies a decomposition of the type (4.8). However, since in general the semigroup in the decomposition is not positive (certainly not if the original semigroup is irreducible), one cannot exploit the equality of the spectral bound and the type for positive semigroups in certain L_p-spaces to reach such a conclusion. Indeed, the answer to this question is unknown.

5. Spectral and compactness properties for kinetic models

In this section we will study positive semigroups related to the operators S, J and K introduced in Section 2. We shall assume that J and K are positive operators, so that the time evolution semigroup is positive and the theory of the previous section applies. For the sake of convenience we shall replace J by cJ, where $c \in \mathbb{R}$. Under the hypotheses of Theorem 2.2, the operator $B_{c,K}$, defined as the operator B_K with J replaced by cJ, is the generator of a positive semigroup $U_{c,K}(t)$, provided $\|K\|<1$. If $\|K\|=1$, we have to take a suitable closed extension of this operator. Similarly, we will define S_K as the operator $B_{c,K}$ (or a suitable closed extension if $\|K\|=1$), for which J=0. In this way we can treat a fixed operator J multiplied by a constant c, as is done in neutron transport theory, where c represents the number of secondaries per collision. In addition to positivity we shall assume certain compactness properties of the collision operator.

We begin with a simple monotonicity property of the semigroups $\{U_{c,K}(t)\}_{t\geq 0}$. Suppose that we have two time dependent kinetic problems on the same phase space, in the same L_p-setting and with the same free streaming operator. Denoting c, K and J for the first problem by c_1, K_1 and J_1 and those for the second problem by c_2, K_2 and J_2, we have

$$U_{c_1,K_1}(t) \leq U_{c_2,K_2}(t), \qquad t\geq 0,$$

whenever $c_1 \leq c_2$, $K_1 \leq K_2$ and $J_1 \leq J_2$. One way to derive the above inequality is to follow step by step the arguments leading to the main existence results in Sections XI.4 and XI.5. We would then obtain this monotonicity result more generally, assuming time dependent phase space, acceleration and boundary conditions, even in spaces of measures. Another way to obtain this result applies to the present semigroup context. On writing the Hille-Dyson-Phillips expansion for both semigroups, one immediately gets the above monotonicity result by term by term comparison. Since the operator norm on a Banach lattice is monotonic on the cone of positive operators, we immediately find

$$\omega_0(U_{c_1,K_1,J_1}) \leq \omega_0(U_{c_2,K_2,J_2})$$

as well as

$$r(U_{c_1,K_1,J_1}) \leq r(U_{c_2,K_2,J_2}),$$

where we have explicitly displayed the dependence of the semigroups on the collision operator. A similar monotonicity property can be derived for the spectral bounds of the semigroups. Using the Laplace transforms of the semigroups, we first prove the monotonicity for the resolvents of the generators on $\{\lambda \in \mathbb{C} : \text{Re } \lambda > \omega_0\}$, where ω_0 is the type of the larger semigroup. Observing that the Laplace transform formula holds for all λ exceeding the spectral bound (cf. (4.6)), we may employ the monotonicity of the resolvents to get the monotonicity of the spectral bounds. In this way we do not have to make assumptions that guarantee the equality of spectral bound and type, as was done in [211] where monotonicity properties of neutron transport semigroups were treated in detail.

For many explicit models it is known that time dependent kinetic equations have a dominant eigenvalue if c becomes sufficiently large. The next results express this fact.

PROPOSITION 5.1. If, for some $n \in \mathbb{N}$ and $\lambda^* \in \mathbb{R}$, the operators $Q_\lambda(c) \equiv ((\lambda I - S_K)^{-1} cJ)^n$ are compact and positivity improving for all $\lambda > \lambda^*$, then $s(B_{c,K}) \to \infty$ and $r(U_{c,K}(t)) \to \infty$ as $c \to \infty$.

Proof: Consider first the limit for the spectral bound. Since $Q_\lambda(c)$ is compact and positivity improving, it follows that $0 < r(Q_\lambda(c)) \in \sigma_p(Q_\lambda(c))$. Then, for every $c > 0$ and for some almost everywhere positive φ_c, one has $Q_\lambda(c)\varphi_c = r(Q_\lambda(c))\varphi_c$. Using the Spectral Mapping Theorem we get, for $R_\lambda(c) \equiv (\lambda I - S_K)^{-1} cJ$, $R_\lambda(c)\varphi_c = \varepsilon_c r(R_\lambda(c))\varphi_c$, where ε_c is an n-th root of unity. Since $R_\lambda(t)$ is positive and φ_c is nonnegative, we have $\varepsilon_c = 1$, whence $\varphi_c \in D(S_K) = D(B_{c,K})$. Since $R_\lambda(c)$ depends analytically on λ and the eigenvalue $r(R_{\lambda,K}(c))$ is algebraically simple, φ_c can be chosen to depend analytically on c. From Theorem VII 1.8 of [213] it follows that

$$(\lambda I - S - \frac{c}{r(R_\lambda(c))} J)\varphi_c = 0,$$

where $\frac{1}{c} r(R_\lambda(c)) = r(R_\lambda(1))$ and thus $\varphi_c = \varphi$ does not depend on c. Therefore,

$$\lambda^* < \lambda \leq s(B_{r(R_\lambda(1))^{-1},K}).$$

Since $r(R_\lambda(1))$ tends to zero as $\lambda\to\infty$, we see that $s(B_{c,K})$ converges to infinity as c does.

To obtain the limit for the spectral radius, we first observe that

$$r(U_{c,K}(t)) \geq e^{\lambda t},$$

since λ is an eigenvalue of $B_{c,K}$ if $c=r(R_\lambda(1))^{-1}$. Hence, if c tends to infinity, so does the corresponding eigenvalue λ (as a result of the fact that the analytic function $r(R_\lambda(1))$ is strictly monotonically decreasing), and hence $r(U_{c,K}(t))$. ∎

The assumption that $Q_\lambda(c)$ is positivity improving for sufficiently large λ is not essential for deriving the above proposition. It is sufficient to assume that $r(Q_\lambda(c))>0$ for some $\lambda>\lambda^*$, some $c>0$ and some $n\in\mathbb{N}$. Analyticity will then guarantee that $r(Q_\lambda(c))>0$ for all $\lambda>\lambda^*$, all $c>0$, and all $n\in\mathbb{N}$. In fact, if $r(Q_\lambda(c))=0$ for some $\lambda>\lambda^*$ and some $n\in\mathbb{N}$, then $r(Q_\lambda(c))\equiv 0$ for all $\lambda>\lambda^*$, all $c>0$ and all $n\in\mathbb{N}$.

So far we have proved that $s(B_{c,K})$ tends to infinity as $c\to\infty$, without linking this property to the existence of a dominant eigenvalue. The key to its existence is the compactness condition in the above proposition. We call J **resolvent compact** with respect to S_K, if for some $n\in\mathbb{N}$ and all $\lambda>\lambda^*$ the operator $Q_\lambda=\{(\lambda I-S_K)^{-1}J\}^n$ is compact. By analyticity the operator Q_λ will then be compact for all λ in the connected component of S_K to which (λ^*,∞) belongs.

The following proposition is an auxiliary result that will enable us to prove the existence of a dominant eigenvalue for sufficiently large c.

PROPOSITION 5.2. If J is resolvent compact with respect to S_K, then the operator $\lambda I-B_{c,K}$ is invertible as a bounded operator for all $\lambda>\mathrm{Re}\,\lambda^*$, except for a discrete set of points $\{\lambda_k\}_{k=1}^\infty$, which are eigenvalues of $B_{c,K}$ of finite algebraic multiplicity.

Proof: If J is resolvent compact with $n=1$, then $\|Q_\lambda\|\to 0$ as $\mathrm{Re}\lambda\to\infty$ so that $I-Q_\lambda$ is invertible as a bounded operator for λ real and sufficiently large. Then the proposition follows from the analytic version of the Fredholm alternative. If J is resolvent compact with $n>1$, then $(I-(\lambda I-S_K)^{-1}J)$ is a Fredholm operator of index zero for all λ with $\mathrm{Re}\,\lambda>\lambda^*$ which is invertible for λ large enough, and the same argument completes the proof. ∎

These two propositions then yield the existence of a dominant eigenvalue for sufficiently large c. Again, one may replace the assumption that Q_λ be positivity improving with the assumption that $r(Q_\lambda)>0$ for some (and hence all) $\lambda>\lambda^*$ and $n\in\mathbb{N}$.

THEOREM 5.3. Let Q_λ be positivity improving for some $\lambda>\lambda^*$ and some $n\in\mathbb{N}$. Then for sufficiently large c the operator $B_{c,K}$ has a dominant eigenvalue, to which corresponds a strictly positive eigenfunction.

Let us define $E_1(t) = U_{0,K}(t)J$ and

$$E_n(t) = \int_0^t E_{n-1}(s)E_1(t-s)ds, \quad n=2,3,...,$$

where $U_{0,K}(t)$ is the semigroup generated by S_K. Here and in the following, the integrals are understood as Riemann integrals in the strong topology. The bounded operator J is called **S_K-smoothing**, if there exists $n\in\mathbb{N}$ such that $E_n(t)$ is a compact operator for all $t>0$, and the mapping $t\to E_n(t)$ is continuous on $(0,\infty)$ in the uniform operator topology. Under the assumption of J being S_K-smoothing, the above results on the existence of a dominant eigenvalue can be extended to results on other points of the asymptotic spectrum. In this way one may derive a more refined picture of the long time behavior of time dependent kinetic systems.

The relevance of smoothing perturbations to kinetic problems was already realized by Jörgens [204], although similar compactness arguments had been previously used by Lehner and Wing [245, 246] in a somewhat different context. Vidav [376] extended Jörgens' result to a quite general setting. For related results involving compact perturbations, holomorphic operator valued functions and positive operators, we refer to Shikhov and Shkurpelov [327] and references therein.

The next theorem generalizes the results of Vidav; the proof improves upon the exposition of Shizuta [328], which was given there for the special case $\partial\Lambda=0$.

THEOREM 5.4. Let J be S_K-smoothing. Then $B_{c,K}$ generates a C_0-semigroup $U_{c,K}(t)$ such that every point of $\sigma(U_{c,K}(t))$ lying outside the circle $|\varsigma|=\exp(\lambda^* t)$ is an isolated eigenvalue of finite algebraic multiplicity. The asymptotic spectrum $\sigma_{as}(B_{c,K})$ consists of isolated eigenvalues with finite algebraic multiplicity, and there are at most a finite number of eigenvalues of $B_{c,K}$ in the half plane $\{\lambda\in\mathbb{C} : \operatorname{Re}\lambda>\lambda^*+\beta\}$ for any $\beta>0$. If the eigenvalues in the set $\sigma_{as}(B_{c,K})$ are ordered such that

Re $\lambda_0 \geq$Re $\lambda_1 \geq \ldots$, then there exists a positive constant M, depending on β, such that

$$\|U_{c,K}(t) - \sum_{i=1}^{n} e^{\lambda_i t} P_i\| \leq Me^{(\lambda^* + \beta)t}, \quad t \geq 0, \tag{5.1}$$

where n is the number of eigenvalues λ_i whose real part is strictly greater than $\lambda^* + \beta$ and P_i is the eigenprojection corresponding to λ_i. The spectrum of $U_{c,K}(t)$ lying outside the circle $|\varsigma| = \exp(\lambda^* t)$ consists of isolated eigenvalues $\mu_k = \exp(\lambda_k t)$ for all $\lambda_k \epsilon \sigma_p(B) \cap \sigma_{as}(B)$.

Proof: Given T>0, one may find M=M(T) such that

$$\|U_{0,K}(t)\| \leq M, \quad 0 \leq t \leq T.$$

By induction, we obtain

$$\|E_n(t)\| \leq \frac{M^n t^{n-1}}{(n-1)!}, \quad 0 \leq t \leq T. \tag{5.2}$$

Define

$$H_n(t) = \int_0^t E_n(s) U_{c,K}(t-s) ds, \quad n = 1,2,\ldots .$$

Then the mapping $t \rightarrow H_n(t)$ is continuous on $(0,\infty)$ in the uniform operator topology. Indeed, let $t > t_0 > 0$. Then

$$H_n(t) - H_n(t_0) = \int_{t_0}^t E_n(t-s) U_{c,K}(s) ds + \int_0^{t_0} [E_n(t-s) - E_n(t_0 - s)] U_{c,K}(s) ds. \tag{5.3}$$

The first integral in (5.3) tends uniformly to zero as $t \rightarrow t_0$, since $\|U_{c,K}(s)\| \leq N = N(T)$ for $0 \leq s \leq T$ and $E_n(t-s)$ can be estimated using (5.2). Since $\|E_n(t-s) - E_n(t_0 - s)\| \rightarrow 0$ for $t \rightarrow t_0$ and for every fixed s, $0 < s < t_0$ and uniformly in t_0 and s on $[\varepsilon, T]$, it follows that the second integral also tends to zero, by the theorem of dominated convergence.

Define

$$R_n^{\varepsilon}(t) = \int_{\varepsilon}^{t} E_n(s)H_n(t-s)ds, \quad n=1,2,... \tag{5.4}$$

We claim that $R_n^0(t)$ is a compact operator on $\mathcal{X}$ for $t\geq 0$. Indeed, let $t>0$ and $\varepsilon>0$ be fixed, consider $0<t\leq T$, and note that the integrand in (5.4) is continuous on $[\varepsilon,T]$ in the uniform operator topology. Then the integral converges in the norm, and $E_n(s)$ is compact by assumption. Therefore, $R_n^{\varepsilon}(t)$ is compact for $0<\varepsilon<t$. From the estimate

$$\|R_n^{\varepsilon}-R_n^{\delta}\| \leq L|\varepsilon-\delta|,$$

which is easily found from (5.2), the definition of $H_n(t)$ and the boundedness of $U_{c,K}(t)$ on $[0,T]$, we obtain that $R_n^0(t)$ is compact for every $t>0$.

Define

$$F_n(t) = \int_0^t E_n(t)U_{0,K}(t-s)ds, \quad n=1,2,... .$$

Then, by applying the Hille-Dyson-Phillips expansion, we write

$$U_{c,K}(t) = U_{0,K}(t) +...+ F_n(t) +...+ F_{2n-1}(t) + R_n^0(t) \equiv Q_n(t) + R_n^0(t), \quad n\in\mathbb{N}.$$

Since $\|U_{0,K}(t)\|\leq Me^{\lambda^* t}$, one easily computes

$$\|Q_n(t)\| \leq e^{\lambda^* t} \sum_{j=0}^{2n-1} \frac{M^{j+1}\|J\|^j}{j!} t^j \equiv e^{\lambda^* t}p(t)$$

for some polynomial function $p(t)$. Since $R_1^0(t)$ is compact, it follows from a stability result for analytic Fredholm operator functions that $U_{c,K}(t)$ and $Q_1(t)$ have the same essential spectrum. This implies that in the spectrum of $U_{c,K}(t)$ lying outside the circle $\{\varsigma\in\mathbb{C} : |\varsigma|=p(t)\exp(\lambda^* t)\}$ there are only isolated eigenvalues μ_k of finite algebraic multiplicity. In fact, due to the polynomial nature of $p(t)$, the same is true for the spectrum of $U_{c,K}(t)$ lying outside this circle.

The characterization of the asymptotic spectrum is a consequence of the Spectral Mapping Theorem for discrete eigenvalues of finite algebraic multiplicity. For the estimate involving the spectral projections P_i, set $P=P_1+...+P_k$. Then we have a decomposition of the underlying space $\mathcal{X}=\mathcal{X}'\oplus\mathcal{X}''$, where $\mathcal{X}'=P\mathcal{X}$ and $\mathcal{X}''=(I-P)\mathcal{X}$. Let us denote by B'' and $U''(t)$ the restrictions of B and $U_{c,K}(t)$ to $\mathcal{X}''$. It is obvious that $\sigma(B'')\subset\{\lambda : \text{Re } \lambda\leq\alpha\}$, where $\alpha=\inf\{\text{Re } \lambda_i\}$ and $\sigma(U''(t)) \subset$

$\{\varsigma : |\varsigma| \le e^{\alpha t}\}$. This means that the spectral radius of $U''(t)$ is $\exp(\alpha t)$. In particular, $r(U''(1))=\exp(\alpha)$. Therefore,

$$\lim_{n\to\infty} \|\{U_{c,K}(1)(I-P)\}^n\|^{1/n} = \lim_{n\to\infty} \|U_{c,K}(n)(I-P)\|^{1/n} = e^{\alpha}.$$

Since $\lambda^*+\beta>\alpha$, we get $\|U_{c,K}(n)(I-P)\| \le e^{(\lambda^*+\beta)n}$ for $n>N$, where N is an integer depending on β. Hence, $\|U_{c,K}(n)(I-P)\| \le C\exp\{(\lambda^*+\beta)n\}$ for $n=0,1,\dots$, where

$$C = \sup\{\|U_{c,K}(n)(I-P)\|e^{-\beta n}+1 : 0\le n\le N\}.$$

Setting

$$M = C\sup\{\|U_{c,K}(s)(I-P)\|e^{-(\lambda^*+\beta)s} : 0\le s\le 1\},$$

we obtain (5.1). ■

According to the above theorem it is always possible to order the eigenvalues belonging to $\sigma_{as}(B)$ in such a way that $\mathrm{Re}\lambda_0 \ge \mathrm{Re}\lambda_1 \ge \dots$. If one denotes by B_i the restriction of B to the finite dimensional subspace $\mathcal{X}_i=P_i\mathcal{X}$, $i=0,1,2,\dots$, then B_i can be expressed as $B_i=\lambda_i I_i+D_i$, where D_i is the nilpotent operator given by

$$D_i = \frac{1}{2\pi i}\int_{\Gamma_i} (\lambda-\lambda_i)(\lambda I-B_{c,K})^{-1}d\lambda.$$

If $m_i=\dim \mathcal{X}_i$, then $D_i^N=0$ for $N\ge m_i$, and we obtain the following expression for the restriction of $U_{c,K}(t)$ to $\mathcal{X}_i$:

$$U_i(t) = e^{\lambda_i t}e^{D_i t} = e^{\lambda_i t}\sum_{k=0}^{m_i-1}\frac{D_i^k}{k!}t^k.$$

Setting

$$U_{c,K}(t) = \sum_{i=1}^{n} U_i(t)P_i + Z_n(t),$$

obviously (5.1) is an estimate of the norm of the semigroup $Z_n(t)$. We have thus indicated a case in which the decomposition (4.8) holds true.

The earlier work of Vidav [376] assumed that there exists $\ell \in \mathbb{N}$ such that the operator

$$U(t_1,t_2,\ldots,t_\ell) = U_{0,K}(t_1)J\ldots JU_{0,K}(t_\ell)J \tag{5.5}$$

is compact for all ℓ-tuples $(t_1,\ldots,t_\ell)$ in $(0.\infty)$ and depends continuously on $(t_1,\ldots,t_\ell)$ in the uniform operator topology. An operator J satisfying these properties was called **semigroup compact** with respect to the generator S_K of $\{U_{0,K}(t)\}_{t\geq 0}$. It is easily seen that this condition implies that J is S_K-smoothing, whence the statements of Theorem 5.4 are true for this case.

All these results have the disadvantage of being formulated in such an abstract way that it is not clear beforehand which kinetic models satisfy these conditions. This difficulty was somewhat alleviated by Voigt [379, 382] in the neutron transport setting. His idea was to show that for some $m \in \mathbb{N}$ the m-th order remainder in the Hille-Dyson-Phillips expansion (4.1a) is compact, which guarantees the invariance of the essential spectrum of the semigroup under the perturbation induced by J. In fact, since the m-th order remainder term $R_m(t)$ can also be written as

$$R_m(t) = \underset{\substack{s_1+\ldots+s_m\leq t \\ s_1\geq 0,\ldots,s_m\geq 0}}{\int ds_1\ldots\int ds_m} \quad U_{0,K}(s_1)J\ldots JU_{0,K}(s_m)JU_K(t-s_1-\ldots-s_m),$$

it is immediate that Voigt's condition is weaker than Vidav's.

Chapter XIII

APPLICATIONS OF THE INITIAL VALUE PROBLEM

1. Kinetic equations in neutron transport

In this chapter, we shall consider specific kinetic models related to the transport of neutrons and electrons, and to cellular growth. The first two sections will be devoted to neutron transport, with special attention to spectral properties of the full transport operator and implications to hydrodynamics. The third and fourth sections deal with electron transport. In the first of these, the Spencer–Lewis equation models the slowing down of electrons by a thermalizing medium. In the following, a linearized Boltzmann equation is presented, which describes the drift of electrons in a weakly ionized gas. Finally, in the last section, we will outline a biological model for the growth of cells, due to Rotenberg, Lebowitz and Rubinow.

Let us consider a linear kinetic equation, which describes neutron transport in an arbitrary (three dimensional) spatial domain with a three dimensional velocity domain which is rotationally invariant. We will allow for a rather general collision term, which may be written as the difference of a gain term containing a cross section or collision frequency and a loss term which has the form of a bounded integral operator. The kernel of the integral operator will, at least initially, not be assumed nonnegative, and both dissipative and conservative boundary conditions will be allowed. Under these assumptions, we obtain a model equation which is applicable to a host of specific problems in radiative transfer and rarefied gas dynamics, as well as in neutron transport theory.

Let the spatial domain Ω be an open region in $\mathbb{R}^3$ with piecewise continuously differentiable boundary surface $\partial\Omega$. The region Ω need be neither bounded nor convex. Let the velocity domain V have the form $V = F \times S$, where the speed domain F is a Borel subset of $(0,\infty)$ endowed with the finite Borel measure ρ and S is the unit sphere in $\mathbb{R}^3$ endowed with the surface Lebesgue measure σ. The phase space will then be the set $\Lambda=\Omega\times V$ endowed with the measure $dxd\lambda$, where $d\lambda = d\rho d\sigma$. Special cases are the monoenergetic problem, where F is a singleton set, and the multigroup problem, where F is a finite set. The linear kinetic equation is then given by

$$\frac{\partial u}{\partial t}(x,v,t) + v \cdot \frac{\partial u}{\partial x}(x,v,t) + \nu(x,v)u(x,v,t) =$$

$$= \int_V k(x,\hat{v}\to v)u(x,\hat{v},t)d\lambda(\hat{v}) + q(x,v,t), \tag{1.1}$$

with initial condition

$$\lim_{t \downarrow 0} u(x,v,t) = u_0(x,v), \tag{1.2}$$

and a boundary condition to be specified later. Here $(x,v) \in \Lambda$. The function $\nu(x,v)$, which is the total cross section in neutron transport, the extinction coefficient in radiative transfer and the collision frequency in gas dynamics, will be assumed bounded and measurable. The inhomogeneous term $q(x,v,t)$ accounts for internal particle or radiation sources, and the kernel $k(x,\hat{v}\to v)$, which is called the scattering kernel in neutron transport, the phase function in radiative transfer, and the redistribution function in gas dynamics, is assumed measurable.

In order to write down the boundary conditions, we consider the characteristic equations

$$\frac{dx}{dt} = v, \qquad \frac{dv}{dt} = 0,$$

of the vector field $X=v\cdot\frac{\partial}{\partial x}$. Along its characteristics, the velocity will be constant, while $x(t)=x(0)+vt$. Thus the projections of the integral curves on the spatial domain Ω are straight lines parallel to v, on which the particle or radiation moves in the direction of v as time increases. The "length" of a trajectory will then be the maximal extent of the trajectory in the interior of Ω, divided by the speed $|v|$. Let us consider the left and right endpoints of the characteristics. Since the boundary of Ω is piecewise continuously differentiable, it can be divided into finitely many components on each one of which the outer unit normal $n(x)$ is continuous in $x \in \partial\Omega$. Since the surface is piecewise C^1, the trajectories beginning or ending at points where the outer unit normal is either not defined or discontinuous form a set in Λ of $dxd\lambda$-measure zero (see Section XI.3). Then the sets D_- and D_+ of left and right endpoints are given by

$$D_- = \{(x,v) : x \in \partial\Omega,\ n(x)\cdot v \le 0\},$$

$$D_+ = \{(x,v) : x \in \partial\Omega,\ n(x)\cdot v \ge 0\}.$$

The intersection of these two sets consists of the left and right endpoints of those trajectories that are tangent to the boundary. This situation can only occur if Ω is not a convex region. As a result of the fact that the surface is piecewise C^1, the trajectories beginning or ending at a point of $D_+\cap D_-$ form a set of $dxd\lambda$-measure zero (see Section XII.3).

Let us specify clearly what is meant by Xu. For this reason we introduce the test function spaces Φ and Φ_0. Here Φ is the linear space of those bounded Borel functions φ on Λ that are continuously differentiable with bounded directional derivative along each characteristic and are such that the lengths of the trajectories meeting the support of φ have a uniform positive lower bound. By Φ_0 we denote the set of those functions in Φ that vanish at the endpoints of the trajectories. For every $u\in L_p(\Lambda,dxd\lambda)$ with $1\le p<\infty$ we define Xu as the following distributional derivative:

$$\int_\Omega\int_V (Xu)(x,v)\varphi(x,v)dxd\lambda = -\int_\Omega\int_V u(x,v)(X\varphi)(x,v)dxd\lambda, \quad \varphi\in\Phi_0.$$

We then obtain the Green's identity

$$\int_\Omega\int_V (Xu)(x,v)dxd\lambda = \int_{D_+} u(r(x,v))d\nu_+ - \int_{D_-} u(\ell(x,v))d\nu_-, \tag{1.3}$$

where $\{u,Xu\}\subset L_p(\Lambda,dxd\lambda)$ for some $1\le p<\infty$ and $\ell(x,v)$ and $r(x,v)$ are the left and right endpoints of the integral curve passing through $(x,v)\in\Lambda$. The restrictions of u to $D_\pm$ belong to $L_{p,loc}(D_\pm,d\nu_\pm)$. Here by $L_{p,loc}(D_\pm,d\nu_\pm)$ we mean the vector space of all measurable functions on $D_\pm$ which are L_p-functions with respect to $d\nu_\pm$ on every bounded Borel subset $D\subset D_\pm$ such that the lengths of the trajectories meeting D have a uniform positive lower bound. It is easily seen that the boundary measures appearing in (1.3) are given by

$$d\nu_\pm(x,v) = |v\cdot n(x)|\,dxd\lambda(v).$$

We then define F_p as the linear space of those $u \in L_p(\Lambda, dxd\lambda)$ such that $Xu \in L_p(\Lambda, dxd\lambda)$ and the restrictions of u to $D_\pm$ belong to $L_p(D_\pm, d\nu_\pm)$.

We shall now formulate the boundary condition to Eq. (1.1). For this purpose we consider an arbitrary bounded linear operator K from $L_p(D_+, d\nu_+)$ into $L_p(D_-, d\nu_-)$ and state the boundary condition in the form

$$u(x,v,t) = (Ku)(x,v,t) + g_0(x,v), \quad (x,v) \in D_-, \tag{1.4}$$

where $g_0 \in L_p(D_-, d\nu_-)$. Well-known examples of boundary operators are those for vacuum boundary conditions and specular reflection. The former is given by K=0, while the latter has the form

$$(Ku)(x,v) = \alpha(x,v)u(x,v - 2(n(x)\cdot v)n(x)), \quad (x,v) \in D_-. \tag{1.5}$$

Here the accommodation coefficient $\alpha(x,v)$ is a measurable function on D_- such that

$$0 \leq \alpha(x,v) \leq 1, \quad (x,v) \in D_-.$$

The first rigorous, thorough and detailed analysis of the linear kinetic problem was performed by Lehner and Wing [246] for monoenergetic neutron transport with isotropic scattering in slab geometry and a Hilbert space setting. Following their method, different boundary conditions, geometries and collision models for neutron transport were subsequently tackled by a number of authors. We mention, in particular, Belleni-Morante [39], Borysiewicz and Mika [50], Marti [258], Angelescu et al. [8, 10], Pimbley [308], and Albertoni and Montagnini [2]. Important landmarks in subsequent developments were Vidav's consideration of the L_1-setting (see [375]), and Voigt's careful analysis of the vector field X for a variety of positive boundary operators [380]. Later developments in an L_1-setting were due to Montagnini [268], Suhadolc [345], Palczewski [298] and Hejtmanek [189].

We shall now apply the theory of Section XII.2 to get the semigroup property of the initial-boundary value problem (1.1)-(1.2)-(1.4). Let S_K and B_K be operators defined on the domain

$$D(S_K) = D(B_K) = \{u \in F_p : u_-(x,v) = (Ku_+)(x,v), \ (x,v) \in D_-\},$$

acting as follows:

$$(S_K u)(x,v) = -(Xu)(x,v) - \nu(x,v)u(x,v),$$

$$(B_K u)(x,v) = -(Xu)(x,v) - \nu(x,v)u(x,v) + (Ju)(x,v).$$

Here we observe that $\nu(x,v)$ is essentially bounded on Λ, while we impose such conditions on $k(x,\hat{v}\rightarrow v)$ that the operator

$$(Ju)(x,v) = \int_V k(x,\hat{v}\rightarrow v)u(x,\hat{v})d\lambda(\hat{v})$$

is bounded on $L_p(\Lambda,dxd\lambda)$. A sufficient condition in order that J is bounded on $L_p(\Lambda,dxd\lambda)$ is the condition

$$\operatorname*{ess\,sup}_{(x,\hat{v})\in\Lambda} \int_V |k(x,\hat{v}\rightarrow v)|\,d\lambda(v) < \infty$$

for $p=1$, and the condition

$$\int_\Omega\int_V \left[\int_V |k(x,\hat{v}\rightarrow v)|^p d\lambda(v)\right]^{1/(p-1)} dxd\lambda(\hat{v}) < \infty$$

for $1<p<\infty$. On applying Theorem XII 2.2 one observes that, for $\|K\| < 1$, S_K and B_K generate the strongly continuous semigroups $\{U_{0,K}(t)\}_{t\geq 0}$ and $\{U_K(t)\}_{t\geq 0}$, respectively. These semigroups are related by the Hille–Dyson–Phillips expansion

$$U_K(t) = U_{0,K}(t) + \sum_{n=1}^{\infty} \int ds_0 \,\cdots \int ds_n \; U_{0,K}(s_0)J\ldots JU_{0,K}(s_n), \quad \begin{matrix} s_0\geq 0,\ldots,s_n\geq 0 \\ s_0+\ldots+s_n=t \end{matrix}$$

which converges absolutely in the operator norm. The free streaming semigroup $\{U_{0,K}(t)\}_{t\geq 0}$ is contractive, and positive if K is positive. If the operator J is positive, that is, if $k(x,\hat{v}\rightarrow v)$ is nonnegative for almost every $(x,v,\hat{v}) \in \Omega\times V\times V$, and if K is positive, then the transport semigroup $\{U_K(t)\}_{t\geq 0}$ is positive. The transport semigroup is contractive if for every $u\in L_p(\Lambda,dxd\lambda)$ the estimate

$$\int_\Omega\int_V \operatorname{sgn}(u(x,v))\,|u(x,v)|^{p-1}\left[\nu(x,v)u(x,v) - \int_V k(x,\hat{v}\rightarrow v)u(x,\hat{v})d\lambda(\hat{v})\right]dxd\lambda(v) \geq 0$$

is satisfied.

Now let us consider the case of a conservative boundary, where $\|K\| = 1$. In general, it is not true that S_K and B_K generate strongly continuous semigroups, even if K is positive, as is exemplified by an example of Voigt [380]. On the other hand, if the spatial region Ω is plane parallel (a half space, a finite slab or the complete three dimensional space) and the speed domain F is bounded, the lengths of the trajectories are bounded below (by infinity for a half space or $\mathbb{R}^3$, and by w/v_{max}, where w is the width and v_{max} the supremum of F, for a finite slab). The lengths of the trajectories are also bounded away from zero, if the speed domain F is bounded and the spatial domain consists of a plane parallel region from which a finite number of disjoint closed and convex "cavities," with piecewise continuously differentiable boundary surfaces and at positive distance from the boundary of the plane parallel region and from each other, are removed. Thus, the trajectories have their lengths bounded away from zero on the exterior region of a sphere or finite or (semi)infinite cylinder if the speed domain F is bounded, while this is no longer true for the interior region of a sphere or cylinder. For all these cases, where the lengths of the integral curves are bounded away from zero, one may prove the above semigroup results, including the positivity and contractivity statements, provided K is positive. Here we apply the paragraph following the proof of Theorem XII 2.3, while observing that it does not matter for the semigroup properties whether or not J is positive.

More generally, it is difficult to state useful conditions which guarantee that S_K and B_K generate the strongly continuous free streaming and transport semigroups. Abstractly, it is clear that S_K and B_K are the generators of the free streaming and transport semigroups if and only if every $u \in F_p$ satisfying the boundary condition $u_- = Ku_+$ has its restrictions to $D_\pm$ in $L_p(D_\pm, d\nu_\pm)$, but there is no general method for determining when such a condition is true. However, if K is positive and has unit norm, we can use the monotonicity argument of Theorem XII 2.3 to prove that suitable closed extensions of S_K and B_K generate the strongly continuous semigroups $\{U_{0,K}(t)\}_{t\geq 0}$ and $\{U_K(t)\}_{t\geq 0}$, which arise as the monotone limits of the semigroups $\{U_{0,\alpha K}(t)\}_{t\geq 0}$ and $\{U_{\alpha K}(t)\}_{t\geq 0}$ as $\alpha \uparrow 1$.

In order to obtain more detailed information about the neutron transport example, let us consider the case of plane parallel symmetry where the spatial domain is either the finite interval $(0,2a)$ or the half line $(0,\infty)$. The model equation then becomes

$$\frac{\partial u}{\partial t}(x,\mu,t) + \mu\frac{\partial u}{\partial x}(x,\mu,t) + u(x,\mu,t) = \tfrac{1}{2}c\int_{-1}^{1} u(x,\hat{\mu},t)\,d\hat{\mu} + q(x,\mu,t), \tag{1.6}$$

where $x\in(0,2a)$ for a parameter $0<a\leq\infty$ and $\mu\in[-1,1]$. This equation describes the transport of neutrons in a plane parallel domain with a width of 2a mean free paths, under the conditions of constant speed and isotropy of the single scattering processes. The direction cosine of propagation of the neutrons is $\mu\in[-1,1]$. The solution $u(x,\mu,t)$ and the term $q(x,\mu,t)$ represent the neutron angular density and the internal neutron sources, as a function of position in units of mean free path, direction and time. The constant c is the average number of secondaries per collision. The spatial medium is plane parallel of width 2a, but because of the invariance of the physical processes under translation parallel to the surface, we may reduce the problem to an equation with one dimensional spatial geometry.

We impose the initial condition

$$\lim_{t\downarrow 0} u(x,\mu,t) = u_0(x,\mu), \tag{1.7}$$

as well as boundary conditions coupling the incident and outgoing neutron "fluxes". As to the latter, we have to distinguish between half space geometry where $a=\infty$, and finite slab geometry where a is finite. For half space domains we require

$$u(0,\mu,t) = (Ku)(0,\mu,t), \quad \mu>0. \tag{1.8}$$

For finite slab media we impose the boundary conditions

$$u(0,\mu,t) = (K_{++}u)(0,\mu,t) + (K_{+-}u)(2a,\mu,t), \quad \mu>0, \tag{1.9a}$$

$$u(2a,\mu,t) = (K_{--}u)(2a,\mu,t) + (K_{-+}u)(0,\mu,t), \quad \mu<0. \tag{1.9b}$$

Here the operators K and

$$\mathbf{K} = \begin{bmatrix} K_{++} & K_{+-} \\ K_{-+} & K_{--} \end{bmatrix}$$

are linear and do not depend on time. They also satisfy a contractivity condition, which will be specified below. In most applications the reflection processes at the two surfaces $x=0$ and $x=2a$ are decoupled so that K_{+-} and K_{-+} vanish. Typical examples are combinations of specular and diffuse reflection, for instance,

$$(K_{++}u)(0,\mu,t) = \alpha_+ u(0,-\mu,t) + \beta_+ \int_0^1 (\hat{\mu}/\mu) u(0,-\hat{\mu},t)\,d\hat{\mu}, \quad \mu>0,$$

$$(K_{--}u)(2a,\mu,t) = \alpha_- u(2a,-\mu,t) + \beta_- \int_{-1}^0 |\hat{\mu}/\mu|\, u(2a,-\hat{\mu},t)\,d\hat{\mu}, \quad \mu<0.$$

Here $\alpha_\pm$ and $\beta_\pm$ are the accommodation coefficients for specular and diffuse reflection, respectively, $0 \le \alpha_\pm \le 1$ and $0 \le \beta_\pm \le 1-\alpha_\pm$. For the half space problem the analogous boundary condition is

$$(Ku)(0,\mu,t) = \alpha u(0,-\mu,t) + \beta \int_0^1 (\hat{\mu}/\mu) u(0,-\hat{\mu},t)\,d\hat{\mu}, \quad \mu>0,$$

where $0 \le \alpha \le 1$ and $0 \le \beta \le 1-\alpha$. It should be observed that there exist physically interesting boundary conditions (1.9) that do not decouple the surfaces $x=0$ and $x=2a$, for example the periodic boundary condition

$$u(2a,\mu,t) = u(0,\mu,t),$$

where K_{++} and K_{--} vanish and K_{+-} and K_{-+} reduce to identity operators.

Let us incorporate the above model in the theory developed in the previous chapters. First of all, the phase space is $\Lambda=(0,2a)\times(-1,1)$ endowed with the Lebesgue measure $dx\,d\mu$. The vector field X is given by

$$(Xu)(x,\mu) = \mu\frac{\partial u}{\partial x}(x,\mu).$$

On using time as a parameter, the characteristic equations of X are given by

$$\frac{dx}{dt} = \mu, \qquad \frac{d\mu}{dt} = 0,$$

whence μ is constant on trajectories and $x=x_0+\mu t$. The integral curves are then given by the lines $\{(x,\mu) : 0<x<2a\}$, where x increases with t for $\mu>0$ and decreases with t for $\mu<0$, as well as the points $\{(x,0) : 0<x<2a\}$. Hence, the sets $D_\pm$ of left and right endpoints are given by

$$D_- = [\{0\}\times(0,1)] \cup [\{2a\}\times(-1,0)],$$

$$D_+ = [\{0\}\times(-1,0)] \cup [\{2a\}\times(0,1)],$$

for the finite slab, and

$$D_- = \{0\} \times (0,1),$$

$$D_+ = \{0\} \times (-1,0),$$

for the half space. The "length" of a trajectory, which is measured with the parameter t, represents the total travel time along the integral curve. It is given by

$$\ell(a,\mu) = \begin{cases} 2a|\mu|^{-1} & \text{for the finite slab,} \\ \infty & \text{for the half space,} \end{cases}$$

whence $\ell(a,\mu) \geq 2a$ for all integral curves of X. In considering the vector field $Y=(\partial/\partial t)+X$ with phase space $\Sigma=\Lambda\times(0,T)$ for some fixed $T>0$, we find for the lengths of the trajectories

$$\ell(a,\mu,T) = \begin{cases} \min\{2a|\mu|^{-1},T\} & \text{for the finite slab,} \\ T & \text{for the half space,} \end{cases}$$

as a result of which $\ell(a,\mu,T) \geq \min\{2a,T\}$ for all integral curves of Y.

Next, we consider the proper definition of the streaming operator

$$S = -\mu\frac{\partial}{\partial x} - 1 \tag{1.10}$$

and the transport operator

$$B = S + J, \tag{1.11}$$

where

$$(Ju)(x,\mu) = \tfrac{1}{2}c\int_{-1}^{1} u(x,\hat{\mu})d\hat{\mu}.$$

We analyze these operators on the (real or complex) space $L_p(\Lambda,dxd\mu)$ with $1\leq p<\infty$. The major issue is the correct definition of the derivative appearing in (1.10). As in Chapter XI, we define the test function space Φ as the linear manifold of all bounded Borel measurable functions on Λ which are continuously differentiable with respect to x with bounded partial derivative, for almost all $\mu\epsilon(-1,1)$. In the definition of Φ we do not include a condition curtailing the minimal length of the trajectories meeting the support, since these lengths have a uniform positive lower bound. By Φ_0 we denote the linear subspace of those $\varphi\epsilon\Phi$ that vanish for x=0 and, in the finite slab case, also for x=2a. For every $u\epsilon L_p(\Lambda,dxd\mu)$ we then define

$$\int_{-1}^{1}\int_{0}^{2a}(Su)\varphi dxd\mu = -\int_{-1}^{1}\int_{0}^{2a}u\varphi dxd\mu + \int_{-1}^{1}\int_{0}^{2a}u\cdot\mu\frac{\partial\varphi}{\partial x}dxd\mu$$

for arbitrary $\varphi\epsilon\Phi_0$.

The integral curves of X have a length satisfying $\ell(a,\mu)\geq\delta>0$. Thus, every $u\epsilon L_p(\Lambda,dxd\mu)$ with $\mu(\partial u/\partial x)\epsilon L_p(\Lambda,dxd\mu)$ has restrictions $u(0,\cdot)$ and, for the finite slab case, $u(2a,\cdot)$ belonging to $L_p([-1,1],|\mu|d\mu)$ such that the Green's identities

$$\int_{-1}^{1}\int_{0}^{2a}\mu\frac{\partial u}{\partial x}dxd\mu = \int_{-1}^{1}\mu\{u(2a,\mu)-u(0,\mu)\}d\mu$$

for the finite slab, and

$$\int_{-1}^{1}\int_{0}^{\infty}\mu\frac{\partial u}{\partial x}dxd\mu = -\int_{-1}^{1}\mu u(0,\mu)d\mu$$

for the half space, are valid. For the precise connection with the abstract theory we refer to Proposition XI 3.1 and the paragraph containing Eq. XI 3.11, in combination with Eq. XII 2.1. In any case, for a pair of Borel sets $E\subset(0,1)$ and $F\subset(-1,0)$ we have

$$\nu_-([\{0\}\times E]\cup[\{2a\}\times F]) = \int_{E\cup F}|\mu|d\mu,$$

$$\nu_+([\{0\}\times F]\cup[\{2a\}\times E]) = \int_{E\cup F}|\mu|d\mu,$$

for the finite slab, and

$$\nu_-(\{0\}\times E) = \int_E|\mu|d\mu,$$

$$\nu_+(\{0\}\times F) = \int_F|\mu|d\mu,$$

for the half space.

We now consider K to be a positive (in lattice sense) contraction

$$K\colon L_p([\{0\}\times(0,1)]\cup[\{2a\}\times(-1,0)], |\mu|\,d\mu) \to$$

$$\to L_p([\{0\}\times(-1,0)]\cup[\{2a\}\times(0,1)], |\mu|\,d\mu)$$

for the finite slab, and

$$K\colon L_p(\{0\}\times(0,1), |\mu|\,d\mu) \to L_p(\{0\}\times(-1,0), |\mu|\,d\mu)$$

for the half space. We define S_K and B_K as the linear operators on the common domain

$$D(S_K) = D(B_K) = \{u\epsilon F_p : u_- = Ku_+\}$$

and acting as in (1.10) and (1.11), where the "traces" u_- and u_+ are the restrictions of $u\epsilon F_p$ to D_- and D_+, respectively.

It is easily seen that

$$\int_{-1}^{1}\int_{0}^{2a} \{u(x,\mu)-\tfrac{1}{2}c\int_{-1}^{1} u(x,\hat{\mu})d\hat{\mu}\}dx\,d\mu = (1-c)\|u\|_1$$

for all nonnegative $u\epsilon L_1(\Lambda,dxd\mu)$. Similarly, one obtains

$$\int_{-1}^{1}\int_{0}^{2a} (\mathrm{sgn}\,u)\,|u|^{p-1}\{u(x,\mu)-\tfrac{1}{2}c\int_{-1}^{1} u(x,\hat{\mu})d\hat{\mu}\}dx\,d\mu \ge (1-c)^p\|u\|_p^p$$

for arbitrary real $u\epsilon L_p(\Lambda,dxd\mu)$. We also observe that J is a bounded positive operator. As a result of Theorems XII 2.2 and 2.3 (including the paragraph following the latter's proof) we arrive at the following results:

(i) For every nonnegative K with $\|K\| \le 1$, the operator S_K generates the strongly continuous semigroup $\{U_{0,K}(t)\}_{t\ge0}$ on $L_p(\Lambda,dxd\mu)$. This semigroup is positive and its type $\omega_0(U_{0,K})\le-1$.

(ii) For every nonnegative K with $\|K\| \le 1$, the operator B_K generates the strongly continuous semigroup $\{U_K(t)\}_{t\ge0}$ on $L_p(\Lambda,dxd\mu)$. This semigroup is positive and its type $\omega_0(U_K)\le-(1-c)$.

(iii) The semigroups $\{U_{0,K}(t)\}_{t\ge0}$ and $\{U_K(t)\}_{t\ge0}$ are related via the

Hille–Dyson–Phillips expansion, which is absolutely norm convergent.

These results are valid for $1 \le p < \infty$.

In general, it is not easy to obtain closed form representations for the semigroup $\{U_{0,K}(t)\}_{t \ge 0}$, let alone for $\{U_K(t)\}_{t \ge 0}$. For vacuum boundary conditions ($K=0$) it is straightforward to obtain

$$(U_{0,K=0}(t)u)(x,\mu) = \begin{cases} e^{-t} u(x-\mu t,\mu), & x-\mu t \in (0,2a), \\ 0, & x-\mu t \notin (0,2a), \end{cases}$$

whence

$$\omega_0(U_{0,K=0}) = s(S_{K=0}) = -1.$$

The spectral bound equality follows by applying

$$((\lambda - S_{0,K=0})^{-1}u)(x,\mu) = \int_0^\infty e^{-(\lambda+1)t} u(x-\mu t,\mu)\chi_{[0,2a]}(x-\mu t)\, dt.$$

On defining $t_0 = t_0(x,\mu) = (x/\mu)$ for $\mu > 0$ and $t_0 = t_0(x,\mu) = (x-2a)/\mu$ for $\mu < 0$, we obtain

$$((\lambda - S_{0,K=0})^{-1}u)(x,\mu) = \int_0^{t_0} e^{-(\lambda+1)t} u(x-\mu t,\mu) dt.$$

For Re $\lambda > -1$ and $u \equiv 1$ the right hand side has the form

$$\frac{1}{\lambda+1}\{1 - \exp[-(\lambda+1)t_0(x,\mu)]\},$$

and for $\lambda = -1$ and $u \equiv 1$ the form $t_0(x,\mu)$, which does not belong to $L_p(\Lambda, dx d\mu)$. Thus $s(S_{K=0}) = -1$.

Finally, since for every $K \ge 0$ with $\|K\| \le 1$

$$0 \le U_{0,K=0}(t) \le U_{0,K}(t),$$

while

$$-1 = s(S_{K=0}) \le s(S_K) \le -1$$

and

$$-1 = \omega_0(U_{0,K=0}) = \omega_0(U_{0,K}) \le -1,$$

we obtain

$$s(S_K) = \omega_0(U_{0,K}) = -1.$$

In particular, the spectral bound and type of the free streaming semigroup $\{U_{0,K}\}_{t\ge 0}$ coincide.

2. Neutron transport (continued): Spectral decomposition and hydrodynamics

The transport operator appearing in a typical time dependent kinetic equation is not a normal operator even in the most simplified situations, such as (1.6). Therefore, its spectral decomposition, in a generalized sense, does not automatically follow from the standard results of the general theory. Historically, spectral decomposability of transport operators was seriously questioned after Lehner and Wing's analysis [245, 246]. Indeed, it turned out that the perfectly absorbing boundary conditions they treated, although the most natural from a physical point of view, were the most difficult for the derivation of spectral decomposition results. Even a modified approach suggested by Lehner [244] has not furnished a complete answer in this respect. Yet, certain special boundary conditions, such as periodic or purely specularly reflecting, allow for a generalized spectral decomposition of certain transport operators, such as the example (1.6). In this section we shall present a method, essentially due to Friedrichs and closely related to the scattering theory of the Boltzmann equation [188, 336, 355], which makes it possible to show that certain transport operators are spectral. This section is an extension by Protopopescu of earlier work [9, 12].

An operator L is called a **spectral operator of scalar type** (cf. [105], [109] vol. III) if there exists a norm bounded, strongly countable additive function E, defined on the Borel sets $\mathcal{B}$ of the complex plane, whose values are bounded linear operators on H, such that

$$E(\Delta_1)E(\Delta_2) = E(\Delta_1 \cap \Delta_2), \quad \Delta_1, \Delta_2 \in \mathcal{B}, \tag{2.1}$$

$$E(\mathbb{C}) = 1_H, \quad E(\emptyset) = 0_H, \tag{2.2}$$

$$E(\Delta_1 \cup \Delta_2) = E(\Delta_1) + E(\Delta_2) - E(\Delta_1 \cap \Delta_2), \tag{2.3}$$

and such that the operator L can be diagonalized in the form

$$L = \int_{\mathbb{C}} \lambda \, dE(\lambda).$$

Friedrichs' idea consists of applying a time independent scattering formulation, i.e., constructing non-unitary "wave operators" which realize the similarity between the transport operator under study and a normal operator. We shall consider the monoenergetic transport equation in slab geometry with periodic boundary conditions. The crucial point is that, due to the particular form of the boundary conditions, one may exploit the diagonalization of the transport operator by the Fourier transform and explicitly construct the similarity (wave) operators [9].

Thus, let us consider the transport operator in slab geometry $x \in (-a,a)$,

$$(Bu)(x,\mu) = -\mu\frac{\partial u}{\partial x}(x,\mu) - u(x,\mu) + \tfrac{1}{2}\int u(x,\hat{\mu})d\hat{\mu}, \quad \mu \in [-1,1], \tag{2.4}$$

on the Hilbert space $H=L_2([-a,a]\times[-1,1])$ with domain as indicated in the previous section and periodic boundary conditions $u(-a,\mu) = u(a,\mu)$. For convenience we have taken $x \in (-a,a)$ rather than $x \in (0,2a)$. Here the average secondaries per collision is taken to be $c=1$, with no loss of generality, as will be apparent. Put

$$(Nu)(\mu) = \tfrac{1}{2}\int_{-1}^{1} u(\hat{\mu})d\hat{\mu}.$$

Then the Fourier transform F in the x variable diagonalizes B in the form

$$B = F\left[\bigoplus_{k \in \mathbb{Z}} B_k\right]F^{-1},$$

where each Fourier component of the transport operator,

$$B_k = -ik\frac{\pi}{a}\mu - 1 + N, \quad k \in \mathbb{Z},$$

acts in the space $\mathcal{L} = L_2[-1,1]$. The original Hilbert space H is identified with the orthogonal direct sum

$$H = \bigoplus_{k \in \mathbb{Z}} L_k,$$

where L_k is identified with $\mathcal{L}$ in a natural way. For convenience, we shall actually decompose the operator

$$A_k = \mu + \frac{i\,a}{k\pi} N,$$

obtained from B_k by deleting the trivial part -1 and dividing by $-ik\frac{\pi}{a}$, $k \neq 0$. (For $k=0$, B_0 is a self adjoint operator whose spectral measure is readily available.) The operator A_k is a bounded non-normal operator obtained by perturbing the self adjoint multiplicative operator μ, which has simple absolutely continuous spectrum on $[-1,1]$, with the skew adjoint operator $\frac{i\,a}{k\pi}N$ which is, up to the constant $ia/k\pi$, a one dimensional projection.

Perturbations of spectral operators have been extensively studied in the literature (e.g., [153, 214, 266, 348]), but the methods are limited by the difficulty to actually verify, in concrete situations, the abstract conditions imposed in order to ensure decomposability. Such conditions typically involve smallness and regularity requirements on the perturbation, and continuity of the integral operator associated with the related eigenvalue problem. In the case at hand, depending on a and k, $\|\frac{i\,a}{k\pi}N\|$ may become very large, and the kernel of the integral operator, identically one for $(\mu,\hat{\mu}) \in [-1,1]\times[-1,1]$ and zero outside, is discontinuous at the boundary.

Direct construction of the similarity operator will obviate these difficulties. A generalized eigenfunction expansion for the operator $A_k:\mathcal{L}\to\mathcal{L}$ will be proved if one can find a Hilbert space $\tilde{\mathcal{L}}$, a normal operator $\tilde{A}_k$: $\tilde{\mathcal{L}}\to\tilde{\mathcal{L}}$ having the same spectrum as A_k, and two bounded operators (wave, similarity or intertwining operators) Ω_k^+: $\tilde{\mathcal{L}}\to\mathcal{L}$ and Ω_k^-: $\mathcal{L}\to\tilde{\mathcal{L}}$, such that

$$\text{(a)} \quad \Omega_k^+\Omega_k^- = 1_{\mathcal{L}}, \tag{2.5}$$

$$\text{(b)} \quad \Omega_k^-\Omega_k^+ = 1_{\tilde{\mathcal{L}}}, \tag{2.6}$$

$$\text{(c)} \quad A_k\Omega_k^+ = \Omega_k^+\tilde{A}_k, \qquad \Omega_k^-A_k = \tilde{A}_k\Omega_k^-. \tag{2.7}$$

Indeed, if this is the case and if $\sup_k \|\Omega_k^{\pm}\| < \infty$, let us denote by $\tilde{E}_k(\cdot)$, $k \neq 0$, the spectral measure of $\tilde{A}_k$. Then the spectral measure of B is given by

$$E(\Delta) = E_0(\Delta) \bigoplus_{\substack{k \in \mathbb{Z} \\ k \neq 0}} \Omega_k^+ \tilde{E}_k\left(\frac{i\,a}{k\pi}(\Delta+1)\right)\Omega_k^-, \qquad \Delta \in \mathcal{B}, \tag{2.8}$$

where $E_0(\cdot)$ is the spectral measure of the self adjoint operator B_0.

By explicit computation, we arrive at two cases:

(i) $|k| > \frac{a}{2}$, when A_k has no eigenvalues;

(ii) $0 < |k| < \frac{a}{2}$, when A_k has the nondegenerate eigenvalue $\lambda_k = i \cot an \frac{k\pi}{a}$.

In case (i) we take $\tilde{\mathcal{L}} \equiv \mathcal{L}$ and $\tilde{A}_k \equiv A^\circ$, where A° is the unperturbed operator of multiplication by μ, and define $\Omega_k^{\pm} = U_k^{\pm}: \mathcal{L} \to \mathcal{L}$. Conditions (a), (b), (c) then take the form

$$(a_1) \quad U_k^+ U_k^- = 1_{\mathcal{L}}, \tag{2.9}$$

$$(b_1) \quad U_k^- U_k^+ = 1_{\mathcal{L}}, \tag{2.10}$$

$$(c_1) \quad A_k U_k^+ = U_k^+ A^\circ, \qquad U_k^- A_k = A^\circ U_k^-. \tag{2.11}$$

In case (ii) we take $\tilde{\mathcal{L}} = \mathcal{L} \oplus \mathbb{C}$, with vectors written as $\begin{bmatrix} u \\ \eta \end{bmatrix}$ for $u \in \mathcal{L}$ and $\eta \in \mathbb{C}$, set

$$\tilde{B}_k = \begin{bmatrix} A^\circ & 0 \\ 0 & \lambda_k \end{bmatrix},$$

and define $\Omega_k^+: \tilde{\mathcal{L}} \to \mathcal{L}$ and $\Omega_k^-: \mathcal{L} \to \tilde{\mathcal{L}}$ by

$$\Omega_k^+ \begin{bmatrix} u \\ \eta \end{bmatrix} = (U_k^+ \quad \varphi_k^+) \begin{bmatrix} u \\ \eta \end{bmatrix} = U_k^+ u + \eta \varphi_k^+, \quad u \in \mathcal{L}, \quad \eta \in \mathbb{C},$$

$$\Omega_k^- u = \begin{bmatrix} U_k^- \\ (\varphi_k^-, \cdot) \end{bmatrix} u = \begin{bmatrix} U_k^- u \\ (\varphi_k^-, u) \end{bmatrix}.$$

Here $U_k^{\pm}$ are bounded operators in $\mathcal{L}$ and $\varphi_k^{\pm} \in \mathcal{L}$. The special form of U_k^- is dictated by its continuity and by the Riesz representation theorem. Applying relation (c), one obtains the four equations

$$A_kU_k^+ = U_k^+A^\circ, \qquad U_k^-A_k = A^\circ U_k^-, \tag{2.12}$$

$$A_k\varphi_k^+ = \lambda_k\varphi_k^+, \qquad A_k^*\varphi_k^- = \bar{\lambda}_k\varphi_k^-. \tag{2.13}$$

Because of (2.13), $\varphi_k^+(\mu) = \overline{\varphi_k^-(\mu)} = c_k(\lambda_k-\mu)^{-1}$, i.e., it is the nonnormalized eigenvector corresponding to the eigenvalue λ_k. Applying condition (b) then yields

$$\begin{bmatrix} U_k^-U_k^+ & U_k^-\varphi_k^+ \\ (U_k^{+*}\varphi_k^-,\cdot) & (\varphi_k^-,\varphi_k^+) \end{bmatrix} = \begin{bmatrix} 1_{\mathcal{L}} & 0 \\ 0 & 1 \end{bmatrix},$$

and therefore we have

(a_2) $U_k^+U_k^- = 1_{\mathcal{L}} - \varphi_k^+(\varphi_k^-,\cdot)$,

(b_2) $U_k^-U_k^+ = 1_{\mathcal{L}}$,

(c_2) $A_kU_k^+ = U_k^+A^\circ, \quad U_k^-A_k = A^\circ U_k^-$.

The vector φ_k is normalized via the relation $(\varphi_k^-,\varphi_k^+)=1$:

$$\varphi_k(\mu) = i\{(\lambda_k-\mu)\sqrt{2}\,\sin\frac{k\pi}{a}\}^{-1}.$$

Observing that the conditions (c_1) and (c_2) are identical, we shall use them to determine $U_k^\pm$ simultaneously for both cases. We then have to verify (a_1), (b_1) and (a_2), (b_2) for the cases (i) and (ii), respectively.

Writing $A_k=A^\circ+\frac{i\,a}{k\pi}N$, condition (c) can be reformulated as

$$[A^\circ, U_k^+] = -i\frac{a}{k\pi}NU_k^+ = R_k^+, \tag{2.14}$$

$$[A^\circ, U_k^-] = i\frac{a}{k\pi}U_k^-N = R_k^-, \tag{2.15}$$

where by square brackets one denotes the commutator $[F, G] = FG-GF$. If $U_k^\pm$ are bounded operators satisfying (2.14)–(2.15), then the Riesz representation theorem yields that $R_k^\pm$ are degenerate integral operators of rank one, namely

$$(R_k^+u)(\mu) = \int_{-1}^{1} r_k^+(\hat{\mu})u(\hat{\mu})d\hat{\mu}, \quad r_k^+ \in \mathcal{L},$$

$$(R_k^-u)(\mu) = r_k^-(\mu) \int_{-1}^{1} u(\hat{\mu})d\hat{\mu}, \quad r_k^- \in \mathcal{L}.$$

Supposing for the moment that $R_k^\pm$ were known, one then has to find a bounded operator Z satisfying

$$[A^\circ, Z] = R, \tag{2.16}$$

where R is a suitable rank one integral operator. If (2.16) has a solution, it certainly has infinitely many, differing from each other by bounded operators which commute with A°. We shall show that it has a bounded solution whenever R has a kernel of the form $r_1(\mu)r_2(\hat{\mu})$, r_1, $r_2 \in L_\infty[-1,1]$, by effectively constructing one solution.

To this end, let us define for $\varepsilon > 0$ the bounded integral operator G_ε on L with continuous kernel

$$G_\varepsilon(\mu,\hat{\mu}) = (\mu-\hat{\mu}+i\varepsilon)^{-1}.$$

As $\varepsilon \downarrow 0$, the operators $G_{\pm\varepsilon}$ have strong limits, denoted by $G_\pm$, which satisfy

$$G_+ - G_- = -2\pi i 1_{\mathcal{L}}. \tag{2.17}$$

The formal kernels of $G_\pm$ will be denoted $[\mu-\hat{\mu}]_\pm^{-1}$, respectively.

LEMMA 2.1. For $k \in \mathbb{Z}$, $|k| \neq \frac{a}{2}$, $k \neq 0$, the intertwining operators $U_k^\pm$ are given by

$$U_k^+ = 1_{\mathcal{L}} + G_+R_k^+, \tag{2.18}$$

$$U_k^- = 1_{\mathcal{L}} + R_k^-G_+. \tag{2.19}$$

Their (formal) kernels are

$$U_k^+(\mu,\hat{\mu}) = \delta(\mu-\hat{\mu}) - [\mu-\hat{\mu}]_+^{-1} \frac{ia}{2k\pi} [1 + \frac{a}{2k} + \frac{ia}{2\pi k}\log\{(1-\hat{\mu})/(1+\hat{\mu})\}]^{-1},$$

$$U_k^-(\mu,\hat{\mu}) = \delta(\mu-\hat{\mu}) + \frac{ia}{2k\pi}\left[1 - \frac{a}{2k} + \frac{ia}{2\pi k}\log\{(1-\mu)/(1+\mu)\}\right]^{-1}[\mu-\hat{\mu}]_+^{-1},$$

and their norms are bounded by

$$\|U_k^\pm\| \le 1 + \frac{a}{2|k|}\left|1 - \frac{a}{2|k|}\right|^{-1}. \tag{2.20}$$

Proof: The proof is a verification of (2.14) and (2.15). The norm boundedness results from (2.17) and from the concrete form of the kernels of $R_k^\pm$. ■

PROPOSITION 2.2. For $|k|>\frac{a}{2}$, A_k is similar to the self adjoint operator A°. The similarity is realized by $U_k^\pm$ given by (2.18)–(2.19). For $|k|<\frac{a}{2}$, A_k is similar to the normal operator $\tilde{A}_k=A^\circ\oplus\lambda_k$ acting on $\tilde{\mathcal{L}}=\mathcal{L}\oplus\mathbb{C}$. The similarity is realized by the operators $\Omega_k^\pm$.

Proof: We sketch the proof in case (ii), which is more complicated. Condition (b_2) reads

$$1_{\mathcal{L}} = U_k^-U_k^+ = 1_{\mathcal{L}} + G_+R_k^+ + R_k^-G_+ + R_k^-G_+G_+R_k^+. \tag{2.21}$$

In calculating the contribution of $R_k^-G_+G_+R_k^+$ we take into account the identity

$$(\mu-\mu'+i\varepsilon)^{-1}(\mu'-\hat{\mu}+i\varepsilon)^{-1} = (\mu-\hat{\mu}+2i\varepsilon)^{-1}[(\mu-\mu'+i\varepsilon)^{-1}+(\mu'-\hat{\mu}+i\varepsilon)^{-1}],$$

which leads to the cancellation of the last three terms in (2.21), proving (b_2). A similar argument proves (a_2). Indeed, when computing the kernel of $G_+R_k^+R_k^-G_+$, one may use the identity

$$\frac{a}{ik}\left[1 + \frac{a}{2k} + \frac{ia}{2k\pi}\log\frac{1-\mu}{1+\mu}\right]^{-1} \cdot \frac{ia}{2k\pi}\left[1 - \frac{a}{2k} + \frac{ia}{2k\pi}\log\frac{1-\mu}{1+\mu}\right]^{-1} =$$

$$= -i\left\{-\frac{ia}{2k\pi}\left[1 + \frac{a}{2k} + \frac{ia}{2k\pi}\log\frac{1-\mu}{2+\mu}\right]^{-1} + \frac{ia}{2\pi k}\left[1 - \frac{a}{2k} + \frac{ia}{2k\pi}\log\frac{1-\mu}{1+\mu}\right]^{-1}\right\}.$$

The contribution of $G_+R_k^+R_k^-G_+$ then leads to the integral

$$(-a/4\pi^2k)\int_C(\mu-\lambda)^{-1}\left(1 + \frac{ia}{2k\pi}\int_{-1}^1\frac{d\mu'}{\mu'-\lambda}\right)^{-1}(\lambda-\hat{\mu})^{-1}d\lambda,$$

where the positively oriented simple Jordan contour C encloses the segment $[-1,1]$, but not the pole λ_k. Applying Cauchy's theorem, this is equal to the same integral on a sufficiently small negatively oriented circle around λ_k, which yields the corresponding projector $\varphi_k^+(\varphi_k^-, \cdot)$ in condition (a_2). ■

We now have the spectral decomposition result, except for the critical values a=2,4,... .

THEOREM 2.3. For $a \neq 2,4,...$ the Boltzmann operator (2.4) with periodic boundary conditions is spectral of scalar type.

Proof: For any Borel set $\Delta \subset \mathbb{C}$ and any $k \in \mathbb{Z}$, $k \neq 0$, we construct

$$E_k(\Delta) = \begin{cases} \Omega_k^+ \hat{\chi}_{[-1,1]\cap\Delta} \Omega_k^-, & |k| > \frac{a}{2}, \\ \Omega_k^+ \hat{\chi}_{[-1,1]\cap\Delta} \Omega_k^- + \chi_\Delta(\lambda_k) \varphi_k^+(\varphi_k^-, \cdot), & |k| < \frac{a}{2}, \end{cases} \tag{2.22}$$

where χ_Δ is the characteristic function of the set Δ and $\hat{\chi}_\Delta$ is the operator corresponding to multiplication by χ_Δ. The non-orthogonal bounded projection valued measure $E_k(\cdot)$ is concentrated on $[-1,1]$ if $|k| > \frac{a}{2}$ and on $[-1,1] \cup \{\lambda_k\}$ if $|k| < \frac{a}{2}$. Using (2.20) one shows that

$$\|E_k(\Delta)\| \leq c[(1 - \frac{2|k|}{a})^2 + d^2]^{-\frac{1}{2}} + \chi_\Delta(\lambda_k)(\frac{2k\pi}{a} \operatorname{cosec} \frac{2k\pi}{a})^{\frac{1}{2}},$$

where c is some constant and $d = \inf\{|z| : z \in \Delta\}$. Therefore, via the similarity, the Spectral Theorem for $\tilde{A}_k$ defines

$$A_k = \int_{\mathbb{C}} \lambda \, dE_k(\lambda)$$

as a norm convergent integral with respect to a countably additive spectral measure. The result for the operator B follows by a trivial rephrasing. ■

Operators of the type considered above restricted to the class of Hölder continuous functions have been studied using the methods of singular integral equations theory [70]. The Banach space X of continuously differentiable functions on $[-1,1]$ with norm $\|f\| = |f(-1)| + \max_\mu |f'(\mu)|$ is a continuously embedded dense subspace of

$\mathcal{L}$, and $A_k|_X$ is a bounded operator on X. For fixed $\hat{\mu}\in[-1,1]$ the formal kernel $U_k^+(\mu,\hat{\mu})$ defines an element of the dual space of X by

$$\varphi \to (U_k^{+*}\varphi)(\hat{\mu}) = (u_k^+(\hat{\mu}),\varphi).$$

Indeed, for $\varphi\in X$ the equivalence class of $U_k^{+*}\varphi\in\mathcal{L}$ has a continuous element, as one may easily see, and $\|(U_k^{+*}\varphi)(\hat{\mu})\|\leq\text{const.}\|\varphi\|$. For fixed $\hat{\mu}$, the formal kernel $U_k^{+*}(\mu,\hat{\mu})$ is a generalized eigenfuction of A_k with eigenvalue $\hat{\mu}$, in the sense that $(u_k^{+*}(\hat{\mu}),A_k^{*}\varphi)=\hat{\mu}(u_k^{+*}(\hat{\mu}),\varphi)$. But this is just the expression $(U_k^{+*}A_k^{*}\varphi)(\hat{\mu}) = (A^{\circ}U_k^{+*}\varphi)(\hat{\mu})$, i.e., the intertwining property of U_k^{+*}. With this interpretation, the relations $\Omega_k^+\Omega_k^- = 1_{\mathcal{L}}$ and $\Omega_k^-\Omega_k^+ = 1_{\tilde{\mathcal{L}}}$ can be viewed as "completeness" and "orthogonality" relations, respectively.

The constructive method based on Friedrichs' wave operators failed to give an answer to the similarity question in the critical cases a=2,4,..., since the L_∞-norm of the integral kernels $r^{\pm}(\mu,\hat{\mu})$ becomes infinite for these particular values of the slab widths. One may ask whether a more clever construction or a completely different method may show similarity in these cases. The question may be settled in the negative by studying the "characteristic function" of the operator A_k (cf. [348]). However, a weaker form of spectral decomposition can be formulated as an easy consequence of Lemma 2.1.

PROPOSITION 2.4. Suppose $a=2|k|$, $k\in\mathbb{Z}$, $k\neq0$. Then for every Borel set $\Delta\subset\mathbb{C}$ at nonzero distance from zero, the operators $E_{\pm a/2}(\Delta)$ given by (2.22) are bounded projections in $\mathcal{L}$ connecting with A_k, and the restriction of A_k to $E_k(\Delta)\mathcal{L}$ is similar to the restriction of A° to $\hat{\chi}_{[-1,1]\cap\Delta}\mathcal{L}$.

The critical behaviour observed for the set of exceptional values a=2,4,..., has many formal similarities with critical phenomena in statistical mechanics (cf. [314]).

With appropriate definitions of the operator B and the Hilbert space H, the infinite medium can be treated analogously, with the difference that $\frac{\pi k}{a}$ $(k\in\mathbb{Z})$ is replaced by real k and the orthogonal direct sum decomposition of H is replaced by an orthogonal direct integral with respect to Lebesgue measure, $\mathcal{H} = \int_{\oplus}\mathcal{L}_k dk$. In terms of subspaces $\mathcal{L}_k$, which may be naturally identified with $L_2[-1,1]$, one may obtain the following result.

THEOREM 2.5. For every $\delta>0$ and every Borel set $\Delta\subset\mathbb{C}$ at nonzero distance from zero, we may construct projectors $E_k(\cdot)$, $k \in \mathbb{R}$, as in (2.22). Then the subspace

$$\mathcal{H}_{\delta,\Delta} = \int_{\oplus |k \pm \frac{\pi}{2}| > \delta} \mathcal{L}_k \, dk \oplus \int_{\oplus |k \pm \frac{\pi}{2}| \le \delta} E_k(\Delta)\mathcal{L}_k \, dk$$

of $\mathcal{H}$ is invariant under the operator A, and A restricted to $\mathcal{H}_{\delta,\Delta}$ is similar to a normal operator.

This similarity method can be applied whenever the Fourier transform (discrete or continuous) diagonalizes the transport operator [9]. Besides the example discussed above, energy dependent [86] and anisotropic scattering [312] models with specularly reflecting boundary conditions in semi-infinite, slab and parallelepiped geometries have been treated. For a similar treatment of the linearized Vlasov equation, see [313].

We conclude this section by considering the long time behavior of the transport equation (1.6). The physical idea behind such an estimate is rather simple and generally accepted. It is assumed that the information contained in the original distribution function $u_0(x,\mu)$ at time $t=0$ can be split into two parts: a "microscopic" part accounting for individual, more chaotic movements, which are rapidly decaying, and a "hydrodynamic" part, related to collective, more regular movements. Then, in the final stage of the kinetic evolution described by a transport equation, one may replace the more complete description of the system via its one particle distribution function $u(x,\mu,t)$ by a "reduced" description involving only the first few moments of $u(x,\mu,t)$. These moments are usually the hydrodynamic quantities of the system, e.g., the local densities of mass, momentum, and energy. The first systematic deduction of hydrodynamics from kinetics originated in the work of Hilbert [193] and, later, Chapman [90, 91] and Enskog [113]. This method, which is essentially a singular perturbation approach, is still a fertile area of research. We quote here – both because of their particular impetus and of the relevance to the present context – the work of Arsenyev [14], Ellis and Pinsky [111] and Papanicolaou [300]. For the model equation (1.6), the only relevant hydrodynamic moment is the local density, which is defined as

$$n(x,t) = \int_{-1}^{1} u(x,\mu,t)d\mu = 2(Nu)(x,\mu,t).$$

Let us consider the evolution equation

$$\frac{\partial u}{\partial t}(x,\mu,t) = -\mu\frac{\partial u}{\partial x}(x,\mu,t) - u(x,\mu,t) + \tfrac{1}{2}\int_{-1}^{1} u(x,\hat{\mu},t)d\hat{\mu} \tag{2.23}$$

in $\mathcal{H} = L_2(\mathbb{R}x[-1,1])$ with initial condition

$$\lim_{t\downarrow 0} u(x,\mu,t) = u_0(x,\mu).$$

Denote by $n^D(x,t)$ the solution of the diffusion equation

$$\frac{\partial n^D}{\partial t}(x,t) = \tfrac{1}{3}\,\frac{\partial^2 n^D}{\partial x^2}(x,t) \tag{2.24}$$

with initial condition

$$\lim_{t\downarrow 0} n^D(x,t) = n_0^D(x) = 2Nu_0(x)$$

in $L_2(\mathbb{R})$. Then the asymptotic equivalence result is contained in the following theorem, which can also be proved for semi-infinite geometry with specularly reflecting boundary conditions [12].

THEOREM 2.6. Let $u_0(x,\mu)\geq 0$, $u_0\neq 0$, $u_0\in L_2(\mathbb{R}\times[-1,1])\cap L_1(\mathbb{R}\times[-1,1])$. Then there exists a constant $C_{u_0}>0$ such that in $L_2(\mathbb{R})$ one has the estimate

$$\frac{\|n-n^D\|}{\|n^D\|} \leq C_{u_0}\, t^{-1/2}. \tag{2.25}$$

Proof: Denoting by $U^D(t)$ the semigroup generated by $\tfrac{1}{3}\,\frac{\partial^2}{\partial x^2}$ in $L_2(\mathbb{R})$, one has to estimate

$$\|n-n^D\|^2 = \|2NU(t)u_0-U^D(t)2Nu_0\|^2.$$

Since the Fourier transform is isometric in L^2, the norms may be estimated in the k-representation:

$$\|n-n^D\|^2 \leq \int_{|k|>\frac{1}{2}\pi+\delta} \left| e^{\lambda_k t}(1,\varphi_k)(\bar{\varphi}_k,u_0)1 - e^{-\frac{1}{3}k^2 t}(1,u_0)1\right|^2 dk +$$

$$+ \int_{\varepsilon<|k|<\frac{1}{2}\pi+\delta} |e^{\lambda_k t}(1,\varphi_k)(\bar{\varphi}_k,u_0)1 - e^{-\frac{1}{3}k^2 t}(1,u_0)1|^2 dk +$$

$$+ \int_{|k|<\varepsilon} |e^{\lambda_k t}(1,\varphi_k)(\bar{\varphi}_k,u_0)1 - e^{-\frac{1}{3}k^2 t}(1,u_0)1|^2 dk,$$

where

$$\lambda_k = -1 + k \text{ cotan} k = -\frac{k^2}{3} - \frac{k^4}{45} + \dots ,$$

and

$$\varphi_k = \{(\sqrt{2} \text{ sin} k)(\text{cotan} k + i\mu)\}^{-1}.$$

The first integral can be exponentially bounded in t, by using the spectral decomposition result (see Theorem 2.5). The second integral contains the "critical" point $|k|=\pi/2$ for which the spectral decomposition per se fails; however, by taking advantage of the concrete form of the operators B_k, one can apply the Spectral Mapping Theorem directly and bound the second integral by const.$e^{-\varepsilon t}$. Then, up to exponentially bounded functions, the square of the difference in local densities $\|n-n^D\|^2$ may be bounded by the term

$$\int_{-\infty}^{\infty} |e^{\lambda_k t}(1,\varphi_k)(\bar{\varphi}_k,u_0)1 - e^{\lambda_k t}(1,u_0)1|^2 dk + \int_{-\infty}^{\infty} |e^{\lambda_k t} - e^{-\frac{1}{3}k^2 t}|^2 (1,u_0)1 dk \sim$$

$$\sim \int_{-\infty}^{\infty} e^{-\frac{2}{3}k^2 t} C_{1,u_0} k^2 dk + \int_{-\infty}^{\infty} e^{-\frac{2}{3}k^2 t} C_{2,u_0} \cdot k^8 t^2 dk \sim C_{3,u_0} \cdot t^{-3/2}$$

Taking into account that, for $u_0 \in L_2 \cap L_1$, $\|n^D\|^2 \sim C_{4,u_0} t^{-\frac{1}{2}}$, one has (2.25). ∎

3. Spencer–Lewis equation and electron deceleration

In this section we shall analyze the time dependent Spencer–Lewis equation [13, 29]

$$\frac{\partial u}{\partial t}(x,\mu,E,t) + \mu\frac{\partial u}{\partial x}(x,\mu,E,t) - \frac{\partial(\beta u)}{\partial E}(x,\mu,E,t) + \sigma(x,E)u(x,\mu,E,t) =$$

$$= \int_{-1}^{1} \sigma_s(x,\mu,\hat{\mu},E)u(x,\hat{\mu},E,t)d\hat{\mu} + f(x,\mu,E,t). \tag{3.1}$$

Here $x\in[0,a]$ is a position variable, $\mu\in[-1,1]$ is the direction cosine of propagation, and $E\in[E_m,E_M]\subset(0,\infty)$ is the energy variable. The equation describes the slowing down of electrons from high energies larger than E_M to low energies smaller than E_m. Its solution $u(x,\mu,E,t)$ is the density distribution of electrons within the given interval and the inhomogeneous term $f(x,\mu,E,t)$ represents the contribution of internal sources. The function $\beta(x,E)$ is the stopping power, $\sigma(x,E)$ the total elastic scattering cross section, and $\sigma_s(x,\mu,\hat{\mu},E)$ the (azimuthally integrated) elastic scattering cross section. As a result, β, σ and σ_s, as well as the functions u and f, are nonnegative. The equation describes the electron distribution in the intermediate energy range $[E_m,E_M]$, where electrons are continuously slowed down from high to thermal energies. Important applications include the slowing down of electrons in an electron microscope, damage to semiconductors by intrusion of high energy electrons, and the use of electron guns for etching semiconductors (microlithography).

Along with (3.1) are given boundary conditions, which specify the incident electron distribution,

$$u(0,\mu,E,t) = g_m(\mu,E,t), \quad \mu>0, \quad E\in[E_m,E_M], \tag{3.2a}$$

$$u(a,\mu,E,t) = g_M(\mu,E,t), \quad \mu<0, \quad E\in[E_m,E_M], \tag{3.2b}$$

as well as the initial condition

$$u(x,\mu,E,0) = g_0(x,\mu,E). \tag{3.3}$$

Here g_m, g_M and g_0 are nonnegative functions.

Equation (3.1) was initially derived in the 1950's by Spencer [343] and Lewis [250] to study the continuous slowing down of electrons of intermediate energy. Instead of the energy derivative of βu, the equation contained the term $\beta\frac{\partial u}{\partial E}$, which greatly facilitated the analysis of the problem. More recently, the topic of interest has been the stationary counterpart of the Spencer–Lewis equation. Nelson [283] proved unique solvability for a simple rod model of the equation and, in cooperation with Seth [284], found strong indications for its unique solvability for the case of a stopping power that is piecewise constant in energy. In this section we shall prove the wellposedness of the above initial–boundary value problem and its stationary counterpart under a simplifying assumption on the stopping power. We shall assume that there exists of partition $E_m=E_0<E_1<...<E_r=E_M$ of $[E_m,E_M]$ such that the stopping power can be written as

$$\beta(x,E) = \beta_i(x), \quad E\in(E_{i-1},E_i), \quad i=1,2,...,r, \tag{3.4}$$

with each β_i a Lipschitz continuous function on $[0,a]$ with at most finitely many zeros. This assumption was made by Nelson and Seth [284] in order to prove the convergence of a finite difference scheme for numerical computation of the solution. Further, since σ represents the total cross section and one must also account for the loss of electrons within $[E_m,E_M]$ by absorption, we assume that σ_s and σ are nonnegative measurable functions satisfying

$$\int_{-1}^{1} \sigma_s(x,\mu,\hat{\mu},E)d\mu \le \sigma(x,E), \qquad (x,\hat{\mu},E) \in [0,a]\times[-1,1]\times[E_m,E_M]. \tag{3.5}$$

This assumption will serve us well if we adopt an L_1–setting for studying Eqs. (3.1)–(3.3). When using an L_p–setting with $1<p<\infty$, we will replace (3.5) by the hypothesis

$$\left[\int_{-1}^{1} \left[\int_{-1}^{1} \sigma_s(x,\mu,\hat{\mu},E)^p d\mu\right]^{1/(p-1)} d\mu\right]^{1/q} \le \sigma(x,E), \tag{3.6}$$

where $q=p/(p-1)$. In addition to one of (3.5) and (3.6), we also assume that $\sigma(x,E)$ is bounded as a function of $(x,E)\in[0,a]\times[E_m,E_M]$.

Let us incorporate Eqs. (3.1)–(3.3) in the abstract theory of the previous chapters. For $i=1,2,...,r$ we define $\Lambda_i=(0,a)\times(-1,1)\times(E_{i-1},E_i)$ and consider as the phase space the set

$$\Lambda = \bigcup_{i=1}^{r} \Lambda_i,$$

endowed with the Lebesgue measure $dx d\mu dE$. We thus have a phase space with r connected components $\Lambda_1,\dots,\Lambda_r$.

Now consider the vector field

$$X = \mu\frac{\partial}{\partial x} - \beta_i(x)\frac{\partial}{\partial E} \quad \text{on } \Lambda_i.$$

Then clearly the vector field is continuous and divergence free on Λ. On using time as a parameter, its characteristic equations have the form

$$\frac{dx}{dt} = \mu, \quad \frac{d\mu}{dt} = 0, \quad \frac{dE}{dt} = \beta_i(x),$$

where $(x,\mu,E)\in\Lambda_i$. Hence, the integral curves of X are the curves

$$(x_0+\mu t,\ \mu,\ E_{0,i} - \int_{E_m}^{E_M} \beta_i(x_0+\mu \xi)d\xi)$$

with x increasing with t for $\mu>0$ and decreasing with t for $\mu<0$, as well as the line segments $(x,\mu,E_{0,i}-\beta_i(x)t)$ for fixed (x,μ). Since the stopping power is nonnegative, the energy E is non increasing in time along trajectories. Because the stopping power is bounded, there is a uniform positive lower bound to the "lengths" of the integral curves. The sets $D_\pm$ of left and right endpoints of integral curves are then given by

$$D_- = \bigcup_{i=1}^{r} \Big[[\{0\}\times(0,1)\times(E_{i-1},E_i)] \cup [\{a\}\times(-1,0)\times(E_{i-1},E_i)] \cup [(0,a)\times(-1,1)\times\{E_i\}]\Big],$$

$$D_+ = \bigcup_{i=1}^{r} \Big[[\{0\}\times(-1,0)\times(E_{i-1},E_i)] \cup [\{a\}\times(0,1)\times(E_{i-1},E_i)] \cup [(0,a)\times(-1,1)\times\{E_{i-1}\}]\Big].$$

One easily obtains the Green's identity

$$\int_0^a\int_{-1}^{1}\int_{E_m}^{E_M} (Xu)(x,\mu,E)dx d\mu dE =$$

$$= \sum_{i=1}^{r} \Bigg[\int_{-1}^{1}\int_{E_{i-1}}^{E_i} \mu\{u(a,\mu,E)-u(0,\mu,E)\}d\mu dE +$$

$$+ \int_0^a\int_{-1}^{1} \beta_i(x)\{u(x,\mu,E_{i-1})-u(x,\mu,E_i)\}dx d\mu\Bigg].$$

Hence, the boundary measures $d\nu_{\pm}$ are given by $|\mu|d\mu dE$ for parts of the boundary within x=0 and x=a, and $\beta_i(x)dxd\mu$ for parts of the boundary of Λ_i within $E=E_{i-1}$ and $E=E_i$.

We may consider the test function space Φ as the linear space of all bounded Borel functions on Λ that are continuously differentiable along trajectories with bounded directional derivative, for $\nu_{\pm}$-almost all integral curves. We write Φ_0 for those functions in Φ that vanish at the endpoints of the trajectories. On $L_p(\Lambda,dxd\mu dE)$ we may then define Xu as the distributional derivative satisfying

$$\int_0^a\int_{-1}^1\int_{E_m}^{E_M}(Xu)v\ dxd\mu dE = -\int_0^a\int_{-1}^1\int_{E_m}^{E_M}u(Xv)\ dxd\mu dE, \qquad v\in\Phi_0,$$

and specify F_p for $1\le p<\infty$ as the linear space of those $u\in L_p(\Lambda,dxd\mu dE)$ such that $Xu\in L_p(\Lambda,dxd\mu dE)$. Since the integral curves have their lengths uniformly bounded away from zero, every $u\in F_p$ has "traces" $u_{\pm}$ belonging to $L_p(D_{\pm},d\nu_{\pm})$.

Because we have imposed the "vacuum" boundary conditions (3.2), we would at first sight expect to define the boundary operator as K=0. However, since we have created a phase space with additional boundaries at the intermediate energies $E_1,\ldots,E_{r-1}$, the boundary operator K must also reflect the continuity of the electron distribution at these intermediate energies. Thus we define the operator $K:L_p(D_+,d\nu_+)\to L_p(D_-,d\nu_-)$ by

$$(Ku)(x,\mu,E) = \begin{cases} 0, & x=0,\quad \mu>0 \\ u(x,\mu,E^+), & E\in\{E_1^-,\ldots,E_{r-1}^-\} \\ 0, & x=a,\quad \mu<0, \end{cases}$$

where, for $i=1,\ldots,r-1$, E_i is written as E_i^+ if $(x,\mu,E_i)\in\Lambda_{i+1}$ and as E_i^- if $(x,\mu,E_i)\in\Lambda_i$. We now define the "free" streaming operator S_0 and the transport operator B_0 on the common domain $D(S_0) = D(B_0) = \{u\in F_p : u_-=Ku_+\}$ by the equations

$$(S_0u)(x,\mu,E) = -(Xu)(x,\mu,E) - \sigma(x,E)u(x,\mu,E)$$

and

$$(B_0u)(x,\mu,E) = -(Xu)(x,\mu,E) - \sigma(x,E)u(x,\mu,E) + (Ju)(x,\mu,E),$$

where

$$(Ju)(x,\mu,E) = \int_{-1}^{1} \sigma_s(x,\mu,\hat{\mu},E)u(x,\hat{\mu},E)d\hat{\mu}.$$

When treating Eqs. (3.1)–(3.3) in the L_p-setting, we shall require (3.5) for p=1 and (3.6) for $1<p<\infty$. Using the boundedness of $\sigma(x,E)$, we then easily obtain that J is bounded on $L_p(\Lambda,dxd\mu dE)$.

Let us now derive the semigroup properties of Eqs. (3.1)–(3.3). Observe that J and K are bounded positive operators on $L_p(\Lambda,dxd\mu dE)$, that $\|K\|=1$, and that the lengths of the integral curves of X have a uniform positive lower bound. Further, for all nonnegative $u\in L_1(\Lambda,dxd\mu dE)$,

$$\int_0^a\int_{-1}^1\int_{E_m}^{E_M} \{\sigma(x,E)u(x,\mu,E) - \int_{E_m}^{E_M} \sigma_s(x,\mu,\hat{\mu},E)u(x,\hat{\mu},E)d\hat{\mu}\}dxd\mu dE \geq$$

$$\geq \int_0^a\int_{-1}^1\int_{E_m}^{E_M} \{\sigma(x,E)-\hat{\sigma}(x,E)\}u(x,\mu,E)dxd\mu dE \geq 0,$$

where

$$\hat{\sigma}(x,E) = \operatorname*{ess\,sup}_{\hat{\mu}\in[-1,1]} \int_{-1}^{1} \sigma_s(x,\mu,\hat{\mu},E)d\mu$$

is dominated by $\sigma(x,E)$ by virtue of (3.5). Likewise, for arbitrary $u\in L_p(\Lambda,dxd\mu dE)$,

$$0 \leq \int_0^a\int_{-1}^1\int_{E_m}^{E_M} \mathrm{sgn}(u)\,|u(x,\mu,E)|^{p-1} \times$$

$$\times \{\sigma(x,E)u(x,\mu,E) - \int_{-1}^{1} \sigma_s(x,\mu,\hat{\mu},E)u(x,\hat{\mu},E)d\hat{\mu}\}dxd\hat{\mu}dE,$$

by virtue of (3.6). We may therefore apply Theorem XII 2.3 and the paragraph following its proof to obtain the following results:

(i) For every $1\leq p<\infty$ the operator S_0 generates the strongly continuous positive contraction semigroup $\{U_0(t)\}_{t\geq 0}$.

(ii) For every $1 \le p < \infty$ the operator B_0 generates the strongly continuous contraction semigroup $\{U(t)\}_{t \ge 0}$.

(iii) The relationship between the two semigroups is given by the Hille–Dyson–Phillips expansion, which is absolutely norm convergent.

We shall now consider the stationary boundary value problem corresponding to the Spencer–Lewis equation. We shall apply the theory of Section XII.3 to derive unique solvability, however, under somewhat more restrictive assumptions on the stopping power.

The boundary value problem for the stationary Spencer–Lewis equation has the form

$$\mu\frac{\partial u}{\partial x}(x,\mu,E) - \frac{\partial(\beta u)}{\partial E}(x,\mu,E) + \sigma(x,E)u(x,\mu,E) =$$

$$= \int_{-1}^{1} \sigma_s(x,\mu,\hat{\mu},E)u(x,\hat{\mu},E)d\hat{\mu} + f(x,\mu,E) \tag{3.7}$$

with boundary conditions

$$u(0,\mu,E) = g_m(\mu,E), \qquad \mu>0, \tag{3.8a}$$

$$u(a,\mu,E) = g_M(\mu,E), \qquad \mu<0. \tag{3.8b}$$

The stopping power $\beta(x,\mu,E)$ has again the form (3.4). At the interfaces $E=E_1,\dots,E=E_{r-1}$ we impose the condition of continuity on the electron distribution $u(x,\mu,E)$. As in Section XII.3 we may formulate the stationary problem (3.7)–(3.8) in $L_p(\Lambda,dxd\mu dE)$ for $1 \le p < \infty$, and unique solvability is equivalent to the property that $0 \notin \sigma(B_0)$. One first constructs $g \in F_p$ whose restrictions $g_\pm$ to $D_\pm$ satisfy $g_- = Kg_+$, where g_- "coincides" with $g_m \in L_p([0,1]\times[E_m,E_M],|\mu|d\mu dE)$ on the set $\{0\}\times[0,1]\times[E_m,E_M]$, "coincides" with $g_M \in L_p([-1,0]\times[E_m,E_M],|\mu|d\mu dE)$ on the set $\{a\}\times[-1,0]\times[E_m,E_M]$, and vanishes on the interfaces $E=E_1,\dots,E=E_{r-1}$. Since g is, in fact, constructed by interpolation along integral curves, it will be nonnegative whenever g_m and g_M are nonnegative. We may then write the unique solution of Eqs. (3.7)–(3.8) as

$$u(x,\mu,E) = -(B_0^{-1}(f+B\hat{g}))(x,\mu,E) + \hat{g}(x,\mu,E),$$

where $B=-(X+\sigma-J)$ is an extension of B_0 to an operator with domain F_p (cf. the proof of Proposition XII 3.1). A sufficient condition for having $0\notin\sigma(B_0)$ and hence the unique solvability of Eqs. (3.7)-(3.8) is $\omega_0(U_0)<0$. This will be the case, for example, if for some constant $\delta>0$,

$$\sigma(x,E) - \int_{-1}^{1} \sigma_s(x,\mu,\hat{\mu},E)d\mu \geq \delta$$

for p=1, and

$$\sigma(x,E) - \left[\int_{-1}^{1} \left[\int_{-1}^{1} \sigma_s(x,\mu,\hat{\mu},E)^p d\mu\right]^{1/(p-1)} d\mu\right]^{1/q} \geq \delta$$

for $1<p<\infty$. Note that this implies an absorptive process in the electron transport.

4. Electron drift in a weakly ionized gas

In this section we shall study existence and uniqueness properties of a time dependent linearized Boltzmann equation describing the free electron distribution of a weakly ionized gas in an external electric field [306, 290]. If one considers only collisions between the electrons and the neutral gas molecules and disregards the interactions between electrons and between electrons and ions (weak ionization), one obtains the linearized Boltzmann equation

$$\frac{\partial u}{\partial t}(x,v,t) + v\cdot\frac{\partial u}{\partial x}(x,v,t) + a\cdot\frac{\partial f}{\partial v}(x,v,t) = \iiint dv_1 d\hat{v}_1 d\hat{v} \times \tag{4.1}$$

$$\times \left[\sigma(\hat{v}_1,\hat{v}\rightarrow v_1,v)(|\hat{v}_1-\hat{v}|)u(x,\hat{v},t)F(x,\hat{v}_1,t)-\sigma(v_1,v\rightarrow\hat{v}_1,\hat{v})(|v_1-v|)u(x,v,t)F(x,v_1,t)\right].$$

Here $F(x,v,t)$ and $u(x,v,t)$ are the distributions of the neutral gas molecules and the free electrons as functions of position $x\in\mathbb{R}^3$, velocity $v\in\mathbb{R}^3$ and time $t\in\mathbb{R}_+$. The former is given, nonnegative and measurable, while the latter is to be calculated. The acceleration a has the form $a=-kE$, where k is the absolute quotient of the electron charge and the electron mass and E is the external electric field. The collision frequency

$$\nu(x,v,t) = \iiint (|v-v_1|)\sigma(v_1,v\rightarrow\hat{v}_1,\hat{v})F(x,v_1,t)dv_1 d\hat{v}_1 d\hat{v}$$

is in general a function of x, v and t. However, since by assumption the molecular distribution is nonnegative and isotropic (i.e., $F=F(x,|v|,t)$), and the cross section $\sigma(v_1,v\to\hat{v}_1,\hat{v})$ is nonnegative and invariant with respect to rigid rotation of the four velocities v, v_1, $\hat{v}$ and $\hat{v}_1$, the collision frequency is nonnegative and depends only on x, $|v|$ and t.

We may now rewrite Eq. (4.1) as

$$\frac{\partial u}{\partial t}(x,v,t) + v\cdot\frac{\partial u}{\partial x}(x,v,t) + a\cdot\frac{\partial u}{\partial v}(x,v,t) =$$

$$= \int \nu(x,\hat{v},t)P(\hat{v}\to v;x,t)u(x,\hat{v},t)d\hat{v} - \nu(x,v,t)u(x,v,t), \tag{4.2}$$

with initial condition

$$u(x,v,0) = u_0(x,v). \tag{4.3}$$

Since the spatial and velocity domains both coincide with $\mathbb{R}^3$, no boundary condition is imposed.

The quantity

$$c(x,\hat{v},t) = \int P(\hat{v}\to v;x,t)dv$$

determines which of the processes, ionization or recombination, is dominant. If there is an equilibrium between these two processes, then $c(x,\hat{v},t)\equiv 1$. We shall assume that the collision frequency is essentially bounded and that the redistribution function $P(\hat{v}\to v;x,t)$ is nonnegative and measurable, so that

$$c = \text{ess sup}\,\{c(x,v,t) : (x,v,t)\in\mathbb{R}^3\times\mathbb{R}^3\times(0,\infty)\} < \infty.$$

We shall also assume that the acceleration is Lipschitz continuous on $\mathbb{R}^3\times\mathbb{R}^3\times(0,\infty)$, partially differentiable with respect to the three velocity components, velocity divergence free,

$$\frac{\partial a}{\partial v} \equiv 0,$$

nonzero whenever the velocity $v=0$, and satisfying the estimate

$$|a(x,v,t)| \leq C\,(1 + |x| + |v|)$$

for some constant C. As a result, multiplication by the collision frequency as well as the operator

$$(J^T u)(x,v,t) = \int \nu(x,\hat{v},t)P(\hat{v}\to v;x,t)u(x,\hat{v},t)d\hat{v}$$

are bounded positive operators on $L_1(\Sigma,dxdvdt)$, which satisfy

$$\int_0^T\!\!\int\!\!\int[\nu(x,v,t)u(x,v,t)-(J^T u)(x,v,t)]dxdvdt \geq (1-c)\int_0^T\!\!\int\!\!\int \nu(x,v,t)u(x,v,t)dxdvdt \quad (4.4)$$

for every nonnegative $u\in L_1(\Lambda,dxdvdt)$.

Now let us consider the vector field

$$X = v\cdot\frac{\partial}{\partial x} + a\cdot\frac{\partial}{\partial v}$$

on the phase space $\Lambda=\mathbb{R}^3\times\mathbb{R}^3$, as well as the vector field $Y = \frac{\partial}{\partial t} + X$ on the extended phase space $\Sigma=\Lambda\times(0,T)$ for some fixed $T>0$. Then the characteristic equations of X are given by

$$\frac{dx}{dt} = v, \qquad \frac{dv}{dt} = a,$$

where t is the parameter. The assumed estimate on the acceleration guarantees that no integral curve of either X or Y runs off to infinity in finite time. The integral curves of X are then given by $\{(x(t),v(t)) : t\in\mathbb{R}\}$, while those for Y are given by $\{(t,x(t),v(t)) : t\in(0,T)\}$. Let us introduce Φ^T as the linear space of bounded Borel functions on Σ that are continuously differentiable with bounded directional derivative along almost all integral curves of Y, and Φ_0^T as the linear space of functions in Φ^T which vanish for $t=0$ and $t=T$. We then define for arbitrary $u\in L_1(\Sigma,dxdvdt)$ the function Yu as the distributional derivative

$$\int_0^T\!\!\int\!\!\int (Yu)(x,v,t)\varphi(x,v,t)dxdvdt = -\int_0^T\!\!\int\!\!\int u(x,v,t)(Y\varphi)(x,v,t)dxdvdt, \qquad \varphi\in\Phi_0^T.$$

For those $u\in L_1(\Sigma,dxdvdt)$ such that $Yu\in L_1(\Sigma,dxdvdt)$ we obtain the Green's identity

$$\int_0^T \iint (Yu)(x,v,t)\,dxdvdt =$$

$$= \int_{\{t=T\}} u(x(T),v(T),T)\,d\nu_-(x(T),v(T)) - \int_{\{t=0\}} u(x(0),v(0),0)\,d\nu_-(x(0),v(0)),$$

where ν_- is a suitable bounded Borel measure on $\{t=0\}$ and the points on $\{t=T\}$ have been propagated from those on $\{t=0\}$ along the integral curves. Similarly, introducing Φ as the linear space of bounded Borel functions on Λ that are continuously differentiable with bounded directional derivative along almost every integral curve of X, we may define for arbitrary $u \in L_1(\Lambda,dxdv)$ the function Xu as the distributional derivative

$$\iint (Xu)(x,v)\varphi(x,v)\,dxdv = -\iint u(x,v)(X\varphi)(x,v)\,dxdv, \quad \varphi \in \Phi.$$

Since every trajectory of X has the real line as its parameter domain, there is no Green's identity associated with X.

Let us first derive the wellposedness of the initial value problem (4.2)–(4.3) using the formalism of Chapter XI. Denote by E_1 the linear space of $u \in L_1(\Sigma,dxdvdt)$ such that $Yu \in L_1(\Lambda,dxdvdt)$. Since all of the integral curves of Y have the same positive length T, the restrictions $u(\cdot,\cdot,0)$ and $u(\cdot,\cdot,T)$ of u to $\{t=0\}$ and $\{t=T\}$ belong to $L_1(\{t=0\},d\nu_-)$ and $L_1(\{t=T\},d\nu_+)$, where ν_+ is obtained from ν_- by propagation along integral curves. Applying Theorems XI 4.3–4.5, it is immediate that for every $u_0 \in L_1(\{t=0\},d\nu_-)$ there exists a unique function $u \in L_1(\Sigma,dxdvdt)$ satisfying the equations

$$(Xu)(x,v,t) + \nu(x,v,t)u(x,v,t) = (J^T u)(x,v,t), \quad (x,v,t) \in \Sigma,$$

$$u(x,v,0) = u_0(x,v), \quad (x,v) \in \Lambda.$$

If u_0 is nonnegative, then so is the solution u. Moreover, if $c \in [0,1]$, then the evolution is contractive in the sense that

$$\|u(\cdot,\cdot,t)\|_{L_1(\Lambda\times\{t\},d\nu_t)} \le \|u_0\|_{L_1(\{t=0\},d\nu_-)}. \tag{4.5}$$

Here ν_t is the measure on $\{(x,v,t) : (x,v) \in \Lambda\}$ obtained from ν_- by propagation along integral curves. In fact, since Λ does not have boundaries, it follows directly from the identity following (XI 4.21) that the contractiveness condition (4.5) is satisfied if and only if $c(x,\hat{v},t) \leq 1$ almost everywhere on Σ. Indeed, this condition is necessary and sufficient in order that the left hand side of (4.4) be nonnegative for all nonnegative $u \in L_1(\Sigma, dxdvdt)$, and the latter is in turn equivalent to contractiveness. The identity following (XI 4.21) also implies that

$$\|u(\cdot,\cdot,t)\|_{L_1(\Lambda\times\{t\},d\nu_t)} = \|u_0\|_{L_1(\{t=0\},d\nu_-)}, \qquad 0\leq t\leq T,$$

if and only if $c(x,\hat{v},t) \equiv 1$ almost everywhere on Σ. In physical terms, the total number of free electrons is independent of time for every initial electron distribution if and only if there is local equilibrium (in position and velocity) between ionization and recombination.

Let us now assume that the acceleration, the cross section, the redistribution function, the distribution of the neutral gas molecules, and the collision frequency are all time independent. Then we may cast the existence and uniqueness results in the semigroup framework of Section XII.2. Let F_1 be the linear space of $u \in L_1(\Lambda, dxdv)$ such that $Xu \in L_1(\Lambda, dxdv)$. We may define the operators S and B on the common domain F_1 by

$$(Su)(x,v) = -(Xu)(x,v) - \nu(x,v)u(x,v),$$

$$(Bu)(x,v) = -(Xu)(x,v) - \nu(x,v)u(x,v) + (Ju)(x,v),$$

with

$$(Ju)(x,v) = \int \nu(x,\hat{v},t)P(\hat{v}\to v;x)u(x,\hat{v})d\hat{v}.$$

Then S and B generate strongly continuous positive semigroups $\{U_0(t)\}_{t\geq 0}$ and $\{U(t)\}_{t\geq 0}$ on $L_1(\Lambda, dxdv)$. The former is contractive. A necessary and sufficient condition for contractiveness of the latter semigroup is that $c(x,\hat{v}) \leq 1$ almost everywhere on Λ, which is satisfied if $c \in [0,1]$. A necessary and sufficient condition in order that $\{U(t)\}_{t\geq 0}$ be a semigroup of isometries is that $c(x,\hat{v}) \equiv 1$ almost everywhere on Λ. These statements are easily derived from the formula following (XI 4.21) in combination with the identity (4.4).

The runaway electron problem corresponds to giving necessary and sufficient conditions in order that the time dependent solution converges as $t\to\infty$. The resulting limit will then be an equilibrium distribution for the electrons. Obviously, this is only an interesting question if $c(x,\hat{v},t) \equiv 1$, i.e., in the case of local equilibrium between ionization and recombination. Cavalleri and Paveri-Fontana [72], from whom we have borrowed the above sketch of the problem, considered the special case in which all relevant data are independent of position. For this case they gave a necessary condition in order that the electrons relax to a steady state distribution. They also showed that if this condition is not satisfied the electron speed will increase indefinitely and hence the electrons will run away to infinity.

5. A transport equation describing growing cell populations

In this section the abstract time dependent theory will be applied to a transport problem modeling growing cell populations. This model was recently developed by Rotenberg [322] as an improvement of a model of Lebowitz and Rubinow [243]. Rotenberg's effort was directed towards a Fokker–Planck approximation of the equation, for which he obtained numerical solutions. For a variety of boundary conditions, van der Mee and Zweifel [370] obtained analytical solutions by eigenfunction expansion. Here we shall examine a Boltzmann type equation rather than its Fokker–Planck approximation.

The transport equation has the form

$$\frac{\partial u}{\partial t}(\mu,v,t) + v\frac{\partial u}{\partial \mu}(\mu,v,t) =$$

$$= \int_0^\infty r(\mu;v,\hat{v})u(\mu,\hat{v},t)d\hat{v} - R(\mu,v)u(\mu,v,t) + S(\mu,v,t), \tag{5.1}$$

where $\mu \in [0,1]$, $v \in [0,\infty)$, and

$$R(\mu,v) = \int_0^\infty r(\mu;v,\hat{v})d\hat{v}. \tag{5.2}$$

We impose the initial condition

$$u(\mu,v,0) = u_0(\mu,v) \tag{5.3}$$

as well as a boundary condition. This equation describes the number density $u(\mu,v,t)$

of a cell population as a function of the degree of maturation μ, the maturation velocity v, and the time t. Here the degree of maturation is defined so that $\mu=0$ at the birth and $\mu=1$ at the death of a cell. The transition rate $r(\mu;v,\hat{v})$ specifies the number of transfers per unit time, unit population and unit velocity interval. By integration one obtains the total transition cross section $R(\mu,v)$. By assumption, $r(\mu;v,\hat{v})$ and $R(\mu,v)$ are nonnegative and measurable. The number density of cells with maturity μ is obtained from $u(\mu,v,t)$ by integration over all v; the total population at time t in turn is the integral of $u(\mu,v,t)$ over both μ and v.

A general reproduction law coupling the velocity distribution before mitosis ($\mu=1$) to the velocity distribution after mitosis ($\mu=0$) has been introduced by Lebowitz and Rubinow [243]. It has the form of a boundary condition,

$$vu(0,v,t) = p\int_0^\infty \hat{v}u(1,\hat{v},t)d\kappa(v,\hat{v}), \tag{5.4a}$$

where $\kappa(v,\cdot)$ is a positive Borel measure on $(0,\infty)$, normalized to unity for all $v\in(0,\infty)$, and the constant $p\in(0,2]$ is the average number of viable offspring per parent cell on mitosis. If $\kappa(v,\cdot)$ is an absolutely continuous measure for all $v\in(0,\infty)$, then one writes its Radon–Nikodym derivative as $k(v,\hat{v})$ and the boundary condition as

$$vu(0,v,t) = p\int_0^\infty k(v,\hat{v})\hat{v}u(1,\hat{v},t)d\hat{v}. \tag{5.4b}$$

In most applications it will be convenient to write the boundary conditions as in (5.4b), even if $\kappa(v,\cdot)$ is not absolutely continuous. We shall do so for the special case of reproduction with "perfect memory", where $k(v,\hat{v})=\delta(v-\hat{v})$ is the formal kernel of the identity operator, the constant $p\le 1$, and the boundary condition is given by

$$u(0,v,t) = pu(1,v,t).$$

Let us incorporate Eq. (5.1) with initial condition (5.2) and either of the boundary conditions (5.4) in the abstract framework of Chapters XI and XII. The phase space will be $\Lambda=(0,1)\times(0,\infty)$ with Lebesgue measure, and the vector field is given by $X = v\frac{\partial}{\partial\mu}$. Then the characteristic equations of X are

$$\frac{d\mu}{dt} = v, \qquad \frac{dv}{dt} = 0,$$

where t is the parameter. The integral curves of X are the lines (μ_0+vt,v), where

$\mu_0 \in (0,1)$, $v \in (0,\infty)$. These integral curves have "length" v^{-1}, while the sets D_- and D_+ of left and right endpoints are given by $D_- = \{0\}\times(0,\infty)$ and $D_+ = \{1\}\times(0,\infty)$. The Green's identity has the form

$$\int_0^1\int_0^\infty (Xu)(\mu,v)d\mu dv = \int_0^\infty v\{u(1,v) - u(0,v)\}d\mu, \tag{5.5}$$

so that $\nu_\pm$ is the weighted Lebesgue measure vdv. One should note that the lengths of the trajectories of X do not have a uniform positive lower bound.

We shall analyze the initial-boundary value problem (5.1)-(5.2)-(5.4) on the space $L_1(\Lambda,d\mu dv)$, where we assume $R(\mu,v)$ to be essentially bounded on Λ. On this space the operator

$$(Ju)(\mu,v) = \int_0^\infty r(\mu;v,\hat{v})u(\mu,\hat{v})d\hat{v}$$

is bounded and positive, while for every nonnegative $u \in L_1(\Lambda,d\mu dv)$

$$\int_0^1\int_0^\infty \{R(\mu,v)u(\mu,v) - \int_0^\infty r(\mu;v,\hat{v})u(\mu,\hat{v})d\hat{v}\}d\mu dv = 0, \tag{5.6}$$

as a corollary of (5.3). With the help of the natural identifications between $L_1(D_\pm,d\nu_\pm)$ and $L_1([0,\infty),vdv)$ one defines the boundary operator K by

$$(Ku)(v) = \frac{p}{v}\int_0^\infty k(v,\hat{v})\hat{v}u(\hat{v})d\hat{v}. \tag{5.7}$$

Then for all nonnegative $u \in L_1([0,\infty),vdv)$ one has

$$\int_0^\infty v(Ku)(v)dv = p\int_0^\infty vu(v)dv,$$

whence K is a bounded positive operator with $\|K\|\leq 1$.

As test function space we choose the linear vector space Φ of bounded Borel functions φ on Λ that are continuously differentiable with respect to μ with bounded partial derivative for almost all $v \in (0,\infty)$, and whose support is contained in the set $[0,1]\times[0,v_{max}]$ for some $v_{max} \in (0,\infty)$. By Φ_0 we denote the linear subspace of functions in Φ which vanish at $\mu=0$ and $\mu=1$ for almost all $v \in (0,\infty)$. It should be noted that the lengths of integral curves of X that meet the support of a function in Φ do in fact have a positive lower bound, namely, some $\delta \geq (1/v_{max})$. For $u \in L_1(\Lambda,d\mu dv)$ we then define Xu as the distributional derivative specified by

$$\int_0^1\int_0^\infty (Xu)(\mu,v)\varphi(\mu,v)d\mu dv = -\int_0^1\int_0^\infty u(\mu,v)(X\varphi)(\mu,v)d\mu dv, \qquad \varphi\in\Phi_0.$$

We define F_1 to be the linear vector space of $u\in L_1(\Lambda,d\mu dv)$ such that $Xu\in L_1(\Lambda,d\mu dv)$ and the restrictions $u(0,\cdot)$ and $u(1,\cdot)$ belong to $L_1([0,\infty),vdv)$. In general, a function u such that $\{u,Xu\}\subset L_1(\Lambda,d\mu dv)$ only has restrictions $u(0,\cdot)$ and $u(1,\cdot)$ in $L_{1,loc}([0,\infty),vdv)$ satisfying (5.5). Here $L_{1,loc}([0,\infty),vdv)$ is the vector space of those Lebesgue measurable functions on Λ that are integrable on every bounded Lebesgue measurable set. In fact, as it will turn out, every solution u of the initial-boundary value problem such that $\{u,Xu\}\subset L_1(\Lambda,d\mu dv)$ has its restrictions $u(0,\cdot)$ and $u(1,\cdot)$ in $L_1([0,\infty),vdv)$.

We now define S_K and B_K as linear operators with the common domain

$$D(S_K) = D(B_K) = \{u\in F_1 : u(0,\cdot)=Ku(1,\cdot)\}$$

satisfying

$$(S_K u)(\mu,v) = -(Xu)(\mu,v) - R(\mu,v)u(\mu,v),$$

$$(B_K u)(\mu,v) = -(Xu)(\mu,v) - R(\mu,v)u(\mu,v) + \int_0^\infty r(\mu;v,\hat{v})u(\mu,\hat{v})d\hat{v},$$

so that $B_K = S_K+J$.

Let us apply Theorem XII 2.2 for the case $0<p<1$. In this case $\|K\|<1$. In view of the positivity of J and K and the identity (5.6) we obtain the following result:

(i) For $p\in(0,1)$ the operators S_K and B_K generate the strongly continuous positive contraction semigroups $\{U_{0,K}(t)\}_{t\geq 0}$ and $\{U_K(t)\}_{t\geq 0}$, respectively.

(ii) The semigroups $\{U_{0,K}(t)\}_{t\geq 0}$ and $\{U_K(t)\}_{t\geq 0}$ are related by the Hille-Dyson-Phillips expansion, which is absolutely norm convergent.

Theorem XII 2.2 cannot be used to obtain these semigroup results for $p\geq 1$, although for $p=1$ a somewhat weaker result can still be derived. Nevertheless, apart from contractiveness, we may extend (i) and (ii) to the case $p\geq 1$. For this purpose we choose some $q>p$ and apply the following transformation to the problem,

$$u(\mu,v,t) \rightarrow w(\mu,v,t) = q^{-\mu}u(\mu,v,t), \tag{5.8}$$

which replaces the initial-boundary value problem by the initial-boundary value problem

$$\frac{\partial w}{\partial t}(\mu,v,t) + v\frac{\partial w}{\partial \mu}(\mu,v,t) =$$

$$= \int_0^\infty r(\mu;v,\hat{v})w(\mu,\hat{v},t)d\hat{v} - \{R(\mu,v)-\log(q)\}w(\mu,v,t) + q^{-\mu}S(\mu,v,t), \tag{5.9}$$

$$w(\mu,v,0) = q^{-\mu}u_0(\mu,v), \tag{5.10}$$

$$vw(0,v,t) = \frac{p}{q}\, v\int_0^\infty k(v,\hat{v})\hat{v}w(1,\hat{v},t)d\hat{v}. \tag{5.11}$$

We have replaced Eqs. (5.1)-(5.2)-(5.4) by a new problem (5.9)-(5.11), where the boundary operator has norm $(p/q)<1$. A straightforward application of Theorem XII 2.2 to the modified problem (5.9)-(5.11) in combination with the invertibility of $(N_q u)(\mu,v)=q^{\mu}u(\mu,v)$ on $L_1(\Lambda,d\mu dv)$ and the strict contractiveness of the modified boundary operator then imply the following results:

(i) For all $p>0$ the operators S_K and B_K generate strongly continuous positive semigroups $\{U_{0,K}(t)\}_{t\geq 0}$ and $\{U_K(t)\}_{t\geq 0}$, respectively.

(ii) For the type and spectral bound of the semigroup generated by B_K we have $s(B_K) = \omega_0(U_K) \leq \log(p)$.

(iii) The semigroups $\{U_{0,K}(t)\}_{t\geq 0}$ and $\{U_K(t)\}_{t\geq 0}$ are related by the Hille-Dyson-Phillips expansion, which is absolutely norm convergent.

The estimate in (ii) requires some elaboration. In the first place, the types of the semigroups $\{U_{0,K}(t)\}_{t\geq 0}$ and $\{U_K(t)\}_{t\geq 0}$ do not change when applying the transformation (5.8). On the other hand, since the condition (5.6) implies the contractiveness of the semigroup generated by B_K, the additional term in (5.9) containing $\log(q)$ requires its type not to exceed $\log(q)$ for every $q>p$. As a result, the type of the latter semigroup does not exceed $\log(p)$. The estimate is then clear, because the spectral bound and the type of a positive semigroup coincide on L_1.

BIBLIOGRAPHY

[1] Abramowitz, M. and Stegun, I.A., **Handbook of mathematical functions**, New York, Dover Publishing Co., 1964.

[2] Albertoni, S. and Montagnini, B., "On the spectrum of neutron transport equation in finite bodies," J. Math. Anal. Appl. **13**, 19–48 (1966).

[3] Aliprantis, C.D. and Burkinshaw, O., **Positive operators**, Pure and Applied Mathematics 119, Orlando, Academic Press, 1985.

[4] Allan, G.R., "One one-sided inverses in Banach algebras of holomorphic vector-valued functions," J. London Math. Soc. **42**, 463–470 (1967).

[5] Altomare, F. and Nagel, R., "Semigroups of positive operators," Report, Universita degli Studi di Bari, 1983 [Italian].

[6] Ambarzumian, V.A., "Diffusion of light by planetary atmospheres," Astron. Zhurnal **19**, 30–46 (1942) [Russian].

[7] Ando, T., "On fundamental properties of a Banach space with a cone," Pac. J. Math. **12**, 1163–1169 (1962).

[8] Angelescu, N., Marinescu, N. and Protopopescu, V., "On the spectrum of the linear transport operator with generalized boundary conditions," Lett. Math. Phys. **1**, 141–146 (1976).

[9] Angelescu, N., Marinescu, N. and Protopopescu, V., "On the spectral decomposition of the linear transport operator with periodic boundary conditions," Lett. Math. Phys. **1**, 329–333 (1976).

[10] Angelescu, N., Marinescu, N. and Protopopescu, V., "Neutron transport with

periodic boundary conditions," Transport Theory Stat. Phys. **5**, 115–125 (1976).

[11] Angelescu, N. and Protopopescu, V., "On a problem in linear transport theory," Rev. Roum. Phys. **22**, 1055–1061 (1977).

[12] Angelescu, N. and Protopopescu, V., "Asymptotic equivalence between the hydrodynamic and kinetic stages for the Lorentz model," Lett. Math. Phys. **3**, 87–92 (1979).

[13] Arkuszewski, J., Kulikowska, T. and Mika, J., "Effect of Singularities on Approximation in S_N methods," Nucl. Sci. Eng. **49**, 20–26 (1972).

[14] Arsenyev, A.A., "The Cauchy problem for the linearized Boltzmann equation," U.S.S.R. J. Comput. Maths. Math. Phys. **5**, 110–136 (1965) = Zh. Vychisl. Mat. Mat. Fiz. **5** (5), 864–882 (1965) [Russian].

[15] Arthur, M.D. and Cercignani, C., "Non-existence of a steady rarefied supersonic flow in a half-space," Z. Angew. Math. Phys. **31**, 634–645 (1980).

[16] Arthur, M.D., Greenberg, W. and Zweifel, P.F., "Vlasov theory of plasma oscillations: linear theory," Phys. Fluids **20**, 1296–1301 (1977).

[17] Asano, K., "Local solutions to the initial and the initial boundary value problem for the Boltzmann equation with external force. I," J. Math. Kyoto Univ. **24**, 225–238 (1984).

[18] Atkinson, F.V., **Discrete and continuous boundary value problems**, New York, Academic Press, 1964.

[19] Azizov, T.Ya. and Iokhvidov, I.S., "Linear operators in spaces with indefinite metric and their applications," J. Soviet Math. **15**, 438–490 (1981) = Itogi Nauki i Techniki, Ser. Math. Analiz **17**, 113–205 (1979) [Russian].

[20] Babovsky, H., "Initial and boundary value problems in kinetic theory. I. The Knudsen gas," Transport Theory Stat. Phys. **13**, 455–474 (1984).

[21] Ball, J. and Greenberg, W., "A Pontrjagin space analysis of the supercritical transport equation," Transport Theory Stat. Phys. **4**, 143–154 (1975).

[22] Baouendi, M.S. and Grisvard, P., "On an evolution equation changing its type," J. Funct. Anal. **2**, 352–367 (1968) [French].

[23] Bardos, C., "Boundary value problems for first order partial differential equations with real coefficients; Approximation theorems; Application to the transport equation," Ann. Scient. École Normale Supérieure, Série **4**, **4**, 185–233 (1970) [French].

[24] Bardos, C., Caflisch, R.E. and Nicolaenko, B., "The Milne and Kramers problems for the Boltzmann equation of a hard sphere gas," Comm. Pure Appl. Math. **39**, 323–352 (1986).

[25] Bart, H., Gohberg, I. and Kaashoek, M.A., **Minimal factorization of matrix and operator functions**, Basel, Birkhäuser OT 1, 1979.

[26] Bart, H., Gohberg, I., Kaashoek, M.A. and van Dooren, P., "Factorizations of transfer functions," SIAM J. Control Opt. **18**, 675–696 (1980).

[27] Bart, H., Gohberg, I. and Kaashoek, M.A., "Wiener–Hopf factorization, inverse Fourier transforms and exponentially dichotomous operators," J. Funct. Anal. **68**, 1–42 (1986).

[28] Bartholomäus, G. and Wilhelm, J., "Existence and uniqueness of the solution for the non–stationary Boltzmann equation for the electrons in a collision dominated plasma by means of operator semigroups," Annalen der Physik **38**, 211–220 (1981).

[29] Bartine, D.E., Alsmiller Jr., R.G., Mynatt, F.R., Engle Jr., W.W. and Barish, J., "Low energy electron transport by the method of discrete ordinates," Nucl. Sci. Eng. **48**, 159–178 (1972).

[30] Batty, C.J.K. and Robinson, D.W., "Positive one parameter semigroups on

ordered Banach spaces," Acta Appl. Math. **1**, 221–296 (1984).

[31] Beals, R., "On an equation of mixed type from electron scattering," J. Math. Anal. Appl. **58**, 32–45 (1977).

[32] Beals, R., "On an abstract treatment of some forward–backward problems of transport and scattering," J. Funct. Anal. **34**, 1–20 (1979).

[33] Beals, R., "Partial range completeness and existence of solutions to two–way diffusion equations," J. Math. Phys. **22**, 954–960 (1981); **24**, 1932 (1983).

[34] Beals, R., "Indefinite Sturm–Liouville problems and half–range completeness," J. Diff. Eqs. **56**, 391–407 (1985).

[35] Beals, R. and Protopopescu, V., "Half–range completeness for the Fokker–Planck equation," J. Stat. Phys. **32**, 565–584 (1983).

[36] Beals, R. and Protopopescu, V., "Half–range completeness and orthogonality for the Fokker–Planck equation with general boundary conditions. II," Transport Theory Stat. Phys. **13**, 43–55 (1984).

[37] Beals, R. and Protopopescu, V., "Abstract time dependent transport equations," J. Math. Anal. Appl., to appear.

[38] Bell, G.I. and Glasstone, S., **Nuclear reactor theory**, Princeton, van Nostrand, 1971.

[39] Belleni–Morante, A., "The initial–value problem for neutron transport in a slab with perfect reflection boundary conditions," J. Math. Anal. Appl. **30**, 353–374 (1970).

[40] Belleni–Morante, A. and Farano, R., "Neutron transport in a slab with moving boundaries," SIAM J. Appl. Math. **31**, 591–599 (1976).

[41] Berežanskii, Y.M., **Expansions in eigenfunctions of self adjoint operators**, Transl. Math. Monographs 17, Providence, R.I., American Mathematical Soc.,

1968 = Kiev, Izdat. "Naukova Nauka," 1965 [Russian].

[42] Bethe, H.A., Rose, M.E. and Smith, L.P., "The multiple scattering of electrons," Proc. Amer. Phil. Soc. **78**, 573–585 (1938).

[43] Bhatnagar, P.L., Gross, E.P. and Krook, M., "Model for collision processes in gases. I. Small amplitude processes in charged and neutral one component systems," Phys. Rev. **94**, 511–525 (1954).

[44] Birkhoff, G., "Reactor criticality in transport theory," Proc. Nat. Acad. Sci. **45**, 567–569 (1959).

[45] Bochner, S. and Phillips, R.S., "Absolutely convergent Fourier expansions for non-commutative normed rings," Ann. Math. **43**, 409–418 (1942).

[46] Bognár, J., **Indefinite inner product spaces**, Berlin, Springer Verlag, 1974.

[47] Bognár, J., "A proof of the spectral theorem for J-positive operators," Acta Sci. Math. Szeged **45**, 75–80 (1983).

[48] Boltzmann, L., "Further studies on the thermal equilibrium of gas molecules," Sitz. Wien. Akad. Wiss. **66**, 275–370 (1872) [German].

[49] Borgioli, G., Frosali, G. and van der Mee, C.V.M., "A stationary criticality problem in general L_p-space for energy dependent neutron transport in cylindrical geometry," Z. Angew. Math. Phys. **35**, 166–180 (1984).

[50] Borysiewicz, M. and Mika, J., "Time behaviour of thermal neutrons in moderating media," J. Math. Anal. Appl. **26**, 461–478 (1969).

[51] Borysiewicz, M. and Mika, J., "Existence and uniqueness of the solution to the critical problem in the multigroup neutron transport theory," Transport Theory Stat. Phys. **2**, 243–270 (1972).

[52] Borysiewicz, M. and Mika, J., "Solution to the critical problem for the neutron transport equation," Transport Theory Stat. Phys. **7**, 139–160 (1978).

[53] Bothe, W., "The scattering and absorption of electron beams," Z. Physik **54**, 161–178 (1929) [German].

[54] Bowden, R.L. and Cameron, W.L., "Constructive solution of the linearized Boltzmann equation: Vector BGK model," Transport Theory Stat. Phys. **8**, 45–62 (1979).

[55] Bowden, R.L. and Garbanati, L.F., "Constructive solution of the linearized Boltzmann equation. Scalar BGK model equation," Transport Theory Stat. Phys. **7**, 1–23 (1978).

[56] Bowden, R.L., Greenberg, W. and Zweifel, P.F., "Resolvent integration techniques for generalized transport equations," J. Math. Phys. **20**, 1099–1108 (1979).

[57] Bowden, R.L., Sancaktar, S. and Zweifel, P.F., "Multigroup neutron transport. II. Half-range," J. Math. Phys. **17**, 82–86 (1976).

[58] Burniston, E.E., Mullikin, T.W. and Siewert, C.E., "Steady-state solutions in the two-group theory of neutron diffusion," J. Math. Phys. **13**, 1461–1465 (1972).

[59] Burschka, M.A. and Titulaer, U.M., "The kinetic boundary layer for the Fokker-Planck equation with absorbing boundary," J. Stat. Phys. **25**, 569–582 (1981).

[60] Burschka, M.A. and Titulaer, U.M., "The kinetic boundary layer for the Fokker-Planck equation: Selectively absorbing boundaries," J. Stat. Phys. **26**, 59–72 (1981).

[61] Busbridge, I.W., **The mathematics of radiative transfer**, Cambridge, Cambridge University Press, 1960.

[62] Busbridge, I.W., "On inhomogeneous stellar atmospheres," Astrophys. J. **133**, 198–209 (1961).

[63] Busbridge, I.W., "On solutions of Chandrasekhar's integral equation," Trans.

Amer. Math. Soc. **105**, 112–117 (1962).

[64] Busoni, G. and Frosali, G., "Streaming operators and semigroups," Meccanica **14**, 119–128 (1979).

[65] Busoni, G. and Frosali, G., "A nonlinear problem in energy and radiation transfer," Aimeta, Sixth National Congress Proc., Genova, 1982.

[66] Caflisch, R.E., "The Boltzmann equation with a soft potential. I.," Comm. Math. Phys. **74**, 71–95 (1980).

[67] Case, K.M., "Plasma oscillations," Ann. Phys. (N.Y.) **7**, 349–364 (1959).

[68] Case, K.M., "Elementary solutions of the transport equation," Ann. Phys. (N.Y.) **9**, 1–23 (1960).

[69] Case, K.M., de Hoffmann, F. and Placzek, G.P., "Introduction to the theory of neutron diffusion," Washington, U.S. Government Printing Office, 1953.

[70] Case, K.M. and Zweifel, P.F., **Linear transport theory**, Reading, Mass., Addison–Wesley, 1967.

[71] Cavalier, J.F. and Greenberg, W., "Analytical solutions of model equations for two phase gas mixtures: transverse velocity perturbations," Phys. Fluids **27**, 1114–1124 (1984).

[72] Cavalleri, G. and Paveri–Fontana, S.L., "Drift velocity and runaway phenomena for electrons in neutral gases," Phys. Rev. A **6**, 327–333 (1972).

[73] Cercignani, C., "Elementary solutions of the linearized gas dynamics Boltzmann equation and their application to the slip–flow problem," Ann. Phys. (N.Y.) **20**, 219–233 (1962).

[74] Cercignani, C., "Flows of rarefied gases supported by density and temperature gradients," Report AS–64–18, University of California, 1964.

[75] Cercignani, C., "Plane Poiseuille flow according to the method of elementary solutions," J. Math. Anal. Appl. **12**, 254-262 (1965).

[76] Cercignani, C., "Plane Couette flow according to the method of elementary solutions," J. Math. Anal. Appl. **11**, 93-101 (1965).

[77] Cercignani, C., "The Kramers problem for a not completely diffusing wall," J. Math. Anal. Appl. **10**, 568-586 (1965).

[78] Cercignani, C., "The method of elementary solutions for kinetic models with velocity dependent collision frequency," Ann. Phys. (N.Y.) **40**, 469-481 (1966).

[79] Cercignani, C., "Unsteady solutions of kinetic models with velocity dependent collision frequency," Ann. Phys (N.Y.) **40**, 454-468 (1966).

[80] Cercignani, C., "On Boltzmann equations with cut-off potentials," Phys. Fluids **10**, 2097-2104 (1967).

[81] Cercignani, C., "Existence and uniqueness in the large for boundary value problems in kinetic theory," J. Math. Phys. **8**, 1653-1656 (1967).

[82] Cercignani, C., "Existence, uniqueness and convergence of the solution of models in kinetic theory," J. Math. Phys. **9**, 633-639 (1968).

[83] Cercignani, C., **Mathematical methods in kinetic theory**, New York, Plenum Press, 1969.

[84] Cercignani, C., **Theory and application of the Boltzmann equation**, New York, Elsevier, 1975.

[85] Cercignani, C., "A nonlinear criticality problem in the kinetic theory of gases," Proc. Oberwolfach Conference on Mathematical Problems in Kinetic Theory, May 1979.

[86] Cercignani, C. and Protopopescu, V., "The spectral decomposition for a class of linear transport operators," J. Stat. Phys. **29**, 363-374 (1982).

[87] Cessenat, M., "L^p–trace theorems for function spaces used in transport," Comptes Rendus Acad. Sci. Paris, Series **1**, **299**, 831–834 (1984) [French].

[88] Chandrasekhar, S., "Stochastic problems in physics and astronomy," Rev. Mod. Phys. **15**, 1–89 (1943).

[89] Chandrasekhar, S., **Radiative transfer**, London, Oxford University Press, 1950 = New York, Dover Publishing Co., 1960.

[90] Chapman, S., "On the law of distribution of molecular velocities, and on the theory of viscosity and thermal conduction, in a non–uniform simple monatomic gas," Phil. Trans. Roy. Soc. **216**, 279–348 (1915).

[91] Chapman, S., "On the kinetic theory of a gas. II. A composit monatomic gas: diffusion, viscosity, and thermal conduction," Phil. Trans. Roy. Soc. **217**, 115–197 (1916).

[92] Chapman, S. and Cowling, T.G., **The mathematical theory of non–uniform gases**, Cambridge, Cambridge University Press, 1970.

[93] Chen Wang, M. and Uhlenbeck, G.E., "On the theory of Brownian motion. II," Rev. Mod. Phys. **17**, 323–342 (1945).

[94] Coddington, E.A. and Levinson, N., **Theory of ordinary differential equations**, New York, McGraw–Hill, 1955.

[95] Crum, M.M., "On an integral equation of Chandrasekhar," Quart. J. Math. Oxford **18**, 244–252 (1946).

[96] Curgus, B., "On the regularity of the critical point infinity of definitizable operators," Int. Eqs. Oper. Theor. **8**, 462–488 (1985).

[97] Davies, E.B., **One parameter semigroups**, London, Academic Press, 1980.

[98] Davison, B., **Neutron transport theory**, Oxford, Clarendon Press, 1957.

[99] Degond, P. and Mas-Gallic, S., "Existence of solutions and diffusion approximation for a model Fokker-Planck equation," Transport Theory Stat. Phys., to appear.

[100] de Haan, J.F., Bosma, P.B. and Hovenier, J.W., "The adding method for multiple scattering calculations of polarized light," Astron. Astrophys., to appear.

[101] de Rham, G., **Differentiable manifolds**, Berlin, Springer Verlag, 1984 = Paris, Hermann, 1970 [French].

[102] Derndinger, R., "On the spectrum of positive generators," Math. Z. **172**, 281-293 (1980) [German].

[103] Derndinger, R. and Nagel, R., "The generator of strongly continuous lattice semigroups on C(X) and its spectrum," Math. Ann. **245**, 159-177 (1979) [German].

[104] Diestel, J. and Uhl Jr., J.J., **Vector measures**, Providence, American Mathematical Soc., 1977.

[105] Dowson, H.R., **Spectral theory of linear operators**, New York, Academic Press, 1978.

[106] Drange, H.B., "The linearized Boltzmann collision operator for cut-off potentials," SIAM J. Math. Anal. **29**, 665-676 (1975).

[107] Drange, H.B., "On the Boltzmann equation with external forces," SIAM J. Appl. Math. **34**, 577-592 (1978).

[108] Duderstadt, J.J. and Martin, W.R., **Transport theory**, New York, Wiley-Interscience, 1979.

[109] Dunford, N. and Schwartz, J.T., **Linear operators**. I, II, III. New York, Wiley-Interscience, 1958.

[110] Dym, H. and Gohberg, I., "On an extension problem, generalized Fourier analysis, and an entropy formula," Int. Eqs. Oper. Theor. **3**, 143–215, 1980.

[111] Ellis, R.S. and Pinsky, M.A., "The first and second fluid approximations to the linearized Boltzmann equation," J. Math. Pures et Appl. **54**, 125–156 (1975).

[112] Enflö, P., "A counterexample to the approximation problem in Banach spaces," Acta Math. **130**, 309–317 (1973).

[113] Enskog, D., "The numerical calculation of reactions in moderately dilute gases," Ark. Mat. Astron. Fys. **16B**, 1–60 (1921) [German].

[114] Erdélyi, A., Magnus, W., Oberhettinger, F. and Tricomi, F.G., **Higher transcendental functions**. I. II, III. New York, McGraw–Hill, 1953.

[115] Everitt, W.N., "Legendre polynomials and singular differential operators." In: W.N. Everitt (Ed.), "Ordinary and partial differential equations," Proc. Dundee Conference 1978, Lecture Notes in Mathematics 827, Berlin, Springer Verlag, 1980, pp. 83–106.

[116] Feldman, I.A., "Wiener–Hopf operator equations and their application to the transport equation," Int. Eqs. Oper. Theor. **3**, 43–61 (1980) = Matem. Issled. **6** (3), 115–132 (1971) [Russian].

[117] Feldman, I.A., "On some projection methods for the solution of the equation of radiative energy transfer," Matem. Issled. **7** (4), 228–236 (1972) [Russian].

[118] Feldman, I.A., "On Wiener–Hopf equations with weakly integrable operator kernels," Matem. Issled. **8**(4), 101–110 (1973) [Russian].

[119] Feldman, I.A., "On an approximation method for the equation of radiative transfer," Matem. Issled. **10** (3), 226–230 (1975) [Russian].

[120] Feller, W., **An introduction to probability theory and applications**. I, II. New York, Wiley, 1966.

[121] Ferziger, J.H. and Kaper, H.G., **Mathematical theory of transport processes in gases**, Amsterdam, North-Holland, 1972.

[122] Fokker, A.D., "The average energy of rotating electrical dipoles in a radiation field," Annalen der Physik **43**, 810-820 (1914) [German].

[123] Frobenius, G., "On matrices with positive elements," Sitz. Preuss. Akad. Wiss., Math.-Phys. Klasse, 471-476 (1908); 514-518 (1909) [German].

[124] Frobenius, G., "On matrices with nonnegative elements," Sitz. Preuss. Akad. Wiss., Math.-Phys. Klasse, 456-477 (1912) [German].

[125] Fry, E.S. and Kattawar, G.W., "Relationships between elements of the Stokes matrix," Appl. Opt. **20**, 2811-2814 (1981).

[126] Ganchev, A.H. and Greenberg, W., "Linear half space transport equations with accretive, invertible collision operators," Transport Theory Stat. Phys., to appear.

[127] Ganchev, A.H., Gong, D., Greenberg, W. and van der Mee, C.V.M., "Perturbation analysis of analytic bisemigroups and applications to linear transport theory," in **Proceedings of International Conference on Operator Theory (Bucharest)**, Operator Theory: Advances and Applications, Basel, Birkhäuser, to appear.

[128] Ganchev, A.H., Greenberg, W. and van der Mee, C.V.M., "Abstract kinetic equations with accretive collision operators," Int. Eq. Oper. Theory, to appear.

[129] Ganchev, A.H., Greenberg, W. and van der Mee, C.V.M., "A class of linear kinetic equations in a Krein space setting," preprint, 1986.

[130] Garcia, R.D.M. and Siewert, C.E., "On the dispersion function in particle transport theory," Z. Angew. Math. Phys. **33**, 801-806 (1982).

[131] Gearhart, L., "Spectral theory for contraction semigroups on Hilbert space," Trans. Amer. Math. Soc. **236**, 385-394 (1978).

[132] Gelfand, I.M., "Normed rings," Mat. Sbornik **9**, 1-20 (1941) [German].

[133] Gelfand, I.M., Minlos, R.A. and Shapiro, Z.Ya., **Representations of the rotation and Lorentz groups and their applications**, Oxford, Pergamon Press, 1963 = Moscow, Fizmatgiz, 1958 [Russian].

[134] Gelfand, I.M., Raikov, D.A. and Shilov, G.E., **Commutative normed rings**, New York, Chelsea, 1964.

[135] Gelfand, I.M. and Shapiro, Z.Ya., "Representations of the group of rotations in three dimensional space and their applications," Amer. Math. Soc. Transl., Series 2, **10**, 207–316 (1956) = Uspehi Matem. Nauk **7**, 3–117 (1952) [Russian].

[136] Germogenova, T.A. and Konovalov, N.V., "The spectrum of the characteristic equation in the theory of transfer with polarization taken into account," Moscow, Institute of Applied Mathematics, AN SSSR, preprint 62, 1978 [Russian].

[137] Gohberg, I., "On linear operators depending analytically on a parameter," Dokl. Akad. Nauk SSSR **78**, 629–632 (1951) [Russian].

[138] Gohberg, I., "The factorization problem for operator functions," Amer. Math. Soc. Transl., Series 2, **49**, 130–161 (1966) = Izvestija Akad. Nauk SSSR, Ser. Matem. **28**, 1055–1082 (1964) [Russian].

[139] Gohberg, I. and Feldman, I.A., **Convolution equations and projection methods for their solution**, Transl. Math. Monographs **41**, Providence, R.I., American Mathematical Soc., 1974 = Moscow, "Nauka," 1971 [Russian].

[140] Gohberg, I. and Heinig, G., "On matrix integral operators on a finite interval with kernels depending on the difference of their arguments," Rev. Roum. Math. Pures et Appl. **20**, 55–73 (1975) [Russian].

[141] Gohberg, I. and Krein, M.G., "Systems of integral equations on a half-line with kernels depending on the difference of arguments," Amer. Math. Soc. Transl., Series 2, **14**, 217–287 (1960) = Uspehi Matem. Nauk **13** (2), 3–72 (1959) [Russian].

[142] Gohberg, I. and Krein, M.G., **Introduction to the theory of linear nonselfadjoint operators in Hilbert space**, Transl. Math. Monographs **18**, Providence, R.I., American Mathematical Soc., 1969 = Moscow, "Nauka," 1965 [Russian].

[143] Gohberg, I. and Krupnik, N., **Introduction to the theory of one dimensional singular integral operators**, Basel, Birkhäuser, 1979 [German] = Kishinev, "Stiinca," 1973 [Russian].

[144] Gohberg, I., Lancaster, P. and Rodman, L., "On self adjoint matrix polynomials," Int. Eqs. Oper. Theor. **1**, 434-439 (1978).

[145] Gohberg, I., Lancaster, P. and Rodman, L., "Spectral analysis of self adjoint matrix polynomials," Research paper 419, University of Calgary, Canada, 1979.

[146] Gohberg, I., Lancaster, P. and Rodman, L., **Matrix polynomials**, New York, Academic Press, 1982.

[147] Gohberg, I., Lancaster, P. and Rodman, L., **Matrix functions and indefinite scalar products**, Basel, Birkhäuser, OT 8, 1983.

[148] Gohberg, I. and Leiterer, J., "Factorization of operator functions with respect to a contour. I. Finite-meromorphic operator functions," Math. Nachrichten **52**, 259-282 (1972) [Russian].

[149] Gohberg, I. and Leiterer, J., "Factorization of operator functions with respect to a contour. II. Canonical factorization of operator functions close to the identity," Math. Nachrichten **54**, 41-74 (1972) [Russian].

[150] Gohberg, I. and Leiterer, J., "Factorization of operator functions with respect to a contour. III. Factorization in algebras," Math. Nachrichten **55**, 33-61 (1973) [Russian].

[151] Gohberg, I. and Semençul, A.A., "On the invertibility of finite Toeplitz matrices and their continuous analogues," Matem. Issled. **7** (2), 201-223 (1972) [Russian].

[152] Gohberg, I. and Sigal, E.I., "An operator generalization of the logarithmic residue theorem and the theorem of Rouché," Math. USSR Sbornik **13**, 603–625 (1971) = Matem. Sbornik **84**, 607–629 (1971) [Russian].

[153] Goldstein, C., "Eigenfunction expansions and similarity for certain nonselfadjoint operators," Bull. Amer. Math. Soc. **75**, 550–553 (1969).

[154] Grad, H., "Principles of the kinetic theory of gases." In: **Handbuch der Physik**, Berlin, Springer Verlag, 1958, pp. 205–294.

[155] Grad, H., "Asymptotic theory of the Boltzmann equation. II." In: J.A. Laurmann (ed.), **Third Symposium on Rarefied Gas Dynamics. I**, New York, Academic Press, 1963, pp. 26–59.

[156] Greenberg, W., "Functional calculus for the symmetric multi-group transport operator," J. Math. Phys. **17**, 159–162 (1976).

[157] Greenberg, W. and Polewczak, J., "Some remarks about continuity properties of local Maxwellians and the BGK model of the Boltzmann equation," J. Stat. Phys. **33**, 307–316 (1983).

[158] Greenberg, W., Polewczak, J. and Zweifel, P.F., "Global existence proofs for the Boltzmann equation." In: E.W. Montroll, J.C. Lebowitz (eds.), **Nonequilibrium phenomena I. The Boltzmann equation**, Studies in Statistical Mechanics 10, North-Holland, New York, 1983, pp. 19–49.

[159] Greenberg, W. and Sancaktar, S., "Solution of the multigroup transport equation in L^p spaces," J. Math. Phys. **17**, 2092–2097 (1976).

[160] Greenberg, W. and van der Mee, C.V.M., "Abstract boundary value problems from kinetic theory," Transport Theory Stat. Phys. **11**, 155–181 (1982/83).

[161] Greenberg, W. and van der Mee, C.V.M., "Stationary transport processes with unbounded collision operators," Lett. Math. Phys. **8**, 7–12 (1984).

[162] Greenberg, W. and van der Mee, C.V.M., "An abstract approach to evaporation models in rarefied gas dynamics," Z. Angew. Math. Phys. **35**, 156–165 (1984).

[163] Greenberg, W. and van der Mee, C.V.M., "Abstract kinetic equations relevant to supercritical media," J. Funct. Anal. **57**, 111–142 (1984).

[164] Greenberg, W. and van der Mee, C.V.M., "An abstract model for radiative transfer in an atmosphere with reflection by the planetary surface," S.I.A.M. J. Math. Anal. **16**, 695–702 (1985).

[165] Greenberg, W., van der Mee, C.V.M. and Walus, W., "Strong solutions of abstract kinetic equations with compact collision operators," S.I.A.M. J. Math. Anal., to appear.

[166] Greenberg, W., van der Mee, C.V.M. and Zweifel, P.F., "Generalized kinetic equations," Int. Eqs. Oper. Theor. **7**, 60–95 (1984).

[167] Greenberg, W. and Walus, W., "Abstract kinetic equations relevant to transport in multiplying media in slab geometry." In: V. Boffi, C. Cercignani (eds.), **Fifteenth Symposium on Rarefied Gas Dynamics**, Stuttgart, Teubner, to appear.

[168] Greenberg, W. and Zweifel, P.F., "Functional analytic treatment of the transport equation," Transport Theory Stat. Phys. **5**, 219–253 (1976).

[169] Greiner, G., "On the Perron–Frobenius theory of strongly continuous semigroups," Math. Z. **177**, 401–423 (1981) [German].

[170] Greiner, G., "Spectrum and asymptotics of strongly continuous semigroups of positive operators," Sitz. Heidelberger Akad. Wiss., 3. Abhandlung, 1982 [German].

[171] Greiner, G., "Spectral properties and asymptotic behaviour of the linear transport equation," Math. Z. **185**, 167–177 (1984).

[172] Greiner, G. and Nagel, R., "On the stability of strongly continuous semigroups of positive operators on $L^2(\mu)$," Ann. Scuola Norm. Sup. Pisa **10**, 257–262 (1983).

[173] Greiner, G., Voigt, J. and Wolff, M., "On the spectral bound of the generator of semigroups of positive operators," J. Operator Theory **5**, 245–256 (1981).

[174] Gross, E.P. and Jackson, E.A., "Kinetic models and the linearized Boltzmann equation," Phys. Fluids **2**, 432–441 (1959).

[175] Grothendieck, A., **Topological tensor products and nuclear spaces**, Memoirs of the Amer. Math. Soc. 8, Providence, R. I., American Mathematical Soc., 1966 [French].

[176] Grothendieck, A., "Resume of the metric theory of topological tensor products," Bol. Soc. Mat. Sao–Paulo **8**, 1–79 (1956) [French].

[177] Guiraud, J.P., "On the Couette problem and the Boltzmann equation," J. de Mécanique **7**, 171–203 (1968) [French].

[178] Guiraud, J.P., "Interior limit problem for the linear Boltzmann equation," J. de Mécanique **9**, 443–499 (1970) [French].

[179] Guiraud, J.P., "The Boltzmann equation in kinetic theory. A survey of mathematical results," Fluid Dynamics Trans. **7** (2), 37–84 (1976).

[180] Hangelbroek, R.J., "A functional analytic approach to the linear transport equation," Ph.D. Thesis, University of Groningen, Netherlands, 1973.

[181] Hangelbroek, R.J., "Linear analysis and solution of neutron transport problems," Transport Theory Stat. Phys. **5**, 1–85 (1976).

[182] Hangelbroek, R.J., "Time independent neutron transport equation with anisotropic scattering in absorbing media," Report ANL 80–60, Argonne Nat. Lab., Argonne, Ill., 1980.

[183] Hangelbroek, R.J., "Monotonicity of slab operators. I," Int. Eqs. Oper. Theor. **9**, 204–217 (1986).

[184] Hangelbroek, R.J. and Lekkerkerker, C.G., "Decompositions of a Hilbert space and factorization of a W–A determinant," SIAM J. Math. Anal. **8**, 458–472 (1977).

[185] Hansen, J.E., "Multiple scattering of polarized light in planetary atmospheres. I. The doubling method," J. Atmosph. Sci. **28**, 120–125 (1971).

[186] Hansen, J.E. and Travis, L.D., "Light scattering in planetary atmospheres," Space Sci. Rev. **16**, 527–610 (1974).

[187] Harris, S., "Steady one dimensional Brownian motion with an absorbing boundary," J. Chem. Phys. **75**, 3103–3106 (1981).

[188] Hejtmanek, J., "Scattering theory of the linear Boltzmann operator," Comm. Math. Phys. **43**, 109–120 (1975).

[189] Hejtmanek, J., "Dynamics and spectrum of the linear multiple scattering operator in the Banach lattice $L_1(\mathbb{R}^3 \times \mathbb{R}^3)$," Transport Theory Stat. Phys. **8**, 29–44 (1979).

[190] Hejtmanek, J., "Time dependent linear transport equations." In: C. Cercignani (ed.), **Kinetic theories and the Boltzmann equation**, Lecture Notes in Mathematics 1048, Berlin, Springer Verlag, 1984, pp. 1–59.

[191] Hejtmanek, J. and Kaper, H.G., "Counterexample to the spectral mapping theorem for the exponential function," Proc. Amer. Math. Soc. **96**, 563–568 (1986).

[192] Herbst, I., "The spectrum of Hilbert space semigroups," J. Operator Theory **10**, 87–94 (1983).

[193] Hilbert, D., "Foundations of kinetic gas theory," Math. Ann. **72**, 562–577 (1912) [German].

[194] Hille, E., **Functional analysis and semi-groups**, Amer. Math. Soc. Colloquium Publ. 31, Providence, R. I., American Mathematical Soc., 1948; Hille, E. and

Phillips, R.S., **Functional analysis and semi-groups**, Amer. Math. Soc. Colloquium Publ. 31 (Revised edition), Providence, R. I., American Mathematical Soc., 1957.

[195] Hille, E., **Lectures on ordinary differential equations**, London, Addison-Wesley, 1969.

[196] Hopf, E., **Mathematical problems of radiative equilibrium**, Cambridge, Cambridge University Press, 1934.

[197] Hovenier, J.W., "Symmetry relationships for scattering of polarized light in a slab of randomly oriented particles," J. Atmosph. Sci. **26**, 488-499 (1969).

[198] Hovenier, J.W., "Principles of symmetry for polarization studies of planets," Astron. Astrophys. **7**, 86-90 (1970).

[199] Hovenier, J.W., "Multiple scattering of polarized light in planetary atmospheres," Astron. Astrophys. **13**, 7-29 (1971).

[200] Hovenier, J.W., "Reduction of the standard problem in radiative transfer for a medium of finite optical thickness," Astron. Astrophys. **82**, 61-72 (1980).

[201] Hovenier, J.W. and van der Mee, C.V.M., "Fundamental relationships relevant to the transfer of polarized light in a scattering atmosphere," Astron. Astrophys. **128**, 1-16 (1983).

[202] Hovenier, J.W. and van der Mee, C.V.M., "Expansion coefficients in polarized light transfer," preprint, 1986.

[203] Jentzsch, P., "On integral equations with a positive kernel," J. Reine Angew. Math. **141**, 235-244 (1912) [German].

[204] Jörgens, K., "An asymptotic expansion in the theory of neutron transport," Comm. Pure Appl. Math. **11**, 219-242 (1958).

[205] Kaniel, S. and Shinbrot, M, "The Boltzmann equation. I. Uniqueness and existence," Comm. Math. Phys. **58**, 65-84 (1978).

[206] Kantorovič, L.V. and Akilov, G.P., **Functional analysis in normed spaces**, Monographs in Pure and Appl. Math. 46, New York, Macmillan, 1964 = Moscow, Fizmatgiz, 1959 [Russian].

[207] Kaper, H.G., "A constructive approach to the solution of a class of boundary value problems of mixed type," J. Math. Anal. Appl. **63**, 691–718 (1978).

[208] Kaper, H.G., "Boundary value problems of mixed type arising in the kinetic theory of gases," SIAM J. Math. Anal. **10**, 161–168 (1979).

[209] Kaper, H.G., Kam Kwong, Man, Lekkerkerker, C.G. and Zettl, A., "Full and partial range eigenfunction expansions for Sturm–Liouville problems with indefinite weights," Proc. Roy. Soc. Edinburgh **98**A, 69–88 (1984).

[210] Kaper, H.G. and Lekkerkerker, C.G., "Spectral analysis of the multigroup transport operator," Transport Theory Stat. Phys. **15**, 411–447 (1986).

[211] Kaper, H.G., Lekkerkerker, C.G. and Hejtmanek, J., **Spectral methods in linear transport theory**, Basel, Birkhäuser OT 5, 1982.

[212] Karlin, S., "Positive operators," J. Math. Mech. **8**, 907–937 (1959).

[213] Kato, T., **Perturbation theory for linear operators**, Berlin, Springer Verlag, 1966.

[214] Kato, T., "Wave operators and similarity for some nonselfadjoint operators," J. Math. Soc. Japan **19**, 123–158 (1967).

[215] Kato, T., "Linear evolution equations of hyperbolic type," J. Fac. Sci. Univ. Tokyo, Sect. 1, **17**, 241–258 (1970).

[216] Kelley, C.T., "Convolution and H–equations for operator–valued functions with applications to neutron transport theory," J. Math. Phys. **18**, 764–769 (1977).

[217] Kittel, C., **Introduction to solid state physics**, New York, Wiley, 1953.

[218] Klaus, M., "Boltzmann collision operator without cut-off," Helv. Phys. Acta **50**, 893-903 (1977).

[219] Klaus, M., van der Mee, C.V.M. and Protopopescu, V., "Solutions of indefinite Sturm-Liouville problems in a half-space," Comptes Rendus Acad. Sci. Paris, Series 1, **300**, 165-168 (1985) [French].

[220] Klaus, M., van der Mee, C.V.M. and Protopopescu, V., "Half-range solutions of indefinite Sturm-Liouville problems," J. Funct. Anal., to appear.

[221] Klinč, T., "On completeness of eigenfunctions of the one speed transport equation," Comm. Math. Phys. **41**, 273-279 (1975).

[222] Kogan, M.N., **Rarefied gas dynamics**, New York, Plenum Press, 1969 = Moscow, "Nauka," 1967 [Russian].

[223] Krasnoselskii, M.A., **Positive solutions of operator equations**, Groningen, Noordhoff, 1964 = Moscow, Fizmatgiz, 1962 [Russian].

[224] Krasnoselskii, M.A., Zabreiko, P.P., Pustylnik, E.I. and Sobolevskii, P.E., **Integral operators in spaces of summable functions**, Leyden, Noordhoff, 1976.

[225] Krein, M.G., "Fundamental properties of normal cones in Banach spaces," Comptes Rendus Acad. Sci. U.R.S.S. **28**, 13-17 (1940) [French].

[226] Krein, M.G., "On the linear approximation of a continuous operator in functional spaces with two norms," Akad. Nauk Ukrain. SSR, Zbirnik Prats Inst. Mat. **9**, 104-129 (1947) [Ukrainian].

[227] Krein, M.G., "On nonlinear integral equations which play a role in the theory of Wiener-Hopf equations." I, II. In: M.G. Krein, **Topics in differential and integral equations and operator theory**, Basel, Birkhäuser OT 7, 1983. First part translated from: Matem. Issled. **42**, 47-90 & 230-231 (1976) [Russian].

[228] Krein, M.G. and Langer, H., "The spectral function of a self adjoint operator in

a space with indefinite metric," Soviet Math. Dokl. **4**, 1236–1239 (1963) = Dokl. Akad. Nauk **4**, 39–42 (1963) [Russian].

[229] Krein, M.G. and Rutman, M.A., "Linear operators leaving invariant a cone in a Banach space," Amer. Math. Soc. Transl., Series 1, **26**, 199–325 (1950) = Uspehi Matem. Nauk **3**, 3–95 (1948) [Russian].

[230] Krein, M.G. and Smuljan, Y.L., "J–polar representations of plus–operators," Mat. Issled. Kishinev **1–2**, 172–210 (1966) [Russian].

[231] Krein, S.G. and Sobolevskii, P.E., "Differential equations with abstract elliptic operators in Hilbert space," Dokl. Akad. Nauk SSSR **110**, 233–236 (1958).

[232] Kriese, J.T., Chang, T.S. and Siewert, C.E., "Elementary solutions of coupled model equations in the kinetic theory of gases," Internat. J. Eng. Sci. **12**, 441–470 (1974).

[233] Kuščer, I. and McCormick, N.J., "Some analytical results for radiative transfer in thick atmospheres." In: J.G. Kuriyan (ed.), **Proceedings of the U.C.L.A. International Conference on Radiation and Remote Probing of the Atmosphere**, N. Hollywood, Cal., Western Periodicals Co., 1974, pp. 196–226.

[234] Kuščer, I. and Ribarič, M., "Matrix formalism in the theory of diffusion of light," Optica Acta **6**, 42–51 (1959).

[235] Kuzmina, M.G., "General functional properties of the transport equation of polarized radiation," Soviet Phys. Dokl. **23**, 26–28 (1978) = Dokl. Akad. Nauk SSSR **238**, 314–317 (1978) [Russian].

[236] Kuzmina, M.G., "The characteristic equation of transport theory of polarized radiation," Soviet Phys. Dokl. **24**, 17–19 (1979) = Dokl. Akad. Nauk SSSR **244**, 62–66 (1979) [Russian].

[237] Langer, H., "Spectral theory of linear operators in J–spaces and some applications to the pencil $L(\lambda)=\lambda^2I+\lambda B+C$," Habilitationsschrift, Technische Universität Dresden, 1965 [German].

[238] Langer, H., "Spectral functions of definitizable operators in Krein spaces." In: D. Butković, H. Kraljević, S. Kurepa (eds.), **Functional analysis**, Lecture Notes in Mathematics 948, Berlin, Springer Verlag, 1982, pp. 1–46.

[239] Larsen, E.W., "Completeness results for linear operators in a Hilbert space and problems in neutron transport theory," Ph.D. Thesis, Rensselaer Polytechnic Institute, Troy, N.Y., 1971.

[240] Larsen, E.W., "A functional analytic approach to the steady, one speed neutron transport equation with anisotropic scattering," Comm. Pure Appl. Math. **27**, 523–545 (1974).

[241] Larsen, E.W. and Habetler, G.J., "A functional analytic derivation of Case's full and half range formulas," Comm. Pure Appl. Math. **26**, 525–537 (1973).

[242] Larsen, E.W. and Zweifel, P.F., "On the spectrum of the linear transport operator," J. Math. Phys. **15**, 1987–1997 (1974).

[243] Lebowitz, J.L. and Rubinow, S.I., "A theory for the age and generation time distribution of a microbial population," J. Math. Biol. **1**, 17–36 (1974).

[244] Lehner, J., "The spectrum of the neutron transport operator for the infinite slab," J. Math. Mech. **11**, 173–181 (1962).

[245] Lehner, J. and Wing, G.M., "Solution of the linearized Boltzmann equation for the slab geometry," Duke Math. J. **23**, 125–142 (1956).

[246] Lehner, J. and Wing, G.M., "On the spectrum of an asymmetric operator arising in the transport theory of neutrons," Comm. Pure Appl. Math. **8**, 217–234 (1955).

[247] Leiterer, J., "Factorization of operator functions with respect to a contour," Habilitationsschrift, Academy of Sciences of the G.D.R., 1974 [German].

[248] Lekkerkerker, C.G., "The linear transport equation. The degenerate case c=1.

I. Full-range theory; II. Half-range theory," Proc. Roy. Soc. Edinburgh **75**A, 259–282 & 283–295 (1975/76).

[249] Lekkerkerker, C.G., "On eigendistributions in linear transport theory,," Proc. Roy. Soc. Edinburgh **83**A, 303–326 (1979).

[250] Lewis, H.W., "Multiple scattering in an infinite medium," Phys. Rev. **78**, 526–529 (1950).

[251] Lotz, H.P., "On the spectrum of positive operators," Math. Z. **108**, 15–32 (1968) [German].

[252] Lucretius Carus, T., **De rerum natura**, Rome, circa 60 B.C. Transl. by W.D. Rouse in W. Durant, **Caesar and Christ**, New York, Simon and Schuster, 1944.

[253] Lumer, G. and Phillips, R.S., "Dissipative operators in a Banach space," Pac. J. Math. **11**, 679–698 (1961).

[254] Luxemburg, W.A.J. and Zaanen, A.C., "Compactness of integral operators in Banach function spaces," Math. Ann. **149**, 150–180 (1962/63).

[255] Magnus, W., Oberhettinger, F. and Soni, R.P., **Formulas and theorems for the special functions of mathematical physics**, Heidelberg, Springer Verlag, 1966.

[256] Marchioro, C., Pellegrinotti, A., Presutti, E. and Pulvirenti, M., "On the dynamics of particles in a bounded region: A measure theoretical approach," J. Math. Phys. **17**, 647–652 (1976).

[257] Marshall, T.W. and Watson, E.J., "A drop of ink falls from my pen ... It comes to earth, I know not when," J. Phys. A **18**, 3531–3559 (1985).

[258] Marti, J., "On initial-value and eigenvalue problems from neutron transport theory," Z. Angew. Math. Phys. **18**, 247–259 (1967) [German].

[259] Maslennikov, M.V., "The Milne problem with anisotropic scattering," Proc. Steklov Inst. Math. **97**, 1969 = Trudy Matem. Instituta im. V.A. Steklova **97**,

1968 [Russian].

[260] Maslova, N.B., "Kramers problem in the kinetic theory of gases," U.S.S.R. Comput. Maths. Math. Phys. **22**, 208–219 (1982) = Zh. Vychisl. Mat. Mat. Fiz. **22**, 700–709 (1982) [Russian].

[261] Maslova, N.B., "Stationary solutions of the linearized Boltzmann equation." In: "Boundary value problems of mathematical physics," Trudy Matem. Instituta im. V.A. Steklova **159**, 1983, pp. 41–60.

[262] Mayya, Y.S. and Sahni, D.C., "One dimensional Brownian motion near an absorbing boundary: Solution to the steady state Fokker–Planck equation," J. Chem. Phys. **79**, 2302–2307 (1983).

[263] McCormick, N.J. and Kuščer, I., "Singular eigenfunction expansions in neutron transport theory," Advances in Nuclear Science and Technology **7**, 181–282 (1973).

[264] Mikusiński, J., **The Bochner integral**, New York, Academic Press, 1978.

[265] Milne, E.A., "Radiative equilibrium in the outer layers of a star: the temperature distribution and the law of darkening," Monthly Not. Roy. Astron. Soc. **81**, 361–375 (1921).

[266] Mochizuki, K., "On the large perturbation of nonselfadjoint operators," J. Math. Soc. Japan **19**, 123–158 (1967).

[267] Molinet, F.A., "Existence, uniqueness and properties of the solutions of the Boltzmann kinetic equation for a weakly ionized gas," J. Math. Phys. **18**, 984–986 (1977).

[268] Montagnini, B., "The eigenvalue spectrum of the linear Boltzmann operator in $L_1(\mathbb{R}^6)$ and $L_2(\mathbb{R}^6)$," Meccanica **14**, 134–144 (1979).

[269] Montagnini, B. and Demuru, M., "Complete continuity of the free gas scattering operator in neutron thermalization theory," J. Math. Anal. Appl. **12**, 49–57

(1965).

[270] Mullikin, T.W., "Neutron branching processes," J. Math. Anal. Appl. **3**, 507–525 (1961).

[271] Mullikin, T.W., "A complete solution of the X and Y equations of Chandrasekhar," Astrophys. J. **136**, 627–635 (1962).

[272] Mullikin, T.W., "Chandrasekhar's X and Y equations," Trans. Amer. Math. Soc. **113**, 316–332 (1964).

[273] Mullikin, T.W., "Wiener–Hopf factorizations," Transport Theory Stat. Phys. **3**, 215–227 (1973).

[274] Mullikin, T.W., "Mathematics of radiative transfer." In: **Proceedings of the Symposium on Theoretical Astrophysics and Relativity**, Chicago, University of Chicago Press, 1975.

[275] Mullikin, T.W. and Victory Jr., H.D., "N–group neutron transport theory: a criticality problem in slab geometry," J. Math. Anal. Appl. **58**, 605–630 (1977).

[276] Muskhelishvili, N.I., **Singular integral equations**, Groningen, Noordhoff, 1953 = Moscow, "Nauka," 1946 [Russian].

[277] Nag, B.R., **Electron transport in compound semiconductors**, Series in Solid State Sciences 11, Berlin, Springer Verlag, 1980.

[278] Nagel, R., "Ergodic and mixing properties of linear operators," Proc. Roy. Irish Acad. **74**A, 245–261 (1974).

[279] Nagel, R., "On the characterization of stable semigroups." In: "Semesterbericht Funktionalanalysis," Wintersemester 1981/1982, Tübingen, 1982 [German].

[280] Narasimhan, R., **Analysis on real and complex manifolds**, Amsterdam, North–Holland, 1968.

[281] Nelson Jr., P., "Subcriticality for transport of multiplying particles in a slab," J. Math. Anal. Appl. **35**, 90–104 (1971).

[282] Nelson Jr., P., "Positive solutions of positive linear equations," Proc. Amer. Math. Soc. **31**, 453–457 (1972).

[283] Nelson Jr., P., "An existence-uniqueness theorem for nonmultiplying rod models subject to an interior jump condition," J. Math. Anal. Appl. **118**, 115–124 (1986).

[284] Nelson Jr., P. and Seth, D.L., "Stability of the upwind discrete energy approximation to the Spencer-Lewis equation of electron transport," Transport Theory Stat. Phys. **15**, 211–243 (1986).

[285] Nelson Jr., P. and Victory Jr., H.D., "Measures on phase space as solutions of the one-dimensional neutron transport equation," Annali di Mat. Pura ed Appl. **62**, 69–90 (1977).

[286] Nikolskii, S., "Linear equations in normed linear spaces," Izvestija Akad. Nauk SSSR **7**, 147–166 (1973) [Russian].

[287] Nishida, T. and Imai, K., "Global solutions to the initial-value problem for the nonlinear Boltzmann equation," Publ. Res. Inst. Math. Sciences, Kyoto University, **12**, 229–239 (1976).

[288] Nonnenmacher, T.M. and Wunderle, R., "Influence of the size effect and surface roughness on phonon transport in bounded systems," Phys. Status Solidi (b) **91**, 147–153 (1979).

[289] Nonnenmacher, T.M. and Zweifel, P.F., "A Boltzmann equation for phonons and electrons," Phys. Status Solidi (b) **96**, 653–658 (1979).

[290] Olaussen, K. and Hemmer, P.C., "The drift velocity of a hard sphere Lorentz gas," J. Phys. A **15**, 3255–3272 (1982).

[291] Pagani, C.D., "On an initial value problem for a singular parabolic equation,"

Rend. Ist. Lombardo, Classe di Scienze (A), **103**, 618–653 (1969) [Italian].

[292] Pagani, C.D., "On some questions connected with the generalized Fokker–Planck equation," Boll. Un. Mat. Ital. **3**, 961–986 (1970) [Italian].

[293] Pagani, C.D., "On the parabolic equation $\mathrm{sgn}(x)|x|^p u_y - u_{xx} = 0$ and a related one," Annali di Mat. Pura ed Appl. **99**, 333–399 (1974).

[294] Pagani, C.D., "On an initial-boundary value problem for the equation $w_t = w_{xx} - xw_y$," Ann. Scuola Norm. Sup. Pisa **2**, 219–263 (1975).

[295] Pagani, C.D. and Talenti, G., "On a forward-backward parabolic equation," Annali di Mat. Pura ed Appl. **90**, 1–57 (1971).

[296] Pahor, S., "A new approach to half-space transport problems," Nucl. Sci. Eng. **26**, 192–197 (1966).

[297] Palczewski, A., "A time dependent linear Boltzmann operator as the generator of a semigroup," Bull. Acad. Pol. Sci., Série Sci. Techn. **25**, 233–237 (1977).

[298] Palczewski, A., "Spectral properties of the space nonhomogeneous linearized Boltzmann operator," Transport Theory Stat. Phys. **13**, 409–430 (1984).

[299] Palczewski, A., "The Cauchy problem for the Boltzmann equation. A survey of recent results." In: C. Cercignani (ed.), **Kinetic theories and the Boltzmann equation**, Lecture Notes in Mathematics 1048, Berlin, Springer Verlag, 1984, pp. 202–206.

[300] Papanicolaou, G.C., "Asymptotic analysis of transport processes," Bull. Amer. Math. Soc. **81**, 330–392 (1975).

[301] Pazy, A., **Semigroups of linear operators and applications to partial differential equations**, New York, Springer Verlag, 1983.

[302] Peebles, G.H. and Plesset, M.S., "Transmission of gamma rays through large thicknesses of heavy materials," Phys. Rev. **81**, 430–439 (1951).

[303] Perron, O., "Fundamentals of a theory of the Jacobi continued fractions algorithm," Math. Ann. **64**, 1-76 (1907) [German].

[304] Phillips, R.S., "Perturbation theory for semi-groups of linear operators," Trans. Amer. Math. Soc. **74**, 343-369 (1953).

[305] Phillips, R.S., "Dissipative operators and hyperbolic systems of partial differential equations," Trans. Amer. Math. Soc. **90**, 193-254 (1959).

[306] Piasecki, J., "The runaway effect in a Lorentz gas," J. Stat. Phys. **24**, 45-58 (1981).

[307] Pietsch, A., **Nuclear Locally Convex Spaces**, Berlin, Springer Verlag, 1972.

[308] Pimbley, G., "Solution of an initial-value problem for the multi-velocity neutron transport with a slab geometry," J. Math. Mech. **8**, 837-866 (1959).

[309] Planck, M., "On a theorem of statistical dynamics and its extension in quantum theory," Sitz. Preuss. Akad. Wiss., Math.-Phys. Klasse, 324-346 (1917) [German].

[310] Pontryagin, L.S., "Hermitian operators in spaces with indefinite metric," Izvestija Akad. Nauk SSSR Ser. Mat. **8**, 243-280 (1944) [Russian].

[311] Prössdorf, S., **Some classes of singular equations**, Amsterdam, North-Holland, 1978 = Berlin, Akademie Verlag, 1974 [German].

[312] Protopopescu, V., "On the spectral decomposition of the transport operator with anisotropic scattering and periodic boundary conditions," Transport Theory Stat. Phys. **14**, 103-115 (1985).

[313] Protopopescu, V., "The spectral decomposition of the linearized Vlasov-Poisson operator," Comptes Rendus Acad. Sci. Paris, Series 1, **302**, 271-274 (1986) [French].

[314] Protopopescu, V., Klein, W. and Keyes, T., "Time dependent transport as critical phenomenon," preprint, 1985.

[315] Ramm, A.G., "Eigenfunction expansions for some nonselfadjoint operators and the transport equation," J. Math. Anal. Appl. **92**, 564–580 (1983).

[316] Reed Jr., K.W., "On a problem in neutron transport theory," J. Math. Anal. Appl. **10**, 161–165 (1965).

[317] Reed Jr., K.W., "On weak and strong solutions for the neutron transport equation," J. d'Anal. Math. **17**, 347–368 (1966).

[318] Ribarič, M., **Functional analytic concepts and structures of neutron transport theory**. I, II. Ljubljana, Acad. Scient. Art. Slovenica, 1973.

[319] Ribarič, M. and Vidav, I., "Analytic properties of the inverses $A(z)^{-1}$ of an analytic linear operator valued function A(z)," Arch. Rat. Mech. Anal. **32**, 298–310 (1969).

[320] Risken, H., **The Fokker–Planck equation. Methods of solution and applications**, Berlin, Springer Verlag, 1984.

[321] Rodewald, B. and Appel, J., "Size effect in the DC electrical conductivity for nearly specular scattering of electrons at rough surfaces," Surface Science **49**, 21–27 (1975).

[322] Rotenberg, M., "Transport theory for growing cell populations," J. Theor. Biol. **103**, 181–199 (1983).

[323] Sakhnovich, L.A., "Equations with kernels depending upon the difference of arguments on a finite interval," Russ. Math. Surveys **35** (4), 81–152 (1980) = Uspehi Matem. Nauk **35** (4), 69–129 (1980) [Russian].

[324] Schaefer, H.H., **Banach lattices and positive operators**, Grundl. Math. Wiss. 215, Berlin, Springer Verlag, 1974.

[325] Scharf, G., "Functional analytic discussion of the linearized Boltzmann equation," Helv. Phys. Acta **40**, 929–945 (1967).

[326] Schnute, J. and Shinbrot, M., "Kinetic theory and boundary conditions for fluids," Can. J. Math. **25**, 1183–1215 (1973).

[327] Shikhov, S.B. and Shkurpelov, A.A., "The analysis of the time dependent kinetic equation for neutron transport in moderating and multiplicative media." In: V.V. Orlov, E.A. Stumbur (eds.), **Theoretical and experimental problems of time dependent neutron transport**, Moscow, Atomizdat, 1972, pp. 97–165 [Russian].

[328] Shizuta, Y., "On the classical solutions of the Boltzmann equation," Comm. Pure Appl. Math. **36**, 705–754 (1983).

[329] Siewert, C.E., "On the equation of transfer relevant to the scattering of polarized light," Astrophys. J. **245**, 1080–1086 (1981).

[330] Siewert, C.E., "On the scattering of polarized light," Z. Angew. Math. Phys. **33**, 807–818 (1982).

[331] Siewert, C.E. and Benoist, P., "The F_N–method in neutron transport theory. Part I: Theory and applications," Nucl. Sci. Eng. **69**, 156–160 (1979).

[332] Siewert, C.E. and Shieh, P.S., "Two group transport theory," J. Nucl. Energy **2**, 383–392 (1967).

[333] Siewert, C.E. and Thomas Jr., J.R., "Half–space problems in the kinetic theory of gases," Phys. Fluids **16**, 1557–1559 (1973).

[334] Siewert, C.E. and Thomas Jr., J.R., "Strong evaporation into a half–space," Z. Angew. Math. Phys. **32**, 421–433 (1981).

[335] Siewert, C.E. and Thomas Jr., J.R., "Strong evaporation into a half–space. II. The three dimensional BGK model," Z. Angew. Math. Phys. **33**, 202–218 (1982).

[336] Simon, B., "Existence of the scattering matrix for the linearized Boltzmann

equation," Comm. Math. Phys. **41**, 99–108 (1975).

[337] Slemrod, M., "Asymptotic behaviour of C_0 semigroups as determined by the spectrum of the generator," Indiana Univ. Math. J. **25**, 782–792 (1976).

[338] Smuljan, Y.L., "Completely continuous perturbation of operators," Amer. Math. Soc. Transl., Series **2**, **10**, 341–344 (1958) = Dokl. Akad. Nauk SSSR **101**, 35–38 (1958) [Russian].

[339] Sobolev, V.V., "The transmission of radiation through an inhomogeneous medium," Soviet Phys. Dokl. **1**, 747–751 (1956) = Dokl. Akad. Nauk SSSR **111**, 1000–1003 (1956) [Russian].

[340] Sobolev, V.V., **A treatise on radiative transfer**, Princeton, van Nostrand, 1963 = **Radiative energy transfer in atmospheres of stars and planets**, Moscow, Gosud. Izdat. Techn.-teoret. Lit., 1956 [Russian].

[341] Sobolev, V.V., "Diffuse reflection and transmission of light by an atmosphere with anisotropic scattering," Astrophysics **5**, 1–7 (1969) = Astrofizika **5**, 5–20 (1969) [Russian].

[342] Sobolev, V.V., **Light scattering in planetary atmospheres**, Oxford, Pergamon Press, 1975 = Moscow, "Nauka," 1972 [Russian].

[343] Spencer, L.V., "Theory of electron penetration," Phys. Rev. **98**, 1597–1615 (1955).

[344] Stokes, G.G., "On the composition and resolution of streams of polarized light from different sources," Trans. Cambridge Phil. Soc. **9**, 399–416 (1852).

[345] Suhadolc, A., "Linearized Boltzmann equation in L_1-space," J. Math. Anal. Appl. **35**, 1–13 (1971).

[346] Suhadolc, A. and Vidav, I., "Linearized Boltzmann equation in spaces of measures," Mathematica Balkanica **3**, 514–529 (1973).

[347] Szegö, G., **Orthogonal polynomials**, Amer. Math. Soc. Coll. Publ. 23, Providence, R. I., American Mathematical Soc., 1939.

[348] Sz.-Nagy, B. and Foias, C., **Harmonic analysis of operators on Hilbert space**, Amsterdam, North-Holland, 1970 = Budapest, Akademiai Kiado, 1967 [French].

[349] Tabata, M., "Estimations for the spectrum of the linearized Boltzmann equation with some external potential field," preprint, 1984.

[350] Thomas Jr., J.R., Chang, T.S. and Siewert, C.E., "Heat transfer between parallel plates with arbitrary surface accommodation," Phys. Fluids **16**, 2116-2120 (1973).

[351] Titulaer, U.M., "The density profile for the Klein-Kramers equation near an absorbing wall," J. Stat. Phys. **37**, 589-607 (1984).

[352] Truesdell, C. and Muncaster, R.G., **Fundamentals of Maxwell's kinetic theory of a simple monatomic gas**, New York, Academic Press, 1980.

[353] Ukai, S., "Solutions of the Boltzmann equation," preprint, 1985.

[354] Ukai, S. and Asano, K., "On the initial-boundary value problem of the linearized Boltzmann equation in an exterior domain," Proc. Japan Acad. **56**A, 12-17 (1980).

[355] Umeda, T., "Smooth perturbations in ordered Banach spaces and similarity for the linear transport operators," preprint, 1985.

[356] van de Hulst, H.C., "A new look at multiple scattering," New York, Techn. Report, Inst. Space Studies, N.A.S.A., 1963.

[357] van de Hulst, H.C., **Multiple light scattering**. I, II. New York, Academic Press, 1980.

[358] van de Hulst, H.C., **Light scattering by small particles**, New York, Wiley, 1957 = New York, Dover Publishing Co., 1981.

[359] van der Mee, C.V.M., "Spectral analysis of the transport equation. I. Nondegenerate and multigroup case," Int. Eqs. Oper. Theor. **3**, 529–573 (1980).

[360] van der Mee, C.V.M., "Semigroup and factorization methods in transport theory," Ph.D. Thesis, Free University, Amsterdam, 1981.

[361] van der Mee, C.V.M., "Linearized BGK and neutron transport equations in finite domains," Transport Theory Stat. Phys. **10**, 75–104 (1981).

[362] van der Mee, C.V.M., "Transport equation on a finite domain. I. Reflection and transmission operators and diagonalization," Int. Eqs. Oper. Theor. **6**, 572–601 (1983).

[363] van der Mee, C.V.M., "Transport equation on a finite domain. II. Reduction to X and Y functions," Int. Eqs. Oper. Theor. **6**, 730–757 (1983).

[364] van der Mee, C.V.M., "Transport theory in L_p–spaces," Int. Eqs. Oper. Theor. **6**, 405–443 (1983).

[365] van der Mee, C.V.M., "Half and finite range completeness for the equation of transfer of polarized light," Math. Meth. Appl. Sci. **6**, 393–416 (1984).

[366] van der Mee, C.V.M., "Albedo operators and H–equations for generalized kinetic models," Transport Theory Stat. Phys. **13**, 341–376 (1984).

[367] van der Mee, C.V.M., "Abstract boundary value problems modeling transport processes in semi–infinite geometry," Transport Theory Stat. Phys. **13**, 29–41 (1984).

[368] van der Mee, C.V.M., "Reflection and transmission of polarized light: The adding method for homogeneous atmospheres," J. Math. Anal. Appl., to appear.

[369] van der Mee, C.V.M. and Protopopescu, V., "Kinetic equations with reflecting boundary conditions," SIAM J. Math. Anal. **17**, 933–944 (1986).

[370] van der Mee, C.V.M. and Zweifel, P.F., "A Fokker-Planck equation for growing cell populations," J. Math. Biol., to appear.

[371] van Kampen, N.G., "On the theory of stationary waves in plasmas," Physica **21**, 949-963 (1955).

[372] van Kampen, N.G., **Stochastic processes in physics and chemistry**, Amsterdam, North-Holland, 1981.

[373] Vekua, N.P., **Systems of singular integral equations**, Groningen, Noordhoff, 1967 = Moscow, "Nauka," 1950 [Russian].

[374] Veling, E.J.M., "Asymptotic analysis of a singular Sturm-Liouville boundary value problem," Int. Eqs. Oper. Theor. **7**, 561-587 (1984).

[375] Vidav, I., "Existence and uniqueness of nonnegative eigenfunctions of the Boltzmann operator," J. Math. Anal. Appl. **22**, 144-155 (1968).

[376] Vidav, I., "Spectra of perturbed semigroups with applications to transport theory," J. Math. Anal. Appl. **30**, 264-279 (1970).

[377] Vladimirov, V.S., "Mathematical problems in the one velocity theory of particle transport," Report AECL-1661, Chalk River, Ontario, 1963 = Trudy Matem. Instituta im. V. A. Steklova **61**, 1961 [Russian].

[378] Voigt, J., "Spectral properties of positivity preserving strongly continuous semigroups," preprint, 1980.

[379] Voigt, J., "A perturbation theorem for the essential spectral radius of strongly continuous semigroups," Monatshefte für Mathematik **90**, 153-161 (1980).

[380] Voigt, J., "Functional analytic treatment of the initial boundary value problem for collisionless gases," Habilitationsschrift, Ludwig Maximilians Universität München, 1980.

[381] Voigt, J., "Positivity in time dependent linear transport," Acta Appl. Math. **2**,

311–331 (1984).

[382] Voigt, J., "Spectral properties of the neutron transport equation," J. Math. Anal. Appl. **106**, 140–153 (1985).

[383] Voigt, J., "On substochastic C_0–semigroups and their generators." In: "Semesterbericht Functionalanalysis," Wintersemester 1984/85. Tübingen, 1985.

[384] Voigt, J., "Interpolation for positive C_0–semigroups on Lp–spaces," Math. Z. **188**, 283–286 (1985).

[385] von Querenburg, B., **Set–theoretic topology**, Berlin, Springer Verlag, 1973 [German].

[386] Walus, W., "Stationary Solutions of Abstract Kinetic Equations," Ph. D. Thesis, Virginia Tech, Blacksburg, 1985.

[387] Wang Chang, C.S. and Uhlenbeck, G.E., "On the propagation of sound in monatomic gases," Report M 999, Eng. Research Inst., University of Michigan, 1952.

[388] Welander, P., "On the temperature jump in a rarefied gas," Arkiv för Fysik **7**, 507–553 (1954).

[389] Wenzel, L., "A Boltzmann equation with inhomogeneous boundary conditions," Math. Nachrichten **117**, 253–262 (1984) [German].

[390] Wiener, N., "Tauberian theorems," Ann. Math. **33**, 1–100 (1932).

[391] Wiener, N., **The Fourier integral and certain of its applications**, Cambridge, Cambridge University Press, 1933 = New York, Dover Publishing Co., 1959.

[392] Wiener, N. and Hopf, E., "On a class of singular integral equations," Sitz. Preuss. Akad. Wiss., Math.–Phys. Klasse, 696–706 (1931) [German].

[393] Wigner, E.P., **Group theory and its application to the quantum mechanics of**

atomic spectra, New York, Academic Press, 1959.

[394] Williams, M.M.R., **Mathematical methods in particle transport theory**, London, Butterworths, 1971.

[395] Williams, M.M.R., "Transport theory in anisotropic media," Proc. Cambridge Phil. Soc. **84**, 549–567 (1978).

[396] Willis, B.L., Zweifel, P.F. and van der Mee, C.V.M., "Whither existence theory?," Transport Theory Stat. Phys. **14**, 669–678 (1985).

[397] Wing, G.M., **An introduction to transport theory**, New York, Wiley, 1962.

[398] Wolff, M., "A remark on the spectral bound of the generator of semigroups of positive operators with applications to stability theory." In: **Functional analysis and approximation**, Basel, Birkhäuser, 1981.

[399] Yosida, K., "On the differentiability and representation of one-parameter semi-groups of linear operators," J. Math Soc. Japan **1**, 15–21 (1948).

[400] Yosida, K., "On the differentiability of semi-groups of linear operators," Proc. Japan Acad. **34**, 337–340 (1958).

[401] Zaanen, A.C., **Integration**, Amsterdam, North-Holland, 1967.

[402] Zaanen, A.C., **Riesz spaces**. II. Amsterdam, North-Holland, 1984.

[403] Zabczyk, J., "A note on C_0 semigroups," Bull. Acad. Pol. Sci., Sér. Sci. Mat. Astron. Phys. **23**, 895–898 (1975).

[404] Ziman, J.M., **Electrons and phonons. The theory of transport phenomena in solids**, New York, Clarendon Press, 1960.

[405] Ziman, J.M., **Principles of the theory of solids**, Cambridge, Cambridge University Press, 1964.

[406] Zweifel, P.F., **Reactor physics**, New York, McGraw–Hill, 1973.

[407] Zweifel, P.F. and Greenberg, W., "Resolvent integration methods in transport theory." In: I. Gohberg (ed.), **Toeplitz Centennial**, Basel, Birkhäuser OT 4, 1982, pp. 525–539.

SUBJECT INDEX